Advances in Wastewater Research

Series Editor

Maulin P. Shah, Environmental Microbiology Consultant, Ankleshwar, Gujarat, India

The Advances in Wastewater Research highlights the prospective future development of wastewater treatment techniques, research and field applications. The series also focuses on the limitations and challenges associated with some generally accepted wastewater strategies and evaluates the possible applications of these corrective strategies to eliminate toxic pollutants from the environment through integrated technologies in wastewater treatment. The series covers the research areas including, but not limited to, toxic pollutants removal from wastewater, methods and procedures, materials, management of wastewater, valorisation possibilities of waste resulting from the removal of toxic pollutants from wastewater, wastewater recycling, environmental impact, and wastewater policies. The series contribute to remove, reduce, or neutralize water contaminants that threaten human health, and ecosystem productivity and integrity. The series includes peer-reviewed monographs, edited volumes, textbooks, and conference proceedings.

Neha Agarwal · Maulin P. Shah ·
Vijendra Singh Solanki · Neetu Singh
Editors

Nanomaterials in Wastewater Research

Progress and Challenges

Springer

Editors
Neha Agarwal
Department of Chemistry
Navyug Kanya Mahavidyalaya (University of Lucknow)
Lucknow, Uttar Pradesh, India

Vijendra Singh Solanki
Department of Chemistry
ISR, IPS Academy
Indore, Madhya Pradesh, India

Maulin P. Shah
Environmental Microbiology Consultant
Industrial Wastewater Research Lab
Anklesvar, Gujarat, India

Neetu Singh
Department of Physics
Maharaja Bijli Pasi Government P.G. College
Lucknow, Uttar Pradesh, India

ISSN 3004-9725 ISSN 3004-9733 (electronic)
Advances in Wastewater Research
ISBN 978-981-96-3907-6 ISBN 978-981-96-3908-3 (eBook)
https://doi.org/10.1007/978-981-96-3908-3

This Springer imprint is published by the registered company Springer Nature Singapore Pte Ltd.
The registered company address is: 152 Beach Road, #21-01/04 Gateway East, Singapore 189721, Singapore

Foreword

The book entitled *Nanomaterials in Wastewater Research: Progress and Challenges* is of extreme importance to all students, professionals, and researchers interested in this area since nanomaterials have shown immense promise in the field of wastewater treatment due to their unique properties like high surface area, reactivity, and the ability to interact with pollutants at the molecular level. This has driven a surge in research over the past few decades, focusing on their potential to enhance traditional treatment methods or introduce entirely new approaches. The readers will find an overview of the progress and challenges in this research area, such as advanced adsorption techniques, nanocatalysts, nanoparticles, membrane technology, antibacterial properties, and hybrid nanocomposites.

The release of nanomaterials into the environment could pose significant risks. The small size of nanoparticles allows them to penetrate biological membranes, potentially causing toxicity to aquatic life and humans. Understanding the long-term effects of these materials on ecosystems and human health is still a significant concern. After wastewater treatment, recovering nanomaterials from water remains a challenge. If left in the treated water, they could contribute to secondary pollution. Developing cost-effective and efficient nanoparticle recovery mechanisms is a priority in this field. While laboratory-scale studies have demonstrated the effectiveness of nanomaterials in treating wastewater, scaling these processes for real-world applications is challenging. The cost of large-scale production, potential risks, and operational challenges make industrial applications complex. Regulatory frameworks governing the use of nanomaterials in wastewater treatment are still underdeveloped. There is a need for standardization in terms of material safety, usage guidelines, and disposal methods. Public concerns regarding such materials' protection and ethical implications must also be addressed. Another obstacle is the high cost of synthesizing nanomaterials and integrating them into existing treatment systems. For widespread adoption, researchers need to focus on cost-effective synthesis methods and ensure that the benefits of using nanomaterials outweigh their costs.

Thinking about future directions, developing eco-friendly and cost-effective methods for synthesizing nanomaterials using biological or waste-based processes is a crucial area of future research. More attention is being given to designing materials that can quickly recovered and reused, such as magnetic nanocomposites, which can be separated from treated water using external magnetic fields. Combining nanomaterials with other wastewater treatment technologies, such as biological processes or electrochemical methods, could lead to more efficient and sustainable systems. Comprehensive LCA studies are needed to evaluate the environmental impact of nanomaterial-based treatments, from production to disposal, to ensure these technologies are truly sustainable.

In this sense, this book shows with quality and great relevance that nanomaterials hold significant potential for advancing wastewater treatment technologies, offering solutions to some of the most pressing challenges in water purification. However, for these technologies to be adopted widely, challenges such as environmental safety, cost-effectiveness, and scalability must be addressed through continued research and development. With appropriate regulatory oversight and technological advances, nanomaterials could be vital in ensuring sustainable water management.

Helen Treichel
Laboratory of Microbiology
and Bioprocess
Federal University of Fronteira Sul
Erechim, RS, Brazil

Preface

The increasing concerns over water pollution and the growing demand for clean water have emphasized the urgent need for efficient, smart, cost-effective, innovative wastewater treatment technologies. Nanotechnology, with its exceptional capabilities to manipulate materials at the molecular scale, has emerged as a potential tool or technology to address the complexities of wastewater remediation. The efficiency of nanomaterials in increasing treatment processes, improving efficiency, and reducing costs has made them a subject of significant interest in environmental science and engineering.

This book, titled *Nanomaterials in Wastewater Research: Progress and Challenges*, deeply explains the latest innovations and developments in the synthesis, characterization, and application of nanomaterials for wastewater treatment. The chapters presented in this comprehensive book are authored by experts in the field of nanotechnology who provide an extensive overview of the several nanotechnological methods that are being utilized to face problems like wastewater pollutants, ranging from heavy metals to organic contaminants.

The book begins with the exploration of the basic aspects of nanomaterials such as their synthesis and characterizations, before delving into their applications in the remediation of wastewater. The chapters cover a distinctive range of topics, from the utilization of nanosensors, carbon quantum dots, and metal oxide nanoparticles to the development of advanced membrane technologies and nano-adsorbents. Some significant parameters such as sustainable and eco-friendly approaches have also been deeply described that aim to minimize the harmful environmental impact of these technologies.

The book also highlights the challenges and opportunities associated with the use of nanomaterials in wastewater treatment which is a very significant part of this book. Because of the continuous development of these technologies, it is very important to understand both their potential benefits and the risks they pose to the environment and human health. The comprehensive discussions on the responsible use of nanomaterials and the future perspectives of their applications in creating a sustainable environment offer valuable insights for researchers, scientists, policymakers, and industry professionals.

We believe that this comprehensive and insightful collection of chapters will surely serve as a valuable resource for researchers, scientists, engineers, students, and environmentalists who are working on the study and application of nanotechnology in the treatment of wastewater. By presenting a wonderful amalgamation of theoretical knowledge and practical applications, this book aims to motivate further research and innovations in developing sustainable solutions for the treatment of wastewater and environmental remediation.

We hope that this extensive research work will not only aid the ongoing advancements in the field but also enhance a deeper understanding of the role of nanotechnology in building a cleaner and more sustainable future.

Lucknow, India — Neha Agarwal
Anklesvar, India — Maulin P. Shah
Indore, India — Vijendra Singh Solanki
Lucknow, India — Neetu Singh

Contents

Chapter 1
Nanomaterials for Wastewater Remediation: Synthesis, Characterization and Applications

Nicholas Yung Li Loh, Wan Ting Tee, Billie Yan Zhang Hiew, Svenja Hanson, and Lai Yee Lee

Abstract The widespread pollution of rivers, lakes, oceans and other water bodies by chemicals, plastics, heavy metals and oil spills, is causing major threats to human wellbeing and aquatic ecosystems. Continuous industrialization and urbanization, and the high population growth rates have aggravated water-related challenges in many countries. This trend indicates the limitations of existing treatment technologies in effectively removing aqueous pollutants. Nanomaterials with unique morphological, mechanical and physicochemical properties, have emerged as promising materials for wastewater remediation. Their outstanding features, including ultra-large surface area, excellent mechanical strength and customizable surface chemistry, have attracted significant research interests in recent years. This chapter covers the different types and classification methods of nanomaterials, including categorization, by dimension, chemical composition and origin. The synthesis techniques for nanomaterials, encompassing the top-down and bottom-up methods, are described in the chapter. It also describes the characterization of nanomaterials based on their physicochemical properties. Additionally, this chapter covers the various applications of nanomaterials in wastewater remediation.

Keywords Nanomaterials · Wastewater · Physicochemical · Pollutants · Remediation

---

N. Y. L. Loh · W. T. Tee · L. Y. Lee (✉)
Department of Chemical and Environmental Engineering, University of Nottingham Malaysia, 43500 Semenyih, Selangor, Malaysia
e-mail: lai-yee.lee@nottingham.edu.my

N. Y. L. Loh · S. Hanson
Department of Chemical and Environmental Engineering, University of Nottingham Ningbo China, Ningbo 315100, China

B. Y. Z. Hiew
School of Engineering and Physical Sciences, Heriot-Watt University Malaysia, 62200 Putrajaya, Malaysia

N. Agarwal et al. (eds.), *Nanomaterials in Wastewater Research*, Advances in Wastewater Research, https://doi.org/10.1007/978-981-96-3908-3_1

1.1 Introduction

Water, a fundamental natural resource essential for sustaining healthy ecosystems and the survival of all organisms, is extensively used in the manufacture of industrial and agricultural goods, as well as in domestic activities (Punia et al. 2021). The escalation of industrial activity and urbanization has led to the generation of considerable volumes of wastewater, which are often discharged without adequate treatment (Dutta et al. 2021). This unsustainable practice has resulted in severe pollution of natural water sources, causing a water shortage crisis and adverse effects on large populations relying on these sources for daily consumption and hygiene. The World Health Organization (WHO, 2023) reported that approximately 2000 million people lived in water-stressed countries in 2021, and over 1700 million people consumed drinking water polluted by fecal matter in 2022. These challenges were intensified in certain regions due to population growth and climate change (Dutta et al. 2021). These data imply that existing wastewater treatment systems are facing bottlenecks, including ineffective treatment, high operational cost and emerging contaminants which are recalcitrant and complex in chemical structure. Hence, effective and innovative treatment methods are critically needed to address these pressing water issues.

Nanomaterials have garnered significance in a number of applications, and these include photovoltaics, semiconductors, catalysis, drug delivery and wastewater remediation. This is due to their tailorable mechanical and physicochemical properties, which are able to enhance their technological performance. Their high surface-to-volume ratio results in unique morphological, chemical, optical and magnetic properties (Dippong 2023). These properties can be tailored for specific applications by modifying parameters during the nanomaterial synthesis. Some of the promising technological applications of nanomaterials in wastewater treatment include adsorption (Al-Kadhi et al. 2022), membrane filtration (Shukla et al. 2018), photocatalysis (Li et al. 2021), disinfection (Yen et al. 2024), coagulation and flocculation (Ribau Teixeira et al. 2024).

Nanomaterials are defined as materials containing at least 50% nanoparticles, with one or more exterior dimensions measuring from 1 to 100 nm (Mekuye and Abera 2023). Due to this overly broad and general definition, a wide variety of materials has been labeled as nanomaterials and classified in several ways, such as by dimensionality, chemical composition and origin. Table 1.1 presents the various classes and types of nanomaterials. According to dimensional categorization, there are zero-dimensional (0D), one-dimensional (1D), two-dimensional (2D) and three-dimensional (3D) nanomaterials. 0D nanomaterials are the group of materials with all their dimensions within the nanoscale (<100 nm) and they are mostly spherical in shape; however, cubic and polygonal shapes of nanoscale can also exist under this class. 1D nanomaterials have one dimension beyond the nanometer range (>100 nm), while the other two dimensions are < 100 nm. These materials usually exhibit a fine wire-shaped or needle-shaped structure, where one dimension is relatively long. In 2D nanomaterials, two of their dimensions are not in the nanoscale, but the other dimension is < 100 nm. Hence, 2D nanomaterials exhibit plate-like or wafer-like configurations. For 3D nanomaterials, all their dimensions are beyond the nanoscale (>100 nm), and they are inter-linked within a bulk matrix to enhance properties, such as mechanical strength, robustness, electronic feature and water adsorption (Mekuye and Abera 2023).

Nanomaterials that are classified according to their major chemical composition can exist as carbon-based, inorganic-based, organic-based and composite nanostructures (Table 1.1). Carbon-based nanomaterials consist of sp^2-hybridized carbon atoms assembled into different dimensions. They exhibit distinct thermal and electrical conductivity, chemical reactivity and stability, as well as exceptional optical and magnetic properties. Organic-based nanomaterials, formed mostly from organic molecules, are held together by covalent interactions, transforming them into special structures like micelles, dendrimers and liposomes. Inorganic-based nanomaterials lack carbon atoms and are mainly composed of metals, metal oxides or semiconducting nanoparticles. Composite nanomaterials are made up of several phases, with at least one phase measuring < 100 nm. In cross-linked nanomaterials, composites with improved properties, such as increased surface area and adsorption capacity, are fabricated (Tee et al. 2024).

Nanomaterials can also be categorized based on their origin, which includes both natural and artificial nanoparticles (Table 1.1). Natural nanomaterials occur in nature, such as viruses and blood, and can also be produced through natural processes, such as nanoparticles present in butterfly wings, volcanic ash, forest fires and ocean spray. Artificial nanomaterials, on the other hand, are produced through human activities. Some are formed for no purpose (incidental nanomaterials), such as those resulting from diesel engine operation and mining. Others are deliberately created with a purpose for certain applications (engineered nanomaterials), such as nanocatalysts and nanosorbents.

This chapter aims to provide an overview of the nanomaterials used in wastewater remediation. It discusses the classification of various nanomaterials, followed by the synthesis methods including the top-down and bottom-up techniques. Common characterization techniques are also covered, along with the notable properties of nanomaterials that make them effective in wastewater treatment. Lastly, the chapter presents the diverse applications of nanomaterials in wastewater remediation.

1.2 Synthesis of Nanomaterials

Nanomaterials can be synthesized using various techniques that enable precise control over their properties for specific applications. These techniques are generally categorized into top-down and bottom-up fabrication processes. Top-down processes create nanostructures from their bulk counterparts through mechanical or physical methods, such as arc discharge, laser ablation, exfoliation and electrospinning. Specifically, these techniques involve breaking down larger material into nanoscale structures. In contrast, bottom-up processes involve assembling individual atoms

Table 1.1 Classification and types of nanomaterials (Mekuye and Abera 2023)

Classification		Type
Dimensionality	Zero-dimensional (0D)	Quantum dots, fullerenes, nanoparticles, nanospheres, nanoclusters
	One-dimensional (1D)	Nanotubes, nanowires, nanofibers, nanorods, nanohorns
	Two-dimensional (2D)	Graphene, graphyne, borophene, germanene, silicone, stanene, plumbene, phosphorene, antimonene, bismuthene, goldene, nanofilms, nanolayers, nanocoatings, nanosheets, transition metal dichalcogenides
	Three-dimensional (3D)	Nanotube bundles, nanowire bundles, nanoparticle dispersion, multi-nanolayers, nanocomposite, aerogels, beads, foams, sponges
Chemical composition	Carbon-based	Carbon nanotubes, carbon quantum dots, fullerenes (buckyballs), graphene, graphene oxide, nanodiamonds
	Inorganic-based	Metal, metal oxide (gold, silver, cerium oxide, iron oxide, manganese oxide, titanium oxide, zinc oxide) and semiconducting (silicon, ceramic) nanoparticles
	Organic-based	Cellulose nanocrystals, chitosan nanoparticles, dendrimers, ferritins, micelles, liposomes
	Composite	Graphene oxide composite, reduced graphene oxide composite, polymer nanocomposite
Origin	Natural	Protein molecules, viruses, minerals (clay), natural colloids (milk, blood), fog (aerosol type), gelatin (gel type), mineralized natural particles (bones, corals, shell, ocean spray, volcanic ash), natural biomolecules (gecko feet, insect wings, spider silk, lotus leaves)
	Artificial	Metal-based nanoparticles, dendrimers, composites

or molecules into well-defined nanostructures. Methods like chemical vapor deposition (CVD), cross-linking self-assembly, co-precipitation, sol–gel, hydrothermal and solvothermal are some examples of bottom-up processes. These methods are generally more effective at regulating the physical and chemical features of nanomaterials, which include antibacterial activity (Yen et al. 2024), high thermal and electrical conductivity (Hassan et al. 2021), surface area (Sebastian et al. 2024), magnetism (Yang et al. 2024) and quantum effects (Yat et al. 2024). As a result, the synthesized nanoparticles can be applied to a variety of fields, including wastewater treatment.

1.2.1 Top-Down Techniques

1.2.1.1 Arc Discharge

Arc discharge is a synthesis technique for creating nanomaterials, especially carbon-based ones like fullerenes and carbon nanotubes (CNTs). In this method, a high-temperature arc is generated between two electrodes, which are typically made of carbon. The intense heat from the arc vaporizes the carbon electrodes, and upon cooling, the vapor condenses to form different nanostructures.

Depending on the arc conditions and the catalyst applied, different forms of CNTs (e.g. single-walled, multi-walled) and fullerenes (e.g. C60, C70) can be synthesized. Generally, higher yields of CNTs are obtained at an arc voltage of 30 V, while a voltage of 40 V tends to produce greater amounts of carbon-soot containing fullerenes (Madhurima et al. 2024). The arc discharge technique is versatile as it allows the production of various forms of nanomaterials by adjusting process parameters. Nevertheless, it requires high precision in process control to synthesize nanomaterials with uniform size and specific nanostructures. In addition, the method is energy-intensive and the resulting product often contains impurities, necessitating further purification (Madhurima et al. 2024).

1.2.1.2 Laser Ablation

Laser ablation involves the use of concentrated laser pulses to deliver a high amount of energy over a very short duration, typically within femtoseconds to nanoseconds. The targeted material undergoes rapid heating, followed by vaporization. The vaporized material forms a plasma plume, which condenses to form nanostructures as it cools.

There are different ways of producing nanomaterials via laser ablation. One common method is pulsed laser ablation in liquid (PLAL), where a target material is immersed in a liquid during which nanoparticles can be directly condensed in the liquid. Another method, pulsed laser deposition (PLD) allows thin films to be formed when a substrate is used for the deposition of ablated material. In the recent years, most nanomaterials have been fabricated through PLAL. For instance, the ablation of a graphite target immersed in dimethylformamide (DMF) results in the synthesis of nitrogen-doped graphene quantum dots (GQDs) (Novoa-De León et al. 2019). The DMF solvent confines the ablated atoms, radicals, ions and clusters, facilitating the formation of GQDs and supporting nitrogen doping in the GQDs. In another study, perovskite and graphene oxide were dispersed in water and irradiated by pulse laser to synthesize a titanium-doped reduced graphene oxide composite (Luo et al. 2022).

In addition, laser ablation can also be used for surface functionalization to enhance properties such as adhesion, wettability and catalytic activity of the nanomaterial. To enhance the catalytic efficiency, CNTs have been functionalized with metal oxides, such as tin (IV) oxide (Al Baroot et al. 2023), nickel (II) oxide (Al-Kadhi et al.

2022) and zinc (II) oxide (Mostafa et al. 2021). Laser ablation offers versatility and simplicity in synthesizing a wide range of nanomaterials. However, its broader application is limited by the high energy consumption and cost associated with the use of high-powered lasers and precision.

1.2.1.3 Exfoliation

Exfoliation is a technique to produce 2D nanomaterials by separating layered materials in their bulk forms to obtain thin, nanoscale layers like graphene. This technique focusses on breaking the weak van der Waals bonds, which link the layers together. Exfoliation can be performed through mechanical, chemical, thermal or electrochemical means.

In mechanical exfoliation, the well-known Scotch Tape method was first applied by Geim and Novoselov to isolate graphene from graphite (Novoselov et al. 2004). Of late, graphene is more popularly made through liquid phase exfoliation where ultrasonic waves are applied to separate the graphitic layers. The bulk material can also be chemically modified, such as the oxidation of graphite to graphite oxide, before exfoliation to produce thin layers of graphene oxide. This can then be reduced to form reduced graphene oxide, which exhibits graphene-like properties. Notably, the oxygen-containing functional groups on the surface of graphene oxide enhance its adsorption capabilities, serving as active sites for pollutant attachment or for further functionalization to target specific contaminants (Tee et al. 2024; Loh et al. 2024).

Electrochemical exfoliation involves the use of an electric field across a bulk material placed in an electrolyte solution. Ions or gases are inserted between the graphitic layers, aiding their separation. For instance, carbon quantum dots (CQDs) can be synthesized using CO_2 nanobubbles and a surfactant, where the nanobubbles wedge between the layers of graphene, enhancing exfoliation of the graphite rods (Yat et al. 2024).

Apart from carbon-based nanomaterials, liquid phase exfoliation is also used to produce 2D transition metal dichalcogenides (TMDs), such as hexagonal-boron-nitride nanosheets. In one approach, tannic acid was introduced into the liquid phase to promote exfoliation while simultaneously functionalizing the nanosheets (Keshebo et al. 2021).

While exfoliation offers several advantages such as functionalization potential and cost-effectiveness, it also has limitations. These include low yield, labor-intensive procedures and the introduction of structural defects due to mechanical or chemcial treatments, which may negatively affect the final product's properties.

1.2.1.4 Electrospinning

Electrospinning is extensively employed for fabricating nanofibers. In this method, a strong electrical field is used to induce a fine jet of polymer solution, which stretches into thin strings. Fibers are formed as the solvent in the jet evaporates and the polymer solidifies upon reaching the collector. The diameter and morphology of the fibers depend on solution properties such as viscosity, surface tension and conductivity, as well as process parameters like applied voltage, flowrate and needle diameter (El-Aswar et al. 2022).

Natural polymers, including chitosan, can be electrospun to form nanofibers. Nanofibrous membranes have been fabricated from a polymeric solution containing chitosan and polyvinyl alcohol (PVA) (Managheb et al. 2021). In one study, porous nanofibers with an average diameter of 99.1 nm and a pore diameter of 0.189 μm were deposited on a polyethersulfone substrate. In another study, graphitic carbon nitride (g-C_3N_4) and titanium dioxide (TiO_2) were incorporated into chitosan-based nanofibers to prevent aggregation of the particles (Li et al. 2021). Interestingly, the addition of g-C_3N_4 and TiO_2 not only produced thinner nanofibers (85 nm diameter) but also enhanced the acid resistance of the chitosan matrix.

Nanofibers can be effectively fabricated by electrospinning, particularly for the synthesis of nanocomposites used in the adsorption (Kim et al. 2021) or photodegradation (Xie et al. 2021) of pollutants. Nevertheless, achieving uniform thickness and surface area of the nanofibers remains challenging and requires precise control of the process parameters.

1.2.2 Bottom-Up Techniques

1.2.2.1 Chemical Vapor Deposition

Chemical vapor deposition (CVD) is commonly used in the synthesis of thin films and coatings. It involves the decomposition or chemical reaction of gaseous precursors to form solid nanomaterials. Graphene nanosheets (GNS) can be repeatably produced by CVD with high productivity. Typically, GNS are grown on transition metal or metal oxide catalysts such as magnesium oxide (MgO). One study employed wastewater sediments containing mainly Mg ions, along with Fe, Mn and Ca ions, as the catalyst precursor for GNS growth (Wang et al. 2024). The type of nanostructures formed by CVD was found to depend on the operating temperature. At 650 °C, GNS with 4 to 6 layers were successfully synthesized, whereas a composite of GNS and CNT was formed at 850 °C.

The effect of temperature was also observed by Chen et al. (2021), where a higher reduction temperature of 700 °C led to the growth of CQDs on CNTs, whereas at 600 °C, defective CNTs were produced. The CQD-decorated CNT was endowed with a larger surface area and more surface functional groups due to the successful attachment of CQDs.

MXenes which are metallic carbides, are 2D nanomaterials commonly explored for their photocatalytic activity. Thirumal et al. (2022) achieved the direct growth of CNTs on the surface of a MXene catalyst through CVD to produce a hybrid nanocomposite.

1.2.2.2 Cross-Linking Self-Assembly

Cross-linking self-assembly is a technique used to synthesize 3D nanomaterials, such as graphene, CNTs and metal–organic frameworks (MOFs). The use of cross-linking agents facilitates the development of an interconnected network through either chemical or physical bonds. Polymers are often used as cross-linking agents to improve mechanical properties and introduce functional groups into the 3D material. For instance, a 3D reduced graphene oxide aerogel grafted with MnO_2 nanoparticles was synthesized using the biopolymer carmellose sodium to support the aerogel network (Hiew et al. 2022). The hydroxyl groups in the cross-linker bonded with the carboxylic and hydroxyl groups of graphene oxide to create a mesoporous aerogel in addition to providing active sites for pollutant attachment. Wu et al. (2024) investigated different polymers, such as polyethyleneimine, glyoxal, polyvinyl chloride and polyvinyl alcohol, for use in chitosan-based aerogels. The authors determined that polyvinyl alcohol generated the most stable aerogel, capable of withstanding acidic conditions.

Metal ions can also be used as cross-linkers as they can coordinate with the chemical functional groups present in the nanomaterials. Loh et al. (2024) employed aluminum ions to provide positively charged anchor points for cross-linking between graphene oxide nanosheets and carboxymethyl cellulose, enhancing the mechanical properties of the network. In another study, Li et al. (2024) constructed an MXene-based aerogel using ferrous ions (Fe^{2+}) as the cross-linking agent. The 2D MXene sheets rapidly formed a dense network, which prevented stacking of MXene layers and improved stability of the structure.

The cross-linking self-assembly method provides a versatile approach to fabricating a wide variety of 3D nanostructures with unique properties and functionalities. However, the choice and concentration of cross-linkers must be carefully considered as they can significantly affect the performance of the as-synthesized nanomaterials.

1.2.2.3 Co-Precipitation

Co-precipitation is an extensively used method for producing mixed metal oxides, hydroxides and composites. In this technique, multiple components in a solution are simultaneously precipitated to form a solid phase. For instance, iron oxide (Fe_3O_4) nanoparticles are commonly synthesized via co-precipitation of a mixture of Fe(II) and Fe(III) solutions. These nanoparticles are then incorporated into other materials, such as chitosan (Asgari et al. 2020), Zn-based MOFs (Ghourchian et al. 2021) and cotton fabric (Yang et al. 2024), through additional synthesis methods.

Co-precipitation can also be performed with mixtures containing three or more metal salts. Silver-doped octahedral zinc ferrites nanoparticles were prepared by adding ammonia solution into nitrate salts of Zn(II), Fe(II) and Ag(I) (Sebastian et al. 2024). The oxalate co-precipitation method was used to synthesize Ni-Mg-Zn ferrites from sulfate salts of Fe(II), Zn(II), Ni(II) and Mg(II), with potassium oxalate acting as the precipitant (Bagade et al. 2023).

1.2.2.4 Hydrothermal and Solvothermal Methods

Hydrothermal and solvothermal techniques rely on the chemical reactions occurring at high temperatures and pressures to produce nanomaterials with precise control over particle size and shape. The main difference between the two methods lies in the type of solvent used: hydrothermal processes use water-based solvents, while solvothermal processes employ organic solvents. As a result, solvothermal methods can operate at higher reaction temperatures and accommodate a broader range of materials.

To prepare a ternary KPCN/GO/$ZnFe_2O_4$ photocatalyst, Kumar et al. (2023) used a hydrothermal process to combine g-C_3N_4 doped with K and P, zinc ferrite and graphene oxide in an autoclave at 180 °C for 12 hours. For the formation of a quaternary Fe_3O_4/CuO/ZnO/rGO catalyst, iron oxide and copper oxide nanoparticles were synthesized in ethylene glycol at 200 °C and in ammonia at 130 °C, respectively, through a solvothermal process (Shen et al. 2023). These nanoparticles were then combined with graphene oxide and zinc oxide in a hydrothermal process to fabricate the co-catalytic system.

Hydrothermal synthesis also offers a multifunctional route, encompassing reduction, doping and self-assembly of 2D graphene oxide to 3D-doped graphene in a single-stage process (Tee et al. 2023a). For instance, an aminated nitrogen-doped graphene hydrogel was synthesized hydrothermally by treating graphene oxide with ethylenediamine at 200 °C for 12 hours. Ethylenediamine served both as the nitrogen source for doping and as a bridging agent for forming the 3D hydrogel (Yongsheng et al. 2021). Another study used thiourea as a sulfur doping source for graphene oxide, along with titanium isopropoxide to develop a photocatalyst through hydrothermal heating at 200 °C for 2 hours (Giang et al. 1271).

Both the hydrothermal and solvothermal processes are effective for producing nanomaterials, where each method offers distinct advantages. The hydrothermal method is more environmentally friendly and straightforward due to its use of water as a solvent. On the other hand, the solvothermal method allows for a wider variety of nanomaterials to be synthesized through the use of diverse organic solvents.

1.2.2.5 Sol–Gel Method

The sol–gel method is a versatile technique to produce various nanomaterials such as thin films, nanoparticles, mesoporous silica, quantum dots and metal-organic frameworks (MOFs). This method typically involves the transition of a colloidal suspension, known as a “sol” into a semi-solid “gel” as condensation reactions occur to form a 3D network.

In synthesizing TiO_2 quantum dots, Mohamed et al. (2023) applied the sol–gel method with titanium tetrachloride ($TiCl_4$) as the precursor. This method is both time-efficient and cost-efficient, requiring relatively low calcination temperatures of 280 °C and 310 °C to produce quantum dots measuring 2.9 and 3.2 nm. In another study, TiO_2 was deposited as a thin coating onto glass beads using the sol–gel method

(Matoh et al. 2023). This coating involved dipping the glass beads into the TiO_2 solution, followed by hot air drying and thermal treatment.

Auto-combustion sol–gel synthesis is a variation of the sol–gel technique, involving an exothermic chemical reaction upon formation of the gel. The addition of a fuel to the precursor solution generates sufficient heat for the decomposition of precursors to form the desired oxides. This technique has been used to synthesize various metal oxide nanomaterials using different fuels. For example, fructose was employed to make Holmium titanate ($Ho_2Ti_2O_7$) nanostructures (Valian et al. 2022), oxalic acid to form double perovskite nanoparticles (Tb_2CoMnO_6) (Dara et al. 2021), banana juice to synthesize $CeO_2/MnCr_2O_4$ nanoparticles (Oroumi et al. 2024) and citric acid to produce copper-doped and dysprosium-doped $NiFe_2O_4$ magnetic nanomaterials (Kotwal et al. 2023).

The flexibility of the sol–gel technique allows the fabrication of a broad range of nanomaterials for various applications. In addition, when combined with a self-sustaining combustion reaction, this method is improved for producing nanomaterials with higher purity and finer particle sizes.

1.3 Characterization of NMs

Characterization of nanomaterials is essential to gain insights into their behavior at the nanoscale. The distinctive properties of nanomaterials render them viable for use in various fields, including energy storage, electronics, optics, catalysis and water pollution remediation. Hence, determining their physicochemical properties and exploring how these properties influence functionality is a critical step in developing new nanomaterials. Generally, the properties of nanomaterials are closely linked to factors such as particle size, surface area, porosity, shape and structural properties. Characterization of nanomaterials is typically performed using high-precision microscopy and spectroscopy techniques, such as field emission scanning electron microscopy (FESEM), transmission electron microscopy (TEM), atomic force microscopy (AFM), scanning tunneling microscopy (STM) and X-ray diffraction (XRD). These tools enable the tailoring of nanomaterial properties such as surface charge density (Ribau Teixeira et al. 2024), magnetism (Yang et al. 2024) and band gap energy (Chegel 2023) to meet specific application requirements through precise manipulation and control over size, shape, synthesis parameters and surface modifications. Additionally, characterization offers information on the quality of the synthesized nanomaterials, and the strengths and limitations of the synthesis methods. In this section, the key properties of nanomaterials will be discussed along with the appropriate characterization instruments.

1.3.1 Particle Size

Particle size has a dominant influence on the properties of nanomaterials. It has been reported that nanoparticles, such as gold, begint to display new properties as their size is reduced to the nanometer scale, as shown in Fig. 1.1. For instance, the melting point, chemical reactivity, optical, electronic, magnetic and adsorption properties of nanomaterials differ significantly from those of their original bulk form (Khan et al. 2019). Therefore, evaluating particle size is a fundamental aspect of nanomaterial characterization. This may involve measuring the overall physical dimensions or the effective size of nanocomposites. High-resolution imaging tools, including FESEM, TEM, AFM and XRD are commonly used for this purpose. FESEM provides size estimations according to the magnifying scale. For example, the length of carbon nanotubes produced by arc discharge was determined to be in the order of several hundred microns using FESEM (Madhurima et al. 2024). TEM offers detailed imaging, and in one study, nitrogen-doped quantum dots were determined to range from 1.5 to 7.5 nm in size, with an average of 4.5 nm (Novoa-De León et al. 2019). XRD can also be used to estimate particle size from diffraction spectrum using Bragg's Law (Madhurima et al. 2024). Another effective technique is AFM which provides surface profiling. For instance, boron nitride nanosheets functionalized with tannic acid were found to have lateral dimensions greater than 500 nm and thickness smaller than 4 nm (Keshebo et al. 2021).

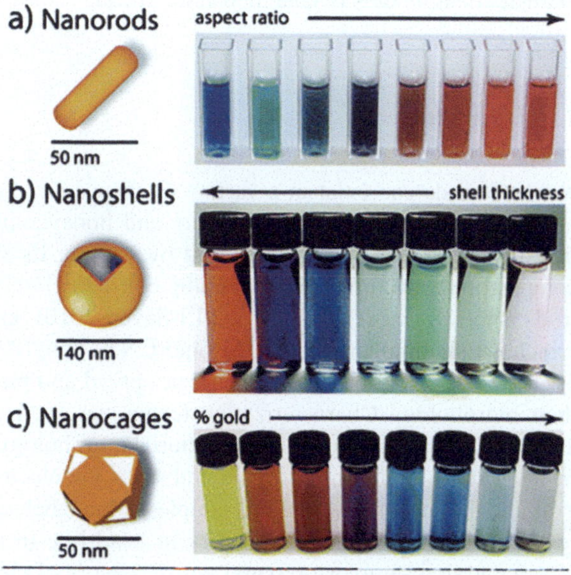

Fig. 1.1 Color dependence of various gold nanoparticles on their size and shape. Modified from Khan et al. (2019) with permission

1.3.2 Surface Area and Porosity

Surface area and porosity are important properties of nanomaterials as they affect surface reactivity and the interactions between surface atoms and surrounding media. The ability to fabricate nanomaterials with large surface area and high porosity has broadened their application spectrum. These properties generally provide a large quantity of functional sites, ample storage voids and improved accessibility to active sites (Hiew et al. 2022). The specific surface area of nanomaterials is commonly determined by nitrogen gas adsorption–desorption analysis using Brunauer–Emmett–Teller (BET) calculations. In addition, t-plot, density functional theory (DFT) and Barrett–Joyner–Halenda (BJH) models are used to determine pore size, pore volume and pore size distribution (Madhurima et al. 2024). For example, anickel-doped $CuBi_2O_4$ photocatalyst was found to be highly mesoporous as it displayed a type-IV nitrogen adsorption isotherm with a hysteresis loop of type H3 between relative pressures of 0.8 and 99 (Haqmal et al. 2024). BET analysis revealed that nickel doping significantly enhanced the surface area from 48.67 to 82.43 m^2/g. The difference in atomic size of the dopant and other composites of the photocatalyst causes crystalline imperfections, which in turn enhance its surface irregularity and surface area for photocatalysis. These enhancements facilitate the breakdown of toxic organic pollutants (Haqmal et al. 2024). In the synthesis of a nanofibrous affinity membrane, the average fiber diameter, porosity and pore diameter were determined from the SEM images, coupled with image processing software such as ImageJ and appropriate mathematical models (Managheb et al. 2021).

1.3.3 Shape

Nanomaterials can exist in a variety of shapes and geometries, including spheres, tubes, helices, ribbons, planes, fibers and honeycombs. Interestingly, the properties of nanomaterials can be altered by varying its shape alone, even when their composition and dimensions remain constant. For example, gold nanoparticles (rods, spheres, cages) (Huang and El-Sayed 2010), graphene nanoribbons (zigzag, armchair) (Majumder et al. 2024) and CNTs (single-walled, multi-walled) (Anzar et al. 2020) exhibit different electrical, optical and plasmonic effects depending on their morphology Characterization techniques such as FESEM, SEM, AFM and TEM enable the detection of nanomaterial morphology at sub-nanometer resolution. In the synthesis of nanofibrous membranes via electrospinning, SEM was used to examine the morphology of the samples (Managheb et al. 2021). The SEM images revealed the presence of impurities in bead-like in some of the samples. It was also found that samples prepared with 3% wt./v of chitosan and chitosan/polyvinyl alcohol in 1:1 ratio exhibited a favorable nanofiber morphology, namely uniform structure with smooth surfaces and the absence of bead-like impurities. In a different study, FESEM images of CNTs produced by arc discharge revealed the presence of graphitic carbon particle impurities, indicating the major drawback of this synthesis method (Madhurima et al. 2024). Despite this, arc discharge remains advantageous for producing CNTs with lengths of several hundred microns high crystallinity. In the preparation of nitrogen-doped quantum dots, TEM was employed for morphology characterization (Novoa-De León et al. 2019). The low-resolution TEM images showed agglomeration of the nanoparticles caused by the drying process, while the high-resolution images depicted isolated nanocrystals with particle sizes varying between 1.5 and 7.5 nm.

1.3.4 Structural Properties

Structural properties of nanomaterials can be broadly categorized into physical (interconnecting frameworks) and chemical (crystallography) characteristics, ranging from the atomic or molecular level to the microstructure level (Munaweera and Madhusha 2023). These structural properties can be analyzed using XRD, Raman spectroscopy, TEM and AFM.

XRD is commonly used to identify and quantify the different crystalline phases of nanomaterials. Each crystalline phase exhibits a unique diffraction pattern, which can be matched against reference databases. Material identification can also be performed by comparing the XRD diffractograms with known standards. For example, in the synthesis of CNTs by Madhurima et al. (2024), the XRD peaks detected are due to stacking of graphitic sheets, hexagonal lattice structures, in-plane regularities and reflection from parallel basal honeycomb planes. The authors also detected a strong (002) peak, which indicated a high degree of order in the stacked layers of CNTs. In composite materials, the XRD pattern typically displays the characteristic peaks of each constituent, provided their crystal structures remain intact during synthesis (Li et al. 2021).

TEM can be used to analyze the interlayer spacing between nanosheets through inverse fast Fourier transform (IFFT) analysis (Keshebo et al. 2021). An interlayer spacing of 0.35 nm was observed for functionalized boron nitride nanosheets which closely matched the 0.34 nm value calculated from XRD using the Bragg's Law (0.34 nm). Additionally, selected-area electron diffraction (SAED) patterns obtained via TEM revealed a hexagonal structure with high crystallinity, indicating that exfoliation using tannic acid did not compromise the material's crystalline integrity (Keshebo et al. 2021).

Raman spectroscopy provides information regarding the structural disorder and crystallinity of carbon-based nanomaterials. The G-peak and D-peak correspond to sp^2 and sp^3 domains, respectively (Novoa-De León et al. 2019). A high D-peak to G-peak intensity ratio (I_D/I_G) indicates the presence of structural defects, edges and dangling sp^4 carbon bonds. Nitrogen doping contributed to great defects that were revealed by the high I_D/I_G. Besides that, a high ratio of 2D-peak to D-peak intensity ratio (I_{2D}/I_D) measures the overall crystalline quality of the graphite network, whereby 2D-peak corresponds to the second order of D-peak (Madhurima et al.

2024). In CNTs fabrication, a high I_{2D}/I_D is desirable as it reflects high quality nanotube formation with long range ordering of the graphite network.

SEM and AFM are useful for evaluating the thickness and lateral dimensions of nanosheets, which are crucial for membrane applications. For example, a membrane synthesized from tannic acid-functionalized hexagonal-boron-nitride nanosheets was found to be relatively thin and dense, as shown in the SEM image of its side view (Keshebo et al. 2021). From AFM analysis, the thickness of the membrane was found to be less than 4 nm, and its lateral size larger than 500 nm. Based on theory, the thickness of a single hexagonal-boron-nitride nanosheet is 0.3 nm. Hence, the AFM results indicate a multilayer form within the synthesized nanosheets.

1.3.5 Elemental and Surface Chemistry

Elemental and surface chemistry analysis allows the determination of elements or components constituting the nanomaterials. From the analysis, further details can be obtained, including the composition, concentration and distribution of elements within the material. The information are crucial for tailoring nanomaterials to suit specific applications. Common techniques used for these properties characterization include X-ray photoelectron spectroscopy (XPS), energy dispersive spectroscopy (EDS), Fourier transform infrared spectroscopy (FTIR) and XRD.

In a recent study, a chitosan/g-C_3N_4/TiO_2 nanofiber intended as adsorbent/ photocatalyst for Cr(VI) removal was analyzed by XPS to assess its chemical composition (Li et al. 2021). The peaks in the XPS spectrum with the respective binding energies confirmed the presence of carbon (C), nitrogen (N), titanium (Ti) and oxygen (O) in the nanofiber. Besides that, binding energy values provided insights into specific chemical bonding between the elements. For example, a peak at 399.3 eV of binding energy indicated N-H_2 bonding in chitosan. The weak signal in the XPS spectrum for Ti 2p was attributed to the low Ti content in the nanofiber. SEM–EDS was then employed to further confirmed the elemental composition of the nanofiber. The EDS elemental mapping showed a homogeneous distribution of C, N, Ti and O throughout the nanofiber, although Ti appeared in significantly lower concentrations.

FTIR spectrum was used to identify chemical functional groups in the material. For example, peaks at 1576 and 1653 cm^{-1} were due to amide groups in the chitosan nanofibers (Li et al. 2021). From XRD, the successful incorporation of raw materials can be deduced from the resultant spectrum, displaying the characteristic peaks of each component, as shown in Fig. 1.2. Moreover, the surface charge of nanomaterials is a crucial property, especially in applications involving pollutant adsorption. Techniques such as zeta potential and point of zero charge measurements are used to determine the surface polarity or charge of the material under varying pH conditions. For instance, carbon-soot and cathode-deposit nanomaterials synthesized by Madhurima et al. (2024), and a graphene oxide composite prepared by Tee et al. (2023b) exhibited a net negative surface charge at neutral pH.

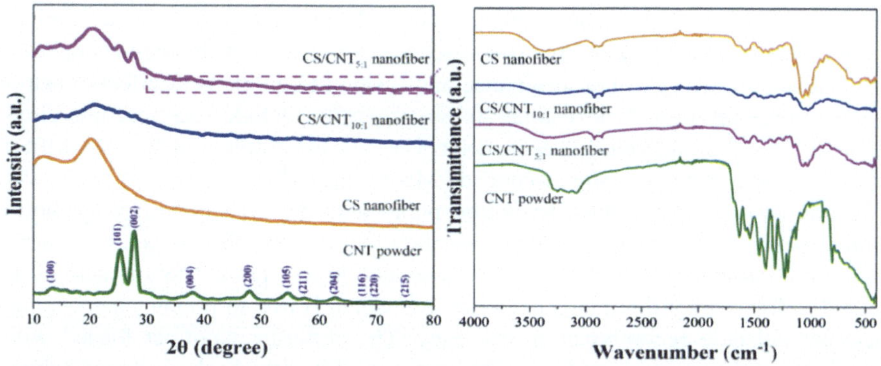

Fig. 1.2 XRD spectrum (left) and FTIR spectrum (right) of g-C_3N_4/TiO_2 (CNT) powder, chitosan (CS) nanofiber and chitosan/g-C_3N_4/TiO_2 (CS/CNT) nanofiber. Modified from Li et al. (2021) with permission.

1.3.6 Magnetic Properties

Magnetic nanomaterials are increasing being used in biomedicine, catalysis, magnetic fluids and environmental remediation, particularly in wastewater decontamination. Their magnetic properties are influenced by several factors, including particle size, shape, composition, crystallographic structure and the synthesis technique used. These properties can be assessed using instruments such as vibrating sample magnetometer (VSM), electron paramagnetic resonance (EPR) and superconducting quantum interference device (SQUID). Notably, the impurity concentration (Punia et al. 2021) and size of nanomaterial (Mekuye and Abera 2023) can negatively affect its magnetic activity up to a certain threshold.

In one study, a nano-catalyst, Fe_3O_4/CuO/ZnO/RGO, was used in a plasma-catalyst system for the degradation of Direct Blue 15 dye. From SQUID-VSM analysis, it was found that the catalyst displayed ferromagnetic behavior (Shen et al. 2023). However, the overall magnetization intensity of the nano-catalyst was lower than that of pure Fe_3O_4. This phenomenon was due to the reduction of exposed crystalline surface from the loading of other nanoparticles. Despite this, the ferromagnetic behavior of the Fe_3O_4/CuO/ZnO/RGO composite allowed the nano-catalyst to be separated from water completely, making it a reusable catalyst for dye pollution remediation. Photocatalysts containing Fe_2O_4 have also shown remarkable magnetic behavior, as determined from VSM studies (Kumar et al. 2023).

1.3.7 Optical Properties

The optical properties of nanomaterials, particularly semiconducting and metallic nanoparticles, are essential for applications in photocatalysis, photoluminescent sensors and photovoltaics. These properties provide information on light absorption, luminescence, reflectivity and phosphorescence, which arise from phenomena such as localized surface plasmon resonance and quantum confinement effects. Instruments commonly used to investigate these properties include spectroscopic ellipsometer, photoluminescence spectrometer, UV–vis spectrophotometer and diffuse reflectance spectrometer (Munaweera and Madhusha 2023).

In one study, a chitosan-based nanofiber photocatalyst was characterized by photoluminescence spectrometry to determine the efficiency of photogenerated charge carrier recombination (Li et al. 2021). Generally, higher photocatalytic activity is related to lower photoluminescence intensity, which is due to poorer photoexcited electron and hole recombination efficiency. The chitosan nanofiber loaded with carbon nitride/titanium dioxide in a 10:1 ratio exhibited the highest photocatalytic activity at the lowest photoluminescence intensity. In addition, the photoluminescence spectrum shows a characteristic peak at 440 nm, indicating that carbon nitride and titanium dioxide were successfully loaded on the chitosan nanofibers. The photocatalyst was further examined using UV–Vis diffuse reflectance spectroscopy to determine its light absorption and band gap properties. The band gap energy was determined using the Kubelka–Munk function (Li et al. 2021). The carbon nitride-loaded and titanium dioxide-loaded chitosan nanofibers exhibited a smaller band gap than pure carbon nitride or titanium dioxide. The result implied that the chitosan composite performed better in degrading the pollutant attributed to its improved photo-absorption ability under visible light irradiation. In another UV–Vis analysis performed by Kumar et al. (2023), the synthesized nanocomposite photocatalyst, KPCN/GO/$ZNFe_2O_4$, was determined to possess higher absorption of visible light than pristine photocatalysts (GCN, PCN, KPCN, KPCN/GO). Furthermore, the band gap analyzed based on the Tauc plot showed that doping of phosphorus and potassium into the nanocomposite has profoundly decreased the band gap of the photocatalyst, enhancing the photocatalytic potential.

A photoluminescence sensor for detecting metal ions and ascorbic acid was developed using nitrogen-doped graphene quantum dots synthesized through pulsed laser ablation in liquid (Novoa-De León et al. 2019). Photoluminescence analysis was performed on the quantum dots under 410 nm of excitation wavelength. The maximum visible light emission was attained at 486 nm corresponding to the relaxation of excited states. However, the initial photoluminescence intensity was relatively low. To enhance the emission, solvothermal treatment was applied, which improved the photoluminescence intensity by tuning the oxygen-containing functional groups of the quantum dots. The photoluminescence emission spectrum was further fitted to the Lorenzian function, confirming that nitrogen was doped into the graphene structure. In addition, the quantum dots were analyzed by UV–Vis-Near-infrared to study the optical absorption characteristics. The results showed that the solvothermal treatment led to a more highly conjugated structure by modifying the oxygenated groups, thereby enhancing the optical properties of the nitrogen-doped quantum dots.

1.4 Application of Nanomaterials in Wastewater Treatment

1.4.1 Types of Water Pollutants

Industrial and domestic wastewater contains a variety of pollutants, which can be broadly categorized into physical, chemical and biological types. Physical pollutants are insoluble solids that remain suspended in water, such as soil, silt, debris and larger particles of plastic, paper and wood. These suspended solids cause turbidity which obstructs sunlight from reaching aquatic plants and animals, thereby disrupting aquatic ecosystems. Mining, construction and metal-processing sectors are the primary sources of suspended solids in industrial wastewater. The suspended solids in domestic wastewater mainly originate from organic debris and food waste discharge from households.

Chemical pollutants can be divided into organic and inorganic types. Organic pollutants such as pesticides, herbicides, oil spills, pharmaceuticals and industrial solvents pose high risks due to their abilities to resist biodegradation and persist in the environment. Some of these contaminants can accumulate in the food chain. Organic pollutants such as organochlorine pesticides and polychlorinated biphenyls, which are toxic and carcinogenic, are widely used in industrial and agricultural sectors (Košnář et al. 2023). In domestic wastewater, organic pollutants are mainly contributed by the use of detergents (Joseph et al. 2021) and pharmaceuticals (Han et al. 2024).

Inorganic pollutants including nitrate, phosphate, chloride, radioactive isotopes (cesium, iodine, uranium) and heavy metals (lead, mercury, cadmium, chromium) are discharged from metal ores processing, mining activities and nuclear facilities. These pollutants can make their way into water resources and the food chain, causing serious health issues (Dhokpande et al. 2024).

Leaching of fertilizers in soils arising from agricultural activities has led to excessive amounts of phosphates and nitrates in the water bodies (Fu et al. 2024). This can cause eutrophication, where the overgrowth of microorganisms like blue-green algae can be promoted by these inorganic pollutants, which in turn causes a reduction in dissolved oxygen. Human and animal waste, and detergents are the main sources of nitrates and phosphates in domestic wastewater (Nancharaiah et al. 2023).

Biological pollutants consist of living organisms or their derivatives, which include pathogens, algae, fungi and biodegradable organic matter. Untreated wastewater containing viruses and bacteria can lead to the spread of diseases, whereas the presence of invasive algae and microorganism species can disrupt local ecosystems and biodiversity. Biological pollutants in domestic wastewater can originate from human and animal waste, and food scraps (Jucherski et al. 2024). The food processing, pharmaceutical and agricultural industries are the major sources of biological pollutants for industrial wastewater (Bôto et al. 2023).

Ultimately, both industrial and domestic wastewater contain diverse pollutants which pose substantial risks to the environment and human health. This necessitates effective treatment, management and monitoring strategies to mitigate their adverse impacts.

1.4.2 Functions of Nanomaterials in Wastewater Treatment Processes

The use of nanomaterials in wastewater treatment provides several advantages over conventional materials, particularly in terms of improved functionality, efficiency and performance. One of the most notable improvements observed in nanomaterials relative to their bulk counterparts is their enhanced adsorption capacity. This is primarily due to their much higher surface area-to-volume ratios which provide abundance of active sites for adsorption. Besides that, many nanomaterials possess porous structures which allow pollutants to access inner surfaces, further increasing their adsorption potential (Tee et al. 2024; Loh et al. 2024; Hiew et al. 2022; Lai et al. 2021).

Nano-photocatalysis typically employs metal-based, metal oxide-based or carbon-based nanomaterials due to their excellent catalytic properties. These materials can absorb a wider spectrum of light wavelengths, facilitate efficient separation of electron–hole pairs, and possess tunable electronic properties for optimum band gap and enhanced reactivity. These features collectively enhance their performance as photocatalysts (Oroumi et al. 2024).

Various strategies have been investigated to incorporate nanomaterials into membrane systems, and these include embedding the materials into membrane supports, grafting them onto the active layer of membranes, coating them as an active layer on a membrane substrate, incorporating them as both active and support layers, or placing them as an interlayer between the support and active layer, as illustrated in Fig. 1.3 (Wang et al. 2021). Furthermore, nanomaterials can act as fillers to enhance membrane performance by improving water permeability and salt rejection rates. Their properties like composition, size, hydrophilicity and porosity, can be tailored to optimize membrane function (Ahmed et al. 2024). The shape of nanomaterials also plays an important role in membrane filtration. Some of the investigated shapes include nanospheres, nanotubes, nanosheets, hollow particles and core–shell nanoparticles (Chew and Yong 2021).

The small size of nanomaterials contributes to their high reactivity, which accelerates coagulation and flocculation kinetics. As a result, nanomaterial-based coagulants/flocculants typically require lower dosages than conventional ones to achieve the same or better coagulation/flocculation efficiencies (Ribau Teixeira et al. 2024). In addition, some nanomaterials can function simultaneously as both coagulant and flocculant, further simplifying the treatment process (Ribau Teixeira et al. 2024). Some nanomaterials possess magnetic properties which facilitate the aggregation of particles and their subsequent separation from water (Afzal et al. 2024). These

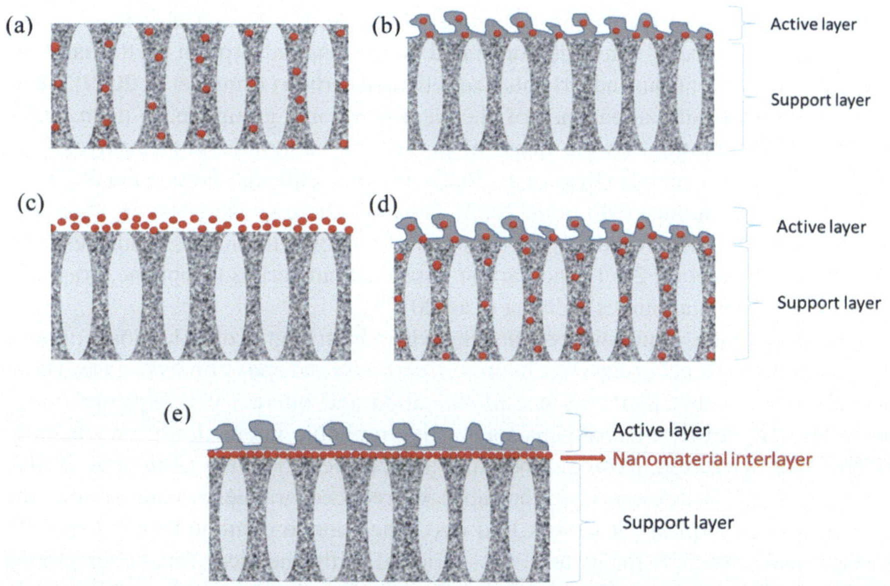

Fig. 1.3 Nanomaterials (a) embedded into membrane supports, (b) grafted onto the active layer of membranes, (c) coated as an active layer on a membrane substrate, (d) incorporated as both active and support layers, or (e) used as an interlayer between support and active layer. Modified from Wang et al. (2021) with permission

features make nanomaterials particularly beneficial for reducing treatment time and chemical usage in coagulation and flocculation applications.

In disinfection, the nanoscale size of materials allows them to effectively penetrate microbial biofilms and cell membranes - protective barriers that conventional disinfectants often fail to infiltrate. Nanomaterials can also be designed to release antimicrobial agents, such as silver ions, in a controlled and sustained manner, enabling long-term and effective disinfection. Unlike conventional disinfectants which often rely on a single mode of action, nanomaterials can offer multiple mechanisms to inactivate microorganisms. This multi-mechanistic approach not only enhances disinfection efficiency but also helps reduce the emergence of antibiotic-resistant microbes (Palanisamy et al. 2023).

1.4.3 Adsorption

Nanomaterials have been extensively studied in recent years as a new generation of adsorbing materials for wastewater treatment. Adsorption is considered an attractive water remediation technique due owing to its relatively simple operation, high selectivity and cost-effectiveness (Zhang et al. 2016). Nanomaterial-derived adsorbents (nano-adsorbents) have demonstrated exceptional adsorption performance as compared to conventional adsorbents like activated carbon (Anjum et al. 2019). Their large surface area and abundance of chemical functional groups make them viable adsorbents for treating a wide range of pollutants, including synthetic dyes (Tan et al. 2015), heavy metals (Kegl et al. 2020), pharmaceuticals (Perumal et al. 2024) and oil-based pollutants (Mir et al. 2022). Generally, nano-adsorbents are classified into two groups: metal oxide-based nanoparticles (iron, zirconium, manganese and titanium) (Gupta et al. 2021) and carbon-based nanomaterials (graphene, graphene oxide and carbon nanotubes) (Zhang et al. 2022a).

Graphene oxide, a monolayer graphene with a high degree of oxidation, contains oxygenated functional groups like hydroxyl, carboxyl and epoxy on its surface. These active groups enable facile surface modification and interact with aqueous pollutants through various mechanisms such as hydrophobic interactions, π-π stacking, electrostatic attraction, hydrogen bonding and covalent bonding (Zhu et al. 2010). Liu et al. (2024) developed a renewable 3D reduced graphene oxide sponge for the efficient adsorption of Congo Red dye. The sponge exhibited a high specific surface area (406.538 m^2/g) and was enriched with chemical functional groups (carboxyl and hydroxyl). The primary adsorption mechanisms included hydrogen bonding, π-π conjugation between the dye and reduced graphene oxide and electrostatic interactions (Liu et al. 2024). In another study, Tee et al. (2024) synthesized a phosphorus-doped 3D graphene oxide composite cross-linked with bentonite and carboxymethyl cellulose for imipramine adsorption. The featured cavities with well-distributed nano-sized openings and chemical functional groups, enabling e effective adsorption of imipramine (458.95 mg/g) via hydrogen bonding, electrostatic attraction, π-π interaction and hydrophobic effects (Tee et al. 2024)

A nanocomposite of CNTs and UiO-66 series metal organic frameworks (MOFs), designated CNT/UiO-66-NH_2, was used to adsorb gallium from aqueous media (Li et al. 2023). The combination of CNTs and MOFs increased the number of active sites, with amine groups playing a major role in gallium adsorption via ionic exchange (Li et al. 2023). Furthermore, mesopores with an average pore size of 3 Å were detected within the adsorbent, contributing to a high specific surface area (419.113 m^2/g) and pore volume (0.740 cm^3/g). These features provided more accessible sites for gallium sorption resulting in a high adsorption capacity (925.44 mg/g) (Li et al. 2023). Moreover, the adsorbent could be regenerated using hydrochloric acid, maintaining an adsorption efficiency of 85.43% after five cycles of regeneration (Li et al. 2023).

Zhang et al. (2022b) innovatively used copper alginate (CA) and CNTs to prepare a nanocomposite for tetracycline adsorption. The CNTs served as a structural support linked to CA, with portions protruding from the surface, enhancing the overall surface area (Zhang et al. 2022b). The incorporation of CNTs significantly improved the adsorption performance, with the CA-CNTs composite achieving a higher tetracycline adsorption capacity (230.2 mg/g) compared to CA alone (134.5 mg/g), demonstrating the synergistic effect of CNTs incorporation (Zhang et al. 2022b).

1.4.4 Membrane Filtration

Membrane filtration is a physical separation technique to filter out pollutants or particulates from a fluid. A sieving medium or membrane with certain pore size is required to selectively retain the pollutants while allowing the fluid to flow through it. There are different types of membrane filtration, including microfiltration (MF), ultrafiltration (UF), nanofiltration (NF) and reverse osmosis (RO). Each of these filtration technique is governed by the size of particulate retained by the membrane (Kyllönen et al. 2005). MF is used to remove particles in the range of 0.1–10 μm, such as suspensions and emulsions, and operates at pressures of 1–4 bar (Anis et al. 2019; Escribá et al. 2021). UF has an exclusion size of 0.002–0.1 μm, or a molecular weight cut-off (MWCO) of 10–100 kDa, which is suitable for filtering humic acids, natural organic matters and biological components at pressures of 2–7 bar (Al Aani et al. 2020). NF is generally used to separate particles around 0.001 μm (1 nm) with an MWCO of 300–500 Da, operating at 6–10 bar (Li et al. 2022). Examples of pollutants for NF includecarbohydrates, synthetic dyes and heavy metals (Elsaid et al. 2023). RO membranes are typically described as having pore sizes of 0.0001–0.001 μm. This membrane type acts as a selective barrier, allowing only water molecules to pass through while rejecting nearly all other substances. This high rejection ability is due to the semipermeable nature of the membrane (Khoo et al. 2022).

Generally, metal oxide NPs (SiO_2, TiO_2, ZnO), carbon nanomaterials and hybrid nanocomposites are often added to existing membrane structures to augment their properties. Properties such as superhydrophilicity, superhydrophobicity, amphiphobicity, anti-fouling and wetting are incorporated into the membrane after the addition of the nanomaterials (Huang et al. 2017). For instance, Sabir et al. (2016) fabricated a desalination membrane from a polymer matrix conjugated with silica nanoparticles (SiO_2 NPs) for the filtration of magnesium sulfate. The incorporation of 0.4 wt% SiO_2 nanoparticles into the membrane resulted in a significant improvement in salt rejection during reverse osmosis operations. Furthermore, the membrane demonstrated enhanced thermal stability and higher permeate flux (Sabir et al. 2016). This superior performance of the desalination membrane was attributed to the surface roughness contributed by SiO_2 nanoparticles, potentially facilitating water transport through the membrane.

Shukla et al. (2018) investigated the use of carbonylated graphene oxide (cGO) to modify the structure of membranes. The fabricated membrane exhibited promising applications in purifying water containing different heavy metal ions, including As(V), Cr(VI), Cd(II), Pb(II) and Zn(II), via nanofiltration. The incorporation of cGO within the membrane structure introduced a significant amount of negative charge due to the abundance of carboxyl groups. This enhanced the membrane's hydrophilicity, leading to an increase in hydraulic permeability from 2.1 $L/m^2{\cdot}h$ (pristine polymer membrane) to 3.5 $L/m^2{\cdot}h$ (Shukla et al. 2018). Furthermore, the negatively charged cGO (zeta potential: − 70 mV) provided ample active sites for the adsorption of cationic metal ions like Cd(II) (82%), Pb(II) (82%) and Zn(II) (87%) (Shukla et al. 2018).

1.4.5 Photocatalysis

Photocatalysis is a green technology that uses light irradiation to activate photocatalysts for use in increasing degradation or transformation rate of organic and inorganic pollutants. Photocatalytic reactions generally occur in four stages: (i) light absorption by the photocatalyst to excite electrons from the valence band (VB) to the conduction band (CB), creating positive holes in the VB, (ii) separation of the photoexcited charges, which are initially bound by Coulombic attraction, (iii) transfer of electrons and holes to the catalyst surface, (iv) use of these charges on the surface for redox reactions (Schneider et al. 2014). In the third stage, a large amount of electrons and holes recombine either on surface sites or during transfer to the surface. In the fourth stage, the photoexcited electrons reduce adsorbed oxygen (O_2) molecules to superoxide radicals ($•O_2^-$) which may further reacts to form highly reactive hydroxyl radicals (•OH). Meanwhile, the VB holes oxidize the surface absorbed hydroxyl groups (OH^-) or water molecules to form hydroxyl radicals (•OH). These potent oxidizing species and free electrons-holes then react or degrade the surface adsorbed pollutants into harmless products.

There are two key challenges in the photocatalytic degradation of pollutants. The photocatalyst needs to possess a narrow bandgap to absorb both visible and ultraviolet light, and to prevent the recombination of photoexcited electron–hole pairs (Pattnaik et al. 2023). This recombination wastes energy and hinders the degradation process. Generally, photocatalysts can be categorized as undoped and doped materials. Undoped photocatalysts typically have low charge recombination rates and are effective for degrading organic pollutants. However, they may struggle to capture electrons for radical formation, limiting their overall efficiency. To address these limitations, doping of nanomaterials is performed to alter the bandgap for visible light absorption, while surface modifications can improve light capture, promote charge separation and increase active sites for pollutant interaction (Pattnaik et al. 2023; Senthil et al. 2017).

Various studies have reported the successful development of nanomaterial-based photocatalysts for pollutant degradation. For instance, Darabdhara and Das (2019) successfully synthesized a dual-responsive magnetic gold-nickel nanostructure-loaded reduced graphene oxide nanocomposite, Au@Ni/rGO, which can be used for colorimetric detection and photodegradation of phenolic compounds. The Au@Ni/rGO nanocomposite exhibited peroxidase-like activity, enabling colorimetric detection of phenolic compounds at a detection limit of 1.68 μM (Darabdhara and Das 2019). Furthermore, the Au nanoparticles generate positive charges in the lower energy 5d level, while presence of Ni enhances light absorption capacity and rGO acts as an electron acceptor, facilitating electron transfer from Au nanoparticles to achieve high photodegradation (> 86.9%) of phenol, 2-chlorophenol and 2-nitrophenol under sunlight irradiation (Darabdhara and Das 2019).

In another study, tin oxide/titanium oxide nanoparticles, SnO_2/TiO_2, were successfully synthesized using a simple solvent-free approach for the photocatalytic degradation of Malachite Green dye (Magdalane et al. 2019). The band gap of the

SnO_2/TiO_2 photocatalyst was found to be 2.8 eV and it was able to degrade the dye under visible light (Magdalane et al. 2019). The photocatalytic mechanism was attributed to the synergistic effect of reduced recombination rate of excitons, enhanced surface area and superior crystallinity, leading to high degradation of Malachite Green dye (~ 96%) (Magdalane et al. 2019).

1.4.6 Disinfection

Disinfection is the process of eliminating or inactivating harmful microorganisms such as bacteria, viruses and algae in wastewater to a level that is safe for release or reuse. Nanomaterials can enhance wastewater disinfection through several mechanisms. Metal oxides nanoparticles, such as titanium dioxide and zinc oxide, can generate reactive oxygen species (ROS) upon ultraviolet or visible light irradiation. ROS, such as hydroxyl radical (•OH) and singlet oxygen ($^1O_2$), damage the cell membranes and proteins of microorganisms, leading to pathogen inactivation. Metal oxide photocatalysts can be immobilized on a support material to minimize particle aggregation and allow easy separation for reuse. Yen et al. (2024) investigated the immobilization of nitrogen-doped titanium dioxide nanoparticles on poly(methyl methacrylate) (PMMA) for the inactivation of *Staphylococcus aureus* and *Escherichia coli (E. Coli)*. The authors concluded that immobilizing the nanoparticles increased ROS formation, thereby improving disinfection efficiency.

Another approach to disinfection involves introducing metal ions with antimicrobial properties, such as silver, copper and zinc, into wastewater. These metal ions specifically bind to the cell membranes and proteins of microorganisms, disrupting cell function and causing cell death. Silver nanoparticle (Ag), in particular, exhibit remarkable antimicrobial activity against a wide range of pathogens. Hydroxyapatite nanowires decorated by Ag nanoparticles were used to construct a porous flow-through reactor for disinfection against various bacteria (Huang et al. 2023). The authors also incorporated materials such as chitosan and tannic acid, both known for their antibacterial properties, to achieve synergism disinfection performance approaching 100%.

Nanomaterials such as graphene oxide and CNTs can cause physical disruption of cell membranes and walls due to their sharp edges. The large surface area and flexibility of graphene oxide sheets also allow them to wrap around bacterial cells, leading to nutrient depletion. Yu et al. (2020) compared different sizes of graphene oxide nanosheets for their antibacterial performance against *Streptococcus mutans*. They discovered that smaller sheets had a greater cutting effect, causing visible cell damage, but minimum cell entrapment. In contrast, the largest graphene oxide sheets exhibited cell entrapment without cutting effects.

1.4.7 Coagulation and Flocculation

Coagulation and flocculation are crucial steps in wastewater treatment plants for removing suspended particles and dissolved substances, with the goal of eliminating turbidity. In coagulation, the process aims to neutralize the charges on suspended particles. This reduces the repulsive forces between particles, enabling them to come together and form flocs.

Flocculation involves the aggregation of these smaller flocs into larger ones, allowing them to settle at the bottom of the tank and be separated from the water. The use of nanomaterials in coagulation and flocculation is versatile, they can assist in charge neutralization, adsorb onto particles to initiate floc formation, act as nuclei for larger floc development or enhance the settling of flocs.

Magnetic nanoparticles outperform conventional coagulants and flocculants due to their high charge density provided by Fe^{2+} and Fe^{3+} ions (Afzal et al. 2024). Their use can shorten the coagulation/flocculation process through magnetic dipole attraction between the magnetic nanoparticles and the suspended or dissolved substances in wastewater. Additionally, magnetic coagulants can act as catalysts for floc development. The incorporation of magnetic nanoparticles into flocs allows magnetic separation, accelerating the settling process. Recovery of magnetic coagulants/flocculants is also possible by applying an external magnetic field. Carboxyl-functionalized magnetic metal-organic framework nanoparticles were employed by Zhu et al. (2020) for the coagulation of dissolved organic matter in wastewater treatment plant effluent. The magnetic coagulant demonstrated improved performance due to its enhanced positive charge and ability to adsorb pollutants through π-π interactions, covalent bonding and hydrogen bonding to form flocs.

Aluminium-based or iron-based coagulants are commonly applied in coagulation/flocculation process. However, they often require pH adjustments and generate large amounts of sludge. Natural coagulants, such as cellulose-based materials, are biodegradable and cost-effective alternatives. Cationic cellulose nanofibers have shown higher removal efficiencies at lower dosages compared to commercial coagulant iron (III) chloride in removing dissolved organic carbon from domestic wastewater (Ribau Teixeira et al. 2024). The nanofiber coagulant primarily facilitates charge neutralization due to its high-density charge. Nevertheless, the positively charged nanofibers also acted as flocculants by inducing floc aggregation through electrostatic interactions.

1.5 Concluding Remarks

The synthesis of nanomaterials involves a variety of techniques which enable precise tuning and control of their properties for specific applications. These techniques are broadly categorized into top-down and bottom-up approaches. Top-down methods, such as arc discharge, laser ablation, exfoliation and electrospinning, break down bulk materials into nanostructures. These processes offer versatility in the types of nanomaterials produced and are widely used despite being energy-intensive and requiring precise operational control.

Conversely, bottom-up methods, including chemical vapor deposition, cross-linking self-assembly, co-precipitation, hydrothermal/solvothermal treatment and sol–gel synthesis, build nanostructures from atomic or molecular precursors. These techniques excel in controlling the chemical and physical characteristics of nanomaterials, making them ideal for applications that demand highly specialized, tailor-made materials.

The choice of synthesis route depends on the desired properties of the nanomaterials and their intended application. With ongoing advancements in production technologies, the ability to tailor nanomaterials for specific uses continues to grow, expanding their applicability across many fields, including wastewater treatment. Characterization techniques including FESEM, TEM, AFM, XRD and FTIR are essential to ensure that the synthesized nanomaterials possess the desired properties and functionalities. These techniques support innovation and the broader application of nanotechnology.

The remarkable properties of nanomaterials, such as ultra-large surface area, high reactivity and potential for functionalization, make them highly effective in removing various pollutants from wastewater. As a result, nanomaterials have demonstrated significant promise in applications including adsorption, membrane filtration, photocatalysis, disinfection and coagulation-flocculation.

Future research should focus on scaling up the synthesis technologies, enhancing the reusability and regeneration of nanomaterials, and ensuring their environmental safety and sustainability. By leveraging the unique properties of nanomaterials, more efficient and effective wastewater treatment solutions can be developed, contributing to cleaner water supplies and improved water recycling rates worldwide.

References

Afzal S, Sherino B, Suppiah DD, Sagadevan S, Julkapli NM (2024) Prospective and potential of magnetic nanoparticles in advanced and sustainable wastewater treatment. J Water Process Eng 63:105540. https://doi.org/10.1016/j.jwpe.2024.105540

Ahmed MA, Mahmoud SA, Mohamed AA (2024) Nanomaterials-modified reverse osmosis membranes: a comprehensive review. RSC Adv 14(27):18879–18906. https://doi.org/10.1039/D4RA01796J

Al Aani S, Mustafa TN, Hilal N (2020) Ultrafiltration membranes for wastewater and water process engineering: a comprehensive statistical review over the past decade. J Water Process Eng 35:101241. https://doi.org/10.1016/j.jwpe.2020.101241

Al Baroot A, Elsayed KA, Haladu SA, Magami SM, Alheshibri M, Ercan F, Çevik E, Akhtar S, Manda AA, Kayed TS, Altamimi NA, Alsanea AA, Al-Otaibi AL (2023) One-pot synthesis of SnO_2 nanoparticles decorated multi-walled carbon nanotubes using pulsed laser ablation for photocatalytic applications. Opt Laser Technol 157:108734. https://doi.org/10.1016/j.optlastec.2022.108734

Al-Kadhi NS, Pashameah RA, Ahmed HA, Alrefaee SH, Alamro, FS, Faqih, HH, Mwafy, EA, Mostafa, AM (2022) Preparation of NiO/MWCNTs nanocomposite for the removal of cadmium ions. J Mater Res Technol 19:1961–1971. https://doi.org/10.1016/j.jmrt.2022.05.149

Anis SF, Hashaikeh R, Hilal N (2019) Microfiltration membrane processes: a review of research trends over the past decade. J Water Process Eng 32:100941 https://doi.org/10.1016/j.jwpe.2019.100941

Anjum M, Miandad R, Waqas M, Gehany F, Barakat MA (2019) Remediation of wastewater using various nano-materials. Arab J Chem 12(8):4897–4919. https://doi.org/10.1016/j.arabjc.2016.10.004

Anzar N, Hasan R, Tyagi M, Yadav N, Narang J (2020) Carbon nanotube—a review on synthesis, properties and plethora of applications in the field of biomedical science. Sens Int 1:100003. https://doi.org/10.1016/j.sintl.2020.100003

Asgari E, Sheikhmohammadi A, Yeganeh J (2020) Application of the Fe_3O_4-chitosan nano-adsorbent for the adsorption of metronidazole from wastewater: optimization, kinetic, thermodynamic and equilibrium studies. Int J Biol Macromol 164:694–706. https://doi.org/10.1016/j.ijbiomac.2020.07.188

Bagade AV, Pund SN, Nagwade PA, Kumar B, Deshmukh SU, Kanagare AB (2023) Ni-doped Mg-Zn nano-ferrites: fabrication, characterization, and visible-light-driven photocatalytic degradation of model textile dyes. Catal Commun 181:106719. https://doi.org/10.1016/j.catcom.2023.106719

Bôto ML, Dias SM, Crespo RD, Mucha AP, Almeida CMR (2023) Removing chemical and biological pollutants from swine wastewater through constructed wetlands aiming reclaimed water reuse. J Environ Manage 326(PartA):116642. https://doi.org/10.1016/j.jenvman.2022.116642.

Chegel R (2023) Magneto-electronic and thermopower properties of B, N and Si doped monolayer graphene. Diam Relat Mater 137:110154. https://doi.org/10.1016/j.diamond.2023.110154

Chen Q, Wang H, Tang X, Ba Z, Zhao X, Wang Y, Deng H (2021) One-step synthesis of carbon quantum dot-carbon nanotube composites on waste eggshell-derived catalysts for enhanced adsorption of methylene blue. J Environ Chem Eng 9(5):106222. https://doi.org/10.1016/j.jece.2021.106222

Chew YT, Yong WF (2021) Recent advances of thin film nanocomposite membranes: effects of shape/structure of nanomaterials and interfacial polymerization methods. Chem Eng Res Des 172:135–158. https://doi.org/10.1016/j.cherd.2021.06.003

Dara M, Hassanpour M, Amiri O, Baladi M, Salavati-Niasari M (2021) Tb_2CoMnO_6 double perovskites nanoparticles as photocatalyst for the degradation of organic dyes: synthesis and characterization. Arab J Chem 14(10):103349. https://doi.org/10.1016/j.arabjc.2021.103349

Darabdhara G, Das MR (2019) Dual responsive magnetic Au@Ni nanostructures loaded reduced graphene oxide sheets for colorimetric detection and photocatalytic degradation of toxic phenolic compounds. J Hazard Mater 368:365–377. https://doi.org/10.1016/j.jhazmat.2019.01.010

Dhokpande SR, Deshmukh SM, Khandekar A, Sankhe A (2024) A review outlook on methods for removal of heavy metal ions from wastewater. Sep Purif Technol. 350:27868. https://doi.org/10.1016/j.seppur.2024.127868

Dippong T (2023) Synthesis, physicochemical characterization and applications of advanced nanomaterials. Materials 16(4) https://doi.org/10.3390/ma16041674

Dutta D, Arya S, Kumar S (2021) Industrial wastewater treatment: current trends, bottlenecks, and best practices. Chemosphere 285:131245. https://doi.org/10.1016/j.chemosphere.2021.131245

El-Aswar EI, Ramadan H, Elkik H, Taha AG (2022) A comprehensive review on preparation, functionalization and recent applications of nanofiber membranes in wastewater treatment. J Environ Manag 301:113908. https://doi.org/10.1016/j.jenvman.2021.113908

Elsaid K, Olabi AG, Abdel-Wahab A, Elkamel A, Alami AH, Inayat A, Chae KJ, Abdelkareem MA (2023) Membrane processes for environmental remediation of nanomaterials: potentials and challenges. Sci Total Environ 879:162569. https://doi.org/10.1016/j.scitotenv.2023.162569

Escribá A, Thome Da Silva BAT, Lourenço SA, Cava CE (2021) Incorporation of nanomaterials on the electrospun membrane process with potential use in water treatment. Colloids Surf A: Physicochem Eng Asp 624:126775. https://doi.org/10.1016/j.colsurfa.2021.126775

Fu W, Zhao Y, Wang Q, Yu X, Song Z, Duan P, Xu M, Zhang X, Rao Z (2024) Characterization of simultaneous removal of nitrogen and phosphorus by novel Raoultella ornithinolytica strain YX-4 and application in real farm wastewater treatment. Bioresour Technol 391:129922. https://doi.org/10.1016/j.biortech.2023.129922

Ghourchian F, Motakef-Kazemi N, Ghasemi E, Ziyadi H (2021) Zn-based MOF-chitosan-Fe_3O_4 nanocomposite as an effective nano-catalyst for azo dye degradation. J Environ Chem Eng 9(6):106388. https://doi.org/10.1016/j.jece.2021.106388

Giang NTH, Tan NN, Huong LM, Hai ND, Thinh NT, Phuc NT, Dat NM, Phong MT, Hieu NH (2023) Photocatalytic degradation of crystal violet on titanium dioxide/graphene aerogel doped sulfur. J Mol Struct 1271:134031 https://doi.org/10.1016/j.molstruc.2022.134031

Gupta K, Joshi P, Gusain R, Khatri OP (2021) Recent advances in adsorptive removal of heavy metal and metalloid ions by metal oxide-based nanomaterials. Coord Chem Rev 445:214100. https://doi.org/10.1016/j.ccr.2021.214100

Han Y, Hu LX, Liu T, Dong LL, Liu YS, Zhao JL, Ying GG (2024) Discovering transformation products of pharmaceuticals in domestic wastewaters and receiving rivers by using non-target screening and machine learning approaches. Sci Total Environ 948:174715 https://doi.org/10.1016/j.scitotenv.2024.174715

Haqmal E, Senthil RA, Ahmed A, Pan J (2024) Uncovering the influence of Ni-doping in $CuBi_2O_4$ photocatalyst on its visible-light-responsiveness for the efficient removal of toxic Alizarin yellow R dye from wastewater. Colloids Surf A: Physicochem Eng Asp 684:133185. https://doi.org/10.1016/j.colsurfa.2024.133185

Hassan K, Hossain R, Sahajwalla V (2021) Novel microrecycled ZnO nanopartcles decorated macroporous 3D graphene hybrid aerogel for efficient detection of NO_2 at room temperature. Sens Actuators B: Chem 330:129278. https://doi.org/10.1016/j.snb.2020.129278

Hiew BYZ, Tee WT, Loh NYL, Lai KC, Hanson S, Gan S, Thangalazhy-Gopakumar S, Lee LY (2022) Synthesis of a highly recoverable 3D MnO_2/rGO hybrid aerogel for efficient adsorptive separation of pharmaceutical residue. J Environ Sci 118:194–203. https://doi.org/10.1016/j.jes.2021.12.036

Huang QL, Huang Y, Xiao CF, You YW, Zhang CX (2017) Electrospun ultrafine fibrous PTFE-supported ZnO porous membrane with self-cleaning function for vacuum membrane distillation. J Membr Sci 534:73–82. https://doi.org/10.1016/j.memsci.2017.04.015

Huang X, El-Sayed MA (2010) Gold NPs: Optical properties and implementations in cancer diagnosis and photothermal therapy. J Adv Res 1(1):13–28. https://doi.org/10.1016/j.jare.2010.02.002

Huang P, Fang W, Yang L, Sun Y, Yang H, Chen XZ, Zeng H (2023) Ultralong hydroxyapatite nanowires-based flow-through reactor with high loading of silver nanoparticles for fast continuous catalytic reduction of organic dyes and disinfection of wastewater. Chem Eng J 475:146305. https://doi.org/10.1016/j.cej.2023.146305

Joseph CG, Farm YY, Taufiq-Yap YH, Pang CK, Nga JLH, Li Puma G (2021) Ozonation treatment processes for the remediation of detergent wastewater: a comprehensive review. J Environ Chem Eng 9(5):106099. https://doi.org/10.1016/j.jece.2021.106099

Jucherski A, Walczowski A, Bugajski P, Operacz A (2024) Reliability of organic and biogenic pollutant removal in selected technologies used in domestic wastewater treatment plants: a comparative analysi. J Environ Manage 354:120381. https://doi.org/10.1016/j.jenvman.2024.120381

Kegl T, Košak A, Lobnik A, Novak Z, Kralj AK, Ban I (2020) Adsorption of rare earth metals from wastewater by nanomaterials: a review. J Hazard Mater 386:121632. https://doi.org/10.1016/j.jhazmat.2019.121632.

Keshebo DL, Hu CC, Hung WS, Wang CF, Tsai HC, Lee KR, Lai JY (2021) Simultaneous exfoliation and functionalization of hexagonal boron nitride in the aqueous phase for the synthesis of high-performance wastewater treatment membrane. J Clean Prod 314:28083. https://doi.org/10.1016/j.jclepro.2021.128083

Khan I, Saeed K, Khan I (2019) Nanoparticles: properties, applications and toxicities. Arab J Chem 12(7):908–931. https://doi.org/10.1016/j.arabjc.2017.05.011

Khoo YS, Goh PS, Lau WJ, Ismail AF, Abdullah MS, Mohd Ghazali NH, Yahaya NKEM, Hashim N, Othman AR, Mohammed A, Kerisnan NDA, Mohamed Yusoff MA, Fazlin Hashim NH, Karim J, Abdullah NS (2022) Removal of emerging organic micropollutants via modified-reverse osmosis/nanofiltration membranes: a review. Chemosphere 305:135151. https://doi.org/10.1016/j.chemosphere.2022.135151

Kim Y, Eom HH, Kim D, Harbottle D, Lee JW (2021) Adsorptive removal of cesium by electrospun nanofibers embedded with potassium copper hexacyanoferrate. Sep Purif Technol 255:117745. https://doi.org/10.1016/j.seppur.2020.117745

Košnář Z, Mercl F, Pierdonà L, Chane AD, Míchal P, Tlustoš P (2023) Concentration of the main persistent organic pollutants in sewage sludge in relation to wastewater treatment plant parameters and sludge stabilization. Environ Pollut 333:122060. https://doi.org/10.1016/j.envpol.2023.122060

Kotwal P, Jasrotia R, Prakash J, Ahmed J, Verma A, Verma R, Kandwal A, Godara SK, Kumari S Maji PK, Fazil M, Ahmad T, Tamboli MS, Sharma N, Kumar R (2023) Magnetically recoverable sol-gel auto-combustion developed Ni1-xCuxDyyFe$_2$-yO$_4$ magnetic nanoparticles for photocatalytic, electrocatalytic, and antibacterial applications. Environ Res 231:116103. https://doi.org/10.1016/j.envres.2023.116103

Kumar R, Sudhaik A, Sonu, Raizada P, Nguyen VH, Van Le Q, Ahamad T, Thakur S, Hussain CM, Singh P (2023) Integrating K and P co-doped g-C_3N_4 with $ZnFe_2O_4$ and graphene oxide for S-scheme-based enhanced adsorption coupled photocatalytic real wastewater treatment. Chemosphere 337:139267. https://doi.org/10.1016/j.chemosphere.2023.139267

Kyllönen HM, Pirkonen P, Nyström M (2005) Membrane filtration enhanced by ultrasound: a review. Desalination 181(1–3):319–335. https://doi.org/10.1016/j.desal.2005.06.003

Lai KC, Hiew BYZ, Tee WT, Thangalazhy-Gopakumar S, Gan S, Lee LY (2021) Usage of a new macro-hierarchical graphene sponge in batch adsorption and packed column configuration for efficient decontamination of cadmium in aqueous environment. J Environ Chem Eng 9(5):106057. https://doi.org/10.1016/j.jece.2021.106057

Li QH, Dong M, Li R, Cui YQ, Xie GX, Wang XX, Long YZ (2021) Enhancement of Cr(VI) removal efficiency via adsorption/photocatalysis synergy using electrospun chitosan/g-C_3N_4/TiO_2 nanofibers. Carbohydr Polym 253 117200. https://doi.org/10.1016/j.carbpol.2020.117200

Li S, Wang X, Guo Y, Hu J, Lin S, Tu Y, Chen Y, Ni Y, Huang L (2022) Recent advances on cellulose-based nanofiltration membranes and their applications in drinking water purification: a review. J Clean Prod 333:130171. https://doi.org/10.1016/j.jclepro.2021.130171

Li W, Zhou C, Li C, Zhu W, Shi J, Liu G (2023) Synthesis of CNT/UiO-66-NH_2 adsorbent and the selective adsorption of gallium in solution. Sep Purif Technol 323:124464. https://doi.org/10.1016/j.seppur.2023.124464

Li G, Liu W, Gao S, Lu H, Fu D, Wang M, Liu X (2024) MXene-based composite aerogels with bifunctional ferrous ions for the efficient degradation of phenol from wastewater. Chemosphere 358:142151. https://doi.org/10.1016/j.chemosphere.2024.142151

Liu Z, Gao B, Han H, Li Y, Fu H, Wei D (2024) A green cross-linking method for the preparation of renewable three-dimensional graphene sponges for efficient adsorption of Congo red dye. Chin J Chem Eng 66:84–93. https://doi.org/10.1016/j.cjche.2023.12.004

Loh NYL, Tee WT, Hanson S, Chiu WS, Hiew BYZ, Khiew PS, Lee LY (2024) Enhanced removal of lead and zinc by a 3D aluminium sulphate-functionalised graphene aerogel as an effective adsorption system. Chemosphere 362:142537. https://doi.org/10.1016/j.chemosphere.2024.142537

Luo P, Liu W, Zhu D, Sun B, Cao Y, Wang Y (2022) One-step synthesis of Ti^{3+} self-doped CTO/rGO composites by pulsed laser ablation in liquid with enhanced photocatalytic activity in water purification. Colloids Surf. A Physicochem. Eng Asp 655:130216. https://doi.org/10.1016/j.colsurfa.2022.130216

Madhurima VP, Kumari K, Jain PK (2024) Synthesis and study of carbon nanomaterials through arc discharge technique for efficient adsorption of organic dyes. Diam Relat Mater 141:110538. https://doi.org/10.1016/j.diamond.2023.110538.

Magdalane CM, Kanimozhi K, Arularasu MV, Ramalingam G, Kaviyarasu K (2019) Self-cleaning mechanism of synthesized SnO_2/TiO_2 nanostructure for photocatalytic activity application for waste water treatment. Surf Interfac 17:100346. https://doi.org/10.1016/j.surfin.2019.100346

Majumder S, Meher A, Moharana S, Kim KH (2024) Graphene nanoribbon synthesis and properties in polymer composites: a review. Carbon 216:118558. https://doi.org/10.1016/j.carbon.2023.118558

Managheb M, Zarghami S, Mohammadi T, Asadi AA, Sahebi S (2021) Enhanced dynamic Cu(II) ion removal using hot-pressed chitosan /poly (vinyl alcohol) electrospun nanofibrous affinity membrane (ENAM). Process Saf Environ Protect 146:329–337. https://doi.org/10.1016/j.psep.2020.09.013

Matoh L, Žener B, Kovačić M, Kušić H, Arčon I, Levstek M, Lavrenčič Štangar U (2023) Photocatalytic sol-gel/P25 TiO_2 coatings for water treatment: degradation of 7 selected pharmaceuticals. Ceram Int 49(14, Part B):24395–24406. https://doi.org/10.1016/j.ceramint.2022.09.204

Mekuye B, Abera B, NMS: An overview of synthesis, classification, characterization, and applications. Nano Select 4(8):486–501. https://doi.org/10.1002/nano.202300038

Mir S, Naderifar S, Rahidi AM, Alaei M (2022) Recent advances in oil/water separation using nanomaterial-based filtration methods for crude oil processing-a review. J Pet Sci Eng 215:110617. https://doi.org/10.1016/j.petrol.2022.110617

Mohamed WAA, Abd El-Gawad HH, Handal HT, Galal HR, Mousa HA, Elsayed BA, Labib AA, Abdel-Mottaleb MSA (2023) TiO_2 quantum dots: Energy consumption cost,germination, and phytotoxicity studies, recycling photo and solar catalytic processes of reactive yellow 145 dye and natural industrial wastewater. Adv Powder Technol 34(1):103923. https://doi.org/10.1016/j.apt.2022.103923

Mostafa AM, Mwafy EA, Toghan A, ZnO (2021) Nanoparticles decorated carbon nanotubes via pulsed laser ablation method for degradation of methylene blue dyes. Colloids Surf A Physicochem Eng Asp 627:127204. https://doi.org/10.1016/j.colsurfa.2021.127204

Munaweera I, Madhusha MLC (2023) Characterization techniques for nanomaterials, CRC Press. https://doi.org/10.1201/9781003354185

Nancharaiah YV, Sarvajith M, Mohan TVK (2023) Pilot-scale aerobic granular sludge reactors with granular activated carbon for effective nitrogen and phosphorus removal from domestic wastewater. Sci Total Environ 894:164822. https://doi.org/10.1016/j.scitotenv.2023.164822

Novoa-De León IC, Johny J, Vázquez-Rodríguez S,. García-Gómez V, Carranza-Bernal S, Mendivil I, Shaji S, Sepúlveda-Guzmán S (2019) Tuning the luminescence of nitrogen-doped graphene quantum dots synthesized by pulsed laser ablation in liquid and their use as a selective photoluminescence on–off–on probe for ascorbic acid detection. Carbon 150:455–464. https://doi.org/10.1016/j.carbon.2019.05.057

Novoselov KS, Geim AK, Morozov SV, Jiang D, Zhang Y, Dubonos SV, Grigorieva IV, Firsov AA (2004) Electric field effect in atomically thin carbon films. Science 306(5696):666–669. https://doi.org/10.1126/science.1102896

Oroumi G, Hemmatzadeh A, Salem KH, Dawi EA, Samimi F, Hameed GF, Taha ZA, Salavati-Niasari M (2024) Green sol-gel auto-combustion synthesis of $CeO_2/MnCr_2O_4$ nanocomposite using banana juice and its potential application for enhanced degradation of methyl orange under visible light. Alex Eng J 102:26–35. https://doi.org/10.1016/j.aej.2024.05.109

AP, Palanisamy G, APL, Albeshr MF, Fahad Alrefaei A, Lee J, Liu X (2023) Photocatalytic degradation of organic pollutants and inactivation of pathogens under visible light via SnO_2/rGO composites. Chemosphere 335:139102. https://doi.org/10.1016/j.chemosphere.2023.139102

Pattnaik A, Sahu JN, Poonia AK, Ghosh P (2023) Current perspective of nano-engineered metal oxide based photocatalysts in advanced oxidation processes for degradation of organic pollutants in wastewater. Chem Eng Res Des 190:667–686. https://doi.org/10.1016/j.cherd.2023.01.014

Perumal MKK, Gandhi D, Renuka RR, Lakshminarayanan A, Thiyagarajulu N, Kamaraj C (2024) Advanced nano-based adsorbents for purification of pharmaceutical residue polluted water: a critical review. Process Saf Environ Protect 186:552–565. https://doi.org/10.1016/j.psep.2024.04.011

Punia P, Naagar M, Chalia S, Dhar R, Ravelo B, Thakur P, Thakur A (2021) Recent advances in synthesis, characterization, and applications of nanoparticles for contaminated water treatment—a review, Ceram. Int 47(2):1526–1550. https://doi.org/10.1016/j.ceramint.2020.09.050

Ribau Teixeira M, Ismail A, Medronho B, Alves L, Pedrosa JFS, Ferreira PJT, Serrão Sousa V, Rosa da Costa AM (2024) Nanofibrillated cationic cellulose derivatives as flocculants for domestic wastewater treatment. J Water Process Eng 58:104817. https://doi.org/10.1016/j.jwpe.2024.104817

Sabir A, Shafiq M, Islam A, Jabeen F, Shafeeq A, Ahmad A, Zahid Butt MT, Jacob KI, Jamil T (2016) Conjugation of silica nanoparticles with cellulose acetate/polyethylene glycol 300 membrane for reverse osmosis using $MgSO_4$ solution. Carbohydr Polym 136:551–559. https://doi.org/10.1016/j.carbpol.2015.09.042

Schneider J, Matsuoka M, Takeuchi M, Zhang J, Horiuchi Y, Anpo M, Bahnemann DW (2014) Understanding TiO_2photocatalysis: mechanisms and materials. Chem Rev 114(19):9919–9986. https://doi.org/10.1021/cr5001892

Sebastian N, Yu WC, Balram D, Hong GT, Alharthi SS, Al-Saidi HM (2024) Ultrasensitive detection and photocatalytic degradation of polyketide drug tetracycline in environment and food samples using dual-functional Ag doped zinc ferrite embedded functionalized carbon nanofibers. Chemosphere 348:140692. https://doi.org/10.1016/j.chemosphere.2023.140692

Senthil RA, Theerthagiri J, Selvi A, Madhavan J (2017) Synthesis and characterization of low-cost g-C_3N_4/TiO_2 composite with enhanced photocatalytic performance under visible-light irradiation. Opt Mater 64:533–539. https://doi.org/10.1016/j.optmat.2017.01.025

Shen Y, Wang Y, Chen Y, Kwang Park J, Fang S, Feng K (2023) Synthesis of Fe_3O_4/CuO/ZnO/RGO and its catalytic degradation of dye wastewater using dielectric barrier discharge plasma. Arab J Chem 16(4):104571. https://doi.org/10.1016/j.arabjc.2023.104571

Shukla AK, Alam J, Alhoshan M, Arockiasamy Dass L, Ali FAA, Muthumareeswaran MR, Mishra U, Ansari MA (2018) Removal of heavy metal ions using a carboxylated graphene oxide-incorporated polyphenylsulfone nanofiltration membrane. Environ Sci Water Res Technol 4(3):438–448. https://doi.org/10.1039/c7ew00506g

Tan KB, Vakili M, Horri BA, Poh PE, Abdullah AZ, Salamatinia B (2015) Adsorption of dyes by nanomaterials: recent developments and adsorption mechanisms. Sep Purif Technol 150:229–242. https://doi.org/10.1016/j.seppur.2015.07.009

Tee WT, Loh NYL, Lai KC, Hiew BYZ, Gan S, Lee LY (2023) Application of 3D heteroatom-doped graphene in adsorptive removal of water pollutants: review on hydrothermal synthesis and its influencing factors. Sep Purif Technol 320:124072. https://doi.org/10.1016/j.seppur.2023.124072

Tee WT, Loh NYL, Hiew BYZ, Chiu WS, Khiew PS, Thangalazhy-Gopakumar S, Gan S, Lee LY (2023) Design and development of a high-performance 3D graphene system for adsorptive removal of organic toxins from wastewater: mechanisms and process optimization. Chem Eng Res Des 195:132-150. https://doi.org/10.1016/j.cherd.2023.05.021

Tee WT, Yong JE, Chua J, Loh NYL, Hiew BYZ, Gan S, Lee LY (2024) A high-performance 3D phosphorus-doped graphene oxide adsorbent for imipramine wastewater treatment. Sep Purif Technol 330:125266. https://doi.org/10.1016/j.seppur.2023.125266

Thirumal V, Yuvakkumar R, Kumar PS, Ravi G, Keerthana SP, Velauthapillai D (2022) Facile single-step synthesis of MXene@CNTs hybrid nanocomposite by CVD method to remove hazardous pollutants. Chemosphere 286:131733. https://doi.org/10.1016/j.chemosphere.2021.131733

Valian M, Salavati-Niasari M, Ganduh SH, Abdulsahib WK, Mahdi MA, Jasim LS (2022) Sol-gel auto-combustion synthesis of a novel chitosan/$Ho_2Ti_2O_7$ nanocomposite and its characterization for photocatalytic degradation of organic pollutant in wastewater under visible illumination. Int J Hydrog Energy 47(49):21146–21159. https://doi.org/10.1016/j.ijhydene.2022.04.221

Wang C, Park MJ, Seo DH, Drioli E, Matsuyama H, Shon H (2021) Recent advances in nanomaterial-incorporated nanocomposite membranes for organic solvent nanofiltration. Sep Purif Technol 268:118657. https://doi.org/10.1016/j.seppur.2021.118657

Wang H, L. Yi, F. Huang, Q. Huang, T. Zhou (2024) Facile synthesis of graphene nanosheets on wastewater sediments for high efficient adsorption of methylene blue. Sep Purif Technol 337:126366. https://doi.org/10.1016/j.seppur.2024.126366

World Health Organization (2023) WHO global water, sanitation and hygiene: annual report 2023. World Health Organization, Geneva. https://www.who.int/publications/i/item/9789240106079. Accessed on 15 May 2025

Wu X, Tang Q, Zhang X, Fang Q, Hua Y, Cai T, Tan W, Kou J, Wang Z, Tian M, Qu J, Li M (2024) Highly acid-stable polyvinyl alcohol/chitosan/phytic acid aerogel for selective separation of uranium from actual wastewater. J Environ Chem Eng 12(5):113358. https://doi.org/10.1016/j.jece.2024.113358

Xie W, Shi Y, Wang Y, Zheng Y, Liu H, Hu Q, Wei S, Gu H, Guo Z (2021) Electrospun iron/cobalt alloy nanoparticles on carbon nanofibers towards exhaustive electrocatalytic degradation of tetracycline in wastewater. Chem Eng J 405:126585. https://doi.org/10.1016/j.cej.2020.126585

Yang H, Zhang P, Zheng Q, Nie G, Hayat A, Bajaber MA, Raza S, Li D, Sui Y (2024) Synergistically active Fe_3O_4 magnetic and EDTA modified cellulose cotton fabric using chemical method and their effective pollutants removal ability from wastewater. Int J Biol Macromol 274(Part1):132996. https://doi.org/10.1016/j.ijbiomac.2024.132996

Yat YD, Foo HCY, Tan IS, Lam MK, Show PL, Ng BWL (2024) Synthesis of carbon quantum dots via electrochemically-induced carbon dioxide nanobubbles exfoliation of graphite for heavy metal detection in wastewater. J Environ Chem Eng 12(3):112715. https://doi.org/10.1016/j.jece.2024.112715

Yen LT, Weng CH, Tzeng JH, Chen YC, Jacobson AR, Lin YT (2024) Substantial improvement in photocatalysis performance of N-TiO_2 immobilized on PMMA: exemplified by inactivation of Staphylococcus aureus and Escherichia coli. Sep Purif Technol 345:127298. https://doi.org/10.1016/j.seppur.2024.127298

Yongsheng X, Xintong L, Hongwei H, Yuexiao S, Qing X, Wenchao P (2021) Aminated N-doped graphene hydrogel for long-term catalytic oxidation in strong acidic environment. J Hazard Mater 401:123742. https://doi.org/10.1016/j.jhazmat.2020.123742.

Yu CH, Chen GY, Xia MY, Xie Y, Chi YQ, He ZY, Zhang CL, Zhang T, Chen QM, Peng Q (2020) Understanding the sheet size-antibacterial activity relationship of graphene oxide and the nano-bio interaction-based physical mechanisms. Colloids Surf B: Biointerfaces 191:111009. https://doi.org/10.1016/j.colsurfb.2020.111009

Zhang Y, Wu B, Xu H, Liu H, Wang M, He Y, Pan B (2016) Nanomaterials-enabled water and wastewater treatment. NanoImpact 3–4:22–39. https://doi.org/10.1016/j.impact.2016.09.004

Zhang Y, Luo J, Zhang H, Li T, Xu H, Sun Y, Gu X, Hu X, Gao B (2022) Synthesis and adsorption performance of three-dimensional gels assembled by carbon nanomaterials for heavy metal removal from water: a review. Sci Total Environ 852:158201. https://doi.org/10.1016/j.scitotenv.2022.158201

Zhang Y, Li Y, Xu W, Cui M, Wang M, Chen B, Sun Y, Chen K, Li Q, Du Q, Pi X, Wang Y (2022) Filtration and adsorption of tetracycline in aqueous solution by copper alginate-carbon nanotubes membrane which has the muscle-skeleton structure. Chem Eng Re Des 183:424–438. https://doi.org/10.1016/j.cherd.2022.05.036

Zhu Y, Murali S, Cai W, Li X, Suk JW, Potts JR, Ruoff RS (2010) Graphene and graphene oxide: synthesis, properties, and applications. Adv Mater 22(35):3906–3924. https://doi.org/10.1002/ADMA.201001068

Zhu G, Bian Y, Hursthouse AS, Xu S, Xiong N, Wan P (2020) The role of magnetic MOFs nanoparticles in enhanced iron coagulation of aquatic dissolved organic matter, Chemosphere 247:125921. https://doi.org/10.1016/j.chemosphere.2020.125921

Chapter 2
Carbon Quantum Dots: From Synthesis to Application in Wastewater Treatment

An-Giang Nguyen

Abstract Carbon quantum dots represent a fascinating category of carbon nanoparticles with sizes around 10 nm. Carbon quantum dots possess several advantageous characteristics, including environmental friendliness, low toxicity, biocompatibility, photo-induced electron transfer, tunable fluorescent properties, and easily achievable synthesis protocols. These attributes make carbon quantum dots well-suited for a variety of applications, such as photocatalysts, biological imaging, sensing, adsorbents, energy storage and conversion, and wastewater treatment. This chapter covers advancements in carbon quantum dots research and development, with an emphasis on carbon quantum dots synthesis and application in wastewater treatment. Finally, the chapter summarizes challenges and outlines future perspectives to accelerate the further application of carbon quantum dots in wastewater treatment.

Keywords Carbon quantum dots · Wastewater · Nanomaterials · Photocatalysts

2.1 Introduction

The increase in industrial activities and population growth has led to a surge in environmental issues, particularly the presence of harmful and toxic substances, which has raised worldwide concerns (Lim et al. 2015; Long et al. 2021; Saravanan et al. 2022). Multidisciplinary efforts are underway to mitigate environmental pollution and safeguard human well-being (Long et al. 2021; Saravanan et al. 2022; Nguyen and Park 2023). In this context, the field of material science plays a significant role. In the recent decades, nanomaterials have been extensively studied for their ability to offer novel approaches to address various critical environmental challenges ranging from air pollution to water contamination (Long et al. 2021; Saravanan et al. 2022). Nanomaterials, such as carbon-based nanomaterials, metallic nanoparticles,

A.-G. Nguyen (✉)
Department of Materials Science and Engineering, Chonnam National University, Gwangju, Republic of Korea
e-mail: giangsce@jun.ac.kr

N. Agarwal et al. (eds.), *Nanomaterials in Wastewater Research*, Advances in Wastewater Research, https://doi.org/10.1007/978-981-96-3908-3_2

and metal–organic frameworks, exhibit diverse applications in purifying the environment (Nasrollahzadeh et al. 2021; Sharma et al. 2019). Notably, carbon-based nanomaterials, particularly carbon quantum dots, possess distinctive characteristics, including exceptional stability, high specific surface area, and pore volume, and have been widely employed in numerous environmental applications such as photocatalysts, sensing, adsorbents, wastewater treatment, pollutant removal, waste separation, catalysis, and energy conversion and storage. (Rani et al. 2020; Guo et al. 2021; Nguyen et al. 2022, 2023a, 2024, 2023b; Moon et al. 2024; Vu et al. 2022; Li et al. 2021a; Zhang et al. 2022; Rasal et al. 2021; Nguyen and Verma 2024).

Carbon quantum dots have been recognized as a noteworthy material since their first discovery in 2004 (Xu et al. 2004). Generally, carbon quantum dots are defined as carbon nanoparticles with a diameter of less than 10 nm, covered by functional groups like carboxyl, hydroxyl, amino, or aromatic rings (Rani et al. 2020; Guo et al. 2021; Zhang et al. 2022; Rasal et al. 2021; Liu et al. 2024). Following their discovery, various methods were proposed to synthesize carbon quantum dots to tailor their properties for specific applications (Lim et al. 2015; Rani et al. 2020; Guo et al. 2021; Zhang et al. 2022; Rasal et al. 2021; Liu et al. 2024). In this chapter, the synthesis methods of carbon quantum dots are reviewed. Moreover, the potential application of carbon quantum dots in wastewater treatment is highlighted. Additionally, the current challenges and future perspectives are provided to realize the full advantages of carbon quantum dots in wastewater treatment. We believe that this chapter can provide valuable strategies for the research and development of carbon quantum dots in wastewater treatment.

2.2 Methods of Synthesis

Carbon quantum dots are typically spherical carbon nanomaterials characterized by abundant functional groups on the surface, with a diameter of less than 10 nm, as shown in Fig. 2.1 (Guo et al. 2021; Rasal et al. 2021). Their structure comprises a carbon core, which is covered by a surface passivation layer. The carbon core serves as the heart of the carbon quantum dots, consisting of a combination of sp^2 and sp^3 carbon (Lim et al. 2015; Guo et al. 2021; He et al. 2022). Due to having more unsaturated bonds, uncoordinated groups, and a high specific surface area, carbon quantum dots tend to bond with other groups to create a passivation layer, reducing their Gibbs free energy. In cases where no passivation exists, the surface of carbon quantum dots is typically linked to numerous functional groups containing oxygen atoms, predominantly carboxyl, and hydroxyl groups. Introducing passivation modifiers such as biomolecules, polymers, or oligomers onto the surface of carbon quantum dots enhances their polarity, optical characteristics, and chemical reactivity (Rani et al. 2020; Guo et al. 2021; Rasal et al. 2021).

Till now, various techniques have been reported for synthesizing carbon quantum dots. Each technique significantly influences the physicochemical properties of the

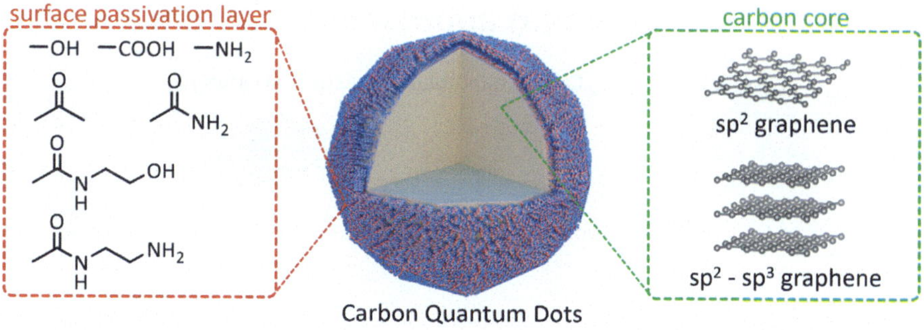

Fig. 2.1 Schematic of the structure of carbon quantum dots

resulting carbon quantum dots. These synthesis techniques can be broadly classified into top-down and bottom-up approaches (Fig. 2.2) (Zhang et al. 2022; Rasal et al. 2021; Liu et al. 2024; Rocco et al. 2023; Yang et al. 2023). Top-down approaches involve physically or chemically stripping large carbon sources (e.g., carbon nanotubes, graphite, graphene, etc.) into smaller particles. In contrast, bottom-up approaches involve building carbon quantum dots from smaller molecular or oligomeric building blocks (Zhang et al. 2022; Rasal et al. 2021; Liu et al. 2024). The following sections will review the synthesis methods that follow these top-down and bottom-up approaches.

2.2.1 *Top-Down Approach*

2.2.1.1 Arc-Discharge Method

In 2004, Xu et al. introduced the arc discharge technique while purifying single-walled carbon nanotubes sourced from arc discharge soot (Xu et al. 2004). This method involves operating within a low-pressure arc chamber in an inert atmosphere. In this system, both the anode and cathode are made of carbon materials, such as graphite. At high temperatures, carbon atoms vaporize from the anode and travel to the cathode to form a mixture of fluorescent nanoparticles and short tubular carbon originating from single-walled carbon nanotubes. The obtained material was functionalized with carboxyl groups by oxidizing with 3.3 N HNO_3 for 48 h to remove amorphous carbon and improve the hydrophilicity of the material. Subsequently, the obtained suspension was extracted by NaOH and water for 2 cycles to collect a stable suspension of single-walled carbon nanotubes. More interestingly, after electrophoresis purification, this suspension was separated into long nanotubes, short irregular tubular material, and fluorescent material. The fluorescent material was later named carbon quantum dots, which exhibited notable fluorescence properties without requiring surface treatment. However, the method suffers from inconsistent

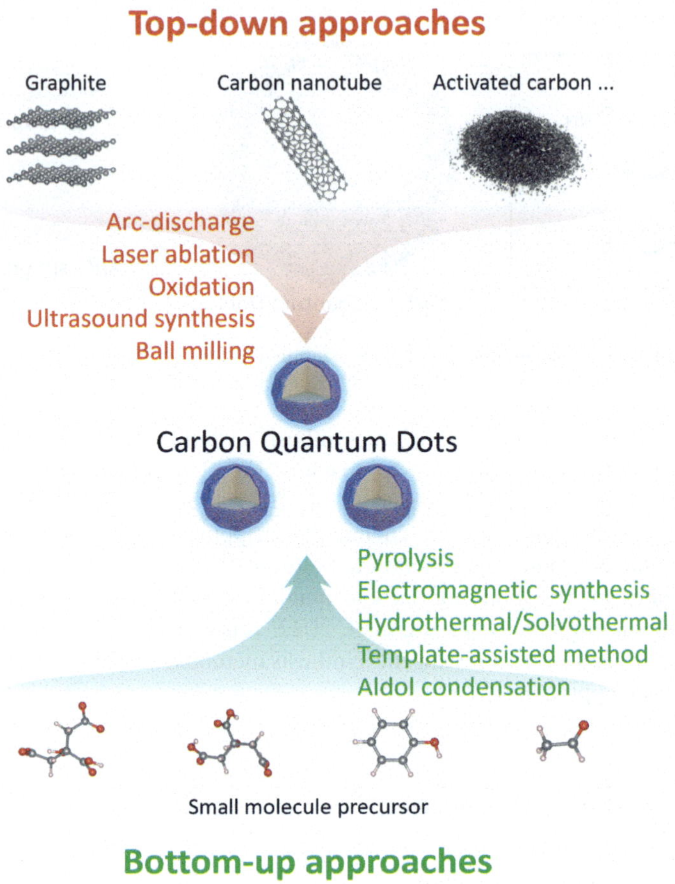

Fig. 2.2 Main approaches to synthesis of carbon quantum dots

particle size distribution and an exceptionally low yield, rendering it impractical for large-scale production (Dey et al. 2014; Biazar et al. 2018).

2.2.1.2 Laser Ablation Method

Laser ablation involves using a high-energy laser to rapidly heat and remove material from a target, with the intense laser pulses vaporizing and consuming the material. This vaporization generates plasma that flows across the surface of the sample, leading to the ionization of the sample material. Sun et al. pioneered the use of laser ablation for synthesizing carbon quantum dots (Sun et al. 2006). They employed a neodymium-doped yttrium aluminum garnet laser source with a wavelength of 1064 nm and a frequency of 10 Hz to irradiate a carbon target. This process occurred in the presence of water vapor, with argon gas serving as the carrier, at a temperature of 900 °C and a pressure of 75 kPa. However, the obtained nanocarbon particles did not demonstrate a photoluminescence effect. Furthermore, they were refluxed in a 2.6 N HNO_3 solution for 12 h. Subsequently, the treated carbon was passivated with several organic materials, such as diamine-terminated oligomeric poly (ethylene glycol) and poly (propionylethylene-imine-co-ethyleneimine), at 120 °C for 72 h. The passivated carbon, namely carbon quantum dots, had a diameter of approximately 5 nm. These carbon quantum dots exhibited photoluminescence in the solid state and suspension stage in the visible and near-infrared regions. Afterward, an amorphous carbon thin film was subjected to a short, near-infrared pulse laser ablation process (Sidorov et al. 2018). During the ablation, it was observed that two distinct carbon-based structures were formed: nanodiamonds and carbon quantum dots with a graphite structure. These structures were found in different regions of the ablation area. The carbon quantum dots were directly generated within the ablation zone and its surrounding plasma region, while the nanodiamonds were primarily formed at the edges of the plasma area, near the laser source. This suggests that different regions of the laser ablation process experience varying conditions, resulting in the formation of diverse reaction byproducts. Consequently, achieving a uniform composition of the desired products becomes challenging due to the varied responses exhibited by distinct regions within the ablation area.

Du's group also fabricated carbon quantum dots by laser irradiating a graphite suspension in several solvents, including poly (ethylene glycol) (Hu et al. 2009). The authors found that photoluminescence was observed only in the carbon quantum dots that had carboxylate groups on their surface. Therefore, by carefully choosing the organic solvents used in the fabrication process, the surface characteristics of the carbon quantum dots can be easily modified and controlled. This allowed them to adjust and tune the light emission properties of the resulting quantum dots as desired. Similarly, Koshizaki et al. fabricated carbon quantum dots by rapidly irradiating 20 mg of nanocarbon material in 50 mL of ethanol, acetone, or water solvent (Li et al. 2011). After ultrasonic dispersion, 4 mL of the obtained suspension was dropped onto glass and irradiated by a neodymium-doped yttrium aluminum garnet pulsed laser with a wavelength of 532 nm under stirring. Subsequently, carbon quantum dots can be separated from the supernatant. However, photoluminescence is only observed in carbon quantum dots prepared in ethanol or acetone solvent. The mechanism behind the observed photoluminescence appears to be related to the surface modification or functionalization of the carbon quantum dots. During the fabrication process, carbon- and oxygen-containing radical species are generated from ethanol or acetone, which subsequently form functional groups on the surface of the carbon quantum dots. These surface functional groups are believed to be the origin and source of the observed photoluminescent properties in the carbon quantum dots.

Hu's research group successfully demonstrated the synthesis of carbon quantum dots with controlled sizes of 3, 8, and 13 nm by regulating the nucleation process by tuning the laser pulse width (Hu et al. 2011). When the laser beam interacts with graphite flakes, it creates an instantaneous high-temperature, high-pressure vapor/plasma at the interface between the liquid medium and the graphite. This leads to the formation of a bubble, which rapidly expands to reach its maximum radius.

Once the laser pulse ends, the bubble starts to contract due to the pressure from the surrounding liquid, resulting in the cooling of its interior and the formation of clusters or nuclei. Since the laser-graphite interaction time is determined by the laser pulse width, bubbles with varying cluster densities can be formed under the same laser beam energy by adjusting the pulse width. The resulting carbon quantum dots from this method display a narrow size distribution, fluorescence characteristics, and good water solubility. Nevertheless, process complexities and high costs limit its widespread application across different fields.

2.2.1.3 Oxidation Method

The oxidation method, including chemical oxidation and electrochemical oxidation, is an approach for producing carbon quantum dots by oxidizing and breaking down larger carbon materials into smaller carbon nanomaterials. For example, Peng et al. used concentrated sulfuric acid to dehydrate carbohydrates (Peng and Travas-Sejdic 2009). The obtained materials were subjected to chemical oxidation using nitric acid and surface passivation by reacting with amine-terminated compounds. This treatment yielded luminescent carbon quantum dots. They also proposed that the surface passivation step plays a crucial role in enhancing the photoluminescent properties of the carbon quantum dots. Specifically, it can be tuned and modulated by varying factors such as the precursors and the duration of the nitric acid oxidation treatment. Another report began with three types of activated carbons, including coconut-activated carbon, wood-activated carbon, and coal-activated carbon (Qiao et al. 2010). Typically, due to their amorphous structure, activated carbons were prone to etching by nitric acid, followed by treatment with an amine-terminated compound, resulting in the formation of carbon quantum dots. The carbon quantum dots obtained from the three different activated carbons exhibited similar properties with good dispersibility and had a size range of 2–6 nm. Their findings corroborated that the size, structure, and properties of carbon quantum dots synthesized through chemical oxidation are largely influenced by the oxidizing agent used and the oxidation process itself. However, the carbon precursor had less impact on the final properties of the carbon quantum dots.

The starch can also be used as a starting material in synthesizing carbon quantum dots (Yan et al. 2015). Initially, starch was pyrolyzed at 300 °C for 4 h under high pressure to produce a carbon material. Subsequently, the carbonized product was dispersed in a mixed oxidant solution consisting of distilled water, acetic acid, and hydrogen peroxide. This dispersion underwent ultrasonic treatment and reflux heating, followed by filtration. The resulting filtrate was then neutralized by adding sodium hydroxide and concentrated to 10 mL. The concentrated solution was dialyzed using a dialysis bag. After dialysis, the resulting solution was evaporated, yielding a light-yellow powder of carbon quantum dots after vacuum drying. This method is versatile and can be applied to a wide variety of carbon-source materials.

Additionally, the process has broad applicability, operational simplicity, and is cost-effective. However, the disadvantage is the use of strong acids, which are corrosive and potentially hazardous in nature, and require neutralizing agents.

The electrochemical oxidation process involves applying a potential to oxidatively strip the carbon electrode to generate carbon quantum dots. Typically, multi-walled carbon nanotubes-covered carbon paper was used as the working electrode, while a platinum wire and $Ag/AgClO_4$ served as the counter electrode and reference electrode, respectively, in a 0.1 M of tetrabutylammonium perchlorate in a degassed acetonitrile solvent as an electrolyte (Zhou et al. 2007). A sweep rate of 0.5 V s^{-1} in the range of –2.0 V to + 2.0 V was applied, resulting in a notable change in the solution color from colorless to yellow and finally to dark brown. It implies that during the electrochemical cycling, the tetrabutylammonium cations intercalated into and disrupted the nanotube structure, leading to the detachment of carbon quantum dots.

In addition, other carbon or graphite electrodes can also serve as working electrodes for the preparation of carbon quantum dots using this electrochemical approach (Xu et al. 2015; Liu et al. 2016). By using an alkaline alcoholic solution (e.g., NaOH-ethanol), electrolytes can offer a relatively cost-effective and environmentally friendly method for generating carbon quantum dots. Liu's group synthesized carbon quantum dots using a three-electrode electrochemical system, comprising a graphite electrode as the working electrode, an Ag/AgCl electrode as the reference electrode, and a platinum foil as the counter electrode in NaOH-ethanol electrolyte (Liu et al. 2016). A constant voltage of 5 V was applied to the graphite electrode for 3 h under a nitrogen atmosphere. Residual NaOH was eliminated through washing and centrifugation, resulting in a homogeneous dispersion of carbon quantum dots. The obtained carbon quantum dots, with high crystallinity, exhibited monodispersity with an average diameter of 4.0 nm. Additionally, the applied potential also affected the obtained carbon quantum dots. The lower applied potential of 3 V required a longer time of 12 h to generate carbon quantum dots, resulting in an average size of 2.9 nm. In contrast, the higher applied potential of 7 V induced electrolyte evaporation. Moreover, the carbon quantum dots produced by applying different potentials during the electrochemical synthesis process exhibited varying quantum yield values. The quantum yields were 9.5%, 11.2%, and 4.6% for applied potentials of 3 V, 5 V, and 7 V, respectively. Furthermore, the authors found that other alkaline alcoholic solutions (e.g., methanol/NaOH, butanol/NaOH) could promote the formation of carbon quantum dots, while solutions of Na_2SO_4/ethanol, acetone/NaOH, or hydrochloric acid/ethanol could not produce any carbon quantum dots. This result indicates that both alcohol molecules and hydroxyl groups play crucial roles in the formation of carbon quantum dots. During electrochemical reactions, gas evolution occurs at both the anode and cathode due to the splitting of water molecules. This water splitting typically takes place at a potential of 1.23 V in moderately concentrated electrolytes. However, at high anodic overpotential (e.g., 5 V), intermediate products such as hydroxyl radicals ($OH^{\bullet}$) can also be generated at the graphite electrode interfaces, as follows:

$$\text{Anode}: 4\,OH^- \rightarrow 2\,H_2O + O_2 + 4\,e^-$$

$$OH^- \rightarrow OH^\bullet + e^-$$

$$\text{Cathode}: \; 2\,H_2O + e^- \rightarrow 2\,OH^- + H_2$$

The hydroxyl radicals generated during the electrochemical process corrode the edges of the graphite electrode. In an alkaline electrolyte without any alcohol addition, these radicals are highly non-selective and reactive, leading to rapid exfoliation of the bulk graphite to form large graphite sheets. However, with the addition of alcohols to the electrolyte, the reactivity of these radicals can be moderately controlled and regulated. This controlled radical activity facilitates the smooth and gradual generation of carbon quantum dots. Consequently, carbon quantum dots with a well-defined hexagonal lattice structure could be electrochemically prepared. Interestingly, it was observed that the initially colorless dispersion of carbon quantum dots turned bright yellow upon exposure to air at room temperature. Characterizations including UV–Vis absorption spectroscopy, fluorescence spectrophotometry, X-ray photoelectron spectroscopy, and Fourier transform infrared spectroscopy revealed that this color change was attributed to the oxidation of the carbon quantum dots.

In an aqueous electrolyte, water was oxidized to highly reactive $OH^\bullet$ and $O^\bullet$ radicals. These radicals act as "electrochemical scissors" capable of exfoliating carbon sources and facilitating the formation of carbon quantum dots. The exfoliation and fragmentation process typically initiates at the edge sites of the carbon material and is further accelerated at defect sites (Rasal et al. 2021). Thus, the solvent plays a crucial role in the electrochemical oxidation process, which can facilitate the stripping or exfoliation of carbon sources.

2.2.1.4 Ultrasound Synthesis Method

In the ultrasound synthesis process, ultrasound serves as a unique energy source that drives reactions. Moreover, ultrasound aids in dispersing particles while preventing their agglomeration. Typically, it employs acoustic cavitation to irradiate a liquid. This process begins with the formation of supersonic liquid jets that create localized regions of extremely low pressure. These low-pressure zones facilitate the formation and expansion of bubbles. As a result, the violent collapse of these bubbles generates intense shockwaves, leading to the fragmentation of larger carbon materials into smaller carbon quantum dots. Hydrodynamic cavitation, propelled by sudden pressure changes and bubble dynamics, plays a crucial role in breaking down the carbon precursors into the desired carbon quantum dots nanostructures through this top-down approach. Typically, polyamide resin precursor was mixed with ethylenediamine as a passivant and immersed in deionized water, followed by ultrasonic treatment for 3 h to prepare carbon quantum dots (Dang et al. 2016). The resulting carbon quantum dots displayed a monodisperse distribution, with particle sizes ranging from 2 to 4 nm. Park et al. utilized food waste to prepare carbon quantum dots with high water dispersibility (Park et al. 2014). Specifically, food waste was mixed with ethanol and treated with ultrasound for 45 min. The supernatant was separated by centrifugation and filtration to obtain carbon quantum dots. Food waste comprises a combination of various organic molecules bonded together by hydrogen bonding. When subjected to ultrasonic irradiation, a series of reactions takes place, including dehydration, polymerization, and carbonization. These reactions lead to a brief, singular nucleation event. Subsequently, the solute molecules diffuse and accumulate on the surface of the carbon nanoparticles, enabling their growth along the generated nucleation. Remarkably, this method could produce approximately 120 g of carbon quantum dots from 100 kg of food waste. Additionally, the ultrasound approach provides the necessary heat for the reaction in a localized manner, rapidly absorbed by the solvent once the reaction finishes. Consequently, macro-level temperature differences are minimal, eliminating the need for high-temperature and high-pressure conditions and equipment.

2.2.1.5 Ball Milling Method

Ball milling approach utilizes mechanical energy to fracture and fragment carbonaceous materials through the impact and friction between the milling balls and the precursor materials. It is a simple and scalable technique that does not require complex equipment or harsh reaction conditions. Additionally, it allows for controlling the size and morphology of the products by adjusting the milling time, milling speed, and the ratio of precursors to milling balls. Furthermore, ball milling can be easily combined with other synthetic approaches, such as functionalization or doping, to tailor the properties of the synthesized materials. For example, carbon black was used as a precursor to produce carbon quantum dots with the assistance of sodium carbonate as a reaction medium and abrasive (Youh et al. 2021). Additionally, the incorporation of sodium carbonate introduces oxygen-containing functional groups onto the surface of the synthesized carbon quantum dots. The resulting carbon quantum dots exhibit irregular shapes with an average size of approximately 3 nm. Wet ball milling processes have also been explored for carbon quantum dots synthesis. Yang's group employed spent coffee grounds as the carbon source, milling them in ethanol at a speed of 150 rpm and room temperature, yielding carbon quantum dots with an average size of ~ 1.6 nm (Jeong et al. 2023). Furthermore, the surface of carbon quantum dots can be functionalized by introducing molecules containing nitrogen and oxygen, such as citric acid. Despite these advantages, the ball milling synthesis of carbon quantum dots may require optimization to achieve high yields and uniform particle size distribution. Additionally, the choice of carbon precursors and milling parameters must be carefully optimized to avoid undesired side reactions or contamination.

2.2.2 *Bottom-Up Approach*

2.2.2.1 Pyrolysis Method

The pyrolysis method is one of the most commonly employed approaches for synthesizing carbon quantum dots, which involves thermally decomposing organic precursors at high temperatures. This method offers several advantages, including simplicity, potential for large-scale carbon quantum dots production, low cost, and a solvent-free process. Additionally, the pyrolysis approach allows for optimizing the properties of the resulting carbon quantum dots by regulating parameters such as the pH of the precursor solution, pyrolysis temperature, and time. In 2008, Giannelis's group pioneered a one-step synthesis of carbon quantum dots through the pyrolysis of various ammonium citrate salts or 4-amino-antipyrine (Bourlinos et al. 2008). The obtained carbon quantum dots exhibited an average size of less than 10 nm. Subsequently, the authors investigated the pyrolysis process in detail and observed that as the degree of carbonization increased, the fluorescence brightness and quantum yield of the carbon quantum dots decreased significantly when the pyrolysis temperature was increased from 180 °C to 400 °C (Krysmann et al. 2012). The pyrolysis of biomass can also produce carbon quantum dots. For example, watermelon peel was pyrolyzed at 220 °C for 2 h in an air atmosphere, followed by filtration and dialysis to produce carbon quantum dots (Zhou et al. 2012). The resulting carbon quantum dots exhibited a small particle size of ~ 2.0 nm, high blue luminescence, favorable fluorescence lifetime, and exceptional stability over a wide pH range. Similarly, Chen and colleagues reported one-step pyrolysis of natural plant leaves (Zhu et al. 2013). The produced carbon quantum dots demonstrated good fluorescent properties without requiring any further surface modification or treatment. Interestingly, the fluorescence intensity of these carbon quantum dots, when dispersed in either aqueous or organic solvent environments, could be further enhanced by employing plasma or microwave-assisted post-treatment processes.

2.2.2.2 Electromagnetic Synthesis Method

The dielectric material, featuring both polar and non-polar bonds, undergoes a transition from a random arrangement to a directional alignment when exposed to an electromagnetic field (Rani et al. 2020; Zhang et al. 2022; Liu et al. 2024). These alignments adjust according to alternating electromagnetic frequencies, inducing molecular movement and friction, thus generating heat, especially in high-frequency electromagnetic fields. Consequently, this heat energy raises the temperature of the medium, promoting the breaking of chemical bonds and facilitating the rapid formation of carbon quantum dots within minutes. Moreover, reaction conditions including temperature and irradiation time can be precisely controlled by electromagnetic-based techniques. This plays a crucial role in ensuring the reproducibility and scalability of carbon quantum dot production, which is vital for large-scale fabrication.

The well-controlled and short reaction time also reduces unwanted side reactions, leading to a more simplified purification procedure. Due to its effectiveness and simple operation, this method proves to be a promising technique for the large-scale production of carbon quantum dots.

For example, Zhu et al. synthesized the carbon quantum dots using a microwave-assisted method, employing poly(ethylene glycol) and saccharides such as glucose and fructose as precursors, heated in a 500 W microwave oven for 5–10 min (Zhu et al. 2009). As the reaction progressed, the solution evolved from colorless to yellow (after 5 min of irradiation) and eventually to dark brown (after 10 min of irradiation), indicating the formation of carbon quantum dots. The size of the carbon quantum dots increased with the reaction time, yielding sizes of 2.75 nm after 5 min and 3.65 nm after 10 min of microwave irradiation. In contrast to the conventional heating approach, the microwave synthesis method promotes fast and more uniform reactions, significantly influencing the purity of synthesized products and yielding high product yields.

2.2.2.3 Hydrothermal/ Solvothermal Synthesis

Hydrothermal/solvothermal method involves sealing a mixture of reactants in water or organic solvents within a sealed Teflon-lined hydrothermal reactor at high temperatures (100 – 200°C) and autogenous pressure, respectively (Zhang et al. 2022). The elevated temperature and pressure conditions facilitate the dehydration and carbonization of the precursors, leading to the formation of carbon quantum dots. Specifically, Hsu et al. proposed that the growth of carbon quantum dots involves four steps: i) dehydration, ii) polymerization, iii) carbonization, and iv) passivation (Hsu and Chang 2012). Subsequently, purified carbon quantum dots can be obtained through subsequent purification and separation steps. Various precursors can be used for this method, such as saccharides or organic acids (e.g., citric acid, glucose, ascorbic acid, etc.) (An et al. 2023; Jia et al. 2012; Liu et al. 2023). For example, ascorbic acid was first dissolved in a mixture of water and ethanol using ultrasonic treatment, followed by a hydrothermal process conducted at 160 °C for 1 h. The carbon quantum dots product was collected by centrifuging the light brown solution and then dialyzed to remove larger particles and residual precursors (Liu et al. 2023). Similarly, Su's group used a solvothermal method to synthesize carbon quantum dots by treating a mixture of glutathione in formamide at 160 °C for 4 h (Su et al. 2023). Both hydrothermal and solvothermal methods offer several advantages, including simplicity, effectiveness, good control over the size and surface functionalization of the synthesized carbon quantum dots, relatively mild reaction conditions compared to other methods, and the potential for scalability. Additionally, a wide range of carbon precursors, such as small organic molecules, polymers, or biomass-derived materials, can be employed in these synthesis routes, providing flexibility in the choice of starting materials.

2.2.2.4 Template-Assisted Method

The template method for synthesizing carbon quantum dots involves the use of a specific removable material as a template. In this approach, carbon quantum dots are first grown or synthesized on the surface of the template material. Subsequently, the template is removed through acid etching or other suitable techniques, allowing for the collection and isolation of the synthesized carbon quantum dots. For example, Liu et al. used surfactant-modified silica spheres as the template for synthesizing carbon quantum dots (Liu et al. 2009). Initially, a satellite-like phenolic resin/F127/silica composite was prepared and carbonized at 900 °C in an argon atmosphere for 2 h, followed by etching with a 2M NaOH solution to eliminate the silica template. The key aspect of this method was the utilization of F127 surfactant-modified silica spheres as a carrier. It not only provided an anchor point for the polymerization of resols in the solution but also hindered the agglomeration of carbon quantum dots during the pyrolysis process. However, complete etching of the silica template is challenging. Other templates were also reported for synthesizing carbon quantum dots such as HKUST-1, ZIF-8, MIL-101 MOFs (Gu et al. 2017; Amali et al. 2014; Yang et al. 2013). However, both the synthesis of the template and its subsequent removal add complexity to the synthesis process. Moreover, the recyclability of the template material also needs to be considered. Therefore, while this may not be the preferred approach for large-scale production, it still offers notable advantages in guiding and controlling the structure of the synthesized carbon quantum dots.

2.2.2.5 Aldol Condensation

This synthesis process involves the aldol condensation of acetone or aldehydes to produce unsaturated ketones or aldehydes under alkaline conditions. Subsequently, these unsaturated carbonyl compounds undergo polymerization, generating polymer-like compounds with extended carbon chains (Liu et al. 2024; Hou et al. 2015; Tu et al. 2023). These elongated chains can twist and intertwine, eventually leading to the formation of numerous carbon quantum dots.

Hou et al. first reported on the aldol reaction of acetone and sodium hydroxide in 2015 (Hou et al. 2015). In the typical synthesis procedure, sodium hydroxide and acetone were mixed together under intense magnetic stirring for 1 h. The resulting mixture was then left in ambient conditions for 120 h. Without any treatment, the mixture gradually turned a dark brown color, and the liquid acetone transformed into a solid product. To neutralize the obtained material, a dilute hydrochloric acid solution was added until a neutral pH was achieved. The resulting product was washed and centrifuged multiple times with deionized water. Finally, carbon quantum dots powder with diameters ranging from 1.5 to 3.0 nm was obtained after drying at 100 °C for 12 h. When acetaldehyde is employed as the precursor, carbon quantum dots with sizes ranging from 3 to 5 nm are synthesized (Hou et al. 2017). More interestingly, this approach could produce up to a kilogram of carbon quantum dots in 2 h (Li et al. 2021b). Furthermore, other heteroatom-containing precursors such as carbamide and cysteine could also be introduced during the synthesis process to generate nitrogen-doped or sulfur, nitrogen-co-doped carbon quantum dots. This method does not require harsh conditions, highly toxic reagents, or specialized equipment, making it a relatively simple and mild approach. Moreover, it is capable of producing high yields of carbon quantum dots, further enhancing its practicality and scalability.

2.3 Wastewater Treatment Applications

Carbon quantum dots exhibit several unique properties that make them highly suitable for various applications in wastewater treatment. Their unique characteristics allow them to be employed for photocatalytic degradation of wastewater pollutants, adsorption-based treatment processes, selective membrane-based separations, and disinfection of wastewater, as illustrated in **Fig. 2.3**.

2.3.1 Photocatalytic Degradation of Wastewater

The photocatalyst can be activated by incident photons, generating electron–hole pairs that subsequently migrate to the photocatalyst's surface. The presence of photoexcited electrons in the conduction band leads to the formation of reactive oxidative species. Additionally, the existence of photo-induced holes contributes to the production of highly oxidative hydroxyl radicals. Currently, the direct utilization of carbon quantum dots as photocatalysts is limited. Nevertheless, most research focuses on the conjunction of carbon quantum dots with other semiconductor materials for photocatalytic applications. This preference stems from well-dispersed

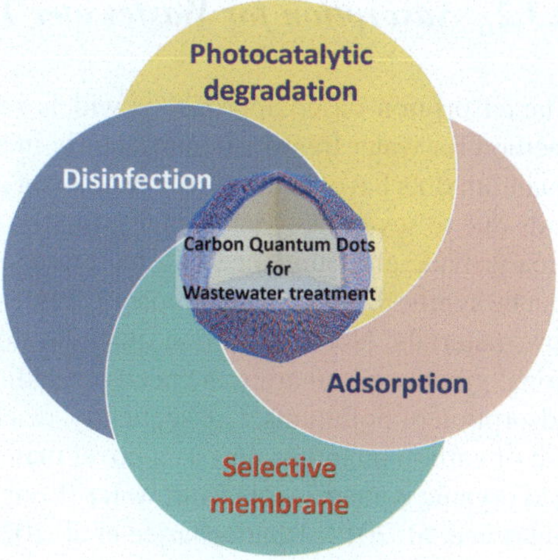

Fig. 2.3 Main application of carbon quantum dots for wastewater treatment

carbon quantum dots onto the surfaces of various other materials. Moreover, the notable electron-transfer characteristics of carbon quantum dots could enhance the charge separation process. Furthermore, the incorporation of carbon quantum dots has the potential to expand the light utilization spectrum of wide band-gap semiconductor photocatalysts, spanning from UV-light to visible light wavelengths (Rani et al. 2020). Typically, carbon quantum dots were incorporated into NH_2-MIL-125 MOF to produce a carbon quantum dots/NH_2-MIL-125 photo-catalyst composite by solvent-deposition method (Wang et al. 2019). Subsequently, this photo-catalyst was introduced into a Rhodamine B (RhB) solution (10 mg/L) and stirred in darkness for 40 min, followed by stirring under Xenon lamp irradiation for 200 min. The RhB was almost removed after 120 min, attributed to the degradation by carbon quantum dots in the presence of the Xenon lamp. The efficiency of photocatalysis increased progressively with rising quantities of carbon quantum dots, suggesting their potential as a luminescent enhancer. The improved photocatalytic performance is thought to be primarily driven by the up-converted photoluminescence behavior and electron-capturing capability of carbon quantum dots. Similarly, 99% RhB was degraded after 90 min exposure to visible light irradiation by carbon quantum dots/$La_2Ti_2O_7$ photocatalyst (Zhang et al. 2018). Carbon quantum dots were also reported as a photocatalytic to promote the degradation of methylene blue (MB). Zhang's group prepared a combination of carbon quantum dots onto the surface of zirconia as a photo-catalyst composite (Guo et al. 2018). The photo-catalyst composite was introduced into a solution containing MB and subjected to sonication for 30 s. Subsequently, the mixture underwent UV irradiation for 60 min. The findings indicate that 95% of MB was degraded after the 60 min UV irradiation. On the other hand, in the absence of carbon quantum dots, only 34% of MB was degraded. This highlights the ability of carbon quantum dots to prevent the recombination of electron–hole pairs, thereby significantly enhancing the degradation efficacy of the photocatalysts.

2.3.2 Adsorption for Wastewater Treatment

The adsorption-based approach is widely recognized as a versatile and inexpensive method for water treatment, particularly in removing pollutants. In addition, carbon quantum dots have emerged as a promising alternative to traditional adsorbent materials due to several key advantages. Firstly, their tiny size and large specific surface area provide an abundance of adsorption sites. Additionally, carbon quantum dots can be synthesized through simple processes using readily available and inexpensive raw materials. Notably, carbon quantum dots inherently possess hydrophilic functional groups, eliminating the need for additional modification steps to facilitate the adsorption of pollutants. Consequently, researchers are actively exploring the potential of carbon quantum dots as a novel material for the removal of heavy metal ions and organic contaminants from water (Long et al. 2021; Nasrollahzadeh et al. 2021; Sharma et al. 2019; Bhattacharjee et al. 2023).

In heavy metal adsorption, the adsorption capacity of carbon quantum dots primarily depends on their chemical composition, specific surface area, pore volumes, and functional groups. The interaction between heavy metals and carbon quantum dots can involve physical adsorption, electrostatic interactions, and chemical chelation. The presence of negatively charged oxygen-containing groups in the carbon quantum dots enhances the chelation of heavy metals through π-π and π-d interactions. Electrostatic interaction is also considered a significant adsorption mechanism between carbon quantum dots and metal ions (Long et al. 2021; Bhattacharjee et al. 2023). For example, Sabet et al. investigated the use of nitrogen-doped carbon quantum dots for the adsorption of Cd^{2+} and Pb^{2+} from wastewater (Sabet and Mahdavi 2019). The study revealed that these nitrogen-doped carbon quantum dots possess a large specific surface area, facilitating the adsorption of heavy metal ions from water. Moreover, the negatively charged nitrogen atoms and oxygen-functional groups on the surface of the nitrogen-doped carbon quantum dots serve as active sites for metal ion adsorption. Thus, nitrogen-doped carbon quantum dots demonstrated the capability to adsorb 37% of Cd^{2+} ions and 75% of Pb^{2+} ions from the aqueous solution. Additionally, the water dispersibility of carbon dots greatly restricts their utility as adsorbents. Therefore, incorporating carbon quantum dots into a host material not only increases their stability but also boosts the adsorption kinetics through synergistic effects. As demonstrated by Zhang's group, carbon quantum dots-modified mesoporous organosilica was reported as an adsorbent for the removal of heavy metal ions, including Hg^{2+}, Cu^{2+}, and Pb^{2+} (Wang et al. 2015). Compared to pristine mesoporous organosilica, the carbon quantum dots-modified counterpart features a variety of functional groups, such as carboxyl and amine groups, which can act as active sites for binding metal ions through electrostatic attraction and n–π stacking interactions. This results in an increased adsorption capacity for a range of metal ions.

Furthermore, carbon quantum dots-modified mesoporous organosilica also demonstrated approximately a 30% capture percentage for 2,4-dichlorophenol at concentration of a 60 mg/L at 10 °C (Wang et al. 2015). Furthermore, the capture percentage noticeably decreases as the pH value of the solution increases from 2 to 8, suggesting the negative impact of protonation on the adsorption behavior. Similarly, carbon quantum dots can serve as effective adsorbents for removing organic pollutants from wastewater. For example, carbon quantum dots-$ZnFe_2O_4$ composite demonstrated a high adsorption performance for methyl orange compared to pristine $ZnFe_2O_4$ (Shi et al. 2018).

2.3.3 Selective Membrane for Wastewater Treatment

Membrane-based water treatment technologies are widely utilized. In addition, carbon quantum dots offer promising functionality in membrane-based separation processes. Typically, carbon quantum dots can be incorporated into membrane substrates due to their abundance of surface functional groups, tiny size, and excellent water dispersibility. They could significantly enhance separation processes without compromising mechanical strength or solute selectivity. Carbon quantum dots can improve membrane permeability, surface hydrophilicity, and stability. For example, sodium-functionalized carbon quantum were introduced into the polyamide layer of hollow fiber membranes through interfacial polymerization reactions for desalination purposes (Gai et al. 2019). Compared to conventional composite membranes, the addition of sodium-functionalized carbon quantum dots could generate a larger effective surface area, a thinner polyamide layer, and an increased presence of hydrophilic oxygen-containing groups within the polyamide layer. Additionally, the interstitial space among the polyamide chains was enlarged due to the inclusion of sodium-functionalized carbon quantum dots. Consequently, incorporating 1 wt% sodium-functionalized carbon quantum dots into the polyamide layer enhanced the pure water permeability of the membranes by 47.1%, without compromising their NaCl rejection rate of 97.7%. Furthermore, incorporating carbon quantum dots into a polyethyleneimine matrix, followed by dip-coating onto polyacrylonitrile membranes, has the potential to enhance filtration performance and membrane selectivity for organic, polar, and non-polar solvents (Yuan et al. 2018). The high surface hydrophilicity of carbon quantum dots, attributed to their hydrophilic groups, can reduce non-specific adsorption. Moreover, carbon quantum dots may promote the diffusion of polar solvents through the membrane by bonding with hydrophilic groups. The resulting composite membrane with incorporated carbon quantum dots exhibited a significant increase in solvent flux, up to 54.3%, and solvent uptake up to 40.5%. Additionally, the incorporation of carbon quantum dots onto the membrane surfaces may hinder the transport of non-polar solvents, thereby improving membrane selectivity. Sun et al. synthesized sulfonic acid-, amino-, and carboxyl group-rich carbon quantum dots (Sun and Wu 2018). By incorporating these carbon quantum dots into the polyamide layer of membranes via interfacial polymerization, the composite membranes acquired various properties attributable to the different functional groups on the carbon quantum dots. Notably, the composite membrane containing negatively charged sulfonic acid group carbon quantum dots exhibited a Na_2SO_4 rejection rate of 93.6% and the highest permeate flux of 42.1 L m^{-2} h^{-1}.

2.3.4 *Disinfection of Wastewater*

Carbon quantum dots possess several desirable properties that make them a potential candidate for the development of antibacterial agents, including precisely controlled size, excellent water dispersibility, and potent antimicrobial capabilities. The antibacterial properties of carbon quantum dots were reported for *Escherichia coli (E. coli)* and *Staphylococcus aureus (S. aureus)* (Yan et al. 2019). The carbon quantum dots were synthesized by an electrochemical method. Two graphite electrodes served as both cathode and anode, and they were immersed in pure water under an applied potential of 15 – 60 V. After 120 h, the obtained material underwent a further purification process to produce carbon quantum dots. Subsequently, a mixture of carbon quantum dots and TiO_2 in ethanol and water was subjected to hydrothermal treatment at 140 °C for 4 h to prepare a quantum dots-TiO_2 composite. During this step, a new chemical bond (Ti–O-C) was generated. Thus, this composite exhibited strong antibacterial characteristics. Suspensions containing carbon quantum dots, TiO_2, and carbon quantum dots-TiO_2 composite were introduced into tubes containing bacterial samples, followed by incubation at 37 °C for 24 h under visible light. Their findings revealed that the antibacterial efficacy of carbon quantum dots-TiO_2 remained at 82%, indicating continued effective antibacterial activity. In contrast, pure TiO_2 exhibited almost no antibacterial activity due to agglomeration. Consequently, carbon quantum dots-TiO_2 can be effectively employed as an antibacterial agent to increase bacterial degradation capabilities. The mechanism of photocatalytic degradation of E. coli using carbon quantum dots-TiO_2 as a photocatalyst involves the photo-generation of electron–hole pairs at the surface of carbon quantum dots, which activate the photocatalytic degradation of *E. coli.*

2.4 Conclusion

This chapter discusses fabrication techniques for carbon quantum dots and their applications in wastewater treatment. Carbon quantum dots can be synthesized through various methods, including arc discharge, laser ablation, oxidation, ultrasound, ball milling, pyrolysis, electromagnetic synthesis, hydrothermal, solvothermal, template-assisted, and aldol condensation methods. However, there is an urgent need to develop innovative synthetic routes, combine existing methods to enhance efficiency, ensure consistent product properties, and streamline purification processes to eliminate residual reagents.

Over the past decade, significant research has focused on exploring how carbon quantum dots and carbon quantum dots-based composites can address environmental challenges. Studies have demonstrated the promising potential of carbon quantum dots in wastewater treatment applications. Nonetheless, further investigations are necessary before practical implementation and large-scale commercial adoption of carbon quantum dots can be realized, as outlined below:

(i). Green synthesis: Explore green chemistry approaches utilizing renewable, eco-friendly, and low-cost precursor materials to reduce environmental impact and improve the sustainability of carbon quantum dots production.

(ii). Consistent properties: The properties of carbon quantum dots can vary considerably depending on the synthesis method, precursor materials, and reaction conditions, making it challenging to obtain consistent and reproducible results. It is crucial to develop standardized synthesis protocols to ensure consistent and reproducible carbon quantum dots properties.

(iii). Surface functionalization and composite materials: Investigating surface functionalization strategies to enhance the affinity of carbon quantum dots toward specific pollutants or contaminants in wastewater. Additionally, developing carbon quantum dots-based composite materials by combining carbon quantum dots with other nanomaterials, polymers, or adsorbents to improve their performance and recyclability in wastewater treatment applications.

(iv). Mechanism: Conducting in-depth mechanistic studies to understand the interactions between carbon quantum dots and various pollutants, as well as the underlying removal mechanisms (e.g., adsorption, photocatalysis, or a combination of processes). Besides, developing and implementing computational models and simulations to predict the behavior and performance of carbon quantum dots in wastewater treatment systems, enabling more efficient design and optimization.

(v). Recyclability: The recyclability of carbon quantum dots in practical wastewater treatment scenarios remains a concern, as they can be affected by factors such as pH, ionic strength, and the presence of other contaminants. Exploring strategies to improve the recyclability of carbon quantum dots, such as surface functionalization or encapsulation in protective matrices, is required.

(vi). Scalability and cost-effectiveness: Existing synthesis methods for carbon quantum dots may not be suitable for large-scale production, and the cost-effectiveness of using carbon quantum dots in wastewater treatment needs to be evaluated. Optimizing synthesis methods, composition, and exploring alternative precursor materials to reduce production costs and improve scalability are critical. Conducting pilot-scale studies and field trials to assess the practical viability of carbon quantum dots-based wastewater treatment technologies is essential. Moreover, exploring the potential integration of carbon quantum dots-based technologies with existing wastewater treatment processes, such as biological treatment, membrane filtration, or advanced oxidation processes, to enhance overall treatment efficiency and versatility.

By addressing these challenges and pursuing the outlined future perspectives, the application of carbon quantum dots in wastewater treatment can be accelerated, leading to more effective and sustainable solutions for water remediation.

References

Amali AJ, Hoshino H, Wu C, Ando M, Xu Q (2014) From Metal–organic framework to intrinsically fluorescent carbon Nanodots, Chem Eur J 20:8279–8282. https://doi.org/10.1002/chem.201402982

An Y, Liu C, Chen M, Yin X, Hou D, Zheng Y, Shi R, He X, Lin X (2023) Solid-state carbon dots with tunable fluorescence via surface substitution: effect of alkyl moieties on fluorescence characteristics. ACS Sustain Chem Eng 11:23–28. https://doi.org/10.1021/acssuschemeng.2c06131

Bhattacharjee T, Konwar A, Boruah JS, Chowdhury D, Majumdar G (2023) A sustainable approach for heavy metal remediation from water using carbon dot based composites: A review. Journal of Hazardous Materials Advances 10:100295. https://doi.org/10.1016/j.hazadv.2023.100295

Biazar N, Poursalehi R, Delavari H (2018) Optical and structural properties of carbon dots/TiO2 nanostructures prepared via DC arc discharge in liquid, p 020033. https://doi.org/10.1063/1.5018965

Bourlinos AB, Stassinopoulos A, Anglos D, Zboril R, Karakassides M, Giannelis EP (2008) Surface Functionalized Carbogenic Quantum Dots. Small 4:455–458. https://doi.org/10.1002/smll.200700578

Dang H, Huang L-K, Zhang Y, Wang C-F, Chen S (2016) Large-Scale Ultrasonic Fabrication of White Fluorescent Carbon Dots. Ind Eng Chem Res 55:5335–5341. https://doi.org/10.1021/acs.iecr.6b00894

Dey S, Govindaraj A, Biswas K, Rao CNR (2014) Luminescence properties of boron and nitrogen doped graphene quantum dots prepared from arc-discharge-generated doped graphene samples. Chem Phys Lett 595–596:203–208. https://doi.org/10.1016/j.cplett.2014.02.012

Gai W, Zhao DL, Chung T-S (2019) Thin film nanocomposite hollow fiber membranes comprising Na+-functionalized carbon quantum dots for brackish water desalination. Water Res 154:54–61. https://doi.org/10.1016/j.watres.2019.01.043

Gu Z, Li D, Zheng C, Kang Y, Wöll C, Zhang J (2017) MOF-Templated synthesis of ultrasmall photoluminescent carbon-nanodot arrays for optical applications. Angew Chem 129:6957–6962. https://doi.org/10.1002/ange.201702162

Guo Y, Cao F, Li Y (2018) Solid phase synthesis of nitrogen and phosphor co-doped carbon quantum dots for sensing Fe3+ and the enhanced photocatalytic degradation of dyes. Sens Actuators B Chem 255:1105–1111. https://doi.org/10.1016/j.snb.2017.08.104

Guo R, Li L, Wang B, Xiang Y, Zou G, Zhu Y, Hou H, Ji X (2021) Functionalized carbon dots for advanced batteries. Energy Storage Mater 37:8–39. https://doi.org/10.1016/j.ensm.2021.01.020

He C, Xu P, Zhang X, Long W (2022) The synthetic strategies, photoluminescence mechanisms and promising applications of carbon dots: Current state and future perspective. Carbon N Y 186:91–127. https://doi.org/10.1016/j.carbon.2021.10.002

Hou H, Banks CE, Jing M, Zhang Y, Ji X (2015) Carbon quantum dots and their derivative 3d porous carbon frameworks for sodium-ion batteries with ultralong cycle life. Adv Mater 27:7861–7866. https://doi.org/10.1002/adma.201503816

Hou H, Shao L, Zhang Y, Zou G, Chen J, Ji X (2017) Large-area carbon nanosheets doped with phosphorus: a high-performance anode material for sodium-ion batteries, Adv Sci 4. https://doi.org/10.1002/advs.201600243

Hsu P-C, Chang H-T (2012) Synthesis of high-quality carbon nanodots from hydrophilic compounds: role of functional groups. Chem Commun 48:3984. https://doi.org/10.1039/c2cc30188a

Hu S-L, Niu K-Y, Sun J, Yang J, Zhao N-Q, Du X-W (2009) One-step synthesis of fluorescent carbon nanoparticles by laser irradiation. J Mater Chem 19:484–488. https://doi.org/10.1039/B812943F

Hu S, Liu J, Yang J, Wang Y, Cao S (2011) Laser synthesis and size tailor of carbon quantum dots. J Nanopart Res 13:7247–7252. https://doi.org/10.1007/s11051-011-0638-y

Jeong G, Park CH, Yi D, Yang H (2023) Green synthesis of carbon dots from spent coffee grounds via ball-milling: Application in fluorescent chemosensors. J Clean Prod 392:136250. https://doi.org/10.1016/j.jclepro.2023.136250

Jia X, Li J, Wang E (2012) One-pot green synthesis of optically pH-sensitive carbon dots with upconversion luminescence. Nanoscale 4:5572. https://doi.org/10.1039/c2nr31319g

Krysmann MJ, Kelarakis A, Dallas P, Giannelis EP (2012) Formation Mechanism of Carbogenic Nanoparticles with Dual Photoluminescence Emission. J Am Chem Soc 134:747–750. https://doi.org/10.1021/ja204661r

Li X, Wang H, Shimizu Y, Pyatenko A, Kawaguchi K, Koshizaki N (2011) Preparation of carbon quantum dots with tunable photoluminescence by rapid laser passivation in ordinary organic solvents. Chem Commun 47:932–934. https://doi.org/10.1039/C0CC03552A

Li S, Li L, Tu H, Zhang H, Silvester DS, Banks CE, Zou G, Hou H, Ji X (2021a) The development of carbon dots: From the perspective of materials chemistry. Mater Today 51:188–207. https://doi.org/10.1016/j.mattod.2021.07.028

Li L, Li Y, Ye Y, Guo R, Wang A, Zou G, Hou H, Ji X (2021b) Kilogram-Scale Synthesis and Functionalization of Carbon Dots for Superior Electrochemical Potassium Storage. ACS Nano 15:6872–6885. https://doi.org/10.1021/acsnano.0c10624

Lim SY, Shen W, Gao Z (2015) Carbon quantum dots and their applications. Chem Soc Rev 44:362–381. https://doi.org/10.1039/C4CS00269E

Liu R, Wu D, Liu S, Koynov K, Knoll W, Li Q (2009) An aqueous route to multicolor photoluminescent carbon dots using silica spheres as carriers. Angew Chem Int Ed 48:4598–4601. https://doi.org/10.1002/anie.200900652

Liu M, Xu Y, Niu F, Gooding JJ, Liu J (2016) Carbon quantum dots directly generated from electrochemical oxidation of graphite electrodes in alkaline alcohols and the applications for specific ferric ion detection and cell imaging. Analyst 141:2657–2664. https://doi.org/10.1039/C5AN02231B

Liu J, Zhan Y, Qiu B, Lin Z, Guo L (2023) Portable Smartphone platform based on aggregation-induced enhanced emission carbon dots for Ratiometric quantitative sensing of fluoride Ions. ACS Sens 8:884–892. https://doi.org/10.1021/acssensors.2c02589

Liu H, Zhong X, Pan Q, Zhang Y, Deng W, Zou G, Hou H, Ji X (2024) A review of carbon dots in synthesis strategy. Coord Chem Rev 498:215468. https://doi.org/10.1016/j.ccr.2023.215468

Long C, Jiang Z, Shangguan J, Qing T, Zhang P, Feng B (2021) Applications of carbon dots in environmental pollution control: A review. Chem Eng J 406:126848. https://doi.org/10.1016/j.cej.2020.126848

Moon C-E, Nguyen A-G, Yang JS, Nazir A, Verma R, Park C-J (2024) Confining SnSe particles in nitrogen-doped carbon nanofibers: A free-standing, binder-free anode for potassium-ion batteries. Carbon N Y 218:118741. https://doi.org/10.1016/j.carbon.2023.118741

Nasrollahzadeh M, Sajjadi M, Iravani S, Varma RS (2021) Carbon-based sustainable nanomaterials for water treatment: State-of-art and future perspectives. Chemosphere 263:128005. https://doi.org/10.1016/j.chemosphere.2020.128005

Nguyen AG, Park CJ (2023) Insights into tailoring composite solid polymer electrolytes for solid-state lithium batteries. J Memb Sci 675:121552. https://doi.org/10.1016/J.MEMSCI.2023.121552

Nguyen AG, Le HTT, Verma R, Vu DL, Park CJ (2022) Boosting sodium-ion battery performance using an antimony nanoparticle self-embedded in a 3D nitrogen-doped carbon framework anode. Chem Eng J 429:132359. https://doi.org/10.1016/J.CEJ.2021.132359

Nguyen AG, Verma R, Song GC, Kim J, Park CJ (2023a) In situ polymerization on a 3D ceramic framework of composite solid electrolytes for room-temperature solid-state batteries. Advanced Science 10:2207744. https://doi.org/10.1002/ADVS.202207744

Nguyen A-G, Lee M-H, Kim J, Park C-J (2024) Construction of a high-performance composite solid electrolyte through in-situ polymerization within a self-supported porous garnet framework. Nanomicro Lett 16:83. https://doi.org/10.1007/s40820-023-01294-0

Nguyen AG, Verma R (2024) Single-atom catalysts for metal-sulfur batteries, in: atomically precise electrocatalysts for electrochemical energy applications, pp 383–402. Springer Nature Switzerland, Cham. https://doi.org/10.1007/978-3-031-54622-8_21

Nguyen AG, Verma R, Didwal PN, Park CJ (2023) Challenges and design strategies for alloy-based anode materials toward high-performance future-generation potassium-ion batteries. Energy Mater https://doi.org/10.20517/energymater.2023.11

Park SY, Lee HU, Park ES, Lee SC, Lee J-W, Jeong SW, Kim CH, Lee Y-C, Huh YS, Lee J (2014) Photoluminescent Green Carbon Nanodots from Food-Waste-Derived Sources: Large-Scale

Synthesis, Properties, and Biomedical Applications. ACS Appl Mater Interfaces 6:3365–3370. https://doi.org/10.1021/am500159p

Peng H, Travas-Sejdic J (2009) Simple Aqueous Solution Route to Luminescent Carbogenic Dots from Carbohydrates. Chem Mater 21:5563–5565. https://doi.org/10.1021/cm901593y

Qiao Z-A, Wang Y, Gao Y, Li H, Dai T, Liu Y, Huo Q (2010) Commercially activated carbon as the source for producing multicolor photoluminescent carbon dots by chemical oxidation. Chem Commun 46:8812. https://doi.org/10.1039/c0cc02724c

Rani UA, Ng LY, Ng CY, Mahmoudi E (2020) A review of carbon quantum dots and their applications in wastewater treatment. Adv Colloid Interface Sci 278:102124. https://doi.org/10.1016/j.cis.2020.102124

Rasal AS, Yadav S, Yadav A, Kashale AA, Manjunatha ST, Altaee A, Chang J-Y (2021) Carbon quantum dots for energy applications: a review. ACS Appl Nano Mater 4:6515–6541. https://doi.org/10.1021/acsanm.1c01372

Rocco D, Moldoveanu VG, Feroci M, Bortolami M, Vetica F (2023) Electrochemical synthesis of carbon quantum dots, Chem Electro Chem 10. https://doi.org/10.1002/celc.202201104.

Sabet M, Mahdavi K (2019) Green synthesis of high photoluminescence nitrogen-doped carbon quantum dots from grass via a simple hydrothermal method for removing organic and inorganic water pollutions. Appl Surf Sci 463:283–291. https://doi.org/10.1016/j.apsusc.2018.08.223

Saravanan A, Kumar PS, Hemavathy RV, Jeevanantham S, Jawahar MJ, Neshaanthini JP, Saravanan R (2022) A review on synthesis methods and recent applications of nanomaterial in wastewater treatment: Challenges and future perspectives. Chemosphere 307:135713. https://doi.org/10.1016/j.chemosphere.2022.135713

Sharma S, Dutta V, Singh P, Raizada P, Rahmani-Sani A, Hosseini-Bandegharaei A, Thakur VK (2019) Carbon quantum dot supported semiconductor photocatalysts for efficient degradation of organic pollutants in water: A review. J Clean Prod 228:755–769. https://doi.org/10.1016/j.jclepro.2019.04.292

Shi W, Guo F, Wang H, Liu C, Fu Y, Yuan S, Huang H, Liu Y, Kang Z (2018) Carbon dots decorated magnetic ZnFe 2 O 4 nanoparticles with enhanced adsorption capacity for the removal of dye from aqueous solution. Appl Surf Sci 433:790–797. https://doi.org/10.1016/j.apsusc.2017.10.099

Sidorov AI, Lebedev VF, Kobranova AA, Nashchekin AV (2018) Formation of carbon quantum dots and nanodiamonds in laser ablation of a carbon film. Quantum Elec (Woodbury) 48:45–48. https://doi.org/10.1070/QEL16456

Su W, Tan M, Wang Z, Zhang J, Huang W, Song H, Wang X, Ran H, Gao Y, Nie G, Wang H (2023)Targeted degradation of PD-L1 and activation of the STING pathway by carbon-dot-based PROTACs for Cancer Immunotherapy, Angew Chem Int Ed 62 (2023). https://doi.org/10.1002/anie.202218128

Sun H, Wu P (2018) Tuning the functional groups of carbon quantum dots in thin film nanocomposite membranes for nanofiltration. J Memb Sci 564:394–403. https://doi.org/10.1016/j.memsci.2018.07.044

Sun Y-P, Zhou B, Lin Y, Wang W, Fernando KAS, Pathak P, Meziani MJ, Harruff BA, Wang X, Wang H, Luo PG, Yang H, Kose ME, Chen B, Veca LM, Xie S-Y (2006) Quantum-sized carbon dots for bright and colorful photoluminescence. J Am Chem Soc 128:7756–7757. https://doi.org/10.1021/ja062677d

Tu H, Liu H, Xu L, Luo Z, Li L, Tian Y, Deng W, Zou G, Hou H, Ji X (2023) Carbon dots from alcohol molecules: principles and the reaction mechanism. Chem Sci 14:12194–12204. https://doi.org/10.1039/D3SC04606K

Vu D-L, Kim D-Y, Nguyen A-G, Park C-J (2022) Stabilizing interface of novel 3D-hierarchical porous carbon for high-performance lithium–sulfur batteries. Electrochim Acta 418:140369. https://doi.org/10.1016/j.electacta.2022.140369

Wang L, Cheng C, Tapas S, Lei J, Matsuoka M, Zhang J, Zhang F (2015) Carbon dots modified mesoporous organosilica as an adsorbent for the removal of 2,4-dichlorophenol and heavy metal ions. J Mater Chem A Mater 3:13357–13364. https://doi.org/10.1039/C5TA01652E

Wang Q, Wang G, Liang X, Dong X, Zhang X (2019) Supporting carbon quantum dots on NH2-MIL-125 for enhanced photocatalytic degradation of organic pollutants under a broad spectrum irradiation. Appl Surf Sci 467–468:320–327. https://doi.org/10.1016/j.apsusc.2018.10.165

Xu X, Ray R, Gu Y, Ploehn HJ, Gearheart L, Raker K, Scrivens WA (2004) Electrophoretic analysis and purification of fluorescent single-walled carbon nanotube fragments. J Am Chem Soc 126:12736–12737. https://doi.org/10.1021/ja040082h

Xu Y, Liu J, Zhang J, Zong X, Jia X, Li D, Wang E (2015) Chip-based generation of carbon nanodots via electrochemical oxidation of screen printed carbon electrodes and the applications for efficient cell imaging and electrochemiluminescence enhancement. Nanoscale 7:9421–9426. https://doi.org/10.1039/C5NR01765C

Yan Z, Shu J, Yu Y, Zhang Z, Liu Z, Chen J (2015) Preparation of carbon quantum dots based on starch and their spectral properties. Luminescence 30:388–392. https://doi.org/10.1002/bio.2744

Yan Y, Kuang W, Shi L, Ye X, Yang Y, Xie X, Shi Q, Tan S (2019) Carbon quantum dot-decorated TiO2 for fast and sustainable antibacterial properties under visible-light. J Alloys Compd 777:234–243. https://doi.org/10.1016/j.jallcom.2018.10.191

Yang Y, Wu D, Han S, Hu P, Liu R (2013) Bottom-up fabrication of photoluminescent carbon dots with uniform morphology via a soft–hard template approach. Chem Commun 49:4920. https://doi.org/10.1039/c3cc38815h

Yang H-L, Bai L-F, Geng Z-R, Chen H, Xu L-T, Xie Y-C, Wang D-J, Gu H-W, Wang X-M (2023) Carbon quantum dots: Preparation, optical properties, and biomedical applications. Mater Today Adv 18:100376. https://doi.org/10.1016/j.mtadv.2023.100376

Youh M-J, Chung M-C, Tai H-C, Chen C-Y, Li Y-Y (2021) Fabrication of carbon quantum dots via ball milling and their application to bioimaging. Mendeleev Commun 31:647–650. https://doi.org/10.1016/j.mencom.2021.09.018

Yuan Z, Wu X, Jiang Y, Li Y, Huang J, Hao L, Zhang J, Wang J (2018) Carbon dots-incorporated composite membrane towards enhanced organic solvent nanofiltration performance. J Memb Sci 549:1–11. https://doi.org/10.1016/j.memsci.2017.11.051

Zhang Z, Wu L, Wang P, Zhang Y, Wan S, Guo X, Jin W, Zhang J (2018) Carbon quantum dots modified La2Ti2O7 nanosheets for visible light photocatalysis. Mater Lett 230:72–75. https://doi.org/10.1016/j.matlet.2018.07.086

Zhang L, Yang X, Yin Z, Sun L (2022) A review on carbon quantum dots: Synthesis, photoluminescence mechanisms and applications. Luminescence 37:1612–1638. https://doi.org/10.1002/bio.4351

Zhou J, Booker C, Li R, Zhou X, Sham T-K, Sun X, Ding Z (2007) An Electrochemical Avenue to Blue Luminescent Nanocrystals from Multiwalled Carbon Nanotubes (MWCNTs). J Am Chem Soc 129:744–745. https://doi.org/10.1021/ja0669070

Zhou J, Sheng Z, Han H, Zou M, Li C (2012) Facile synthesis of fluorescent carbon dots using watermelon peel as a carbon source. Mater Lett 66:222–224. https://doi.org/10.1016/j.matlet.2011.08.081

Zhu L, Yin Y, Wang C-F, Chen S (2013) Plant leaf-derived fluorescent carbon dots for sensing, patterning and coding. J Mater Chem C Mater 1:4925. https://doi.org/10.1039/c3tc30701h

Zhu H, Wang X, Li Y, Wang Z, Yang F, Yang X (2009) Microwave synthesis of fluorescent carbon nanoparticles with electrochemiluminescence properties, Chem Commun. p 5118. https://doi.org/10.1039/b907612c.

Chapter 3
Nanosensors to Monitor Wastewater Pollutants

Amrit Krishna Mitra, Souvik Dey, Vijendra Singh Solanki, and Neha Agarwal

Abstract A serious concern for the environment and life itself is water pollution. Anthropogenic and natural factors both contribute to water contamination. The biota is facing major health issues as a result of large volumes of industrial waste in fresh and marine water due to industrial revolution. The growing human population has also resulted in an increase in waste materials, such as cosmetics, hazardous chemicals, organic garbage, and biological pollutants in water. Many water pollutants are difficult to detect and quantify using conventional methods. The detection technology is costly, not easily portable and requires a lot of maintenance. In order to make the equipment more compact, durable, economical and highly selective, researchers have looked into modern approaches for the detection of water pollutants. Nanostructured devices are an effective strategy due to their low energy consumption, excellent precision, and great efficiency. In the realm of wastewater monitoring, nanosensors are essential as they are used to detect contaminants from various natural and man-made sources. They provide significant benefits over traditional techniques and satisfy the demand for portable analytical instruments that possess stability and selectivity. As nanotechnology has progressed, a great deal of study has been done on how to use the special chemical and physical characteristics of nanostructures and nanomaterials (NMs) to create nanosensors. Because of their superior interaction with nanoscale analytes, high sensitivity, selectivity, portability, quick response and simplicity of use, these nanosensors are far superior to the conventional sensors. In order to monitor wastewater pollutants, nanosensors can be

A. K. Mitra (✉)
Department of Chemistry, Government General Degree College, Singur, Hooghly, West Bengal, India
e-mail: ambrosia12june@gmail.com

S. Dey
Department of Chemistry, Rani Rashmoni Green University, Tarakeswar, Hooghly, West Bengal, India

V. S. Solanki
Department of Chemistry, Institute of Science and Research, IPS Academy, Indore, India

N. Agarwal
Department of Chemistry, Navyug Kanya Mahavidyalaya, University of Lucknow, Lucknow, India

N. Agarwal et al. (eds.), *Nanomaterials in Wastewater Research*, Advances in Wastewater Research, https://doi.org/10.1007/978-981-96-3908-3_3

classed depending on the NMs employed as the sensing material, transduction mechanism and also on the nature of analytes. More effective and selective nanobiosensors have also been designed in the recent past as a result of the fusion of biotechnology and nanotechnology. Many different types of sensor materials, including those based on metal, metal oxide, carbon, and polymers, have been generated. High-efficiency, high-flexibility and multi-functionality sensing applications are being built with NM-enabled sensors. Several nanosensors currently in use possess the innate ability to accomplish these objectives; yet, they necessitate additional advancements to become user and consumer-friendly instruments capable of identifying analytes in previously unreachable domains. Using contemporary nanosensors, this chapter focuses on the detection of pesticides, industrial waste, chemical oxygen demand, biological oxygen demand, biological pollutants, heavy metals and new pollutants in wastewater.

Keywords NMs · Nanosensors · Wastewater · Environment · Nanoscale · Pollutant

3.1 Introduction

NMs and nanotechnology are widely acknowledged today, both in research and in our daily lives. The concept of nanoparticles dates back to the early history of the universe, following the Big Bang. In recent years, scientists and engineers around the world have shown tremendous interest in this field and its related topics. NMs refer to materials with external dimensions or internal structures at the nanoscale. These particles exhibit unique properties, such as optical, mechanical, magnetic, conductive, and adsorptive characteristics, which differ significantly from those of larger-scale particles (Chakraborty et al. 2021).

Nanoscience is a branch of science focused on the synthesis, characterization, and exploration of extremely small particles that can only be measured at the nanometer (a nanometer is one billionth of a meter) scale. To put this into perspective, this length is equivalent to the size of approximately 10 hydrogen atoms or 5 silicon atoms aligned in a row. The prefix 'nano' comes from the Greek word 'nanos,' meaning dwarf or very small. Let us discuss the clear and concise definitions of NM and nanotechnology. NMs are materials with one or more dimensions in the nanoscale range (1 to 100 nm). Examples include quantum wells (1-dimensional), quantum wires (2-dimensional), and quantum dots (3-dimensional) (Mustafa et al. 2021).

Norio Taniguchi, a Japanese scientist, first introduced the term 'nanotechnology' in 1974. This field involves the application of nanoscience theory into practical uses, focusing on observing, measuring, manipulating, assembling, controlling, and manufacturing matter at the nanometer scale.

The National Nanotechnology Initiative in the United States defines nanotechnology as 'a science, engineering, and technology conducted at the nanoscale where unique phenomena enable novel applications in a wide range of fields, from chemistry, physics, and biology to medicine, engineering, and electronics.' (Kaur et al. 2019).

To differentiate between nanotechnology and nanoscience: nanoscience involves the study of the behavior of nanoscale materials and provides theoretical explanations based on this study. In contrast, nanotechnology focuses on the design, production, characterization, and application of NMs. Essentially, nanoscience is the theoretical foundation of NMs, while nanotechnology applies this theory to practical uses (Malik et al. 2020). The primary application of these NMs is in wastewater remediation and water quality monitoring. Both large-scale and small-scale efforts in these areas are challenging due to the variability and complexity of the water matrix and the low concentrations of pollutants that need to be detected (Abdel-Karim et al. 2020). Water pollution is one of the most serious environmental problems. According to recent studies, at least 71% of the world's population, equivalent to 4.3 billion people, faces some degree of water scarcity during certain months of the year. Water shortages are particularly severe in Southeast Asia, North Africa, and the Middle East (Saylan et al. 2020). Wastewater, which contains various harmful materials, is produced and discharged from domestic, industrial, commercial, and agricultural activities (Gaviria-Arroyave et al. 2020). Heavy metals, which are metallic compounds with densities higher than that of water, are a significant concern (Sha and Bhattacharyya 2020). The characteristics of wastewater are determined by its physical appearance, chemical parameters, and the abundance of microorganisms. The composition and concentration of substances in wastewater can be broadly classified into three categories: (i) microorganisms, (ii) inorganic chemicals, and (iii) organic pollutants. These components can cause adverse environmental effects, such as disruption of endocrine and hormone systems, induction of cytotoxicity and/or genotoxicity, and carcinogenic effects (Tuteja et al. 2020; Martins et al. 2022).

This chapter primarily offers an overview of nanosensors, detailing their operation in wastewater monitoring, applications, types, design, fabrication, and the pollutants that contribute to water pollution (Patel et al. 2020).

3.2 Types of Wastewater Pollutants

Understanding the types of pollutants present in wastewater is crucial for effective monitoring and pollution control efforts, as wastewater contains a diverse array of pollutants originating from residential, industrial, and agricultural activities. These pollutants can be broadly categorized into several types: chemical pollutants, biological pollutants, and nutrient pollutants (Table 3.1). Each category has its unique characteristics, sources, and environmental impacts, which are discussed below (Vikesland 2018).

Table 3.1 Categories of wastewater pollutants: understanding their sources and impacts

No.	Pollutants	Sources	Examples
1	Microorganisms	Domestic sewage, domestic waste water, dung heap	Bacteria, virus
2	Organic waste	Industrial effluent, animal excreta, food processing factory waste, detergents and decayed animals and plants	Oil, gasoline, plastic
3	Plant nutrients	Chemical fertilizers	Phosphate and ammonia
4	Heavy metal	Heavy metal producing factories	Cd, Hg, Pb
5	Sediments	Soil erosion by agriculture and strip-mining	Soil, silt
6	Pesticides	Chemicals used for killing insects, fungi and weeds	Urea, Phosphate
7	Radioactive Substance	Mining of uranium containing minerals	Uranium, thorium

3.2.1 Chemical Pollutants and Heavy Metal Pollutants

Chemical pollutants in wastewater encompass a wide range of substances, including heavy metals, organic compounds, biological pollutants, and synthetic chemicals. These pollutants originate from various sources such as industrial processes, urban runoff, and household activities. They pose significant risks to water quality and ecosystem health (Graboski et al. 2020).

Heavy metals are metallic elements characterized by high atomic weights and densities, such as lead, mercury, cadmium, and chromium. These metals are persistent in the environment and have the ability to bioaccumulate in aquatic organisms. This accumulation poses significant risks to both human health and wildlife (Dimitrievska et al. 2023). Each heavy metal exhibits unique properties and toxicity levels, which require specific monitoring and pollution control measures. For example, cadmium (Cd) is naturally found in ores along with zinc (Zn), copper (Cu), and lead (Pb). It is used in various applications including stabilizers in PVC products, coloring pigments, numerous alloys, and rechargeable Ni–Cd batteries. Metallic cadmium has historically been used as an anticorrosion agent. Cadmium can also be found as impurities in phosphate fertilizers. Mining of zinc (Zn) and lead (Pb) ores can be a source of cadmium contamination in groundwater. Tobacco is a significant source of cadmium exposure for smokers, as tobacco plants, like other plants, can absorb cadmium from the soil. High concentrations of cadmium can lead to renal dysfunction, kidney damage, and gastrointestinal disorders. Long-term exposure to cadmium has been associated with an increased risk of cancer. Exposure to elevated levels of cadmium during pregnancy can lead to premature birth and reduced birth weights (Vunain and Mishra 2018).

Again, lead (Pb) is known for its neurotoxic effects, which are particularly harmful to children's development. Mercury bioaccumulates in aquatic ecosystems, posing risks to fish and other wildlife. The main sources of lead in the environment include storage batteries, electroplating, construction materials, ceramics, dyes, and paints. Lead is non-biodegradable and a teratogen, making it extremely harmful to the environment. Recently a label-free colorimetric nanosensor has been developed for detecting Pb^{2+} in water (Ejeian et al. 2018). The release of heavy metals into wastewater can have serious adverse effects on human health and the environment. Chronic exposure to heavy metals through contaminated water sources can lead to various health problems, such as neurological disorders, kidney damage, and cancer. Moreover, heavy metals can accumulate in soil and sediment, posing long-term risks to ecosystems and aquatic life.

3.2.2 Pollutants Based on Organic Compounds

Organic contaminants in wastewater include a wide range of chemicals, including petroleum-derived hydrocarbons, volatile organic compounds (VOCs) and medicines. These contaminants come from a variety of sources, including industrial effluents, urban runoff, and home sewage, creating considerable issues for wastewater treatment and environmental management. Organic pollutants are classified into several different chemical classes, each with its own set of sources and features. Polycyclic aromatic hydrocarbons (PAHs), for example, are by-products of incomplete combustion and can be detected in urban runoff and industrial effluent. VOCs such as benzene and toluene come from industrial operations and automobile emissions, and they contribute to the organic pollutant load in wastewater.

A group of compounds known as Persistent Organic Pollutants (POPs) persist in the environment due to their resistance to chemical, photochemical, and biological degradation processes. These compounds do not undergo significant elimination in any environmental compartment. POPs include substances such as polychlorinated biphenyls (PCBs), polychlorinated dibenzo-*p*-dioxins, organochlorine pesticides (OCPs), perfluorooctanoate (PFOA), polybrominated diphenyl ethers (PBDEs), and perfluorooctane sulfonate (PFOS).

These pollutants have been found in rivers and coastal areas, indicating widespread distribution and possible environmental effect (Ejeian et al. 2018; Jain et al. 2021).

Many organic molecules survive in the environment, preventing decomposition by typical wastewater treatment processes. Their complex chemical structures and microbial resistance make efficient removal from wastewater difficult. As a result, organic contaminants can remain in aquatic ecosystems, posing dangers to both aquatic organisms and human health. This persistence can lead to bioaccumulation and biomagnification, exacerbating the potential effects on ecosystems and human populations (He and Zhang 2019).

3.2.3 Biological Pollutants

Biological contaminants in wastewater include predominantly microorganisms such as pathogens, bacteria, viruses, and parasites. Pathogens are disease-causing microorganisms found in sewage, such as bacteria, viruses, protozoa, and helminths. These microbes can enter wastewater via a variety of routes, notably through human and animal feces. They represent a substantial danger of transmitting waterborne illnesses (Manasa and Mehta 2020). Wastewater can include a wide variety of pathogens, each with its own distinct properties and modes of transmission. Bacterial and viral pollutants are prevalent in wastewater, posing serious threats to water quality and public health. These microbes may live and grow in aquatic environments, increasing the likelihood of waterborne illness epidemics. Bacterial pathogens such as *Escherichia coli* and *Salmonella spp.* are commonly found in sewage and can cause gastrointestinal illnesses. Viral viruses like norovirus and rotavirus are extremely contagious and can cause gastroenteritis and other watery disorders. Pathogenic microorganisms in contaminated water can cause a variety of waterborne diseases, such as gastroenteritis, cholera, hepatitis, and dysentery. Waterborne infections are transmitted through ingestion, inhalation, or direct contact with polluted water, emphasizing the significance of proper wastewater treatment and pathogen monitoring.

Bacteria and viruses are commonly found in human and animal fecal matter, making sewage a primary source of contamination. Once introduced into wastewater, these microorganisms can persist for extended periods, especially in environments with favorable conditions for their growth and survival. Detecting and quantifying bacterial and viral contaminants in wastewater presents significant challenges due to their small size, diverse characteristics, and complex interactions with environmental matrices. Traditional methods of microbial analysis, such as culture-based techniques, are time-consuming and may underestimate microbial concentrations. This underscores the need for the development of rapid and sensitive detection methods (Amin et al. 2014).

3.2.4 Nutrient Pollutants

Nutrient pollutants in wastewater primarily consist of nitrogen and phosphorus compounds, which originate from sources such as human waste, agricultural runoff, and industrial effluents. Excessive levels of these nutrients in wastewater can lead to eutrophication of aquatic ecosystems, thereby posing significant risks to water quality and ecosystem health. Nitrogen contaminants in wastewater include ammonia, nitrate, and nitrite (Steffens et al. 2017). These chemicals are results of organic matter breakdown, synthetic fertilizer application, and industrial operations. If not handled effectively, they can represent serious threats to aquatic ecosystems. Nitrogen compounds enter wastewater via a variety of processes, including human and animal waste, agricultural runoff, and atmospheric deposition (Garcia-Segura et al. 2020). Excess nitrogen levels in water bodies can cause eutrophication, which is characterized by algal blooms, hypoxia, and fish mortality. Additionally, nitrogen compounds can endanger human health by contaminating drinking water and leading to the development of disinfection by-products. Phosphorus contaminants in wastewater are generally composed of orthophosphate, polyphosphate, and organic phosphorus compounds. These substances come from several sources, including human and animal waste, agricultural runoff, and industrial emissions. If not managed appropriately, they can cause eutrophication and ecosystem deterioration (Abd El-Ghany et al. 2023).

Monitoring nitrogen levels in wastewater is essential for assessing water quality and implementing effective pollution control measures. Traditional methods of nitrogen analysis, such as colorimetric assays and spectrophotometry, are widely used but may lack the sensitivity and specificity needed for complex wastewater matrices. Advanced analytical techniques, including chromatography and mass spectrometry, offer enhanced sensitivity and selectivity for detecting nitrogen compounds in wastewater. Phosphorus compounds are essential nutrients for plant growth and development, but excessive phosphorus levels in water bodies can lead to eutrophication and algal blooms. This enrichment of aquatic ecosystems with phosphorus can cause oxygen depletion, habitat degradation, and loss of biodiversity, adversely affecting ecosystem services and human well-being. Monitoring phosphorus levels in wastewater is critical for identifying pollution sources and implementing remediation strategies. Traditional methods of phosphorus analysis, such as colorimetric assays and spectrophotometry, are widely used but may suffer from interference and limited sensitivity. Advanced analytical techniques, including high-performance liquid chromatography (HPLC) and inductively coupled plasma mass spectrometry (ICP-MS), offer improved accuracy and precision for detecting phosphorus compounds in complex wastewater matrices (Zhang et al. 2021; Shmeis 2022).

3.3 Challenges in Wastewater Monitoring

Wastewater tracking plays a crucial role in assessing water quality, identifying pollution sources, and guiding pollution control measures. However, several challenges hinder the effectiveness of wastewater monitoring efforts, including the limitations of traditional methods, the need for real-time monitoring, and specific challenges associated with detecting different types of pollutants. Traditional strategies of wastewater monitoring, such as grab sampling and laboratory analysis, have several drawbacks that can compromise the accuracy, reliability, and timeliness of data collection (Shmeis 2022).

3.3.1 *Time-Consuming Sampling Procedures*

Traditional wastewater monitoring typically involves manual collection of grab samples at discrete time points, which can be labor-intensive, time-consuming, and subject to sampling bias. Additionally, grab sampling fails to capture temporal variations in pollutant concentrations, limiting the ability to assess dynamic changes in water quality.

3.3.2 *Infrequent Sampling Frequency*

Due to logistical constraints and resource limitations, traditional wastewater monitoring programs often rely on infrequent sampling intervals, which may not capture short-term fluctuations or episodic pollution events. Consequently, there is a risk of underestimating pollutant loads and overlooking critical pollution sources (Singh et al. 2022).

3.3.3 *Need for Real-Time Monitoring*

The dynamic nature of wastewater pollution needs real-time monitoring technologies that give immediate data for informed decision-making and quick response to pollution situations.

Real-time monitoring allows for continuous observation of water quality parameters, resulting in early detection of pollution events, timely mitigation measures, and proactive control of pollution sources. Real-time monitoring improves pollution control efforts by providing timely and actionable information, while also reducing threats to human health and the environment.

Instantaneous monitoring supplements traditional grab sample techniques by recording short-term fluctuations, diurnal trends, and pollution spikes that would go undetected with occasional sampling. Real-time monitoring provides high-resolution data on pollutant concentrations and water quality metrics, allowing for more focused pollution prevention and control techniques (Mao et al. 2021).

3.4 Specific Challenges for Different Pollutants

Because of variations in concentrations, detection limits, and interactions with environmental matrices, different types of contaminants pose distinct challenges for wastewater monitoring. Chemical contaminants, such as heavy metals and organic chemicals, may necessitate specialized analytical procedures for identification and quantification due to their various physicochemical features and low concentrations in wastewater. Biological pollutants, such as pathogens and microbial contaminants, present difficulties in microbial detection and enumeration. Consequently, they necessitate quick and sensitive detection technologies capable of identifying low quantities of microbial contaminants in complex environmental matrices. Many pollutants of significance in wastewater, such as emerging contaminants and trace contaminants, exist at quantities that are below the detection limits of conventional monitoring technologies. To achieve reliable detection and quantification of these low-level contaminants, modern analytical techniques, sensitive sensors, and stringent quality assurance protocols must be used to ensure data correctness and precision.

3.5 Nanosensors for Wastewater Monitoring

Nanosensors represent a cutting-edge technology that offers tremendous potential for revolutionizing wastewater monitoring efforts. These miniaturized devices leverage the unique properties of NMs to detect and quantify pollutants with unprecedented sensitivity, selectivity, and speed. Table 3.2 showcases examples of a few representative nanosensors.

Nanosensors hold promise for addressing the limitations of traditional monitoring methods and enabling real-time, continuous surveillance of water quality parameters in wastewater.

A crucial issue that is sometimes disregarded in the hurry to develop new and revolutionary nanosensors is the expectation that they would fulfill existing unmet needs or surpass conventional techniques. While nanosensors have great sensitivity and specificity in detecting toxins or pollutants at very low concentrations, their use must be carefully examined. They are especially useful for detecting heavy metals, chemical compounds, and other contaminants in water. Given the historically conservative nature of the water treatment community, these concerns must be addressed from the very beginning of nanosensor design. Continued growth of applied nanosensor technology necessitates a focus on areas where nanosensors truly deliver innovative capabilities or increased utility in water quality monitoring. Nanosensors encompass a diverse array of sensor platforms and detection mechanisms specifically designed for wastewater monitoring applications. These sensors can detect a variety of pollutants such as chemical contaminants, biological pathogens, and nutrient species with exceptional precision and accuracy (Hooshmand et al. 2023). Nanosensors are versatile in their design and operation, making them suitable for integration into existing wastewater treatment systems, portable field devices, and distributed sensor networks. This versatility enhances their utility in providing real-time monitoring capabilities and improving overall water quality management strategies. Nanosensors are applied across various stages of wastewater treatment, spanning from influent characterization and process monitoring to effluent quality assessment and environmental monitoring. These sensors facilitate real-time monitoring of essential water

Table 3.2 Examples of a few representative nanosensors (Chakraborty et al. 2021)

Types of nanosensors		
Based on detection mechanism	Chemical nanosensors	These nanosensors detect the chemical properties
	Physical nanosensors	These nanosensors measure the physical parameters
Based on sensing element	Nanoparticle-based nanosensors	Utilize nanoparticles, such as quantum dots
	Nanowire-based nanosensors	Carbon nano tube, nano wire are this type of nanosensors
Based on application	Biological nanosensors	It helps detect pathogens, biomolecules, and more
	Environmental nanosensors	It is used to monitor water and air pollution, i.e., environmental pollution
	Industrial nanosensors	It is used to monitor the quality control of industrial waste
Based on transduction principle	Field effect transistor (FET) nanosensors	It helps detect electrical conductance and capacitance
	Surface plasmon resonance (SPR) nanosensors	It operates through a thin metal film, which is influenced by binding events between the sensing element and the analyte
	Surface enhanced Raman scattering (SERS) Nanosensors	It helps detect trace elements
Based on transduction method	Electrochemical nanosensors	It detects changes in electrical properties
	Optical nanosensors	It utilizes changes in optical properties
	Mechanical nanosensors	It detects changes in mechanical properties, such as mass
Based on scale	It identifies molecules to detect specific analytes	It recognizes molecules to detect specific analytes
	It is an extremely sensitive nanosensor	It is a highly sensitive nanosensor

quality parameters such as pollutant concentrations, microbial contamination levels, and nutrient dynamics. This capability offers valuable insights into treatment performance and aids in the development of effective pollution control strategies (Das et al. 2015).

3.5.1 Targeted Pollutant Detection

Nanosensors can be customized to detect specific pollutants present in wastewater, including heavy metals, organic compounds, pathogens, and nutrient species. Through the functionalization of NMs with selective receptors or recognition elements, nanosensors can achieve high sensitivity and specificity for target analytes. This capability enables rapid and reliable detection even within complex wastewater matrices (Gangar et al. 2021).

Nanosensors used for wastewater monitoring are classed according to the contaminants they are meant to detect, which include chemical, biological, and nutritional species. Each type of nanosensor employs unique sensing mechanisms and detection principles adapted to the target analytes. This classification allows for thorough monitoring of water quality parameters in wastewater, covering specific monitoring requirements with enhanced detection capabilities (Gangar et al. 2021).

- **Chemical Nanosensors**

Chemical nanosensors are especially intended to detect and quantify chemical contaminants such as organic molecules, heavy metals, and emerging pollutants in wastewater. These sensors use NMs' unique qualities, including as high surface-to-volume ratio, surface reactivity, and optical properties, to detect target analytes in a sensitive and selective manner. Chemical nanosensors can improve the accuracy and efficiency of wastewater pollution monitoring by utilizing these features (Jahn et al. 2020).

- **Biological and Microbial Nanosensors**

Biological and microbial nanosensors are widely used for detecting and.

quantifying microbial pollutants such as bacteria, viruses, and protozoa in wastewater. These sensors make use of biomolecular recognition components including antibodies, aptamers, and enzymes that have been mounted on NM surfaces. This enables them to specifically capture and identify infections with great sensitivity and specificity. Biological and microbial nanosensors use biomolecular interactions to help monitor and manage microbial pollution in wastewater systems (Li et al. 2015a).

- **Nutrient-Specific Nanosensors**

Nutrient-specific nanosensors are intended to track the movement of nutrients such as nitrogen and phosphorus species in wastewater. These sensors use selective receptors or catalytic compounds mounted on NM surfaces. This design allows them to identify and quantify target nutrient species in real time, providing vital information about nutrient cycling, eutrophication processes, and the efficacy of wastewater treatment systems. Nutrient-specific nanosensors help to improve management strategies for sustaining water quality and ecosystem health by enabling continuous monitoring (Gangar et al. 2021; Jahn et al. 2020; Li et al. 2015a).

3.5.2 Advances in NMs

Recent advances in NMs have resulted in the development of sophisticated nanosensors with improved performance characteristics for monitoring wastewater. These advancements include developing new NMs with tailored physical and chemical properties, using techniques to change surfaces and bioconjugate functionalities, and incorporating nanosensors into microfluidic platforms and wireless communication systems (Li et al. 2015a).

- **Materials Used in Sensor Fabrication**

Nanosensors are made from a variety of NMs, including nanoparticles, nanotubes, nanowires, graphene, and other materials. These NMs offer significant advantages such as high surface area, tunable optical and electrical properties, and biocompatibility, making them ideal for the growth of sensor technology (Oliveira et al. 2015).

- **Enhancement of Sensitivity and Selectivity**

Recent advances in NM synthesis and functionalization have produced nanosensors with increased sensitivity, selectivity, and stability for detecting contaminants in wastewater. Researchers can improve sensor performance and detect target analytes at trace levels in complex environmental circumstances by precisely controlling the surface chemistry and shape of NMs. These advances are critical for improving wastewater monitoring capabilities and implementing effective pollution control methods (Li et al. 2015a).

Optical nanosensors, as defined by rely on NMs' unique optical properties when interacting with light signals, a process known as transduction (Scheme 1). The sensitivity of these nanosensors is mostly dependent on the approaches used to detect optical events. They respond to changes in optical signal transduction and can be further classified according to specific optical features such as absorption and emission, surface plasmon resonance (SPR), light scattering, fluorescence, and others (Gangar et al. 2021; Jahn et al. 2020; Li et al. 2015a).

The primary operating mechanism of such nanosensors includes using fluorescence processes, including both emission and quenching. When a quencher interacts with fluorescein isothiocyanates (FITCs) in their excited state, it reduces the fluorescence emitted by FITCs (Shmeis 2022; Jahn et al. 2020). Various types of nanosensors have been listed in Fig. 3.1.

3.5.3 Design and Fabrication of Nanosensors for Wastewater

The design and manufacture of nanosensors for wastewater monitoring require advanced engineering tactics aimed at generating sensitive, selective, and long-lasting sensor systems capable of detecting pollutants with pinpoint precision and

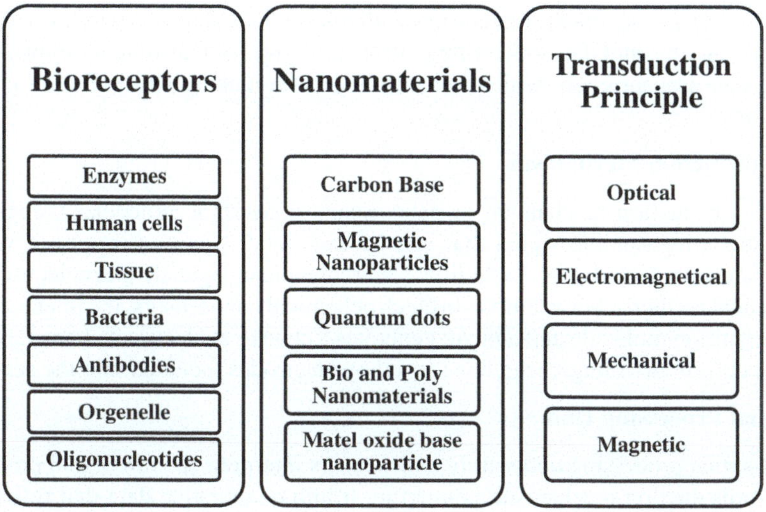

Fig. 3.1 Main components of nanosensors and their various types according to types of NMs used, transduction principle and recognition elements (for nanobiosensors) for environmental application (Chakraborty et al. 2021)

accuracy. These nanosensors use the unique features of NMs and novel sensor architectures to improve performance in challenging wastewater conditions.

- **Nanosensor Components**

Nanosensors are made up of several key components, each of which contributes significantly to sensor performance and pollutant detection. These components are precisely designed and combined to maximize sensor performance and reliability in harsh wastewater settings (Zhang et al. 2017b).

- **Sensor Structure and Functional Elements**

Nanosensors typically have a substrate or support material that is functionalized with NMs and sensing elements. NMs, such as nanoparticles, nanowires, and nanotubes, serve as sensing transducers, providing a large surface area and reactivity for detecting pollutants. Functionalized coatings or recognition elements, such as antibodies, aptamers, or molecularly imprinted polymers, are applied to the NM surface in order to selectively capture target analytes and facilitate signal transduction (Jahn et al. 2020; Li et al. 2015a).

- **Integration with Existing Wastewater Infrastructure**

Nanosensors are designed to seamlessly connect with existing wastewater treatment systems and monitoring platforms, making them easier to incorporate into operational workflows. These sensors can be used at several stages of the wastewater treatment process, including influent characterization, process monitoring, and effluent quality assessment. Integration with microfluidic devices, wireless communication systems, and data collecting platforms allows for real-time monitoring and remote data transmission, which improves decision-making and pollution control operations.

- **Transduction Mechanism**

The transduction mechanism in nanosensors converts a molecular recognition event into a measurable signal that can be examined. Nanosensors use a variety of transduction techniques, including electrochemical, optical, piezoelectric, and mechanical methods. For example, in electrochemical nanosensors, the interaction of the recognition molecule and analyte generates a change in electrical characteristics, such as current or voltage, which is sensed by electrodes included into the sensor.

- **Signal Processing Unit**

The signal processing unit within nanosensors interprets the raw signal produced by the transduction mechanism, converting it into quantitative data that reflect the concentration of the target analyte. This unit typically incorporates analog or digital circuitry, microcontrollers, and signal processing algorithms. These components are designed to amplify, filter, and analyze the sensor output effectively.

Advanced signal processing techniques, including machine learning algorithms and data fusion methods, may also be utilized to improve sensor accuracy, reliability, and overall robustness. These techniques help in extracting meaningful information from complex sensor signals, enabling more precise detection and analysis of pollutants in wastewater monitoring applications (Jahn et al. 2020).

- **Tailoring for Specific Pollutants**

Nanosensors are intended to identify specific pollutants in wastewater, such as chemical contaminants, biological pathogens, and nutrient species. Their design and manufacturing procedures are meticulously adjusted to increase sensitivity, selectivity, and stability while detecting these specific analytes. This improvement assures that nanosensors can detect and measure contaminants even in complex environmental situations like wastewater (Orebiyi et al. 2010).

- **Customization Based on Pollutant Characteristics**

Nanosensors are meant to focus on the physicochemical properties of the contaminants they detect, such as molecular size, charge, and reactivity. For example, in the detection of heavy metals, nanosensors may use functionalized NMs with a high affinity for metal ions. Microbial detection sensors, on the other hand, may include biomolecular recognition components that can selectively bind to specific infections. This modification guarantees that nanosensors are properly tailored to capture and detect their intended targets in wastewater and other environmental matrices (Iqbal and Shah 2013).

- **Optimization of Transduction Mechanism**

Customizing the transduction mechanism entails optimizing the sensor's pathway for signal transduction to effectively convert the binding event between the recognition molecule and pollutant into a measurable signal. This process includes selecting suitable transduction techniques like electrochemical, optical, or piezoelectric methods that provide high sensitivity and selectivity for the target pollutant.

For instance, electrochemical nanosensors might employ specific redox reactions or changes in conductivity to precisely detect pollutant concentrations. This approach ensures that nanosensors are tailored to reliably detect and quantify specific pollutants in wastewater and other environmental samples.

3.5.4 Multi-Sensor Systems for Comprehensive Monitoring

Multi-sensor systems integrate numerous sensing modalities or sensor arrays onto a single platform, allowing for simultaneous monitoring of multiple water quality indicators. These systems use synergies between several sensor types to improve sensitivity, specificity, and reliability in identifying contaminants in wastewater. Multi-sensor systems provide a complete wastewater monitoring solution by combining chemical, biological, and physical sensing elements. They give significant information on pollutant behavior, treatment effectiveness, and environmental impact, which improves overall water resource management and protection.

- **Integration and Deployment**

Once validated, customized nanosensors can be seamlessly integrated into monitoring systems and deployed for routine environmental monitoring tasks. This integration often includes embedding the sensors into portable devices, field-deployable platforms, or interconnected sensor arrays to monitor pollutant levels continuously in real-time. These nanosensors are designed to work in concert with data acquisition systems and communication networks, allowing for remote monitoring and data sharing. This capability enhances the ability to make timely decisions and implement effective pollution control strategies based on accurate and up-to-date information gathered from the sensors.

- **Integration with Treatment Systems**

Nanosensors are integrated into wastewater treatment systems to enhance treatment efficiency, reduce environmental impact, and ensure compliance with regulations. These sensors offer continuous monitoring and real-time feedback on treatment processes, which enables operators to optimize process parameters, detect operational faults promptly, and implement adaptive control strategies. By monitoring critical water quality indicators such as pollutant concentrations, microbial activity, and nutrient levels, nanosensors support proactive management and targeted interventions to mitigate pollution sources effectively. This integration helps wastewater treatment facilities achieve higher operational efficiency and maintain water quality standards more effectively.

- **Enhancing Efficiency of Wastewater Treatment**

Nanosensors improve the effectiveness of wastewater treatment processes by offering prompt information regarding the efficiency of pollutant removal, treatment rates, and opportunities for process enhancement. Through continuous monitoring of essential parameters like dissolved oxygen, pH, and nutrient levels, nanosensors empower operators to adjust treatment procedures efficiently, reduce energy usage, and optimize overall treatment efficacy.

- **Minimizing Environmental Impact**

Nanosensors play a crucial role in reducing the environmental impact of wastewater discharges through early detection of pollutant spills, leaks, and contamination events. Their ability to detect pollutants rapidly and with high sensitivity allows for swift response actions, implementation of containment measures, and mitigation strategies aimed at preventing pollution incidents and mitigating environmental harm (Li et al. 2016).

3.5.5 Properties of NMs

Some essential properties of nanoparticles include: (1) surface effects, which can be explained by the fact that nanoparticles have more surface atoms than inner atoms; (2) the availability of more free energy in the surface, which can increase the rate of a chemical reaction. (3) The other attribute is superparamagnetism, which happens when the particle is smaller than the magnetic domain of a material. (4) In a free electron model, average energy spacing increases as the number of atoms decreases, improving the catalytic characteristics of nanoparticles; (5) the aggregation state, form, and fractal dimension; (6) solubility; and (7) the chemical composition and crystal structure (Iwase et al. 2018).

3.6 Applications of Nanosensors in Wastewater Monitoring

Nanosensors are highly versatile tools for monitoring wastewater, capable of detecting and measuring pollutants in real-time with exceptional sensitivity, selectivity, and dependability. These advanced sensors offer significant promise for enhancing the evaluation of water quality, implementing effective pollution control measures, and managing environmental concerns across various wastewater treatment scenarios. In recent times, nanosensors have emerged as potent and efficient instruments for detecting water pollution and facilitating its remediation (Li et al. 2015b; Chandran et al. 2018).

The fundamental operating principle of nanosensors involves their responsiveness and sensitivity to changes in environmental stimuli, which act as transduction mechanisms. These stimuli include thermal factors (such as temperature, heat flow, entropy, specific heat), chemical factors (concentration, pH, composition, oxidation/reduction potential, reaction rate), magnetic factors (magnetic moment, flux density, permeability, field intensity), optical factors (absorbance, transmittance, fluorescence), mechanical factors (volume, area, length, pressure, mass flow, acoustic intensity, acoustic wavelength), and electrical factors (current, voltage, resistance, charge).

When exposed to target analytes, nanosensors detect these variations and transduce them into signals, typically electrical, which are then analyzed to provide an output. In biosensors, key components include bio-recognition elements like enzymes, oligonucleotides, and antibodies, which facilitate specific binding to target molecules (Li et al. 2015b).

The production methods for nanosensors primarily involve three approaches: (1) top-down lithography, (2) bottom-up methods, and (3) self-assembly, each tailored to optimize sensor performance and functionality (Chandran et al. 2018).

3.6.1 Monitoring of Chemical Contaminants

Nanosensors play an important role in the monitoring and detection of chemical contaminants in wastewater, including heavy metals, organic chemicals, inorganic compounds, and emerging pollutants. These sensors are distinguished by their sensitivity and specificity to target analytes, enabling rapid detection and precise quantification in complex environmental situations (Jiang et al. 2018).

- **Heavy Metal Detection**

  Nanosensors are used to detect heavy metals like lead, mercury, cadmium, and chromium in wastewater. These sensors employ functionalized NMs, such as nanoparticles and nanotubes, which are designed to selectively bind to metal ions. This selective binding enables sensitive detection through different transduction methods, including electrochemical, optical, and piezoelectric sensing (Low et al. 2017).

- **Organic Compound Analysis**

  Nanosensors are effective and efficient in detecting organic compounds such as polycyclic aromatic hydrocarbons (PAHs), volatile organic compounds (VOCs), and pharmaceuticals in wastewater. Surface-functionalized NMs serve as sensing elements, selectively interacting with target analytes to produce measurable signals that indicate pollutant concentrations (Miao et al. 2017).

3.6.2 Monitoring of Biological Contaminants

Nanosensors play a crucial role in detecting biological contaminants, including bacteria, viruses, protozoa, and other microbial pathogens, in wastewater. They provide fast and sensitive detection, allowing for early warning of microbial contamination and timely measures to prevent waterborne disease outbreaks (Zhou et al. 2017).

- **Pathogen Detection**

  Nanosensors are used to detect bacterial and viral pathogens in wastewater, such as Escherichia coli, Salmonella spp., norovirus, and rotavirus. Surface-functionalized NMs, combined with biomolecular recognition elements, selectively capture these target pathogens, allowing for sensitive detection through various transduction methods, including impedance spectroscopy, fluorescence, and surface plasmon resonance (Wu et al. 2018).

- **Microbial Activity Monitoring**

  Nanosensors are used to monitor microbial activity and metabolic processes in wastewater treatment systems. They detect changes in microbial biomass, metabolic byproducts, and enzyme activity, offering valuable information on treatment efficiency, microbial ecology, and process optimization.

3.6.3 Nutrient Monitoring and Management

Nanosensors are essential for monitoring nutrient dynamics, such as nitrogen and phosphorus, in wastewater. They provide real-time assessments of nutrient concentrations, aiding in the optimization of nutrient removal, control of eutrophication, and protection of the environment (Zhu et al. 2017).

- **Nitrogen Species Detection**

  Nanosensors detect nitrogen species in wastewater, including ammonia, nitrate, and nitrite. Functionalized nanoparticles, when paired with selective receptors or catalytic materials, selectively absorb and measure nitrogen species, enabling for quick evaluation of nutrient loading and nitrogen removal effectiveness in treatment systems (Yuan et al. 2017).

- **Phosphorus Species Analysis**

  Nanosensors are used to detect phosphorus species, such as orthophosphate, polyphosphate, and organic phosphorus compounds, in wastewater. Surface-functionalized NMs, integrated with specific binding ligands or enzymes, selectively capture and quantify these phosphorus species, offering valuable insights into nutrient cycling, eutrophication processes, and the performance of wastewater treatment (Tu et al. 2015).

- **Nutrient Removal Technologies in Wastewater Treatment**

Wastewater treatment uses a variety of processes to remove nutrients, including biological, physical, and chemical approaches. Biological methods such as activated sludge, trickling filters, and sequencing batch reactors use microorganisms to degrade organic materials and extract nutrients from wastewater. Advanced approaches, such as membrane bioreactors (MBRs) and nutrient removal systems (NRS), improve nutrient removal efficiency through filtration, adsorption, and precipitation (Wu et al. 2017).

- **Enhanced Nutrient Removal Strategies**

Wastewater treatment plants use enhanced nutrient removal (ENR) technologies such as biological nutrient removal (BNR), increased coagulation, and tertiary treatment procedures to satisfy rigorous nutrient removal targets while still protecting delicate aquatic habitats. BNR approaches, which include denitrification, nitrification, and phosphorus removal, optimize microbial processes to increase nutrient removal efficiency. Chemical precipitation, adsorption, and ion exchange techniques supplement biological treatments to reduce nutrient levels in treated wastewater (Liang et al. 2018).

3.7 Real-Time Monitoring and Process Control

Nanosensors enable real-time monitoring and control of wastewater treatment systems, resulting in increased operational efficiency, pollutant management, and regulatory compliance. They continually monitor water quality data, allowing for timely reactions to pollution incidents, adaptive control tactics, and treatment process optimization (Zhang et al. 2017a).

- **Continuous Surveillance**

Nanosensors provide ongoing monitoring of essential water quality indicators such as pollutant levels, microbial contamination, and nutrient fluctuations in wastewater treatment processes. By offering real-time data collection and analysis, they facilitate proactive pollution management, early identification of treatment issues, and the enhancement of treatment efficiency (Ou et al. 2018).

- **Adaptive Control Strategies**

Nanosensors enable adaptive control in wastewater treatment systems by providing real-time data that allows for dynamic adjustments to treatment processes. By tracking essential parameters such as dissolved oxygen, pH, and nutrient levels, nanosensors help operators enhance treatment effectiveness, reduce energy use, and meet regulatory requirements (Zeng et al. 2017).

3.8 Integration with Smart Water Management Systems

Nanosensors are incorporated into advanced water management systems, allowing for remote monitoring, data analysis, and decision-making support in managing water quality and controlling pollution. These systems utilize cutting-edge sensor technology, wireless communication networks, and data visualization tools to optimize resource use, boost environmental sustainability, and enhance public health (Zeng et al. 2017).

- **Remote Monitoring and Data Transmission**

  Nanosensors allow for the remote monitoring of water quality parameters, providing stakeholders with access to real-time data and alerts through web-based platforms or mobile apps. Wireless communication networks, including IoT (Internet of Things) systems, support effortless data transfer and remote control of sensor networks, enabling quick responses to pollution events and adaptive management approaches. IoT-enabled devices, such as wireless sensor nodes, data loggers, and cloud storage solutions, support real-time data transmission, remote monitoring, and centralized data management, converting raw sensor information into actionable insights available to stakeholders through web and mobile applications (Lin et al. 2018).

- **Data Analytics and Decision Support**

  Nanosensors generate extensive datasets on water quality parameters, which can be analyzed with advanced data analytics techniques, such as machine learning and artificial intelligence, to uncover valuable insights and patterns. Decision support tools, including predictive models and risk assessment frameworks, leverage this sensor data to guide decision-making, prioritize interventions, and optimize resource allocation for pollution control and environmental management (Lin et al. 2018).

- **Smart Control and Automation**

  Smart water management systems utilize nanosensor data and predictive analytics to enable intelligent control and automation of water treatment processes, distribution networks, and irrigation systems. Automated control algorithms dynamically adjust parameters like flow rates, chemical dosages, and valve positions in real-time, based on sensor feedback and predefined optimization goals. This approach maximizes efficiency, reduces energy consumption, and ensures adherence to regulatory standards (Banerjee et al. 2016).

- **Resilience and Sustainability**

  Integrating nanotechnology with smart water management systems allows communities and industries to strengthen the resilience and sustainability of water resources amidst increasing water scarcity, pollution risks, and climate change impacts. The proactive monitoring, predictive modeling, and adaptive control offered by these smart systems enable stakeholders to anticipate and address emerging challenges, optimize resource usage, and protect water quality for current and future generations (Llorente et al. 2017).

3.9 Innovative and Advanced Techniques for Water Treatment

In recent years, advancements in nanotechnology have introduced innovative, efficient, and cost-effective approaches to wastewater treatment. These advancements stem from the unique properties of NMs, such as their high reactivity, precision, and environmentally friendly, cost-efficient production methods. The following sections will explore these developments in more detail (Wang et al. 2017a).

- **Photocatalysis**

Photocatalysis is a cutting-edge method for water purification, which uses a light-responsive nanostructured catalyst to break down various pollutants in water. This process involves modifying the rate of chemical reactions through the application of UV, visible, or infrared light, with the reaction occurring in the presence of a photocatalyst. The core principle behind this process is that the photocatalyst absorbs light, leading to the chemical transformation of the reaction's participants. In a photocatalytic system, a semiconductor acts as the catalyst, and when it absorbs light with energy greater than its bandgap, it generates electron–hole (e–h) pairs (Fig. 3.2). These pairs create highly reactive radicals, such as superoxides and hydroxyl ions, in the water. These radicals then undergo secondary reactions to degrade the organic and inorganic pollutants present in the contaminated water. An overview of this process is depicted in the diagram below. Examples of nanostructured photocatalysts include titanium dioxide (TiO_2), zinc oxide (ZnO), ferric oxide (Fe_2O_3), zinc sulfide (ZnS), and cadmium sulfide (CdS). For semiconductors to function effectively as photocatalysts, they must possess a large bandgap (Arslan et al. 2017; Leong et al. 2018).

- **Nanofiltration**

Nanofiltration (NF) is a membrane separation technique driven by pressure and is rapidly gaining prominence in water purification and wastewater treatment due to its charge-based repulsion capabilities. This method utilizes size exclusion and electrostatic interactions to selectively remove ions, organic compounds, and fine particles from water and other aqueous solutions (Lee et al. 2017a; Zhang et al. 2017b).

- NF membranes are composed of thin, semi-permeable films with pore sizes typically ranging from 1 to 10 nm (nm). These pores are larger than those in reverse osmosis (RO) membranes but smaller than those in ultrafiltration (UF) membranes (Liu et al. 2017).

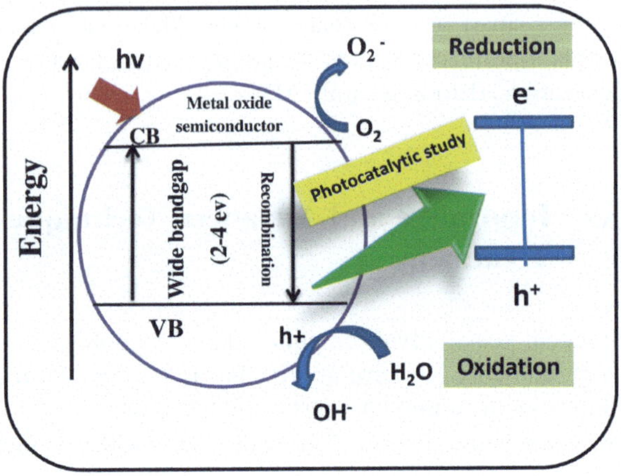

Fig. 3.2 Schematic representation of photocatalysis occurring on the surface of the nanostructured metal oxide semiconductor photocatalyst

- The NF separation process relies on both size exclusion and electrostatic interactions. Smaller particles and ions are more effectively retained, while larger molecules and solvents pass through the membrane (Zhang et al. 2018).
- The selectivity of NF membranes is determined by factors such as pore size distribution, surface charge, and molecular weight cut-off (MWCO), allowing for the precise separation of contaminants based on size and charge (Matthews et al. 2015).

3.10 Common Applications of Nanofiltration

1. **Water and Wastewater Treatment:** NF is extensively used in processes like desalination, softening, and the removal of dissolved organic compounds, heavy metals, and microorganisms. It is particularly effective for treating brackish water, groundwater, and industrial effluents (Wang et al. 2017b).
2. **Food and Beverage Industry:** In this sector, NF is used for concentrating, clarifying, and fractionating juices, dairy products, and beverages. It effectively removes color, odor, and undesirable components while preserving valuable nutrients and flavors.
3. **Pharmaceutical and Chemical Manufacturing:** NF is critical in pharmaceutical and chemical production, aiding in the purification of active ingredients, separation of by-products, and recovery of valuable solvents and intermediates (Wang et al. 2017b).

NF membranes are generally made from thin-film composite (TFC) or thin-film nanocomposite (TFN) materials, which consist of multiple layers with distinct properties (Fig. 3.3). These layers typically include a porous support layer, an active filtration layer, and sometimes a selective rejection layer. Commonly used materials for NF membranes include polyamide, polyethersulfone, cellulose acetate, and

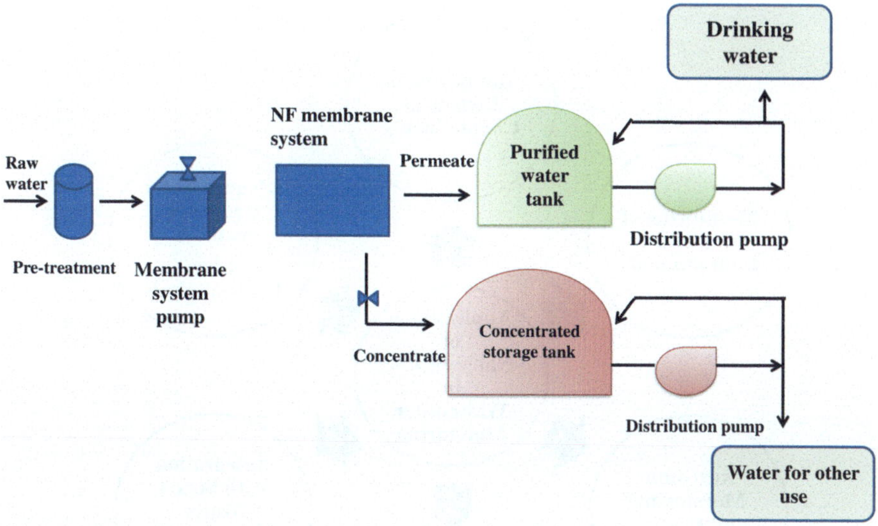

Fig. 3.3 Schematic representation of nanofiltration technique

ceramic nanoparticles, all of which provide high chemical resistance, mechanical strength, and permeability, making them suitable for a broad range of applications (Reineck et al. 2016).

The operating conditions for NF systems are influenced by factors such as feedwater composition, membrane characteristics, and the desired separation efficiency. Key operating parameters include pressure, temperature, crossflow velocity, and feedwater pH. NF systems typically function at lower pressures compared to RO, ranging from 5 to 30 bar (500 to 3000 kPa), which helps reduce energy consumption and operational costs. However, higher pressures may be necessary for specific applications or particular feedwater conditions (Li et al. 2017a).

3.11 Application of Nanosensors in Industrial Wastewater Treatment

In this study, we explore the application of nanosensors in industrial wastewater treatment, focusing on a specific facility to showcase the practical implementation and advantages of nanosensor technology in monitoring and managing wastewater pollution (Fig. 3.4) (Song et al. 2018).

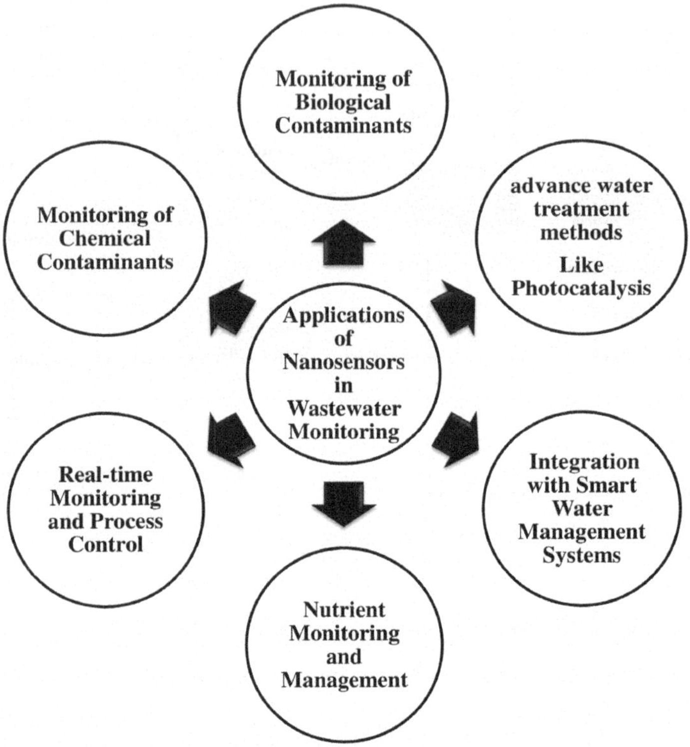

Fig. 3.4 Schematic representation of the application of nanosensors in waste water monitoring

3.11.1 Industrial Wastewater Characterization

The study begins with a detailed overview of the industrial facility, including its manufacturing processes, sources of wastewater generation, and characteristics of pollutant discharge. By analyzing the composition and quality of the wastewater, we identify key pollutants, such as heavy metals, organic compounds, and specific process-related contaminants (Li et al. 2017b).

- **Identification of Pollutants**

Through thorough sampling and analysis of the wastewater, specific pollutants are identified, including heavy metals like chromium and nickel, organic solvents such as benzene and toluene, and process byproducts like phenols and cyanides. These pollutants pose significant environmental risks and present regulatory compliance challenges, necessitating robust monitoring and control measures (Zhao et al. 2016a).

- **Quantification of Pollutant Loads**

Using advanced analytical techniques like chromatography, spectrometry, and atomic absorption spectroscopy, we measure the concentrations of these pollutants in the wastewater. By assessing pollutant loads and their variability over time, insights are gained into pollutant sources, treatment effectiveness, and potential environmental impacts.

3.11.2 Implementation of Nanosensor Technology

This section examines the integration of nanosensor technology into the wastewater treatment infrastructure to enhance pollutant monitoring, process control, and regulatory compliance. Nanosensors offer high sensitivity, selectivity, and real-time detection capabilities, making them ideal for addressing the complexities of industrial wastewater management (Sun et al. 2017).

- **Selection of Nanosensor Platforms**

We assess various nanosensor platforms suitable for detecting the identified pollutants in the wastewater stream. Nanosensors designed for heavy metal detection, organic compound analysis, and microbial monitoring are evaluated based on their performance, compatibility with industrial wastewater, and ease of integration into existing systems.

- **Customization for Industrial Applications**

To optimize the performance of nanosensors for industrial wastewater monitoring, sensor designs and functionalization strategies are tailored to meet the facility's specific requirements. Surface modifications, receptor coatings, and sensor arrays are customized to enhance sensitivity, selectivity, and stability in challenging industrial environments (Zhang et al. 2017c).

3.11.3 Real-World Deployment and Monitoring

This section details the deployment of nanosensor technology within the industrial facility for real-time monitoring of wastewater quality. Nanosensors are installed at key points along the treatment process—such as influent characterization, treatment unit monitoring, and effluent assessment—to provide comprehensive surveillance of pollutant levels and treatment efficiency (Lv et al. 2017).

- **Continuous Surveillance**

Nanosensors provide continuous monitoring of crucial water quality parameters, such as heavy metal concentrations, organic compound levels, and microbial contamination in the wastewater stream. Real-time data collection and analysis enable early detection of pollution incidents, process optimization, and regulatory compliance (Zhao et al. 2016b).

- **Adaptive Process Control**

Nanosensors facilitate adaptive control strategies in the wastewater treatment system, allowing for dynamic adjustments to operations based on real-time data. By monitoring critical parameters like pH, dissolved oxygen, and nutrient levels, nanosensors enable timely interventions, optimize treatment efficiency, and minimize environmental impact (Yu et al. 2017).

3.11.4 Performance Evaluation and Outcomes

The study concludes with an evaluation of the performance and outcomes following the deployment of nanosensors in the wastewater treatment facility. By comparing data from before and after deployment, we assess the effectiveness of nanosensors in improving wastewater quality, reducing pollutant discharge, and enhancing regulatory compliance (Aslan et al. 2017).

- **Reduction in Pollutant Levels**

The deployment of nanosensors results in a significant reduction in pollutant concentrations in the wastewater stream, demonstrating the effectiveness of real-time monitoring and adaptive control in reducing pollution sources and optimizing treatment performance. This contributes to better environmental stewardship and sustainable industrial practices (Gao et al. 2018).

- **Operational Efficiency Gains**

Nanosensor-enabled optimization leads to operational efficiency gains for the facility, including reduced energy consumption, minimized chemical use, and improved resource utilization. By fine-tuning processes based on real-time data, the facility achieves cost savings, regulatory compliance, and excellence in wastewater management (Li et al. 2019).

- **Enhanced Environmental Compliance**

Nanosensor technology supports improved environmental compliance for the facility, allowing for timely and accurate reporting of wastewater quality data to regulatory authorities and stakeholders. By providing transparent monitoring data, nanosensors underscore the facility's commitment to environmental responsibility and regulatory adherence, building trust and credibility within the community (Li et al. 2020).

3.12 Challenges and Solutions in Nanosensor Implementation

Implementing nanosensors in wastewater treatment presents various challenges related to sensor design, fabrication, deployment, and data interpretation. Addressing these challenges is crucial for unlocking the full potential of nanosensor technology and overcoming obstacles to its widespread adoption in environmental monitoring (Guo et al. 2019).

3.12.1 Sensor Design and Fabrication Challenges

Designing and fabricating nanosensors involves several hurdles related to performance, stability, and reproducibility, which must be addressed to ensure reliable and effective operation in wastewater environments.

- **Sensitivity and Selectivity Optimization**

Achieving optimal sensitivity and selectivity for detecting target analytes in wastewater is a significant challenge. Enhancing sensor performance requires careful optimization of surface functionalization, sensor coatings, and recognition elements to reduce interference from background substances (Li and Wei 2017).

- **Stability and Durability Enhancement**

Nanosensors must endure harsh conditions such as fluctuating pH, temperature changes, and chemical exposure typical in wastewater treatment. Improving their stability and durability through advanced materials engineering, encapsulation methods, and surface passivation is essential for ensuring long-term performance and reliability (He et al. 2020).

3.12.2 Deployment and Integration Challenges

Deploying nanosensors in real-world wastewater monitoring presents logistical, operational, and technical challenges that need to be addressed for successful implementation and seamless integration with existing systems (Marieeswaran and Panneerselvam 2020).

- **Sensor Compatibility and Interfacing**

Integrating nanosensors with existing wastewater treatment systems and monitoring platforms involves overcoming compatibility issues and interface requirements. Standardized sensor interfaces, communication protocols, and data formats are crucial for ensuring smooth integration into operational workflows (Wu et al. 2020).

- **Sensor Calibration and Validation**

Accurate calibration and validation of nanosensors for measuring target analytes in complex wastewater samples are challenging. Establishing robust calibration procedures, reference standards, and quality control measures is critical for ensuring precision, accuracy, and traceability of sensor measurements (Stock and Biswas 2012).

3.12.3 Data Interpretation and Analysis Challenges

Interpreting data from nanosensors and deriving meaningful insights from complex wastewater datasets pose significant challenges. Addressing these challenges is vital for informed decision-making and effective pollution control and environmental management (Bo et al. 2016).

- **Data Fusion and Integration**

Combining nanosensor data with traditional monitoring methods, remote sensing, and modeling approaches enhances the reliability and completeness of wastewater monitoring information. Techniques for data fusion, statistical analysis, and machine learning facilitate the integration of diverse datasets and extraction of actionable insights.

- **Uncertainty Quantification and Risk Assessment**
- Quantifying measurement uncertainties and assessing associated risks in nanosensor data interpretation are crucial for ensuring the credibility of monitoring results. Frameworks for uncertainty quantification, sensitivity analysis, and probabilistic modeling support robust risk assessment and decision-making under uncertainty (Garcia-Rodríguez et al. 2014).

3.12.4 Regulatory and Ethical Considerations

Navigating regulatory requirements, ethical considerations, and stakeholder engagement processes is essential for the responsible deployment of nanosensor technology in wastewater monitoring.

- **Compliance with Regulatory Standards**

Adhering to environmental regulations, quality standards, and data reporting requirements is vital for the acceptance and adoption of nanosensor technology.

Collaborating with regulatory agencies, standardization bodies, and industry stakeholders helps align with regulatory expectations and demonstrate sensor reliability and performance (Ibáñez et al. 2013).

- **Ethical and Societal Implications**

Addressing ethical concerns, privacy issues, and societal implications related to nanosensor deployment requires proactive engagement with stakeholders, including community members, policymakers, and environmental groups. Ensuring transparency, accountability, and inclusivity in decision-making fosters trust and responsible innovation in nanosensor applications (Gaudin 2017).

3.13 Future Directions and Innovations in Nanosensor Technology for Wastewater Monitoring

Nanosensor technology is rapidly evolving, with emerging innovations addressing current limitations and enhancing sensor performance. Future developments promise to revolutionize wastewater monitoring and environmental management, paving the way for sustainable water resource management and environmental stewardship.

3.13.1 Advanced Sensor Platforms and Materials

Future advancements in nanosensor technology are expected to focus on improving sensor platforms and materials to enhance sensitivity, selectivity, and stability for pollutant detection in wastewater.

- **Next-Generation NMs**

Developing novel NMs with tailored properties, such as improved surface reactivity, adjustable optical and electronic characteristics, and biocompatibility, holds promise for enhancing sensor performance. Innovations in NM synthesis, including top-down, bottom-up, and self-assembly approaches, allow precise control over material composition and structure, leading to highly sensitive and selective nanosensors.

- **Multifunctional Sensor Platforms**

Advancements in sensor design enable the creation of multifunctional platforms capable of detecting multiple water quality parameters simultaneously. By integrating different sensing modalities—such as electrochemical, optical, and piezoelectric sensing—within a single platform, multifunctional nanosensors offer comprehensive monitoring of pollutants, microbial contamination, and nutrient dynamics, facilitating holistic water quality assessment.

3.13.2 Miniaturization and Integration

Future nanosensor technologies are likely to focus on miniaturization and integration to create compact, portable, and cost-effective sensor platforms for widespread use in wastewater monitoring.

- **Lab-on-a-Chip Devices**

The development of lab-on-a-chip devices and microfluidic platforms enables the integration of nanosensors into portable, handheld devices for on-site wastewater monitoring. These systems offer benefits such as reduced sample volume requirements, rapid analysis, and automation capabilities, making them suitable for field use in remote or resource-limited settings (Wang et al. 2020).

- **Wearable and Implantable Sensors**

Innovations in wearable and implantable sensor technologies facilitate continuous real-time monitoring of water quality parameters, providing insights into individual exposure risks and health outcomes. Wearable sensors integrated into clothing or accessories and implantable sensors embedded in the body offer personalized monitoring solutions for assessing water quality-related health risks (Bilal et al. 2019).

3.13.3 Smart Sensor Networks and IoT Integration

Integrating nanosensors with smart sensor networks and Internet of Things (IoT) platforms enables real-time data acquisition, communication, and decision support for effective wastewater management and pollution control (Tian et al. 2020).

- **Distributed Sensor Networks**

Deploying distributed sensor networks equipped with nanosensors provides extensive coverage of wastewater treatment systems and environmental monitoring sites. Utilizing wireless communication and cloud-based data analytics, these networks enable real-time monitoring, predictive modeling, and adaptive control strategies to optimize treatment operations and reduce environmental impacts.

- **IoT-Enabled Monitoring Systems**

IoT-enabled monitoring systems combine nanosensors with data acquisition, communication networks, and cloud computing to facilitate remote monitoring, data visualization, and decision support for water quality management. These systems offer scalable, interoperable solutions for collecting, analyzing, and disseminating sensor data, enhancing collaborative governance and stakeholder engagement in water resource management (Lee et al. 2018).

3.13.4 Artificial Intelligence and Data Analytics

Integrating artificial intelligence (AI) and data analytics with nanosensor data enables advanced processing, pattern recognition, and predictive modeling for improved wastewater monitoring and pollution control.

- **Machine Learning Algorithms**

  Machine learning techniques, such as neural networks, support vector machines, and random forests, analyze large datasets from nanosensors to identify patterns, correlations, and anomalies related to water quality and pollution sources. These models support predictive analytics, anomaly detection, and decision-making to enhance wastewater treatment processes and mitigate pollution risks (Lee et al. 2017b).

- **Data-Driven Decision Support**

  AI and data analytics tools transform raw sensor data into actionable insights, recommendations, and visualizations for stakeholders and decision-makers. Integrating nanosensor data with environmental models and risk assessment frameworks enables evidence-based decision-making, adaptive management strategies, and timely interventions to protect water quality and public health (Chen et al. 2018).

3.14 Conclusion and Future Perspectives

- **Impact of Nanosensors on Wastewater Monitoring**

  This chapter has explored the significant role of nanosensors in wastewater monitoring, highlighting their exceptional sensitivity, selectivity, and real-time data acquisition capabilities. Nanosensors are transforming our approach to evaluating water quality and managing pollution by detecting a diverse array of contaminants, such as heavy metals, organic compounds, and microbial pathogens. Their contributions enhance our understanding of water quality dynamics and enable proactive measures to protect public health and maintain ecosystem integrity (Wang et al. 2012).

- **Addressing Environmental Challenges and Enhancing Water Quality**

  Nanosensors are crucial in tackling urgent environmental issues and improving global water quality. By providing precise and timely data on pollutant levels, nanosensors enable decision-makers to implement effective pollution control strategies, optimize wastewater treatment processes, and ensure regulatory compliance. The integration of nanosensor technology into wastewater management significantly boosts our ability to monitor, evaluate, and mitigate water pollution, leading to cleaner and safer water resources for current and future generations (Singh et al. 2020).

- **Encouraging Further Research and Development**

While we celebrate the progress achieved with nanosensor technology, it is important to continue advancing research and development in this area. Investing in innovative sensor designs, novel NMs, and sophisticated data analytics will propel the next breakthroughs in nanosensor technology. These advancements will help address emerging water quality challenges and promote environmental sustainability.

- **Promoting Integration in Wastewater Management**

To fully leverage the benefits of nanosensor technology, widespread integration into wastewater management practices is essential. Building collaborative partnerships between academia, industry, government, and non-profit organizations will facilitate knowledge sharing, technology transfer, and capacity building. By fostering interdisciplinary collaboration and stakeholder engagement, we can accelerate the adoption of nanosensors in wastewater monitoring and pollution control efforts, addressing global water challenges more effectively (Dolati et al. 2018).

- **Insights and Lessons from Case Studies and Applications**

The case studies and applications discussed in this chapter offer valuable insights into the practical use and impact of nanosensors in wastewater monitoring. Nanosensors provide versatile solutions for detecting chemical contaminants, monitoring microbial pathogens, and assessing nutrient levels. By reviewing real-world examples and success stories, we gain a deeper understanding of the capabilities, limitations, and best practices related to nanosensor technology in environmental monitoring.

- **Future Advancements and Innovations**

As we conclude this chapter, it is clear that the journey with nanosensor technology is just beginning. The knowledge gained here lays the groundwork for future advancements and innovations in wastewater monitoring. By embracing emerging technologies, fostering interdisciplinary collaborations, and engaging stakeholders, we can drive progress toward more sustainable and resilient water management systems. Harnessing the transformative potential of nanosensor technology will help safeguard water resources, protect public health, and preserve ecosystems for future generations.

Acknowledgements Amrit Krishna Mitra would like to acknowledge the financial assistance provided by the Department of Science & Technology and Biotechnology, Government of West Bengal, India (Memo No. 860(Sanc.)/STBT-11012(25)/5/2019-ST SEC dated 03/11/2023). The authors would like to acknowledge Prof. Ashutosh Ghosh, Vice-Chancellor of Rani Rashmoni Green University, for his guidance and encouragement throughout the process of preparing this manuscript.

References

Abd El-Ghany, M. N., Yahia, R. A., & Fahmy, H. A. (2023) Nanosensors for agriculture, water, environment, and health. In: Handbook of Nanosensors: Materials and Technological Applications pp 1–29. Springer Nature Switzerland Cham

Abdel-Karim R, Reda Y, Abdel-Fattah A (2020) Nanostructured materials-based nanosensors. J Electrochem Soc 167(3):037554

Amin MT, Alazba AA, Manzoor U (2014) A review of removal of pollutants from water/wastewater using different types of NMs. Adv Mater Sci Eng 2014(1):825910

Arslan O, Topuz F, Eren H, Biyikli N, Uyar T (2017) Pd nanocube decoration onto flexible nanofibrous mats of core–shell polymer–ZnO nanofibers for visible light photocatalysis. New J Chem 41(10):4145–4156

Aslan E, Gonce MK, Yigit MZ, Sarilmaz A, Stathatos E, Ozel F, Patir IH (2017) Photocatalytic H_2 evolution with a Cu2WS4 catalyst on a metal free D-π-A organic dye-sensitized TiO_2. Appl Catal B 210:320–327

Banerjee T, Sulthana S, Shelby T, Heckert B, Jewell J, Woody K, Santra S (2016) Multiparametric magneto-fluorescent nanosensors for the ultrasensitive detection of Escherichia coli O157: H7. ACS Infect Dis 2(10):667–673

Bilal M, Adeel M, Rasheed T, Zhao Y, Iqbal HM (2019) Emerging contaminants of high concern and their enzyme-assisted biodegradation–a review. Environ Int 124:336–353

Bo L, Shengen Z, Chang CC (2016) Emerging pollutants-part II: Treatment. Water Environ Res 88(10):1876–1904

Chakraborty U, Kaur G, Chaudhary, GR (2021) Development of environmental nanosensors for detection monitoring and assessment. New Front NMs Environ Sci 91–143

Chandran RB, Breen S, Shao Y, Ardo S, Weber AZ (2018) Evaluating particle-suspension reactor designs for Z-scheme solar water splitting via transport and kinetic modeling. Energy Environ Sci 11(1):115–135

Chen K, Zhang W, Zhang Y, Huang L, Wang R, Yue X, Wang J (2018) Label-free fluorescence aptasensor for sensitive determination of bisphenol S by the salt-adjusted FRET between CQDs and MoS2. Sens Actuators, B Chem 259:717–724

Das S, Sen B, Debnath N (2015) Recent trends in NMs applications in environmental monitoring and remediation. Environ Sci Pollut Res 22:18333–18344

Dimitrievska I, Paunovic P, Grozdanov A (2023) Recent advancements in nano sensors for air and water pollution control. Mater Sci Eng 7:113–128

Dolati S, Ramezani M, Nabavinia MS, Soheili V, Abnous K, Taghdisi SM (2018) Selection of specific aptamer against enrofloxacin and fabrication of graphene oxide based label-free fluorescent assay. Anal Biochem 549:124–129

Ejeian F, Etedali P, Mansouri-Tehrani HA, Soozanipour A, Low ZX, Asadnia M, Razmjou A (2018) Biosensors for wastewater monitoring: A review. Biosens Bioelectron 118:66–79

Gangar T, Satyam S, Patra S (2021) Monitoring/sensing techniques to address pollutant heterogeneity assessment in wastewater. In: Microbial ecology of wastewater treatment plants, pp 279–314. Elsevier

Gao C, Chen S, Wang Y, Wang J, Zheng X, Zhu J, Xiong Y (2018) Heterogeneous single-atom catalyst for visible-light-driven high-turnover CO_2 reduction: the role of electron transfer. Adv Mater 30(13):1704624

Garcia-Rodríguez A, Matamoros V, Fontàs C, Salvadó V (2014) The ability of biologically based wastewater treatment systems to remove emerging organic contaminants—a review. Environ Sci Pollut Res 21:11708–11728

Garcia-Segura S, Qu X, Alvarez PJ, Chaplin BP, Chen W, Crittenden JC, Westerhoff P (2020) Opportunities for nanotechnology to enhance electrochemical treatment of pollutants in potable water and industrial wastewater–a perspective. Environ Sci Nano 7(8):2178–2194

Gaudin V (2017) Advances in biosensor development for the screening of antibiotic residues in food products of animal origin–A comprehensive review. Biosens Bioelectron 90:363–377

Gaviria-Arroyave MI, Cano JB, Peñuela GA (2020) NM-based fluorescent biosensors for monitoring environmental pollutants: A critical review. Talanta OpEn 2:100006

Graboski AM, Martinazzo J, Ballen SC, Steffens J, Steffens C (2020) Nanosensors for water quality control. In: Nanotechnology in the beverage industry. pp 115–128 Elsevier

Guo H, Li J, Li Y, Wu D, Ma H, Wei Q, Du B (2019) Exciton energy transfer-based fluorescent sensor for the detection of Hg^{2+} through aptamer-programmed self-assembly of QDs. Anal Chim Acta 1048:161–167

He X, Zhang C (2019) Recent advances in structure design for enhancing photocatalysis. J Mater Sci 54(12):8831–8851

He ZJ, Kang TF, Lu LP, Cheng SY (2020) An electrochemiluminescence sensor based on CdSe@ CdS-functionalized MoS 2 and a GOD-labeled DNA probe for the sensitive detection of Hg (ii). Anal Methods 12(4):491–498

Hooshmand S, Kassanos P, Keshavarz M, Duru P, Kayalan CI, Kale İ, Bayazit MK (2023) Wearable nano-based gas sensors for environmental monitoring and encountered challenges in optimization. Sensors 23(20):8648

Ibáñez M, Gracia-Lor EMMA, Bijlsma L, Morales E, Pastor L, Hernández F (2013) Removal of emerging contaminants in sewage water subjected to advanced oxidation with ozone. J Hazard Mater 260:389–398

Iqbal J, Shah MH (2013) Health risk assessment of metals in surface water from freshwater source lakes, Pakistan. Hum Ecol Risk Assess Int J 19(6):1530–1543

Iwase Y, Tomita O, Naito H, Higashi M, Abe R (2018) Molybdenum-substituted polyoxometalate as stable shuttle redox mediator for visible light driven Z-scheme water splitting system. J Photochem Photobiol, A 356:347–354

Jahn IJ, Mühlig A, Cialla-May D (2020) Application of molecular SERS nanosensors: where we stand and where we are headed towards? Anal Bioanal Chem 412:5999–6007

Jain K, Patel AS, Pardhi VP, Flora SJS (2021) Nanotechnology in wastewater management: a new paradigm towards wastewater treatment. Molecules 26(6):1797

Jiang Z, Wan W, Li H, Yuan S, Zhao H, Wong PK (2018) A hierarchical Z-scheme α-Fe2O3/g-C3N4 hybrid for enhanced photocatalytic CO2 reduction. Adv Mater 30(10):1706108

Kaur R, Sharma SK, Tripathy, SK (2019). Advantages and limitations of environmental nanosensors. In Advances in nanosensors for biological and environmental analysispp. 119–132). Elsevier.

Lee JE, Bera S, Choi YS, Lee WI (2017a) Size-dependent plasmonic effects of M and M@ SiO2 (M= Au or Ag) deposited on TiO_2 in photocatalytic oxidation reactions. Appl Catal B 214:15–22

Lee EH, Lim HJ, Lee SD, Son A (2017b) Highly sensitive detection of bisphenol A by NanoAptamer assay with truncated aptamer. ACS Appl Mater Interfaces 9(17):14889–14898

Lee ES, Kim GB, Ryu SH, Kim H, Yoo HH, Yoon MY, Kim YP (2018) Fluorescing aptamer-gold nanosensors for enhanced sensitivity to bisphenol A. Sens Actuators, B Chem 260:371–379

Leong KH, Abd Aziz A, Sim LC, Saravanan P, Jang M, Bahnemann D (2018) Mechanistic insights into plasmonic photocatalysts in utilizing visible light. Beilstein J Nanotechnol 9(1):628–648

Li C, Wei C (2017) DNA-templated silver nanocluster as a label-free fluorescent probe for the highly sensitive and selective detection of mercury ions. Sens Actuators, B Chem 242:563–568

Li W, Shi Y, Gao L, Liu J, Cai Y (2015a) Occurrence, fate and risk assessment of parabens and their chlorinated derivatives in an advanced wastewater treatment plant. J Hazard Mater 300:29–38

Li H, Zhou Y, Tu W, Ye J, Zou Z (2015b) State-of-the-art progress in diverse heterostructured photocatalysts toward promoting photocatalytic performance. Adv Func Mater 25(7):998–1013

Li H, Tu W, Zhou Y, Zou Z (2016) Z-Scheme photocatalytic systems for promoting photocatalytic performance: recent progress and future challenges. Advanced Science 3(11):1500389

Li K, Hogan NJ, Kale MJ, Halas NJ, Nordlander P, Christopher P (2017a) Balancing near-field enhancement, absorption, and scattering for effective antenna–reactor plasmonic photocatalysis. Nano Lett 17(6):3710–3717

Li X, Sun Y, Xiong T, Jiang G, Zhang Y, Wu Z, Dong F (2017b) Activation of amorphous bismuth oxide via plasmonic Bi metal for efficient visible-light photocatalysis. J Catal 352:102–112

Li X, Yun W, Guo W, Zhang W, Yang L (2019) Simple, one step and sensitive fluorescent nanoprobe for simultaneous detection of Pb^{2+} and Cu^{2+} based on pincer enzyme strand and signal amplification strategy. Sens Actuators, B Chem 301:127170

Li D, Yuan X, Li C, Luo Y, Jiang Z (2020) A novel fluorescence aptamer biosensor for trace Pb (II) based on gold-doped carbon dots and DNAzyme synergetic catalytic amplification. J Lumin 221:117056

Liang S, Han B, Liu X, Chen W, Peng M, Guan G, Lin Z (2018) 3D spatially branched hierarchical Z-scheme CdS-Au nanoclusters-ZnO hybrids with boosted photocatalytic hydrogen evolution. J Alloy Compd 754:105–113

Lin W, Cao E, Zhang L, Xu X, Song Y, Liang W, Sun M (2018) Electrically enhanced hot hole driven oxidation catalysis at the interface of a plasmon–exciton hybrid. Nanoscale 10(12):5482–5488

Liu J, Zhang C, Ma B, Yang T, Gu X, Wang X, Hu C (2017) Rational design of photoelectron-trapped/accumulated site and transportation path for superior photocatalyst. Nano Energy 38:271–280

Llorente VB, Dzhagan VM, Gaponik N, Iglesias RA, Zahn DR, Lesnyak V (2017) Electrochemical tuning of localized surface plasmon resonance in copper chalcogenide nanocrystals. J Phys Chem C 121(33):18244–18253

Low J, Jiang C, Cheng B, Wageh S, Al-Ghamdi AA, Yu J (2017) A review of direct Z-scheme photocatalysts. Small Methods 1(5):1700080

Lv D, Zhang D, Pu X, Kong D, Lu Z, Shao X, Dou J (2017) One-pot combustion synthesis of $BiVO_4$/BiOCl composites with enhanced visible-light photocatalytic properties. Sep Purif Technol 174:97–103

Malik R, Tomer VK, Joshi N, Chaudhary V, Lin L (2020) Nanosensors for monitoring indoor pollution in smart cities. In Nanosensors for smart cities, pp 251–266. Elsevier

Manasa RL, Mehta A (2020) Wastewater: sources of pollutants and its remediation. Environ Biotechnol 2:197–219

Mao K, Zhang H, Pan Y, Yang Z (2021) Biosensors for wastewater-based epidemiology for monitoring public health. Water Res 191:116787

Marieeswaran M, Panneerselvam P (2020) A magnetic nanoscale metal–organic framework (MNMOF) as a viable fluorescence quencher material for ssDNA and for the detection of mercury ions via a novel quenching–quenching mechanism. RSC Adv 10(7):3705–3714

Martins R, Figueiredo J, Sushkova A, Wilhelm M, Tedim J, Loureiro S (2022) "Smart" nanosensors for early detection of corrosion: Environmental behavior and effects on marine organisms. Environ Pollut 302:118973

Matthews JR, Payne CM, Hafner JH (2015) Analysis of phospholipid bilayers on gold nanorods by plasmon resonance sensing and surface-enhanced Raman scattering. Langmuir 31(36):9893–9900

Miao X, Shen X, Wu J, Ji Z, Wang J, Kong L, Song C (2017) Fabrication of an all solid Z-scheme photocatalyst g-C3N4/GO/AgBr with enhanced visible light photocatalytic activity. Appl Catal A 539:104–113

Mustafa G, Khan MS, Asif MI, Ullah A, Khan I, Ullah I (2021) Water pollutants and nanosensors. In Aquananotechnology, pp 105–133 Elsevier

Oliveira TS, Murphy M, Mendola N, Wong V, Carlson D, Waring L (2015) Characterization of pharmaceuticals and personal care products in hospital effluent and waste water influent/effluent by direct-injection LC-MS-MS. Sci Total Environ 518:459–478

Orebiyi EO, Awomeso JA, Idowu OA, Martins O, Oguntoke O, Taiwo AM (2010) Assessment of pollution hazards of shallow well water in Abeokuta and environs, southwest, Nigeria

Ou M, Wan S, Zhong Q, Zhang S, Song Y, Guo L, Xu Y (2018) Hierarchical Z-scheme photocatalyst of g-C3N4@ Ag/BiVO4 (040) with enhanced visible-light-induced photocatalytic oxidation performance. Appl Catal B 221:97–107

Patel G, Pillai V, Bhatt P, Mohammad S (2020) Application of nanosensors in the food industry. In Nanosensors for smart cities, pp 355–368. Elsevier

Reineck P, Brick D, Mulvaney P, Bach U (2016) Plasmonic hot electron solar cells: the effect of nanoparticle size on quantum efficiency. J Phys Chem Lett 7(20):4137–4141

Saylan Y, Yılmaz F, Özgür E, Derazshamshir A, Denizli A (2020). Plasmonic smart nanosensors for the determination of environmental pollutants. Emerg Carbon-Based Nanocomposites Environ Appl 237–279

Sha R, Bhattacharyya TK (2020) MoS2-based nanosensors in biomedical and environmental monitoring applications. Electrochim Acta 349:136370

Shmeis, RMA (2022) Nanotechnology in wastewater treatment. In Comprehensive analytical chemistry, vol 99, pp 105–134. Elsevier

Singh S, Nigam P, Pednekar A, Mukherjee S, Mishra A (2020) Carbon quantum dots functionalized agarose gel matrix for in solution detection of nonylphenol. Environ Technol 41(3):322–328

Singh M, Yadav A, Kaushik S, Gupta N, Singh G, Gupta, P (2022) Air and water pollution monitoring and control through bioNM-based sensors. In: Bionanotechnology towards sustainable management of environmental pollution, pp 183–208. CRC Press

Song H, Meng X, Dao TD, Zhou W, Liu H, Shi L, Ye J (2018) Light-enhanced carbon dioxide activation and conversion by effective plasmonic coupling effect of Pt and Au nanoparticles. ACS Appl Mater Interfaces 10(1):408–416

Steffens C, Steffens J, Graboski AM, Manzoli A, Leite FL (2017) Nanosensors for detection of pesticides in water. In New pesticides and soil sensors, pp 595–635. Academic Press

Stock N, Biswas S (2012) Synthesis of metal-organic frameworks (MOFs): routes to various MOF topologies, morphologies, and composites. Chem Rev 112(2):933–969

Sun Y, Zhao Z, Zhang W, Gao C, Zhang Y, Dong F (2017) Plasmonic Bi metal as cocatalyst and photocatalyst: the case of Bi/(BiO) $2CO_3$ and Bi particles. J Colloid Interface Sci 485:1–10

Tian H, Xu X, Qu J, Li H, Hu Y, Huang L, Li B (2020) Biodegradation of phenolic compounds in high saline wastewater by biofilms adhering on aerated membranes. J Hazard Mater 392:122463

Tu W, Zhou Y, Feng S, Xu Q, Li P, Wang X, Zou Z (2015) Hollow spheres consisting of Ti0. 91O2/ CdS nanohybrids for CO_2 photofixation. Chem Commun 51(69):3354–13357

Tuteja SK, Arora D, Dilbaghi N, Lichtfouse E (eds) (2020) Nanosensors for environmental applications. Springer International Publishing

Vikesland PJ (2018) Nanosensors for water quality monitoring. Nat Nanotechnol 13(8):651–660

Vunain E, Mishra AK (2018) Nanosensors as tools for water resources. Nanotechnol Sustain Water Resour. 177–198

Wang QH, Kalantar-Zadeh K, Kis A, Coleman JN, Strano MS (2012) Electronics and optoelectronics of two-dimensional transition metal dichalcogenides. Nat Nanotechnol 7(11):699–712

Wang D, Wang W, Wang Q, Guo Z, Yuan W (2017a) Spatial separation of Pt and IrO_2 cocatalysts on SiC surface for enhanced photocatalysis. Mater Lett 201:114–117

Wang DD, Ge CW, Wu GA, Li ZP, Wang JZ, Zhang TF, Luo LB (2017b) A sensitive red light nano-photodetector propelled by plasmonic copper nanoparticles. J Mater Chem C 5(6):1328–1335

Wang J, Sui Q, Lyu S, Huang Y, Huang S, Wang B, Yu G (2020) Source apportionment of phenolic compounds based on a simultaneous monitoring of surface water and emission sources: A case study in a typical region adjacent to Taihu Lake watershed. Sci Total Environ 722:137946

Wu F, Li X, Liu W, Zhang S (2017) Highly enhanced photocatalytic degradation of methylene blue over the indirect all-solid-state Z-scheme g-C3N4-RGO-TiO2 nanoheterojunctions. Appl Surf Sci 405:60–70

Wu Y, Wang H, Tu W, Liu Y, Tan YZ, Yuan X, Chew JW (2018) Quasi-polymeric construction of stable perovskite-type $LaFeO_3$/g-C3N4 heterostructured photocatalyst for improved Z-scheme photocatalytic activity via solid pn heterojunction interfacial effect. J Hazard Mater 347:412–422

Wu F, Ye J, Cao Y, Wang Z, Miao T, Shi Q (2020) Recent advances in fluorescence sensors based on DNA–MOF hybrids. Luminescence 35(4):440–446

Yu L, Zhang S, Zhang M, Chen J (2017) Superhydrophobicity construction with dye-sensitised TiO_2 on fabric surface for both oil/water separation and water bulk contaminants purification. Appl Surf Sci 425:46–55

Yuan YJ, Chen D, Yang S, Yang LX, Wang JJ, Cao D, Zou ZG (2017) Constructing noble-metal-free Z-scheme photocatalytic overall water splitting systems using MoS2 nanosheet modified CdS as a H 2 evolution photocatalyst. J Mater Chem 5(40):21205–21213

Zeng X, Wang Z, Wang G, Gengenbach TR, McCarthy DT, Deletic A, Zhang X (2017) Highly dispersed TiO_2 nanocrystals and WO_3 nanorods on reduced graphene oxide: Z-scheme photocatalysis system for accelerated photocatalytic water disinfection. Appl Catal B 218:163–173

Zhang C, Yu K, Feng Y, Chang Y, Yang T, Xuan Y, Liu S (2017a) Novel 3DOM-$SrTiO_3$/Ag/Ag_3 PO_4 ternary Z-scheme photocatalysts with remarkably improved activity and durability for contaminant degradation. Appl Catal B 210:77–87

Zhang G, Miao H, Hu X, Mu J, Liu X, Han T, Wan J (2017b) A facile strategy to fabricate Au/TiO_2 nanotubes photoelectrode with excellent photoelectrocatalytic properties. Appl Surf Sci 391:345–352

Zhang P, Wang T, Zeng H (2017c) Design of Cu-Cu_2O/g-C3N_4 nanocomponent photocatalysts for hydrogen evolution under visible light irradiation using water-soluble Erythrosin B dye sensitization. Appl Surf Sci 391:404–414

Zhang X, Ke X, Yao J (2018) Recent development of plasmon-mediated photocatalysts and their potential in selectivity regulation. J Mater Chem A 6(5):1941–1966

Zhang L, Guo Y, Hao R, Shi Y, You H, Nan H, Fang J (2021) Ultra-rapid and highly efficient enrichment of organic pollutants via magnetic mesoporous nanosponge for ultrasensitive nanosensors. Nat Commun 12(1):6849

Zhao Z, Zhang W, Lv X, Sun Y, Dong F, Zhang Y (2016a) Noble metal-free Bi nanoparticles supported on TiO_2 with plasmon-enhanced visible light photocatalytic air purification. Environ Sci Nano 3(6):1306–1317

Zhao H, Zhang Y, Li G, Tian F, Tang H, Chen R (2016b) Rhodamine B-sensitized BiOCl hierarchical nanostructure for methyl orange photodegradation. RSC Adv 6(10):7772–7779

Zhou FQ, Fan JC, Xu QJ, Min YL (2017) $BiVO_4$ nanowires decorated with CdS nanoparticles as Z-scheme photocatalyst with enhanced H_2 generation. Appl Catal B 201:77–83

Zhu B, Xia P, Li Y, Ho W, Yu J (2017) Fabrication and photocatalytic activity enhanced mechanism of direct Z-scheme g-C_3N_4/Ag_2WO_4 photocatalyst. Appl Surf Sci 391:175–183

Chapter 4
Nanomaterials and Wastewater Treatment: A Step Towards Sustainable Future

Swarupa Bhattacharjee, Rajdeep Das, Sanchita Boruah, and Shyama Prasad Biswas

Abstract Pure water is the basic need of all living organisms. But nowadays, it has become a challenge to get good quality water, also growing need and scarcity of water makes waste water treatment a talked about issue in the field of applied sciences. Presently, some techniques for the primary waste water treatment although are in use but use of nanomaterials (NMs) in this field has attracted the attention. A high efficiency, low-cost, and sustainable of waste water treatment way is the need of the hour. NMs having various important properties such as catalysis, adsorption, greater surface area, and many more make them a probable solution. Many NMs are being explored for the removal of contaminants from waste water. Heavy metals are yet another matter of concern when it comes to contamination of water. Various NMs and assisted techniques are also being used for the purpose of their removal. It is also a favourable way when it comes to energy conservation, environmental remediation and creation of an ecofriendly environment which again is in line with the blooming concept of green technology. Various recyclable nanoparticles their sustainable use and future prospects are an important and fast-growing domain of research and application. The present study deals with the concrete and collected idea of NMs in the field of waste water treatment and their future prospects also taking into account the environmental sustainability. Wastewater treatment presents a pressing global challenge as population growth and industrialization exacerbate water pollution. In recent years, nanotechnology has emerged as a promising avenue for addressing this issue, offering innovative solutions through the application of NMs in wastewater treatment processes. This review explores the latest advancements in

S. Bhattacharjee (✉)
Department of Zoology, Bahona College, Jorhat, India
e-mail: bhattacharjeeswarupa@gmail.com

R. Das
Department of Zoology, Dudhnoi College, Dudhnoi, India

S. Boruah
Department of Zoology, DHSK College, Dibrugarh, India

S. P. Biswas
Department of Life Sciences, Dibrugarh University, Dibrugarh, India

N. Agarwal et al. (eds.), *Nanomaterials in Wastewater Research*, Advances in Wastewater Research, https://doi.org/10.1007/978-981-96-3908-3_4

utilizing NMs for wastewater treatment, highlighting their efficacy, mechanisms of action, and potential environmental impacts. Additionally, the article discusses the role of nanotechnology in advancing towards a sustainable future by improving water quality and resource recovery.

Keywords Wastewater treatment · NMs · Sustainability · Ecofriendly

4.1 Introduction

Since water is the most essential component of all living things, it is estimated that 800 million people globally lack access to clean drinking water for home use (Vallino et al. 2020). Also, as the water resources that are presently available are not free from various anthropogenic influences of inorganic pollutants like salts, metals, and radioactive materials and also the ranges of organic pollutant which includes a vast array of viruses/bacteria, pesticides, anhydrous phase liquids, water scarcity and its quality regulation have grown in importance for sustainable development. The addition of new contaminants to the soil–plant–environment continuum (mostly through water pollution) as a result of steady population increase and industrial development and expansion has raised concerns for human health (Korashy et al. 2017; Kumari et al. 2019; Sharma et al. 2016; Wu et al. 2019). The health of the community, water ecosystems, environmental sustainability, societal economics, and social well-being are all significantly impacted by contaminated water. For example, it is estimated that 30,000 fatalities per day worldwide are caused by a lack of clean water supplies and inadequate hygiene (Mishra et al. 2020). The majority of all the reported cases—80%—occur in village settings and involve new-borns who are very much prone to such problems. The need for clean, sanitary water is rising swiftly throughout these days. But the widely employed treatment techniques are insufficient to provide the necessary volume of high-quality water. For this reason, it's critical to search for innovative energy-saving water purifying technologies. The involvement of NMs in the field of water treatment sector has grown in importance among the existing technologies for the quick removal of impurities through adsorption or breakdown processes. Toxic trace metals, phenols, and colours are just a few of the things that the wastewater treatment industry uses nanotechnology. Magnetic nanoparticles also have the added benefit of being easily recovered from mixed reactions by applying an external magnetic field, and they can be reused many a times (Kefeni et al. 2017). Some nanoparticles (NP) may be hazardous and constitute a new class of pollutants that are not only detrimental to the environment as a whole but also have significant and serious impact on the human health, but relevant concerns are not addressed (Ying et al. 2017). In recent times, magnetic nanoparticles (MNPs) have garnered significant interest due to their unique superparamagnetic characteristics, their high adsorption capacity, and the correlation between their size and surface area (Tshikovhi et al. 2020). Particularly, one of the most significant and popular choice when it comes to nanoparticles are MNPs is ferrites, which are transition metal oxides with spinel structures (Amiri et al. 2019). Ferrites are categorized as garnet ($M3Fe_5O_{12}$, where M = rare cations), hexaferrite ($SrFe_{12}O_{19}$ and $BaFe_{12}O_{19}$), spinel (MFe_2O_4, where M = Co, Mn, Zn, Fe, Ni, etc.), and orthoferrite ($MFeO_3$, M = rare earth cations) based on their crystalline structures and magnetic characteristics. Among these, spinel ferrite nanoparticles (SFNPs) received particular interest. This may be attributed to their straightforward chemical makeup, superior magnetic characteristics, and extensive use in a variety of industries, such as biomedicine, catalysis, and water and sewage treatment (Ciocarlan et al. 2018). The wastewater produced by cities and industries is regarded as a significant pollution carrier and has to be appropriately managed and disposed of. Research has been directly focussed on the need to manage the hazardous contaminants (both organic and inorganic) and complex chemicals found in wastewater. One of the primary methods for improving water quality is the application of efficient technology for the treatment of harmful substances.

With the escalating demand for clean water and the increasing pollution of water bodies, effective wastewater treatment has become imperative for sustaining human health and environmental integrity. Conventional treatment methods often fall short in removing emerging contaminants and achieving stringent water quality standards. Consequently, there is a growing interest in exploring alternative technologies, among which nanotechnology holds significant promise. NMs, due to their unique physicochemical properties, offer novel opportunities for enhancing the efficiency and sustainability of wastewater treatment processes.

This chapter aims at providing a broad overview of various use of nanotechnology in water treatment, including the processes of adsorption, analyte extraction, membrane modification, photocatalysis, pathogen disinfection, and metal ion removal. A wide variety of readers may easily comprehend the qualities, benefits, and drawbacks of various NMs as well as their function in the water treatment process, which has been researched and presented graphically. Wide range of studies have been carried out on various water pollutants, their sources of origin and their adverse effects both on the environment and on public health which are very much essential for understanding the deep routed problem of waste and polluted water. This in itself also advocates the need of treating the waste water. Table 4.1 briefly illustrates the different water pollutants with their sources and adverse effects.

4.2 NMs in Wastewater Treatment

NMs encompass a diverse array of engineered structures with dimensions ranging from 1 to 100 nm. These materials exhibit distinctive properties such as high surface area-to-volume ratio, surface reactivity, and tenable surface chemistry, making them exceptionally suitable for various wastewater treatment applications.

Table 4.1 Indicates different water pollutants with their sources and adverse effects

Water pollutants	Source of pollutants	Effect of pollutants	References
Pathogens	Viruses and bacteria	Waterborne diseases	Umar et al. 2019a)
Agricultural pollutants	Agricultural chemicals	Directly affect the fresh water resources	Tang et al. 2016)
Sediments and suspended solids	Land cultivation, demolition, mining operations	Damaging fish spawning, affecting aquatic environment of insects and fishes	Richter and Ayers 2018)
Inorganic pollutants	Metals compounds, trace elements, inorganic salts, heavy metals, mineral acids,	Aquatic flora and fauna, public health problems	Sizmur et al. 2017)
Organic pollutants	Detergents, insecticides, herbicides	Aquatic life problems, cacogenic	Wang et al. 2019)
Industrial pollutants	Municipal pollutant water	Caused water and air pollution	Liu et al. 2018)
Radioactive pollutants	Different isotopes	Affect bones, teeth, and skin	Bayoumi and Saleh 2018)
Nutrients pollutants	Plant debris, fertilizer	Effect on eutrophication process	Ma et al. 2018)
Macroscopic pollutants	Marine debris	Plastic pollution	Longwane et al. 2019)
Sewage and contaminated water	Domestic wastewater	Waterborne diseases	Rajasulochana and Preethy 2016)

4.2.1 Adsorption

One of the primary mechanisms through which NMs contribute to wastewater treatment is adsorption. Nanoparticles with tailored surface functionalization can effectively adsorb pollutants, including heavy metals, organic dyes, and pharmaceuticals, from aqueous solutions. For instance, graphene-based NMs, such as graphene oxide and reduced graphene oxide, have shown remarkable adsorption capacities for a wide range of contaminants due to their large surface area and π-π interactions.

4.2.2 Catalysis

Nanocatalysts exhibit enhanced catalytic activity and selectivity, enabling the degradation of recalcitrant organic pollutants through advanced oxidation processes (AOPs). Metal and metal oxide nanoparticles, such as titanium dioxide (TiO_2), iron oxide (Fe_2O_3), and zinc oxide (ZnO), are extensively employed as photocatalysts for the degradation of organic compounds under ultraviolet (UV) or visible light irradiation. Additionally, nanocomposites combining catalytic nanoparticles with support matrices enhance stability and recyclability, facilitating practical implementation in wastewater treatment systems.

4.2.3 Membrane Separation

NMs have revolutionized membrane-based separation technologies by enhancing membrane properties such as permeability, selectivity, and antifouling characteristics. Incorporating nanoparticles, such as carbon nanotubes, zeolites, and metal–organic frameworks, into polymeric membranes improves their performance in removing contaminants, including microorganisms, suspended solids, and nanoparticles, from wastewater streams. Furthermore, nanocomposite membranes exhibit enhanced fouling resistance and durability, prolonging operational lifespan and reducing maintenance costs.

4.3 Nanotechnological Processes in Wastewater Treatment

Smaller size, increased surface area, increased mobility in solution, high reactivity (Wu et al. 2019) hydrophobicity, dispersibility, hydrophilicity, porosity properties, and strong mechanical properties are just a few of the remarkable qualities that NMs have (Daer et al. 2015; Tang et al. 2014). Recently, sophisticated breakthroughs in NMs—such as nanosorbents, nanomembranes, nanomotors, nanophotocatalysts, etc.—have been made to remediate contaminated water.

4.3.1 Nanophotocatalyst

The term "photocatalysis," which refers to the destruction of substances in the presence of light, is derived from the combination of the Greek terms "photo" and "catalysis" (Yaqoob et al. 2020). There is not a widely accepted definition of photocatalysis in the scientific community (Saravanan et al. 2017). Nevertheless, the process of stimulating or activating the material by light (visible, solar, or ultraviolet) is referred to as photocatalysis. The primary distinction between a thermocatalyst and a photocatalyst is that the latter is activated by heat, while the former is activated by photons of light energy (Gomes et al. 2019). Because of its shape-dependent properties and greater surface ratio, nanophotocatalyst is frequently employed to remediate polluted water (Chen et al. 2019). Because of their unique surface characteristics and distinct quantum effects, nanoparticles respond differently than bulk materials

(Yaqoob et al. 2020). According to Ong et al. (2018), it also improves their mechanical, magnetic, electric, optical, and also the chemical reactivity. According to a study, a nanophotocatalyst boosted the oxidizing species generation on the material surface, expanding the oxidation capacity and improving the pollutant degradation from the contaminated water (Gómez-Pastora et al. 2017). Semiconductors, zero valance metal bases, and some bimetallic nanoparticles are mostly used to remediate environmental contaminants including chlorpyrifos, azo dyes, nitroaromatics, organochlorine insecticides, etc. (Loeb et al. 2018; Samanta et al. 2016). TiO_2 nanotubes have been used by several researchers to remove organic contaminants (such as chlorinated ethylene, toluene, Congo red, azo dyes, dichlorophenol, trichlorobenzene, and aromatic phenol) from contaminated water (Bhatia et al. 2017; Liang et al. 2019). Al_2O_3, ZnO, SiO_2, TiO_2, and other oxides are, however, the most widely utilized nanophotocatalysts (Ali et al. 2019; Bhanvase et al. 2017). TiO_2 is regarded as one of the best photocatalysts out of all of them because of its inexpensive cost, nontoxic characteristics, ease of availability, and chemical stability. It naturally exists in three states: anatase, rutile, and brookite. Anatase is thought to be a good photocatalyst material among them (Yamakata and Vequizo 2019).

ZnONPs may be efficiently reused and are also utilized as a photocatalyst to remediate wastewater from other materials (Das et al. 2018; Hassan and Elhadidy 2019). Li et al. (2014) investigated how well composite NMs CdS/TiO_2 worked as a photocatalyst to break down the reference material, dimethyl sulfoxide, in the presence of visible light (Li et al. 2014). Because of their ferromagnetism, NMs doped with iron may be recycled and reused with ease (Phokha et al. 2016; Serrà et al. 2020). Berekaa (2016) reported the use of Pd-incorporated ZnONPs' photocatalytic ability to eradicate Escherichia coli from polluted water (Berekaa 2016). To improve metal oxides' photocatalytic properties, however, a lot of effort has been put into altering them with other substances such metal ions and metals (Umar et al. 2019b), dye sanitizers, and carbonaceous compounds (Malik et al. 2014). Still, further adjustment is required.

4.3.2 Nano- and Micromotors

Recent years have seen enormous advancements in the science of nanotechnology, which provide several methods for treating water. These days, energy-storing materials such as nano- and micromotors are being used to provide machine-driven force (Yaqoob et al. 2020). They can use many processes to achieve different objectives. These cutting-edge motors may run on energy or without energy sources (acoustics, magnetic fields, or electric fields) and have intriguing and important applications (Jurado-Sánchez and Wang 2018). These motors have many features including power, high speed, self-mixing ability, and specific control movement. Eliminating contaminants from polluted resources is crucial for the environment's stability and sustainability. Global demand for clean, filtered water led to an increase in water treatment and purification operations. In the past, several conventional methods have been employed to treat freshwater, wastewater from sediments, polluted groundwater, etc. (Yaqoob et al. 2020). In order to improve water treatment, traditional technologies require external agitation due to their poor diffusion. Nano/micromotors, however, can cross the diffusion boundary easily and effectively through energy mixing because of their capabilities and self-propulsion (Yaqoob et al. 2020). By integrating with materials that have a nano/microstructure, or increased workability and surface area, the capability of self-propulsion of nano/micromotors greatly improves the efficiency of treatment of water (Chi et al. 2018; Pacheco et al. 2019). Additionally, where conventional methods are ineffective in microscale detentions, nano/micromotors can function as programmed cleansers. These motors may be utilized to clean up wastewater contaminants in an efficient manner.

4.3.3 Nanomembranes

Nanomembranes are a special kind of membrane made of various nanofibers that are used to extract unwanted tiny particles from aqueous medium (Figs. 10.1 and 10.2). This process develops at a very fast rate of exclusion and may be used as a reverse osmosis pretreatment technique (Jhaveri and Murthy 2016). A water porous membrane can be used for nanofiltration, reverse osmosis, and ultrafiltration (UF). Nanomembranes fall into three categories: inorganic, organic, and hybrids of organic and inorganic materials. According to a study, polymeric materials such as cellulose acetates, polyethersulfone, cellulose nitrates, polyacrylonitrile, polyvinyl alcohol, and biomacromolecules are the building blocks of organic nanomembranes (Soyekwo et al. 2014; Wei et al. 2017). On the other hand, metals, metal oxides (Kanakaraju et al. 2014; Brady-Estévez et al. 2010), carbon nanotubes (CNTs) (Rahaman et al. 2012), and graphene include inorganic nanomembranes. Inorganic–organic hybrid nanomembranes are created when the inorganic material is added to a polymetric system (Goh et al. 2015). Sun et al. (2013) used the water channel protein aquaporin to create a nanomembrane for the purpose of treating water (Sun et al. 2013). An aquaporin Z-based nanomembrane dramatically decreased the content of NaCl (66.2%) and MgC12 (88.1%) in water, respectively, and increased the water flow by 65% when comparison is done with the membrane which is lagging aquaporin Z. In another study, CNTs were used as nanofillers and polyvinylidene fluoride-co-hexafluoropropylene to create a hydrophobic nanomembrane in a different investigation (Tijing et al. 2016). It was stated that the effectiveness of salt removal was measured at > 99.99%, and the water flow was enhanced by 33%–59%. In a similar vein, (Liu et al. 2015) synthesized a zirconium (VI)-based metal–organic framework nanomembrane for the desalination of water. The membrane effectively eliminates multivalent ions, such as Al^{+3} (99.3%), Ca^{+2} (86.3%), and Mg^{+2} (98%) from water, according to test data. A test utilizing several saline solutions that lasted up to 170 h revealed no deterioration in performance. In a different work, arsenic (arsenate) was extracted from polluted water using membranes based on amyloid fibrils by Bolisetty et al. (2017). As a result, 99.6% of the arsenic in the water was removed.

Owing to their tiny size and increased reactivity, nanoparticles (NPs) may readily enter the human bodies the medium can be food chains, such as drinking water, and can lead to oxidative stress and other issues. Thus, it is crucial to get rid of NPs from water. Zhang et al. used cellulose nanofibrils to create an ultra-thin nanomembrane for this purpose. The membrane's thickness was around 30 nm, while its pore width ranged from 2.5 to 12 nm. Significant NP elimination and a water flow of 1.14 L/h m^2 bar were recorded (Zhang et al. 2016). Li et al. created a CNT sponge-based nanomembrane in a different investigation. It has been observed that the CNT nanomembrane entirely eliminates methyl orange and rhodamine B from the water, as well as 100% of AuNPs and TiO_2 and roughly 80% of CdSNPs.

Pesticides, oils, hydrocarbons, phenols, and medicines are examples of organic pollutants that can upset the ecosystem and endanger aquatic life. Using nanomembranes, several tests have been conducted to remove organic contaminants from water. Karim et al. (2014) used cellulose nanocrystals in a chitosan (CS) matrix to create a composite nanomembrane (Karim et al. 2014). The membrane's pore size varied from 10 to 13 nm. The elimination of methyl violet 2B (84%), victoria blue 2B (98%), and rhodamine 6G (70%), according to the results, was effective. Conversely, Lee and Baik (2010) created a nanomembrane by vertically aligning CNTs over steel mesh in order to separate the oil from the water. The membrane is both hydrophobic and oleophilic. Both the high-viscosity lubricating oil and the low-viscosity diesel were successfully removed. Similarly, Li et al. (2014) employed aminated polyacrylonitrile (APAN) nanofibrous membrane immobilized Ag nanoclusters to separate oil from water (Li et al. 2014). APAN nanofibrous membrane effectively cleared oil across a wide pH range and in a hypersaline environment. It has also been shown that other organic pollutants, including methyl blue, direct red dye, dimethyl carbonate, ethanol, dyes, and humic acid may be removed using various nanomembranes (Gao et al. 2013).

Generally speaking, biological pollutants (bacteria, viruses, plankton, and algae) cause waterborne illnesses. Membranes are capable of several functions, including simultaneous separation, photocatalytic oxidation, antimicrobial activity, and antifouling properties. In order to filter the viruses, a highly selective nanoporous membrane employing copolymers was created. At 2 bars, the membrane's pore size diameter, which was around 17 nm, remained highly stable. The membrane screened the human rhinovirus type 14, which has a diameter of around 30 nm. New nanofibrous scaffolds considerably reduce the amount of bacteria and viruses in the water. In a similar vein, titania nanocrystals and ɣ-alumina were coated by Shayesteh et al. (2016) to create a nanomembrane. The membrane was able to remove 100% of the bacteria and around 25% of the ions, according to the results.

4.3.4 Nanoadsorbents

Adsorption is a commonly employed and researched method for aqueous medium decontamination. It is advantageous due to its high efficiency and simplicity. When it

Table 4.2 Most commonly used nanosorbents and their functions

Different nanosorbents	Treatment function	References
Carbon-based nanosorbents	Nickel ions present in water	Rodovalho et al. 2016)
Graphite oxide	Removal of dyes	Sun et al. 2015)
Regenerable polymeric Nanosorbents	Organic pollutants, inorganic contaminants of wastewaters	Lee et al. 2012)
Nanoclay	Hydrocarbons, dyes	Kyzas and Matis 2015)
Nano-metal oxides	Different heavy metals	Wang et al. 2016)
Nano-Aerogels	To remove the uranium from drinking water	Krstić and Pesovski 2018)
Nano-iron oxides	To eliminate the hormones and toxic pharmaceuticals material from water	Unuabonah and Taubert 2014)
Polymer fibres	To remove the arsenic and other toxic metals	Yadav et al. 2019)

comes to cleaning polluted water of a variety of pollutants, the adsorption technique is preferred over other approaches due to its efficiency, ease of use, affordability, and adaptability (Kumari et al. 2019). Nanoadsorbents are NPs (organic/inorganic) with high porosity, larger surface area, and higher active surfaces that have a high adsorption and isolating capacity of a wide spectrum of pollutants (Santhosh et al. 2016). In addition to their capacity to regenerate, nanoadsorbents function quickly and effectively. It indicates that after NPs are separated from the pollutant, they may be employed once more. Many kinds of nanoadsorbents, including metal oxide-based, carbon-based, metal NPs, and composites of NPs, are employed to clean up the environment. These days, there are many different kinds of contaminants in water that fall into three main categories: inorganic, organic, and biological pollutants. The literature indicates that inorganic, organic, and biological contaminants may all be adsorbed equally by nanoadsorbents. Table 4.2 shows the most commonly used nanosorbents and their functions in waste water management.

4.4 Benefits of Nanotechnology Over the Traditional Treatment Methods

A new WHO research estimates that a total of 783 million people worldwide is affected by the low and compromised quality of their drinking water. Additionally, it was said that waterborne illnesses account for 1.87 million child fatalities annually (Heijden 2018). Conventional treatment procedures are not appropriate in developing nations because of high costs, upkeep issues, etc. For drinking purposes, people in many developing nations gather and conserve water from various locations. Water contamination during collection and transportation is a possibility. Treating such issues using ceramic water filters (CWFs) is an effective solution. CWFs are altered and coated with silver nanoparticles these days (Adeleye et al. 2016). Filtered and clean water can be created with CWFs, although this approach cannot purify water at a higher level. Because of silver's antibacterial qualities, using silver NMs will eradicate the infections (Pandey et al. 2017). Many contaminants, including those derived from organic materials, are found in tiny levels in water and can seriously endanger human life. Typically, flocculation and chlorination are employed to eliminate contaminants from water. These traditional treatments have shortcomings, including limited efficacy and sludge generation in water, which also present a number of environmental risks (Umar et al. 2013). Nanomembranes, which prevent any solid particles from entering the water, have been utilized to lessen the amount of sludge that contaminates it. Nanomembranes are seen to be the most appropriate technology because of these advantages, although further advancements are still required to make it flawless. Researchers have documented the photodegradation of several contaminants, including medicines, insecticides, dyes, and metal oxide, utilizing both NP-doped and NP-undoped metals. According to a paper (Umar et al. 2016), NP-doped TiO_2 exhibits more pronounced photodegradation outcomes than undoped TiO_2. Nanomechanics has a number of benefits over conventional methods. When compared to traditional methods, the tiny particles can reduce the total cost of the treatment and be utilized for both in situ and ex situ decontamination (Safdar et al. 2017). At 25 °C, nanophotocatalysis is a crucial method for degrading pollutants (Tahir et al. 2019). Nanophotocatalysis has several benefits, including reduced toxicity, cost-effectiveness, increased stability, and environmental friendliness (Ciambelli et al. 2019). Water pollution may also be effectively eliminated from water via nanoadsorption. Because of its larger surface area, faster adsorption rate, and shorter equilibrium time than traditional sorbents, nanoadsorbents remove a noticeably larger volume of contaminants (Vunain et al. 2016).

4.5 Environmental Implications and Challenges

While nanotechnology offers unprecedented opportunities for advancing wastewater treatment, concerns regarding the environmental fate and toxicity of NMs persist. The potential release and the aftermath of nanoparticles into the environment during manufacturing, application, and disposal stages raises questions regarding their prolonged and chronic impacts in the long run not only on the aquatic ecosystems but also to the public health as a whole. Therefore, comprehensive risk assessments and regulatory frameworks are essential to ensure the safety as well as the responsible and accountable utilization of NMs in wastewater treatment.

4.6 Potential Ecotoxicity of Materials and Procedures Based on Nanotechnology that Are Used in Water Applications

Although there is no direct link between NPs in water and human health, humans may still absorb NPs by eating fish, thus it's critical to consider how NPs affect the aquatic population. According to reports, NPs in water, such as TiO_2, Ag, and CNTs, have a deleterious impact on aquatic life (Asghari et al. 2012; Jackson et al. 2013).

The US Environmental Protection Agency (EPA) released a comprehensive research in 2010 that detailed the impacts of TiO_2 on fish, microorganisms, plants, invertebrates, and algae (EPA (Environmental Protection Agency) 2010; Gehrke et al. 2015). The study assessed the impact of various TiO_2NP forms and entrance mechanisms on the environment and organisms. According to reports, nano-TiO_2 decreased daphnia reproduction and also caused behavioural abnormalities, pathological alterations in the intestines and gills, and respiratory disturbances (EPA (Environmental Protection Agency) 2010). According to reports, nano-TiO_2 can have detrimental impacts on algae, mostly based on its kind and concentration.

Another significant element that determines the toxicity level of NPs is the duration of their exposure. According to reports, daphnia are extremely harmful to nano-TiO_2 when exposed for longer than 72 h, although the effects are negligible within 48 h (Zhu et al. 2010). Bai et al. (2010) assessed ZnONPs' impact on zebrafish embryos. It was observed that zebrafish embryos exposed to ZnONPs at concentrations of 50 and 100 ppm perished, but embryos exposed to concentrations as low as 25 ppm persisted longer before hatching. The fatal concentration of ZnONPs and CuONPs in protozoa and crustacea was shown to be greatly influenced by the quantity of organic matter dissolved in water, as reported by Blinova et al. (2010). Higher quantities of CdSe/ZnS quantum dots were harmful to daphnia in an aqueous environment (Pace et al. 2010). As compared to nano-TiO_2, CdTe quantum dots shown a greater potential for toxicity against green algae (Chlamydomonas reinhardtii). In a similar vein, freshwater mussels (Elliptio complanata) exposed to CdTe quantum dots demonstrated oxidative stress and DNA damage. According to Sanders et al. (2008), exposure to CdS quantum dots resulted in hepatocellular nuclear pleomorphism in the liver of Gasterosteus aculeatus.

When rainbow trout were given CNTs (single-walled carbon nanotubes), there was an increase in Na + K + -ATPase activity in their intestines and gills. Due to precipitation in the gills, it resulted in respiratory impairment (Smith et al. 2007). It has also been observed that double-walled CNTs are hazardous to frog larvae. Fullerene nanoparticle exposure caused oxidative damage in daphnia. Additionally, it was noted that C60 NPs cause daphnia cells to undergo necrotic and apoptotic death (Gaiser et al. 2011).

NPs in the environment can originate from point sources like landfills, wastewater treatment facilities, and other point sources as well as nonpoint sources including clothing, washing machines, and other NPs-based items. For instance, depending on the washing circumstances, a significant quantity of AgNPs may be discharged into the environment when washing textiles. The solubility in water, chemical reactivity (such as redox potential), and binding to other substances are some of the properties of NPs that have a major impact on their mobility. These characteristics along with the NPs' surface charge might make them more immobile. Because of their larger specific surface area, NPs quickly adsorb and accumulate after coming into touch with water. The surface properties of polymers in an aqueous solution facilitate the aggregation of particulate matter and also improve its interaction with certain proteins. For instance, polyethyleneimine (polymeric NPs) have surface properties that promote agglomeration in the water system. NPs can remain in sediments in water bodies like rivers, lakes, etc., for a longer amount of time when they are adsorbed or in the form of agglomerates. While NPs are considered inactive in this form, changes in the ecological conditions, such as a pH decrease or temperature rise, might facilitate NP dissolution. Because of their increased reactivity, metal oxide nanoparticle stability is constrained (Matlochova et al. 2013). Nonetheless, a number of scientific teams are attempting to improve metal oxide NPs' stability.

NPs can degrade in an aqueous system in a variety of ways depending on the environmental circumstances. Depending on the chemical makeup of the NPs and the turbidity of the water, photoreaction with sunlight can alter the NPs' reactivity. It is possible for NPs to degrade via both biotic and abiotic processes. Few metallic and organic NMs have been shown to change anaerobic conditions [91].

Municipal wastewater treatment plants have used a variety of technologies, including filtration, flocculation, and sedimentation, during the wastewater treatment process. Biological treatment is typically used to remove NPs, such as by interacting with the microbiological population, which can improve NP degradation, integration, and sedimentation. Because larger particles settle more quickly during the sedimentation process, the early agglomeration phase is required to increase the NPs' settling rate.

4.7 Future Perspectives

Nanotechnology can play a significant role in the fields of wastewater treatment and water reuse. It can get beyond the risks and worries that are now present about toxic by-products being produced by traditional water treatment methods. The capacitive deionization technique may effectively leverage the unique qualities of NMs, such as catalysis, high facets ratio, antibacterial activities, improved surface attributes, and greater conductivity, for the detection and desalination of high fluorescence. Using freely accessible solar energy and designing NMs as a visible light photocatalyst can help enhance water processing. Nanotechnology can also be utilized to mitigate the deterioration of water quality as it travels through the distribution networks. It is possible to construct solar-powered water treatment devices based on nanotechnology, which could be the most suitable for disaster situations. By using nanotechnology, alternative water sources for drinking and agriculture may be used with minimal energy use. This method holds great significance for underdeveloped nations who are struggling with serious water pollution issues. Adopting cutting-edge, highly effective water treatment technology is now essential. Future water purification systems will probably be built on nanotechnology, which will be affordable, simple to use, and effective in eliminating a wide range of water pollutants.

4.8 Conclusion

Nanotechnology holds immense potential for revolutionizing wastewater treatment processes, offering innovative solutions to address the escalating challenges when it comes to water pollution and also when there is the apparent scarcity of resources. By harnessing the unique properties of NMs, such as adsorption capacity, catalytic activity, and membrane performance, sustainable and efficient wastewater treatment systems can be developed. However, the widespread implementation of nanotechnology in wastewater treatment necessitates interdisciplinary collaboration, rigorous research, and proactive risk management strategies to mitigate potential environmental and health risks. Ultimately, integrating nanotechnology into wastewater treatment practices represents a significant step towards achieving global water sustainability and securing a better future for generations to come. The usage of wastewater in various industries and the possible ramifications of using wastewater untreated are covered at the conclusion of this chapter. There is a vast volume of wastewater produced worldwide that is contaminated with various organic, inorganic, and biological pollutants. Negative effects of the polluted wastewater might be felt directly or indirectly by people, animals, and plants. Adding upon, it is the water scarcity which is growing severely and needs to be addressed at the earliest. The microbial population in the soil is seriously threatened by potentially hazardous substances found in wastewater, and these elements also contribute to the deterioration of the soil. In this chapter, several treatment philosophies have been covered. The adsorption method is a time-honoured and conventional approach for eliminating contaminants. However, this method's effectiveness has multiplied considerably when bigger surface area and higher reactivity nanoadsorbents are used. The application of nanomembrane filtration is a very effective method. It is a pricey and energy-intensive procedure, though. Nanocatalysts can be used in photocatalysis processes to breakdown pollutants found in wastewater. When it comes to organic pollutants, this method is critical. Additionally, photocatalysis has several disadvantages that limit its widespread use. Using antimicrobial NMs, such as AgNPs, is another popular method for ridding the water of microorganisms. Nevertheless, while NMs have the ability to address the issue of contaminated water, they may also be harmful to the ecosystem. More extensive research in this field is needed to come to a concrete solution when it comes to addressing various issues with regard to the environment. But to meet the need of the hour, the use of nanomaterial comes up as a prominent solution. In any case, it is anticipated that the NMs will offer a workable remedy for handling tainted wastewater soon. Also, the nanomaterial use seems to be a compatible and a ready solution when it comes to the future of the waste water treatment. With more and more research and main streaming as well as the commercialization of the use of NMs, new and comparatively unexplored discipline of waste treatment technology with promising future is waiting ahead.

References

Adeleye AS, Conway JR, Garner K, Huang Y, Su Y, Keller AA (2016) Engineered NMs for water treatment and remediation: costs, benefits, and applicability. Chem Eng J 286:640–662

Ali I, Al-Ghamdi K, Al-Wadaani FT (2019) Advances in iridium nano catalyst preparation, characterization and applications. J Mol Liq 104461278

Amiri M, Salavati-Niasari M, Akbari A (2019) Magnetic nanocarriers: evolution of spinel ferrites for medical applications. Adv Colloid Interface Sci 265:29–44

Asghari S, Johari SA, Lee JH, Kim YS, Jeon YB, Choi HJ, Yu IJ (2012) Toxicity of various silver nanoparticles compared to silver ions in *Daphnia magna*. J Nanobiotechnol 10(1):14

Bai W, Tian W, Zhang Z, He X, Ma Y, Liu N, Chai Z (2010) Effects of copper nanoparticles on the development of zebrafish embryos. J Nanosci Nanotechnol 10:8670–8676

Bayoumi TA, Saleh HM (2018) Characterization of biological waste stabilized by cement during immersion in aqueous media to develop disposal strategies for phytomediated radioactive waste. Prog Nucl Energy 107:83–89

Berekaa MM (2016) Nanotechnology in wastewater treatment; influence of NMs on microbial systems. Int J Curr Microbiol App Sci 5(1):713–726

Bhanvase BA, Shende TP, Sonawane SH (2017) A review on graphene–TiO_2 and doped graphene–TiO_2 nanocomposite photocatalyst for water and wastewater treatment. Environ Technol Rev 6(1):1–14

Bhatia D, Sharma NR, Singh J, Kanwar RS (2017) Biological methods for textile dye removal from wastewater: a review. Crit Rev Environ Sci Technol 47(19):1836–1876

Blinova I, Ivask A, Heinlaan M, Mortimer M, Kahru A (2010) Ecotoxicity of nanoparticles of CuO and ZnO in natural water. Environ Pollut 158(1):41–47

Bolisetty S, Reinhold N, Zeder C, Orozco MN, Mezzenga R (2017) Efficient purification of arsenic contaminated water using amyloid–carbon hybrid membranes. Chem Commun 53(42):5714–5717

Brady-Estévez AS, Schnoor MH, Kang S, Elimelech M (2010) SWNT- MWNT hybrid filter attains high viral removal and bacterial inactivation. Langm 26(24):19153–19158

Chen W, Liu Q, Tian S, Zhao X (2019) Exposed facet dependent stability of ZnO micro/nano crystals as a photocatalyst. Appl Surf Sci 470:807–816

Chi Q, Wang Z, Tian F, You JA, Xu S (2018) A review of fast bubble-driven micromotors powered by biocompatible fuel: Low-concentration fuel, bioactive fluid and enzyme. Micromach. 9(10):537

Ciambelli P, La Guardia G, Vitale L (2019) Nanotechnology for green materials and processes. Studies in Surface Science and Catalysis, vol 179, pp 97–116. Elsevier

Ciocarlan RG, Seftel EM, Mertens M, Pui A, Mazaj M, Tusar NN, Cool P (2018) Novel magnetic nanocomposites containing quaternary ferrites systems Co0. 5Zn0. 25M0. $25Fe_2O_4$ (M= Ni, Cu, Mn, Mg) and TiO_2-anatase phase as photocatalysts for wastewater remediation under solar light irradiation. Mater Sci Eng B 230:1–7

Daer S, Kharraz J, Giwa A, Hasan GW (2015) Recent applications of NMs in water desalination: a critical review and future opportunities. Desalination 367:37–48

Das P, Ghosh S, Ghosh R, Dam S, Baskey M (2018) *Madhuca longifolia* plant mediated green synthesis of cupric oxide nanoparticles: a promising environmentally sustainable material for waste water treatment and efficient antibacterial agent. J Photochem Photobiol B 189:66–73

EPA (Environmental Protection Agency) (2010) RTP Division.NM case studies:nanoscale titanium dioxide in water treatment and in topical sunscreen. EPA/600/R-09/057F

Gaiser BK, Biswas A, Rosenkranz P, Jepson MA, Lead JR, Stone V, Fernandes TF (2011) Effects of silver and cerium dioxide micro-and nano-sized particles on *Daphnia magna*. J Environ MOnit 13(5):1227–1235

Gao W, D'Agostino M, Garcia-Gradilla V, Orozco J, Wang J (2013) Multi-fuel driven janus micromotors. Small 9(3):467–471

Gehrke I, Geiser A, Somborn-Schulz A (2015) Innovations in nanotechnology for water treatment. Nanotechnol Sci Appl 8:1–17

Goh PS, Ng BC, Lau WJ, Ismail AF (2015) Inorganic NMs in polymeric ultrafiltration membranes for water treatment. Sep Purificat Rev 44(3):216–249

Gomes J, Domingues E, Quinta-Ferreira RM, Martins RC (2019) N-TiO_2 photocatalysts: a review of their characteristics and capacity for emerging contaminants removal. Water 11(2):373

Gómez-Pastora J, Dominguez S, Bringas E, Rivero MJ, Ortiz I, Dionysiou DD (2017) Review and perspectives on the use of magnetic nano photocatalysts (MNPCs) in water treatment. Chem Eng Sci 310:407–427

Hassan AF, Elhadidy H (2019) Effect of Zr+4 doping on characteristics and sonocatalytic activity of TiO_2/carbon nanotubes composite catalyst for degradation of chlorpyrifos. J Phys Chem Solids 129:180–187

Jackson P, Jacobsen NR, Baun A, Birkedal R, Kühnel D, Jensen KA, Vogel U, Wallin H (2013) Bioaccumulation and ecotoxicity of carbon nanotubes. Chem Cent J 7(1):154

Jhaveri JH, Murthy ZVP (2016) A comprehensive review on anti-fouling nanocomposite membranes for pressure driven membrane separation processes. Desalination 379:137–154

Jurado-Sánchez B, Wang J (2018) Micromotors for environmental applications: A review. Environ Sci: Nano 5 (7):1530–1544

Kanakaraju D, Glass BD, Oelgemöller M (2014) Titanium dioxide photocatalysis for pharmaceutical wastewater treatment. Environ Chem Lett 12(1):27–47

Karim Z, Mathew AP, Grahn M, Mouzon J, Oksman K (2014) Nanoporous membranes with cellulose nanocrystals as functional entity in chitosan: removal of dyes from water. Carb Poly 112:668–676

Kefeni KK, Mamba BB, Msagati TA (2017) Application of spinel ferrite nanoparticles in water and wastewater treatment: a review. Sep. Puri. Technol. 188:399–422

Korashy HM, Attafi IM, Famulski KS, Bakheet SA, Hafez MM, Alsaad AM, Al-Ghadeer ARM (2017) Gene expression profiling to identify the toxicities and potentially relevant human disease outcomes associated with environmental heavy metal exposure. Environ Poll 221:64–74

Krstić V, Pesovski TUB (2018) A review on adsorbents for treatment of water and wastewaters containing copper ions. Chem Eng Sci 192:273–287

Kumari P, Alam M, Siddiqi WA (2019) Usage of nanoparticles as adsorbents for waste water treatment: an emerging trend. Sustain Mater Technol 22:e00128

Kyzas GZ, Matis KA (2015) Review Nano adsorbents for pollutants removal: A review. J Mol Liq 203:159–168

Lee XJ, Foo LPY, Tan KW, Hassell DG, Lee LY (2012) Evaluation of carbon-Based nanosorbents synthesized by ethylene decomposition on stainless steel substrates as potential sequestrating materials for nickel ions in aqueous solution. J Environ Sci 24:1559–1568

Li X, Xia T, Xu C, Murowchick J, Chen X (2014) Synthesis and photoactivity of nanostructured CdS– TiO_2 composite catalysts. Cat Today 225:64–73

Liang X, Cui S, Li H, Abdelhady A, Wang H, Zhou H (2019) Removal effect on stormwater runoff pollution of porous concrete treated with nanometer titanium dioxide. Transp Res D Transp Environ 73:34–45

Liu X, Demir NK, Wu Z, Li K (2015) Highly water-stable zirconium metal–organic framework UiO-66 membranes supported on alumina hollow fibers for desalination. J Am Chem Soc 137(22):6999–7002

Liu C, Hong T, Li H, Wang L (2018) From club convergence of per capita industrial pollutant emissions to industrial transfer effects: An empirical study across 285 cities in China. Ener Pol 121:300–313

Loeb SK, Alvarez PJ, Brame JE, Cates EL, Choi W, Crittenden J, Dionysiou DD, Li Q, Li-Puma J, Quan X, Sedlak DL (2018) The technology horizon for photocatalytic water treatment: sunrise or sunset? Environ Sci Technol 53(6):2937–2947

Longwane GH, Sekoai PT, Meyyappan M, Moothi K (2019) Review Simultaneous removal of pollutants from water using nanoparticles: A shift from single pollutant control to multiple pollutant control. Sci Total Environ 656:808–833

Ma H, Guo Y, Qin Y, Li YY (2018) Review Nutrient recovery technologies integrated with energy recovery by waste biomass anaerobic digestion. Biores Technol 269:520–531

Malik A, Hameed S, Siddiqui MJ, Haque MM, Umar K, Khan A, Muneer M (2014) Electrical and optical properties of nickel-and molybdenum-doped titanium dioxide nanoparticle: improved performance in dye-sensitized solar cells. J Mater Eng Perform 23(9):3184–9192

Matlochova A, Placha D, Rapantová N (2013) The application of nanoscale materials in groundwater remediation. Polish J Environ Stud 22(5):1401–1410

Mishra BK, Chakraborty S, Kumar P, Saraswat C (2020) Water Quality Restoration and Reclamation. Sustainable Solutions for Urban Water Security, pp 59–81. Springer, Cham

Pace HE, Lesher EK, Ranville JF (2010) Influence of stability on the acute toxicity of CdSe/ZnS nanocrystals to *Daphnia magna*. Environ Toxicol Chem 29(6):1338–1344

Pacheco M, López MA, Jurado-Sánchez B, Escarpa A (2019) Self-propelled micromachines for analytical sensing: a critical review. Anal Bioanal Chem 411:6561–6573

Pandey N, Shukla SK, Singh NB (2017) Water purification by polymer nanocomposites: an overview. Nanocomp. 3(2):47–66

Phokha S, Klinkaewnarong J, Hunpratub S, Boonserm K, Swatsitang E, Maensiri S (2016) Ferromagnetism in Fe-doped MgO nanoparticles. J Mater Sci: Mater Elect 27 (1):33–39

Rahaman MS, Vecitis CD, Elimelech M (2012) Electrochemical carbon-nanotube filter performance toward virus removal and inactivation in the presence of natural organic matter. Environ Sci Technol 46(3):1556–1564

Rajasulochana P, Preethy V (2016) Comparison on efficiency of various techniques in treatment of waste and sewage water-A comprehensive review. Resour Effic Technol 4:175–184

Richter KE, Ayers JM (2018) An approach to predicting sediment microbial fuel cell performance in shallow and deep water. Appl Sci 8:2628

Rodovalho FL, Capistrano G, Gomes JA, Sodre FF, Chaker JA, Campos AFC, Bakuzis AF, Sousa AH (2016) Elaboration of magneto-Thermally recyclable nanosorbents for remote removal of toluene in contaminated water using magnetic hyperthermia. Chem Eng J 15:725–732

Safdar M, Simmchen J, Jänis J (2017) Correction: Light-driven micro-and nanomotors for environmental remediation. Environ Science: Nano 4(11):2235

Samanta HS, Das R, Bhattachajee C (2016) Influence of nanoparticles for wastewater treatment-a short review. Austin Chem Eng 3(3):1036

Sanders MB, Sebire M, Sturve J, Christian P, Katsiadaki I, Lyons BP, Feist SW (2008) Exposure of sticklebacks (*Gasterosteus aculeatus*) to cadmium sulfide nanoparticles: biological effects and the importance of experimental design. Marine Environ Res 66(1):161–163

Santhosh C, Velmurugan V, Jacob G, Jeong SK, Grace AN, Bhatnagar A (2016) Role of NMs in water treatment applications: a review. Chem Eng J 306:1116–1137

Saravanan R, Gracia F, Stephen A (2017) Basic principles, mechanism, and challenges of photocatalysis. Nanocomposites for Visible Light-Induced Photocatalysis, pp 19–40. Springer, Berlin/Heidelberg, Germany

Serrà A, Zhang Y, Sepúlveda B, Gómez E, Nogués J, Michler J, Philippe L (2020) Highly reduced ecotoxicity of ZnO-based micro/nanostructures on aquatic biota: Influence of architecture, chemical composition, fixation, and photocatalytic efficiency. Water Res 169:115210

Sharma A, Kaur M, Katnoria JK, Nagpal AK (2016) Heavy metal pollution: A global pollutant of rising concern. Toxicity and waste management using bioremediation, pp 1–26. IGI Global

Shayesteh M, Samimi A, Shafiee Afarani M, Khorram M (2016) Synthesis of titania–γ -alumina multilayer nanomembranes on performance-improved alumina supports for wastewater treatment. Desal Water Treat 57(20):9115–9122
Sizmur T, Fresno, T Akgül G, Frost H, Jiménez EM (2017) Review Biochar modification to enhance sorption of inorganics from water. Biores Technol 246:34–47
Smith CJ, Shaw BJ, Handy RD (2007) Toxicity of single walled carbon nanotubesto rainbow trout, (*Oncorhynchus mykiss*): respiratory toxicity, organ pathologies, and other physiological effects. Aquat Toxicol 82(2):94–109
Soyekwo F, Zhang QG, Deng C, Gong Y, Zhu AM, Liu QL (2014) Highly permeable cellulose acetate nanofibrous composite membranes by freeze-extraction. J Memb Sci 454:339–345
Sun G, Chung TS, Jeyaseelan K, Armugam A (2013) Stabilization and immobilization of aquaporin reconstituted lipid vesicles for water purification. Coll Surf b: Biointerfaces 102:466–471
Sun X, Liu Z, Zhang G, Qiu G, Zhong N, Wu L, Cai D, Wu Z (2015) Reducing the pollution risk of pesticide using nano networks induced by irradiation and hydrothermal treatment. J Environ Sci Health C 50:901–907
Tahir MB, Kiran H, Iqbal T (2019) The detoxification of heavy metals from aqueous environment using nano-photocatalysis approach: a review. Environ Sci Poll Res 26(11):10515–10528
Tang WW, Zeng G, Gong JL, Liang J, Xu P, Zhang C, Haung BB (2014) Impact of humic/fulvic acid on the removal of heavy metals from aqueous solutions using NMs: a review. Sci Total Environ 468:1014–1027
Tang K, Gong C, Wang D (2016) Reduction potential, shadow prices, and pollution costs of agricultural pollutants in China. Sci Total Environ 541:42–50
Tijing LD, Woo YC, Shim WG, He T, Choi JS, Kim SH, Shon HK (2016) Superhydrophobic nanofiber membrane containing carbon nanotubes for high-performance direct contact membrane distillation. J Memb Sci 502:158–170
Tshikovhi A, Mishra SB, Mishra AK (2020) Nanocellulose-based composites for the removal of contaminants from wastewater. Int J Biol Macromol 152:616–632
Umar K, Haque MM, Muneer M, Harada T, Matsumura M (2013) Mo, Mn and La doped TiO_2: synthesis, characterization and photocatalytic activity for the decolourization of three different chromophoric dyes. J Alloys Comp 578:431–438
Umar K, Aris A, Ahmad H, Parveen T, Jaafar J, Majid ZA, Reddy AVB, Talib J (2016) Synthesis of visible light active doped TiO2 for the degradation of organic pollutants—methylene blue and glyphosate. J. Anal Sci Technol 7(1):29
Umar K, Parveen T, Khan MA, Ibrahim MNM, Ahmad A, Rafatullah M (2019a) Degradation of organic pollutants using metal-Doped TiO_2 photocatalysts under visible light: A comparative study. Desal. Water Treat. 161:275–282
Umar K, Ibrahim MNM, Ahmad A, Rafatullah M (2019b) Synthesis of Mn-doped TiO_2 by novel route and photocatalytic mineralization/intermediate studies of organic pollutants. Res Chem Intermed 45(5):2927–2945
Unuabonah EI, Taubert A (2014) Review article Clay-Polymer nanocomposites (CPNs): Adsorbents of the future for water treatment. Appl Clay Sci 99:83–92
Vallino E, Ridolfi L, Laio F (2020) Measuring economic water scarcity in agriculture: a cross-country empirical investigation. Environ Sci Pol 114:73–85
Van Der Heijden AEDM (2018) Developments and challenges in the manufacturing, characterization and scale-up of energetic NMs–A review. Chem Engineer J 350:939–948
Vunain E, Mishra AK, Mamba BB (2016) Dendrimers, mesoporous silicas and chitosan-based nanosorbents for the removal of heavy-metal ions: a review. Int J Biol Macromol 86:570–586
Wang Y, Zhang Y, Hou C, Liu M (2016) Mussel-Inspired synthesis of magnetic polydopamine–Chitosan nanoparticles as bio sorbent for dyes and metals removal. J Taiwan Inst Chem E 61:292–298
Wang J, Wang Z, Carolina LZV, Wolfson JM, Pingtian G (2019) Review on the treatment of organic pollutants in water by ultrasonic technology. Ultrason Sonochem 55:273–278

Wei G, Su Z, Reynolds NP, Arosio P, Hamley IW, Gazit E, Mezzenga R (2017) Self-assembling peptide and protein amyloids: from structure to tailored function in nanotechnology. Chem Soc Rev 46(15):4661–4708

Wu Y, Pang H, Liu Y, Wang X, Yu S, Fu D, Wang X (2019) Environmental remediation of heavy metal ions by novel-NMs: a review. Environ Pollut 246:608–620

Yadav VB, Gadi R, Kalra S (2019) Clay based nanocomposites for removal of heavy metals from water: A review. J Environ Manag 232:803–817

Yamakata A, Vequizo JJM (2019) Curious behaviors of photogenerated electrons and holes at the defects on anatase, rutile, and brookite TiO_2 powders: A review. J Photochem Photobiol c: Photochem Rev 40:234–243

Yaqoob AA, Parveen T, Umar K, Mohamad Ibrahim MN (2020) Role of NMs in the treatment of wastewater: a review. Water 12(2):495

Ying Y, Ying W, Li Q, Meng D, Ren G, Yan R, Peng X (2017) Recent advances of NM-based membrane for water purification. Appl Mater Today 7:144–158

Zhang QG, Deng C, Soyekwo F, Liu QL, Zhu AM (2016) Sub-10 nm Wide Cellulose Nanofibers for Ultrathin Nanoporous Membranes with High Organic Permeation. Adv Funct Mat 26(5):792–800

Zhu X, Chang Y, Chen Y (2010) Toxicity and bioaccumulation of TiO_2 nanoparticle aggregates in *Daphnia magna*. Chemos 78(3):209–215

Chapter 5
Nano-Catalyzed Heavy Metal Remediation from Wastewater

Alisha and Sudesh Kumar

Abstract In the current generation of shortage of water reservoirs, the efficient management of polluted water stands as a fundamental necessity for the progressing society. It is imperative to innovate and adopt novel technologies for water purification characterized by both minimal investment and high efficacy. Amidst different management methods, contemporary advancements in nanotechnology have garnered global research interest. This chapter provides an overview of advancements in nano-material science concerning the treatment of contaminated water. It discusses the application of diverse categories of nano-scale materials in the management of polluted water. These encompass four primary categories. Firstly, sorbent materials like graphene, carbon nanotubes, activated carbon, ZnO, MgO, and iron oxides are commonly employed for the elimination of trace metal ions from polluted water. Following that, catalyst materials, i.e., oxidizing agents, light-assisted catalysts, electrochemical catalysts, and Fenton-based catalysts have demonstrated their capability to eliminate both metal ions and organic pollutants. In addition, nanoporous membranes have been employed for the efficient removal of synthetic dyes, trace metal ions, and impurities. The encouraging outlook of nanotechnology in wastewater treatment along with the possible pathways for future endeavors is also discussed.

Keywords Wastewater treatment · Catalyst · Fenton · Carbon · Metal oxide · Heavy metal

5.1 Introduction

Swift commercialization and the surge in the human populace have contributed to widespread urban expansion, leading to a rapid surge of environmental contamination. Augmenting the quality of soil, water, and air poses a substantial challenge in the modern era. Identifying and alleviating water pollutants, coupled with their proactive

---

Alisha · S. Kumar (✉)
Department of Chemistry, Banasthali Vidyapith, Tonk, Rajasthan, India
e-mail: sudeshneyol@gmail.com

N. Agarwal et al. (eds.), *Nanomaterials in Wastewater Research*, Advances in Wastewater Research, https://doi.org/10.1007/978-981-96-3908-3_5

prevention, are imperative actions for protecting the environment (Wu et al. 2019). Nanotechnology plays a pivotal role in attaining the goal of a sustainable environment, and this technology has made notable progress over the past decade, especially in the realm of nanomaterials (NMs). The availability of clean water is decreasing due to commercial activities, leading to a global scarcity of pure water, particularly in emerging nations. Water contamination can occur due to various pollutants including organics, micro-organisms, pathogens, dyes, and trace metals like cadmium (Cd^{2+}), lead (Pb^{2+}), nickel (Ni^{2+}), zinc (Zn^{2+}), mercury (Hg^{2+}), and chromium (Cr^{3+}). These pollutants do not break down naturally and pose serious threats to both living organisms and the surroundings. Trace metals can cause cancer, kidney damage, liver inflammation, miscarriage, nephrotic syndrome, anemia, and neurological disorders (Madima et al. 2020). Pb^{2+} ions are often released into the environment from industries such as metallurgy, paper manufacturing, and glass production. Electrodeposition in battery manufacturing, solar cell production, mining operations, and textile industries are common sources of cadmium contamination in wastewater (Nimibofa et al. 2018). Contact with items such as jewelry, watches, zippers, and coins containing nickel (Ni^{2+}) ions can lead to skin disorders. Chromium (Cr^{6+}) ions can cause liver damage, kidney inflammation, and abdominal discomfort, and are also a primary cause of mucosal lesions. Due to the severe health risks associated with these trace metals, it is crucial to remove them from water to protect human health. Various methods can be employed for water purification such as charge transfer, bioaccumulation, flocculation, extraction, and reverse osmosis. Among these, adsorption is the most commonly utilized technique due to its low cost, high efficiency, and ease of implementation for removing trace metals at low concentrations. Figure 5.1 illustrates various origins of trace metal ion pollution. Various materials have been employed for remediating wastewater, encompassing both organic and plant-based substances, with humic acid being particularly noteworthy for its effectiveness in wastewater treatment and the elimination of heavy metal ions (Zhang et al. 2017). Material science plays a pivotal role across various sectors, including ecological studies, medical applications, supercapacitors/batteries, wastewater treatment plants, catalyst development, power generation, and energy storage (Baby et al. 2019). NMs offer distinct advantages for treating contaminated water, capitalizing on their large surface area for adsorption, capacity for chemical transformations, and ease of recyclability. NMs are increasingly utilized for removing various contaminants from polluted water, including inorganic compounds, micro-organisms, and organic pollutants (Baby Shaikh et al. 2018).

Various forms of carbon materials like graphene, carbon nanotubes, graphene oxide, fullerenes, and activated carbons have found wide-ranging uses in detectors, energy storage devices, electrical items, wastewater treatment, and pharmaceuticals. This chapter delves into the recent advancements in nanocatalysts used for heavy metals eradication from wastewater. Different materials employed for wastewater treatment are shown in Fig. 5.2.

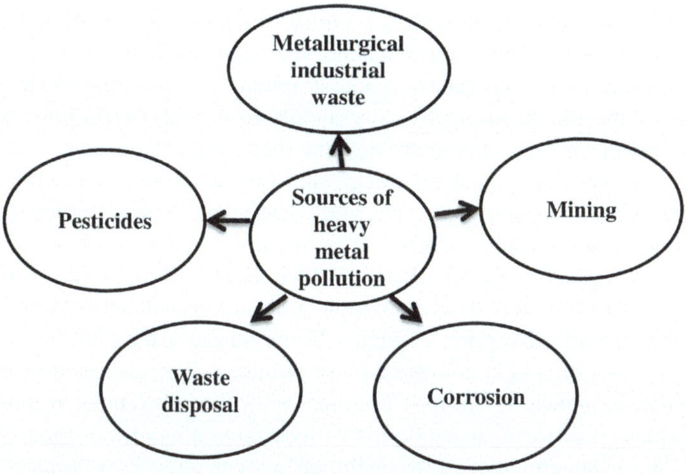

Fig. 5.1 Origins of trace metal ion pollution (Baby et al. 2019)

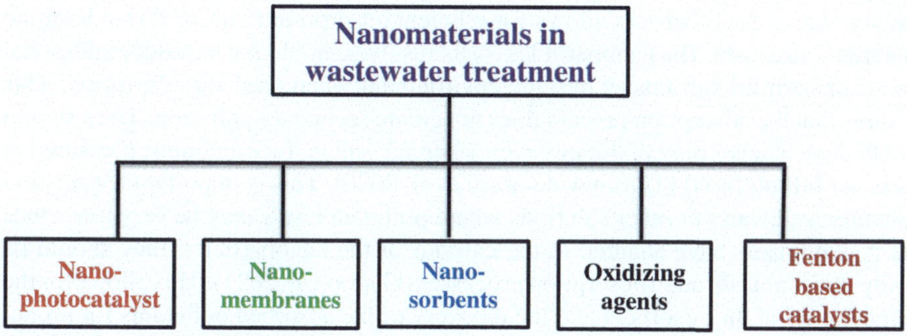

Fig. 5.2 Nanocatalysts for elimination of heavy metal ions from wastewater

5.2 Nanocatalysts for Wastewater Treatment

5.2.1 Nano-Adsorptive Materials

Lately, researchers have become increasingly interested in investigating the possibilities offered by NMs as adsorptive materials (Alayan et al. 2021). The reduced size of these particles leads to a greater area, increasing their activity and sorbing efficiency for metals. The adsorption mechanism relies on factors such as the adsorptive capacity factor and the segregation of contaminants, such as trace metals or organics, under stable conditions. Additionally, oxidative-reductive reactions help in the conversion of ionic forms into inorganic toxins but, some scientists argue that alterations in the oxidation-reduction environment can impact the harmfulness of these contaminants. The nanoparticles include carbon nanotubes, activated carbon, zinc oxide, iron oxide, magnesium oxide, manganese oxide, graphene, and titanium oxide which act as sorbents for the purification of heavy metals from water (Dhillon and Kumar 2019).These materials have characteristics like their surface area and exterior modifications. The chemical, physical, chemical, and characteristics of these materials are related to their surface morphology, size, and constituents. The variables influencing the adsorption are surface area, reactivity, atomic arrangement, absence of diffusion barriers, and strong surface interaction (Iftekhar et al. 2018). As the surface area available for adsorption increases, more molecules or ions can adhere to the surface, increasing the overall adsorption capacity. Some surfaces have higher adsorption activity due to their chemical composition or structure. The chemical properties of the surface can influence its ability to interact with the substances in the solution. Reactive surfaces may form strong bonds with certain molecules or ions, enhancing adsorption. The arrangement of atoms on the surface can affect how molecules or ions interact with it. Different surface structures may have varying affinities for certain substances. In some cases, the adsorption process may be limited by the dispersion of ions within the porous structure of the solid material. In an aqueous environment, the absence of such barriers allows for efficient adsorption to occur throughout the material's structure. The nanoparticles (s) themselves should not introduce additional toxins or harmful substances into the environment when used for adsorption. This ensures that the adsorption process does not create secondary pollution. The s should be effective at adsorbing pollutants even when present in trace amounts, measured in parts per billion (ppb) (Thirunavukkarasu et al. 2020). This is important for treating contaminated water or other solutions where pollutant levels may be very low. Once the contaminants have adhered to the exterior of the nanoparticles, they should be easily removable through desorption processes (Gupta et al. 2021). This allows for the regeneration of the adsorbent and the recovery of the adsorbed pollutants for proper disposal or reuse. The s should be durable and capable of being reused multiple times without significant loss of adsorption capacity or effectiveness. This enhances the efficiency and viability of the adsorption process. Overall, these characteristics ensure that the s function effectively as adsorbents for removing heavy metals from solutions while minimizing environmental impact and maximizing efficiency and recyclability.

5.2.1.1 Iron Oxide

Ferric oxide is an economic-friendly NM that occurs naturally and can be synthesized easily. Its use as an adsorbent is environmentally friendly, with a reduced risk of secondary pollution when applied directly to polluted areas (Sharma et al. 2021). The effectiveness of adsorption of different heavy metal ions onto ferric oxide particles is contingent upon various parameters, comprising acidity, time, temperature, and adsorption level. These parameters need to be optimized for the efficient elimination of toxic metal ions from contaminated environments (Latif et al. 2020). Researchers have conducted surface alterations of ferric particles to enhance their sorption capability (Wang et al. 2013). Surface treatments involve treating the ferric particles with different chemicals or compounds like aminopropyltrimethoxysilane to alter their surface properties and improve their affinity for heavy metal ions.

5.2.1.2 Manganese Oxide

Manganese oxides exhibit strong sorption capability due to their high surface area and multiform composition. These particles have been utilized for eliminating toxic metals like arsenic from polluted water due to their effective adsorption properties. Surface-modified forms of manganese, such as hydrous and nanoporous manganese oxide, are commonly employed for water purification (Husnain et al. 2020). Hydrous manganese oxides can be synthesized by adding $MnSO_4{\cdot}H_2O$ into a sodium hypochlorite solution. The adsorption of heavy metals onto hydrous manganese oxides typically occurs through a surface complexation mechanism, which involves an ion transfer. Adsorption of bivalent metals occurs in two steps: initially, metals adsorb on the outer layer of hydrous manganese oxides, followed by internal diffusion, where the ions diffuse into the interior of the particles.

5.2.1.3 Zinc Oxide

Zinc oxide (ZnO) possesses a nanoscale porous matrix with a large area, rendering it appropriate for adsorbing toxic metals from wastewater. Various nanoforms of ZnO, like nano-plates, nano assemblies, nano-sheets, and nano-rods, are utilized as adsorbents for toxic metal elimination. These surface-modified forms of ZnO exhibit a higher elimination rate for toxic metals compared to industrial-grade ZnO due to their unique porous framework. For example, zinc nano-plates show high removal efficiency for copper (Cu(II)) from aqueous medium. Zinc nano-assemblies were employed for the removal of trace metal ions, including arsenic, cobalt, copper, nickel, lead, cadmium, and mercury (Bharti et al. 2022). Nanostructured micropores, in particular, demonstrate strong attraction for the adsorption of mercury, lead, and arsenic due to their positively charged character. Zinc nano-rods exhibit a large removal rate for cadmium and lead from polluted water.

5.2.1.4 Magnesium Oxide

Magnesium microspheres possess a unique design that improves the adsorption affinity for the elimination of trace metals from polluted areas. These microspheres provide a high surface area for adsorption to occur. Various surface modifications have been done to enhance the removal rate of trace metals of magnesium oxide particles (Ahmed and Kumar 2024). These modifications include altering the morphology of MgO particles into different structures such as nanobelts, nanocubes, nano-rods, nanotubes, and nanowires.

5.2.1.5 Carbon Nanotubes (CNTs)

CNTs are cylindrical nanoforms patterned in a tubular configuration. They are categorized into two classes, namely, single-walled carbon nanotubes (SWCNT) and multi-walled carbon nanotubes (MWCNT). SWCNTs consist of carbon monolayers and MWCNTs consist of multiple layers of carbon. CNTs have found extensive applications in the domain of wastewater treatment, biosensors, stimulators, and membranes due to their distinctive features, including extensive specific area, microporous structure, enclosed structure, layered composition, and robust bonding with contaminants. Various harmful contaminants have been eliminated from water through the utilization of CNTs, leveraging their unique features. Furthermore, CNTs can be synergistically mixed with other carbon NMs to enhance their performance in removing a broad spectrum of contaminants from water (Das et al. 2018). Despite the extensive utilization of CNTs in the area of water purification, these materials may experience suboptimal activity because of adulteration, which may be carbon components used in the synthesis process. The adulteration leads to the modification of their surface characteristics by obstructing the reactive regions for the bonding of CNTs with the contaminants (Sarkar et al. 2018). Nevertheless, the majority of the research studies indicated that surface modification can be averted by tailoring CNTs in acidic or basic solutions. This process also leads to the addition of unique functionality on the CNTs surface, which proved to be useful for contaminants removal. The addition of novel functionality to the CNTs surface, along with the characteristics of the CNTs helps in assessing the removal rate of various pollutants. CNTs have effectively demonstrated as an exceptional NM in eliminating contaminants from water. This contributes to creating greener surroundings, because of its remarkable adsorption efficiency and outstanding restoration capacities (Thines et al. 2017). The photocatalytic process stands as a novel technique employed in water purification that employs semiconductors, specifically iron oxide, zinc oxide, and titanium oxide but the photocatalytic efficacy of these nano-oxides is relatively low, and their response to UV light is sluggish. CNTs are novel materials for catalytic processes due to their enhanced photocatalytic efficacy, nanoscale dimensions, large durability, tubular hollow design, and expanded photo absorption capabilities resulting from their extensive specific area. SWCNTs represent 1-D materials characterized by a tubular hollow structure with walls of just singular atoms in thickness (Xu and Wang 2017). These carbon NMs demonstrate extraordinary physical and chemical characteristics owing to their distinctive design. SWCNTs are extensively utilized for various applications, including semiconductor materials, electric devices, biomedicine, and detectors. They are also commonly employed in the remediation of ecological contamination due to their porosity, large area, facile usage, and nanoscale dimensions. These characteristics make SWCNTs highly promising for water purification. Alijani and co-authors engineered a composite based on SWCNTs by incorporating them with cobalt ferrite sulfide and the resultant materials were employed for the elimination of mercury; findings demonstrated a remarkable adsorption rate exceeding 99.56% in a brief duration of 7 min (Alijani and Shariatinia 2018). In contrast to this, the adsorption rate for solely SWCNTs was observed to be 45.39% for mercury removal. Anitha and co-authors performed molecular modeling of pristine SWCNTs and their modified variants, such as hydroxy SWCNTs, amino SWCNTs, and carboxyl SWCNTs for the elimination of trace metals like lead, cadmium, and mercury from water (Anitha et al. 2015). The results showed that the carboxy SWCNTs exhibited a significantly higher adsorption rate approximately 150–230% greater compared to pristine SWCNTs. The hydroxy SWCNTs and amino SWCNTs were identified to exhibit lower adsorption strength, displaying only 10 to 47% greater adsorption rate in contrast to pristine SWCNTs. The carboxy SWCNTs have also been explored for the removal of heavy metal ions like lead, copper, and cadmium. The results revealed the adsorption efficacies for metal ions of 96.02, 55.89, and 77.00, respectively. In contrast to this, unmodified SWCNTs showed adsorption efficacies of 24.29, 33.55, and 24.07 mg/g for copper, lead, and cadmium, respectively. Zazouli and co-authors engineered composite materials of SWCNT with L-cysteine. They utilized these materials for extracting Hg^{2+} ions from aqueous media with adsorption efficacy of 95% (Zazouli et al. 2014). Gupta and co-authors developed a carbon nanotube-reinforced polysulfone membrane, which was utilized for the elimination of trace metal ions. Integration of SWCNTs led to a decrease in the porosity of the membrane and a more even surface. The engineered polysulfone membrane exhibited significant adsorption capacity of 96.8, 87.6, and 94.2% for chromium, arsenic, and lead, respectively. The sole membrane exhibited adsorption capacities of 28.5, 30.3, and 28.3% for arsenic, chromium, and lead, respectively (Gupta et al. 2015). These findings demonstrate an enhancement in the membrane's efficacy attributed to the integration of SWCNTs. Dehghani and co-authors utilized SWCNTs in the elimination of chromium from aqueous media and assessed the influence of various parameters like time, acidity, and amount of chromium ions on the adsorption capability. It was noted that the removal rate varied with the acidity. The peak efficacy was identified at a pH value of 2.5, and the adsorption conforms to the Langmuir model (Dehghani et al. 2015). These investigations indicated that SWCNTs are effective for water purification. The MWCNTs demonstrate distinctive features, including large areas, elevated heat, and electrical conductivity, as well as exceptional durability. Due to these physical and chemical characteristics, they find extensive applications in electric devices, photovoltaic cells, detectors, and biomedicine. MWCNTs have also been broadly utilized in water purification, particularly trace metals are removed through bonding with the surface moieties of MWCNTs. The oxygen-functionalized MWCNTs have a large adsorption capability and efficacy for the removal of metal ions like chromium, lead, and cadmium from aqueous media. The removal of metals relies on acidity, and this characteristic can be utilized for ion desorption by adjusting the initial pH value, allowing MWCNTs to be reused. Certain investigations have indicated that plasma-treated MWCNTs exhibit superior adsorption characteristics compared to acid-treated MWCNTs; this can be attributed to an abundance of oxygen-functionalized moieties on CNTs. Additionally, reports indicated that plasma-treated MWCNTs can be readily upcycled and reutilized. The MWCNTs composites with iron oxide, zirconium oxide, and aluminum oxide have been employed for the removal of trace metals like chromium, arsenic, nickel, lead, and copper from aqueous media. The adsorption properties of MWCNTs can be influenced by variables such as acidity and the amount of metallic ions. Their research findings align well with the Freundlich isotherm. The removal rate of modified MWCNTs exhibited a notable increase when compared to other metal oxides. Furthermore, the modified MWCNTs showed 20-fold more efficacies compared to unmodified MWCNTs in the removal of trace metal ions. It is widely accepted that the interaction between metal ions and the polarized surface of CNTs is generally considered the primary mechanism of adsorption. Functionalized MWCNTs have demonstrated elevated adsorption capability and efficacy for metal ions like lead, cadmium, and chromium from the aqueous media. The adsorption efficacy of acid-treated MWCNTs enhances the ability to eliminate Pb^{2+} and Cd^{2+} ions with oxygen-containing moieties, resulting in the precipitation of salts or ion complexes. The adsorption capacity of nitric acid-treated MWCNTs experiences a substantial increase primarily because of the presence of oxygen-containing moieties interacting with metallic ions, leading to the formation of ion complexes on the nanotube surface. The composite materials of MWCNTs with oxides, such as Fe_2O_3, ZnO, and Al_2O_3 were synthesized using a coprecipitation technique. The resulting composites have been effectively employed for the adsorption of Cu^{2+}, Ni^{2+}, and Pb^{2+} ions. The adsorption efficacy of these materials was observed to vary with the acidity and temperature. Based on the acidity and temperature, the adsorption capability of MWCNT composite materials ranged from 10 to 31 mg/g and the adsorption conforms to the Langmuir isotherm. The MWCNT composites with iron and manganese oxide were regarded as promising materials for their ability to eliminate chromium ions. The highest adsorption capability of 186.9 mg/g was observed with a peak elimination efficiency of 85% at the optimal pH value of 2.1. The impressive adsorption performance can be attributed to the polarized surface of the nanotubes. Additionally, research indicates that plasma-treated MWCNTs exhibit superior adsorption capabilities in comparison to MWCNTs having oxygen functionalities. The plasma-oxidation method has also been documented in the creation of CNTs incorporating manganese dioxide and titanium oxide for the extraction of Pb^{2+} ions from the aqueous media. On a heterogeneous adsorption site, certain regions combined twice align with the Langmuir–Freundlich model. This model was employed to distinguish between two types of adsorbent surfaces with enhanced and lower electron affinities for nickel ions. The adsorption of Ni^{2+} ions predominantly takes place at modified surfaces of MWCNTs composites, and the functionalization results in a 20% rise in the adsorption capability at low amounts of adsorbate. MWCNTs can also be modified with hydroxyquinoline, and this modification has been employed for the extraction of numerous toxic metal ions like Cu^{2+}, Pb^{2+}, and Cd^{2+}. CNTs, whether in their pristine state or with oxygen-bearing groups and in complex with others, exhibit remarkable capacity for adsorbing trace metals and extensive studies are currently underway for their utilization in water treatment. Elsehly and co-authors utilized available MWCNTs for eliminating Mn^{2+} and Fe^{2+} ions, achieving a removal efficiency of 71.5% and 52%, respectively, at

Table 5.1 Application of modified and raw carbon nanotubes for toxic metals removal (Ouni et al. 2019)

Sorbent material	Pollutant	Highest adsorptive capability ($mg.g^{-1}$)	Type of bonding	Adsorption isotherm
Modified MWCNTs	Zn^{2+}	1.05	Coulombic	Freundlich, Langmuir
SWCNTs	Hg^{2+}	40.1	Chemisorption	Freundlich, Langmuir
SWNTs with thiol groups	Hg^{2+}	131.5	Chemisorption	Langmuir
MWCNTs	As^{5+}	2.9	Chemisorption, coulombic	Combined Langmuir–Freundlich
Oxidized MWCNTs	As^{5+}	3.61	Specific adsorption, chemisorption, coulombic	Combined Langmuir–Freundlich
Ethylenediamine MWCNTs	As^{5+}	12.17	Specific adsorption, chemisorption, coulombic	Freundlich
MWCNTs	Cu^{2+}	3.19×10^{-5} mol/g	Coulombic, deposition, surface bonding	Langmuir
SWCNT	Cr^{6+}	20.3	Chemisorption, coulombic, physiosorption	Langmuir
MWCNTs	Cr^{6+}	2.4	Chemisorption, coulombic, physiosorption	Langmuir
MWCNTs-hydroxyapatite	Co^{2+}	16.2	Surface binding	Langmuir

a 50 ppm concentration of metal ions (Elsehly et al. 2016). Another group utilized CNT nanocomposite materials for extracting Fe^{2+} and Mn^{2+} from aqueous media. Table 5.1 shows the adsorption capacities of modified and unmodified CNTs for various toxic metal ions.

5.2.1.6 Fullerenes

Fullerenes, first identified in 1985 from cosmic particles, possess a unique structure composed of both 5 and 6-membered rings. They are described by the chemical formula C20 + m, where m is an integer. Fullerenes display characteristics such as hydrophobicity, strong electron binding, and a large surface area. These distinct physicochemical properties make them versatile carbon materials suitable for different uses in fields such as electronics, biomedical sciences, photovoltaic cells, detectors, beauty products, photocatalysis, and coatings (Maduraiveeran and

Jin 2021). Fullerenes have found various applications, including the remediation of pollutants from groundwater or surface water through light-induced catalysis (Park et al. 2018). They are also valuable for hydrogen fixation due to their ability to convert C–C bonds to C–H bonds. With an optimal hydrogen storage capacity of 6.1% attributed to their chemical properties and 3-D framework, fullerenes can be easily reverted back due to the stronger bond energies of C–C bonds. In capacitive electrode surfaces, carbon coatings play a crucial role, where their capability depends on area, porosity, and conductance. Carbon NMs, including fullerenes, offer greater conductance compared to traditional ones because of their extensive surface area (Clancy et al. 2018). Fullerenes in composite form have been shown to exhibit superior capacitance and stability compared to pristine graphene. Moreover, fullerenes have been utilized in batteries as anodes, offering increased efficiency while replacing traditional metal electrodes, thus presenting benefits in terms of efficiency and sustainability (Sweetman et al. 2017). They possess physicochemical properties that enable the elimination of organics from aqueous media. Soluble fullerene complexes (SFCs) act as catalysts to generate oxygen-free radicals upon exposure to ultraviolet irradiation, which can decompose contaminants present in water photochemically, while also serving as antiradicals. Fullerenes can adsorb pollutants by infiltrating adsorbed species into the gaps between carbon materials, and their reduced accumulation property and large surface area make them valuable for the elimination of organics from solution. Studies have demonstrated the effectiveness of fullerenes in eliminating metal ions like copper, conforming to the Langmuir theory. However, their high-cost limits real-life applications, although small quantities can increase the absorptivity of materials like zeolites, lignin, and carbon. Fullerene synthesis enhances hydrophobicity, and adsorption efficiency, and improves the recycling of materials (Gusain et al. 2020). Polymer-bound fullerenes, such as those bonded with polyvinylpyrrolidone, have shown value in extracting water contaminants. The membrane filtration method for the purification of water is gaining prominence, and fullerenes hold promise for enhancing its efficacy due to their ease of use, strong electron bonding, high strength, and ability to customize sizes. They can also enhance the adsorption efficacy of nano-adsorbents when bound together.

5.2.1.7 Graphene

Graphene is regarded as a cutting-edge material and has garnered immense interest at the time of its invention in the year 2004. It is a 2D single-layer carbon nanostructure structured with carbon atoms in a planar arrangement in a hexagonal crystal lattice. It has attracted substantial interest as innovative carbon NMs for ecological uses owing to its large area of around 2630 m^2/g, large heat transfer capability, and fast electron mobility (Xu and Wang 2017). Moreover, their exceptional tensile strength and certain remarkable attributes like numerous functional moieties, large electron charge density, and high water-attracting features) render them attractive materials in polluted water treatment. Graphene and reduced graphene oxide

(RGO) can be easily synthesized by graphene oxide reduction through an exfoliation process. Furthermore, graphene can undergo functionalization to produce graphene oxide using the Hummers technique, resulting in a significant amount of oxygen functionalities (such as carboxyl acid group, hydroxy group, carbonyl functional group, and epoxide group) within the carbon framework (Jilani et al. 2018). This characteristic is especially noteworthy for the removal of diverse contaminants from water. Graphene oxide can be produced through facile and cost-effective procedures, like the oxidative transformation from graphite to graphene oxide, followed by ultrasonic separation (Wang et al. 2013). Because of its water-attracting characteristics, large surface area, and abundant density of functional moieties, it can serve as an adsorbent and function as a catalyst for eliminating various contaminants from polluted water. Utilization of graphene in the area of wastewater treatment is presently impeded by limited recyclability and isolation, challenges that can be surmounted by hybridization and functionalization. Moreover, graphene is generally combined with other materials for appropriate uses because of its no band energy difference and vulnerability to oxidation. When paired with other materials, it can boost the photocatalysis of coupling materials by serving as an electron receptor and carrier. Graphene is composed of hexagonal layers, forming a material with walls of just one atom thick, and is recognized as the finest NM, possessing a tensile strength 200 times greater than that of steel. Graphene was identified in 2004 by Konstantin Novoselov and Andre Geim, who were given a Nobel Prize in 2010 for their innovation. Graphene, a two-dimensional NM, is extensively employed in nearly every sector, including touch panels, mobile devices, liquid crystal displays, semiconductor devices, microprocessors, energy storage devices, energy production, water purification, supercapacitor devices, solar panels, biomedicine, and ecological research. The increasing focus on graphene NMs in water purification is driven by their distinctive physical and chemical properties, particularly conductivity, large area, heat conductivity, tensile strength, and adjustable surface properties. Tabish and co-authors developed graphene NMs and employed them as an adsorbent for effectively eliminating pollutants and trace metals from aqueous media (Tabish et al. 2018). Moreover, this graphene NM was utilized for the elimination of arsenic from aqueous media, demonstrating a removal efficacy of 80%. The material maintained its water purification efficacy even after repeated cycles and reutilization. Guo and co-authors engineered a graphene oxide nanocomposite using iron oxide through an on-site co-precipitation technique. Subsequently, they employed this composite for the elimination of lead ions from aqueous media (Guo et al. 2018). The formulated material demonstrated outstanding adsorptive performance in extracting lead ions from water, exhibiting an outstanding adsorption capability of 373.14 mg/g. Zhang and co-authors modified graphene oxide with 4-sulfophenylazo and utilized it to eliminate various trace metals from water (Zhang et al. 2018). The engineered NM exhibited the highest adsorption capability of 59, 689, 267, 66, and 191 mg/g for copper, lead, cadmium, nickel, and chromium ions, respectively. Diana and co-authors engineered an autonomous microscopic robot based on graphene, comprising a layered composition made of nickel, graphene oxide, and platinum. Each layer served a distinct usage; for instance, graphene oxide extracted trace metals, while the intermediate layer of nickel facilitated microscopic robot control through exterior magnetic force. Additionally, the inside platinum layer contributed to the motor in an autonomous system. The devised system successfully eliminated 80% of lead ions from the aqueous media. Yang and co-authors formulated hydrogen beads employing sodium alginate and graphene oxide and effectively employed hydrogen beads for extracting manganese ions from water, demonstrating an outstanding adsorption capacity of 56.49 mg/g (Yang et al. 2018). Zheng and co-authors engineered a material composed of tea polyphenol, zinc oxide, and RGO. The developed NM was utilized for the elimination of trace metals, offering the additional benefit of antimicrobial characteristics. They employed this NM for extracting lead ions from water, achieving an adsorption capacity of 98.9%, and the nanocomposite was discovered to exhibit antimicrobial characteristics against Streptococcus mutans, resulting in a 99% removal rate (Zheng et al. 2018). Mousavi and co-authors engineered graphene oxide nanocomposites with Fe_3O_4 and utilized them for extracting lead ions from an aqueous solution. The NM demonstrated a 98% adsorption efficacy with an adsorption capability of 126.6 mg/g. Modified graphene NM can be used as an adsorbent for eliminating lead ions from water, the maximum adsorption capacity recorded for the removal of lead ions by graphene is 406.6 mg/g at a pH value of 5.0 in a duration of 40 min (Mousavi et al. 2018). Graphene-hydrogel nanocomposite modified with lingo sulfonate with oxygen-bearing functional moieties that make the highly polarizable surface was identified to enhance the speed of lead ion adsorption. The highest adsorption efficiency obtained was 1308 mg/g in duration of 40 min.

5.2.1.8 Activated Carbon (AC)

AC demonstrates promise as a NM for adsorption, boasting a significant surface area, ample porosity, and an easily achievable synthesis method utilizing various precursors. Its exceptional physicochemical properties make it widely applicable for removing pollutants originating from diverse sources such as pharmaceutical, agricultural, oil and petroleum, cosmetics, transportation, and textile industries. It is mainly employed for adsorbing air pollutants, recovering solvents, and purifying water, especially in the removal of dyes and other organics. It serves as a cost-effective carbon material for removing pollutants such as suspended particulate matter, biochemical oxygen demand, and chemical oxygen demand from water while maintaining the acidity for biochemical uses. In a study conducted by Maguana et al., they produced AC from pear seed cake and utilized it to remove methylene blue, achieving an adsorption capacity of 260 mg/g (Maguana et al. 2019). Similarly, Antonio et al. synthesized AC derived from the kenaf plant and successfully applied it for managing and removing medicinal contaminants from water. AC is commonly known as activated charcoal (Macías-García et al. 2019). This material is synthesized through specific procedures, resulting in high porosity and a surface area exceeding 3000 m^2/g. It is widely produced using materials such as timber, agricultural waste, and coal. Apart from its porous nature, this nanostructure also exhibits remarkable durability, making it suitable for diverse uses including catalyst supports, capacitors, batteries, hydrogen storage, and adsorption of organic contaminants. Thanks to its high tensile strength, AC can be recycled and reused. A study by Abeer et al. involved synthesizing AC from apricot stones and utilizing it to remove Zn^{2+} and Al^{3+} ions, achieving an adsorption efficiency of 92% (El-Saharty et al. 2018). Ibrahim and co-authors produced AC using sewage sludge, utilizing it to extract Cu^{2+} ions from water. They observed that the NM exhibited a removal efficiency exceeding 50% (El-Saharty et al. 2018). In another study, Li et al. synthesized AC derived from sludge and modified it with sulfur groups (Li et al. 2018). They used the sulfur-modified AC to remove lead, cadmium, nickel, and copper ions from polluted water. The adsorption capacities were determined to be 238.1, 96.2, 87.7, and 52.4 mg/g for lead, cadmium, copper, and nickel ions, respectively. Additionally, Cao et al. developed highly porous activated carbon from long-root Eichhornia crassipes and used it for extracting toxic metal ions, including zinc, lead, cadmium, copper, and nickel (Cao et al. 2019). Dong et al. investigated the use of spent AC for removing toxic metal ions from water, achieving maximum adsorption capacities of 95% for lead ions and 86% for cadmium ions (Dong et al. 2018). In another study, M. Bali and co-authors employed industrial AC for the removal of trace metals, achieving complete removal of cadmium ions within 15 min and attaining adsorption equilibrium for lead, zinc, and copper ions in 45 min (Bali and Tlili 2019). Cadmium ions exhibited the highest removal percentage. AC produced from Moso and Ma bamboo has demonstrated high efficacy in removing various metals, including lead, copper, chromium, and cadmium ions, with the highest adsorption capacity exceeding 90%. In another study, microwave-prepared AC derived from olive stones were utilized for removing metals, i.e., iron, zinc, copper, cadmium, and nickel from water, achieving an adsorption efficacy of over 98% (Alslaibi et al. 2013). AC is easy to produce, economically friendly, and stands out as a promising material for extracting organics. It can be synthesized using a wide range of carbon precursors, particularly agricultural waste. In addition to its straightforward synthesis method, AC can be easily functionalized.

5.2.2 Adsorption Pathways

The adsorption technique is broadly recognized as highly efficient for eliminating diverse pollutants from water, with preference given to adsorbents possessing high adsorption capabilities. Achieving optimal adsorption efficiency for contaminant removal requires a deeper understanding of the adsorption pathways between the adsorbent material and the adsorbed species. Mechanical interactions play a significant role in determining these pathways, with key interactions including water-repelling interactions, pi-stacking, coulombic forces, and hydrogen bonding. Different adsorbent materials bond with adsorbed species in various ways, depending on factors such as the composition and properties of contaminants and the functional groups present on the surface of the adsorbent material. For instance, when considering adsorption involving functional groups like carboxyl acid, hydroxy, amino, and carbonyl groups, hydrogen bonding plays a significant role. On the other hand, for adsorbent materials with electrically charged moieties, coulombic forces are predominant, while non-polar or hydrophobic forces typically dominate for adsorbents containing aliphatic compounds (Tong et al. 2019). Carbon-based NMs have shown significant effectiveness in removing pollutants from wastewater. CNTs demonstrate high removal efficacy for pollutants due to their large surface area and strong affinity for organic pollutants (Madhura et al. 2019). The adsorption pathways between CNTs and pollutants typically involve interactions such as hydrophobic forces, pi bonding, London dispersion forces, coulombic forces, and hydrogen bonding, which can occur independently or concurrently. Additionally, oxidative functionalization of CNTs can significantly enhance their adsorptive behavior for inorganic contaminants by introducing functional groups such as carboxyl acid, hydroxy, and phenolic groups onto the surface of the CNTs. In the case of unfunctionalized CNTs, adsorption primarily occurs at the surface of the adsorbent. However, for oxidized CNTs, adsorption occurs both at the material's surface and through the functionalities present on the surface (Mashkoor et al. 2020). These functional groups act as the main adsorption sites for many inorganic contaminants through coulombic forces and molecular bonding. When removing organic contaminants using graphene, adsorption is governed by the interaction between the graphene materials and the pollutants. Therefore, a higher surface area leads to a greater elimination rate of pollutants. Elimination typically occurs when acidic or basic solutions are introduced, depending on the properties and suitability of the materials. The predominant bonding between graphene and organic contaminants is pi-stacking. In the removal of inorganic contaminants, the presence of functional groups plays a crucial role in determining the removal efficiency. The functionalized graphene and graphene oxide have a high capability to adsorb inorganic contaminants compared to untreated graphene. The mechanism of adsorption between these graphene and inorganic pollutants typically involves coulombic forces, indicating that electrostatic interactions play a significant role in the adsorption process (Crini et al. 2019). The prevalence of iron is a main choice for researchers developing heterogeneous catalysts due to its widespread availability and cost-effectiveness. Iron's strong activity in the Fenton reaction, a process used in environmental remediation and organic synthesis, further reinforces its utility in catalysis. Iron catalysts are commonly used in Fenton reactions, where they act as catalysts by transitioning between Fe^{3+} and Fe^{2+} states. Some researchers have also suggested the presence of ferryl species (Fe^{4+}), particularly in alkaline environments.

5.3 Photocatalysts

Photocatalytic processes refer to chemical reactions where metals interact with photonic energy to catalyze the reactions. These NPs, often composed of metals possess powerful photocatalytic characteristics (Thongam and Chaturvedi 2021).

In the context of wastewater treatment, these photocatalysts are particularly useful for breaking down organic contaminants. These pollutants, which include pesticides, dyes, and organic compounds, are commonly found in wastewater and can be harmful to the surroundings and human well-being if not properly treated. By harnessing the energy from light, NP photocatalysts can facilitate the decomposition of these pollutants into harmless by-products, thereby aiding in the purification of wastewater. Semiconductor metals are versatile and can target a wide range of pollutants, making them valuable for environmental remediation efforts. These catalysts operate effectively even at low amounts, meaning that small amounts can have a significant impact on pollutant degradation. They can work under relatively mild conditions, which can be advantageous for practical applications and energy efficiency. Titanium oxide is extensively used in photocatalyst reactions because of its high chemical reactivity when exposed to UV) radiation with a wavelength shorter than 390 nm (Chen et al. 2020). Similar to titanium oxide, zinc oxide is also widely used for its photocatalytic properties. Various researches have demonstrated the efficacy of these synthesized catalysts in degrading pollutants and facilitating various photocatalytic processes. The efficacy of photocatalysts like zinc and titanium oxides depends on various parameters including:

1. Band gap energy: The energy difference between the valence band and the conduction band of the semiconductor material affects its ability to absorb light and initiate photocatalytic reactions.
2. Particle size: Smaller particle sizes generally result in higher surface area-to-volume ratios, which can enhance photocatalytic activity by providing more active sites for reactions to occur.
3. Dose: The amount or concentration of the photocatalyst used in the reaction can influence the rate and extent of photocatalytic degradation.
4. Pollutant concentration: The concentration of pollutants in the solution or environment being treated can impact the efficiency of the photocatalytic process, with higher pollutant concentrations potentially leading to reduced efficiency due to factors such as mass transfer limitations.
5. pH: The pH level of the solution can affect the surface charge and reactivity of the photocatalyst, thereby influencing its efficiency in degrading contaminants.

5.4 Heterogeneous Fenton-Like Reaction

The Fenton-like reaction is initiated when ferrous ions are substituted with ferric ions or other metals in the Fenton's solution. While heterogeneous Fenton-like reaction also occurs alongside homogeneous type, its effectiveness is constrained by the limited pH variation and tends to disperse rapidly when ideal situations are disrupted. A heterogeneous Fenton-like reaction offers a viable solution to the operational and financial limitations linked with a Fenton reaction that is homogeneous type, like the large chemical usage, catalyst depletion, and extensive sludge accumulation. The Fe^{3+} ion is primarily utilized in the insoluble form, either by utilizing indigenous minerals like γ-Fe_2O_3, Fe_3O_4, FeS_2, and α-Fe_2O_3 or by embedding it onto appropriate support materials in heterogeneous Fenton reaction to increase the specific surface area. The heterogeneous Fenton reaction differs significantly from the homogeneous type because adsorption primarily determines the efficacy of the reaction (Jain et al. 2018). The heterogeneous Fenton reaction occurs in 3 stages: the adsorption of organic materials onto the exterior surface of the catalyst, the on-site production and attack of hydroxyl radicals on organic substances, and finally, the release of oxidation compounds as shown in Fig. 5.3 (Zhu et al. 2019). To elucidate the impelling factor behind adsorption in heterogeneous Fenton-like reactions, the Langmuir adsorption model can be employed. This model indicates that the adsorption efficiency primarily relies on the single-layer adsorption capability of the catalyst.

Consequently, a larger area of the catalyst is open in the water medium, leading to increased adsorption. While multilayer adsorption may also occur, it is typically observed only when organic substances exist in very large amounts. When the organic substances are adsorbed onto the catalyst, hydroxyl radicals are produced via reactive regions on the catalyst and commence oxidative breakdown of the organic substances. The oxidized compounds either go through subsequent oxidation or release from the catalyst, thereby concluding the catalysis process. The heterogeneous Fenton reaction was initially created using Fe^{3+} ions to address the drawbacks of the homogeneous type, aiming to reduce issues like excessive sludge production and high chemical usage, which in turn minimizes prices but a significant challenge remains in optimizing and regulating pH, as low pH environment is preferred for large efficacy but can cause metal ions to leach from the active site of the catalyst. This leaching reduces catalyst performance, making the method less ecologically viable. Consequently, scientists are exploring different metals like Fe, Ce, Cu, Mn, and Ag to create heterogeneous catalysts with improved durability while maintaining adequate activity levels. Iron is unique in that it forms strong complexes with the decomposition by-products generated during the oxidation process. Copper also shows a

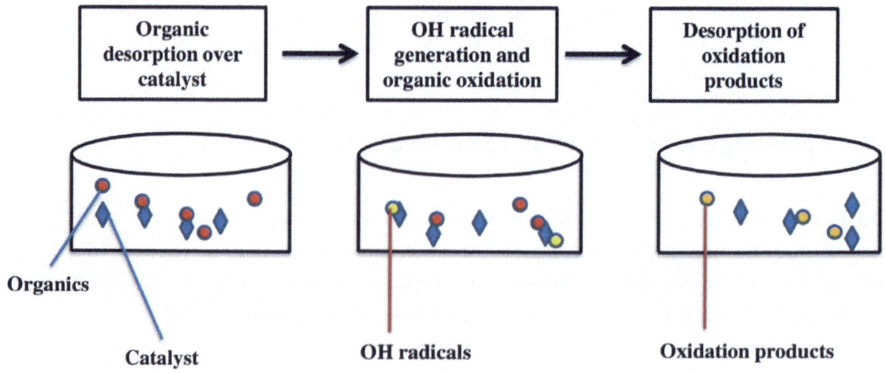

Fig. 5.3 Heterogeneous Fenton process

similar oxidation-reduction process to iron. However, unlike iron, copper generates transient complexes with the decomposition by-products. Silver is being considered for its distinctive electron transfer cycle containing both metallic silver and silver ions. Cerium is of interest because it forms complexes with the oxidizing agents in the Fenton process (Kurian 2020). Manganese is of interest because it has two distinct electron transfer cycles contingent on the characteristics of manganese substances.

5.5 Functions of NMs in Oxidizing Contaminants

The utilization of nano-catalysts in oxidation for wastewater treatment offers several advantages compared to conventional methods (Zhang et al. 2022). Catalysts, particularly those derived from metals like platinum, gold, and palladium, have demonstrated great catalytic capability for degrading different pollutants. These catalysts can effectively target and break down recalcitrant substances that are hard to manage with traditional approaches. The catalytic activity of NMs significantly accelerates the oxidation reactions, leading to shorter treatment times. This rapid degradation of pollutants reduces the overall processing time required for wastewater treatment, thereby increasing operational efficiency. Chemical oxidation facilitated by nano-catalysts can transform organic pollutants into less harmful or even useful by-products (Lu and Astruc 2020). This aspect is particularly beneficial as it not only aids in pollutant removal but also promotes resource recovery and potential reuse of treated wastewater or its by-products. While NMs offer numerous benefits for wastewater treatment, such as high catalytic activity and selectivity, they also present challenges, including the elevated price of noble metals like platinum and the problem of recovering and reusing the catalysts (Hodges et al. 2018). To address these difficulties and make catalytic processes more cost-effective, recent research has focused on improving the reactivity and sustainability of noble metals by creating mixed-metal alloys by mixing them with other metallic elements. The strategy of forming bimetallic alloys involves combining platinum, gold, and palladium with less expensive or more abundant metals to enhance catalytic performance while reducing costs. By leveraging the synergistic effects between different metal components, bimetallic nanoparticles can show increased catalytic performance, specificity, and robustness in contrast to their monometallic counterparts.

5.6 Nano-Membranes

The membrane filtration technique integrated with nano-catalysts is identified as one of the most efficient approaches among modern wastewater purification methods (Obotey Ezugbe and Rathilal 2020). It surpasses traditional methods by incorporating nanotechnology concepts to achieve superior performance in terms of water quality improvement, disinfection efficiency, and compact design of treatment plants.

NMs integrated into membrane filtration systems offer new functionalities such as catalytic activity, high porosity, and resistance to fouling. These attributes increase the efficacy of the treatment method (Bera et al. 2022). Additionally, it is noted for its economic feasibility, efficiency, and simplicity in contrast to alternative management technologies. This technique is particularly efficient in removing various contaminants from wastewater, including dyes; trace metals, and other pollutants. The integration of NMs into novel membrane designs plays an important role not only in physically separating particles but also in chemically decomposing contaminants, thereby enhancing the overall separation efficiency. A variety of membranes are employed for filtration and elimination of NPs. These membranes primarily consist of 1-D materials, which include nanofibers, nanotubes, and nanoribbons. One such membrane, synthesized with carbon nanofibers, has demonstrated exceptional elimination capability at high pressures (Chadha et al. 2022). Additionally, incorporating beta-cyclodextrins into carbon fibers via a straightforward separation method shows promise for the elimination of fuchsin acid and phenolphthalein. Osmotic filtration refers to a process where a membrane is utilized to separate solutes from a solution based on their osmotic pressure gradients. Nanoscale zeolite membranes can be used for this purpose. Zeolites are crystalline aluminosilicate minerals with a porous structure, allowing for the selective separation of molecules based on their size and charge. Common zeolites employed in osmotic filtration include MFI-type, sodalite, and Linde Type-A (Alsayed and Ashraf 2021).

5.6.1 *Membrane Fouling and Tailoring*

Membranes are widely available in the market and are used in various uses. However, the increasing demand for producing novel water sources from polluted water necessitates membranes with enhanced efficiency and resistance to fouling at a cheap price (Dharupaneedi et al. 2019). Fouling of membranes occurs when contaminants in polluted water interact with a water-repellent film, leading to the settling of particles either on the exterior surface or within the membrane. This fouling phenomenon results in the production of impure treated water and diminishes the dependability of filtration systems, thereby constraining further advancement in this field. Additionally, under low-pressure conditions, the flow rate of nano-membranes is significantly reduced. To mitigate the reduction in flow rate due to membrane fouling, several approaches can be employed. These include chemical cleaning which involves using specific chemicals to dissolve or dislodge fouling substances from the membrane surface or pores. Chemical cleaning can effectively remove organic and inorganic deposits, restoring the membrane's performance. Mechanical methods, such as physical scrubbing or backwashing, can help dislodge and remove fouling materials from the membrane surface (Abdel-Karim et al. 2021). These techniques are particularly useful for removing larger particles or debris. Membrane replacement is used in severe cases of fouling where cleaning is ineffective, or if the membrane is damaged beyond repair, complete replacement of the membrane may be necessary.

This ensures the restoration of optimal filtration performance. The development of various technologies over the previous years is to minimize membrane fouling. One such approach is membrane alteration, which involves altering the surface properties of the membrane to make it more resistant to fouling (Saffarimiandoab et al. 2021). This can include changes in surface chemistry, morphology, or the incorporation of fouling-resistant coatings. Overall, these techniques aim to prolong membrane lifespan, maintain high flux rates, and improve the reliability of membrane filtration systems. The primary focus has been on modifying membranes by applying a water-loving polymer coating like chitosan and polyvinyl alcohol but there are some limitations associated with these techniques. The process of coating membranes with hydrophilic polymers can be expensive, particularly if high-quality materials are used. This cost factor may limit the widespread adoption of these modification methods. This process can be complex and labor-intensive. It may require specialized equipment and expertise, adding to the overall complexity of the membrane modification process. Some of the methods may involve the use of chemicals or solvents that can potentially generate pollutants or waste products. This can have environmental implications and may not align with sustainability goals. The research into the synthesis of composite membranes incorporating carbonaceous nanotubes as additives has yielded promising results. Specifically, the resulting carbon nanotube-mixed matrix membrane has demonstrated significant capability for achieving large flux rates.

5.7 Conclusion

Lately, there has been a pressing demand for novel wastewater techniques to guarantee clean water and eradicate various contaminants. In this context, nanoscience emerges as one of the optimal solutions for advancing effluent management practices. Numerous nano-catalysts have been efficiently innovated and implemented for the purification of polluted water. Nanotechnology offers several advantages and benefits for wastewater treatment. Nanotechnology enables the development of highly efficient treatment processes that can effectively remove various pollutants, including chemical and biological contaminants, from wastewater. NMs have distinctive characteristics like large surface area and reactivity, which enhance the efficiency of wastewater treatment processes by facilitating the adsorption, degradation, or filtration of pollutants. Nanotechnology can be applied across a wide range of wastewater treatment applications, including drinking water treatment, industrial wastewater treatment, and wastewater reuse, making it a versatile and adaptable solution. Nanotechnology-based wastewater treatment processes often require fewer resources, such as chemicals and energy, compared to traditional treatment methods, leading to cost savings and environmental benefits. Nanotechnology enables the development of innovative management techniques, such as adsorption, oxidation, photocatalysis, and nano-enabled membranes, which offer novel approaches to addressing complex wastewater treatment challenges.

References

Abdel-Karim A, Leaper S, Skuse C, Zaragoza G, Gryta M, Gorgojo P (2021) Membrane cleaning and pretreatments in membrane distillation–a review. Chem Eng J 422:129696. https://doi.org/10.1016/j.cej.2021.129696

Ahmed S, Kumar A (2024) Research progress in synthesis strategies of magnesium oxide NPs for water treatment applications. Environ Sci Water Res Technol. https://doi.org/10.1039/D3EW00528C

Alayan HM, Aljumaily MM, Alsaadi MA, Mjalli FS, Hashim MA (2021) A review exploring the adsorptive removal of organic micropollutants on tailored hierarchical carbon nanotubes. Toxicol Environ Chem 103(4):282–325. https://doi.org/10.1080/02772248.2021.1932888

Alijani H, Shariatinia Z (2018) Synthesis of high growth rate SWCNTs and their magnetite cobalt sulfide nanohybrid as super-adsorbent for mercury removal. Chem Eng Res des 129:132–149. https://doi.org/10.1016/j.cherd.2017.11.014

Alsayed AF, Ashraf MA (2021) Modified nanofiltration membrane treatment of saline water. In: Water engineering modeling and mathematic tools. Elsevier, pp 25–44. https://doi.org/10.5004/dwt.2020.25302

Alslaibi TM, Abustan I, Ahmad MA, Foul AA (2013) Application of response surface methodology (RSM) for optimization of Cu^{2+}, Cd^{2+}, Ni^{2+}, Pb^{2+}, Fe^{2+}, and Zn^{2+} removal from aqueous solution using microwaved olive stone activated carbon. J Chem Technol Biotechnol 88(12):2141–2151. https://doi.org/10.1002/jctb.4073

Anitha K, Namsani S, Singh JK (2015) Removal of heavy metal ions using a functionalized single-walled carbon nanotube: a molecular dynamics study. J Phys Chem A 119(30):8349–8358. https://doi.org/10.1021/acs.jpca.5b03352

Baby R, Saifullah B, Hussein MZ (2019) Carbon NMs for the treatment of heavy metal-contaminated water and environmental remediation. Nanoscale Res Lett 14:1–17. https://doi.org/10.1186/s11671-019-3167-8

Baby Shaikh R, Saifullah B, Rehman FU (2018) Greener method for the removal of toxic metal ions from the wastewater by application of agricultural waste as an adsorbent. Water 10(10):1316. https://doi.org/10.3390/w10101316

Bali M, Tlili H (2019) Removal of heavy metals from wastewater using infiltration-percolation process and adsorption on activated carbon. Int J Environ Sci Technol 16(1):249–258. https://doi.org/10.1007/s13762-018-1663-5

Bera SP, Godhaniya M, Kothari C (2022) Emerging and advanced membrane technology for wastewater treatment: a review. J Basic Microbiol 62(3–4):245–259. https://doi.org/10.1002/jobm.202100259

Bharti JJS, Kumar SS, Kumar V, Kumar A, Kumar D (2022) A review on the capability of zinc oxide and iron oxides NMs, as a water decontaminating agent: adsorption and photocatalysis. Appl Water Sci 12(3):46. https://doi.org/10.1007/s13201-021-01566-3

Cao F, Lian C, Yu J, Yang H, Lin S (2019) Study on the adsorption performance and competitive mechanism for heavy metal contaminants removal using novel multi-pore activated carbons derived from recyclable long-root Eichhornia crassipes. Biores Technol 276:211–218. https://doi.org/10.1016/j.biortech.2019.01.007

Chadha U, Selvaraj SK, Thanu SV, Cholapadath V, Abraham AM, Manoharan M, Paramsivam V (2022) A review of the function of using carbon NMs in membrane filtration for contaminant removal from wastewater. Mater Res Express 9(1):012003. https://doi.org/10.1088/2053-1591/ac48b8

Chen D, Cheng Y, Zhou N, Chen P, Wang Y, Li K, Huo S, Cheng P, Peng P, Zhang R, Wang L (2020) Photocatalytic degradation of organic pollutants using TiO_2-based photocatalysts: a review. J Clean Prod 268:121725. https://doi.org/10.1016/j.jclepro.2020.121725

Clancy AJ, Bayazit MK, Hodge SA, Skipper NT, Howard CA, Shaffer MS (2018) Charged carbon NMs: redox chemistries of fullerenes, carbon nanotubes, and graphenes. Chem Rev 118(16):7363–7408. https://doi.org/10.1021/acs.chemrev.8b00128

Crini G, Lichtfouse E, Wilson LD, Morin-Crini N (2019) Conventional and non-conventional adsorbents for wastewater treatment. Environ Chem Lett 17:195–213. https://doi.org/10.1007/s10311-018-0786-8

Das R, Leo BF, Murphy F (2018) The toxic truth about carbon nanotubes in water purification: a perspective view. Nanoscale Res Lett 13:1–10. https://doi.org/10.1186/s11671-018-2589-z

Dehghani MH, Taher MM, Bajpai AK, Heibati B, Tyagi I, Asif M, Agarwal S, Gupta VK (2015) Removal of noxious Cr (VI) ions using single-walled carbon nanotubes and multi-walled carbon nanotubes. Chem Eng J 279:344–352. https://doi.org/10.5772/intechopen.84225

Dharupaneedi SP, Nataraj SK, Nadagouda M, Reddy KR, Shukla SS, Aminabhavi TM (2019) Membrane-based separation of potential emerging pollutants. Sep Purif Technol 210:850–866. https://doi.org/10.1016/j.seppur.2018.09.003

Dhillon A, Kumar D (2019) New generation nano-based adsorbents for water purification. In: Nanoscale materials in water purification. Elsevier, pp 783–798. https://doi.org/10.1016/B978-0-12-813926-4.00036-7

Dong L, La H, Wang Z, Gu P, Chen G, Jiang R (2018) A new function of spent activated carbon in BAC process: removing heavy metals by ion exchange mechanism. J Hazard Mater 359:76–84. https://doi.org/10.1016/j.jhazmat.2018.07.030

El Maguana Y, Elhadiri N, Bouchdoug M, Benchanaa M, Jaouad A (2019) Activated carbon from prickly pear seed cake: optimization of preparation conditions using experimental design and its application in dye removal. Int J Chem Eng. https://doi.org/10.1155/2019/8621951

El-Saharty A, Mahmoud SN, Manjood AH, Nassar AAH, Ahmed AM (2018) Effect of apricot stone activated carbon adsorbent on the removal of toxic heavy metals ions from aqueous solutions. Int J Ecotoxicol Ecobiol 3(2):51. https://doi.org/10.11648/j.ijee.20180302.13

Elsehly E, Chechenin N, Bukunov K, Makunin A, Priselkova A, Vorobyeva E, Motaweh H (2016) Removal of iron and manganese from aqueous solutions using carbon nanotube filters. Water Sci Technol Water Supply 16(2):347–353. https://doi.org/10.2166/ws.2015.143

Guo T, Bulin C, Li B, Zhao Z, Yu H, Sun H, Ge X, Xing R, Zhang B (2018) Efficient removal of aqueous Pb (II) using partially reduced graphene oxide-Fe_3O_4. Adsorpt Sci Technol 36(3–4):1031–1048. https://doi.org/10.1177/0263617417744402

Gupta S, Bhatiya D, Murthy C (2015) Metal removal studies by composite membrane of polysulfone and functionalized single-walled carbon nanotubes. Sep Sci Technol 50(3):421–429. https://doi.org/10.1080/01496395.2014.973516

Gupta K, Joshi P, Gusain R, Khatri OP (2021) Recent advances in adsorptive removal of heavy metal and metalloid ions by metal oxide-based NMs. Coord Chem Rev 445:214100. https://doi.org/10.1016/j.ccr.2021.214100

Gusain R, Kumar N, Ray SS (2020) Recent advances in carbon NM-based adsorbents for water purification. Coord Chem Rev 15(405):213111. https://doi.org/10.1016/j.ccr.2019.213111

Hodges BC, Cates EL, Kim JH (2018) Challenges and prospects of advanced oxidation water treatment processes using catalytic NMs. Nat Nanotechnol 13(8):642–650. https://doi.org/10.1038/s41565-018-0216-x

Husnain SM, Asim U, Yaqub A, Shahzad F, Abbas N (2020) Recent trends of MnO_2-derived adsorbents for water treatment: a review. New J Chem 44(16):6096–6120. https://doi.org/10.1039/C9NJ06392G

Iftekhar S, Ramasamy DL, Srivastava V, Asif MB, Sillaää M (2018) Understanding the factors affecting the adsorption of Lanthanum using different adsorbents: a critical review. Chemosphere 204:413–430. https://doi.org/10.1016/j.chemosphere.2018.04.053

Jain B, Singh AK, Kim H, Lichtfouse E, Sharma VK (2018) Treatment of organic pollutants by homogeneous and heterogeneous Fenton reaction processes. Environ Chem Lett 16:947–967. https://doi.org/10.1007/s10311-018-0738-3

Jilani A, Othman MHD, Ansari MO, Hussain SZ, Ismail AF, Khan IU, Inamuddin (2018) Graphene and its derivatives: synthesis, modifications, and applications in wastewater treatment. Environ Chem Lett 16:1301–1323. https://doi.org/10.1007/s10311-018-0755-2

Kurian M (2020) Cerium oxide based materials for water treatment–a review. J Environ Chem Eng 8(5):104439. https://doi.org/10.1016/j.jece.2020.104439

Latif A, Sheng D, Sun K, Si Y, Azeem M, Abbas A, Bilal M (2020) Remediation of heavy metals polluted environment using Fe-based NPs: mechanisms, influencing factors, and environmental implications. Environ Pollut 264:114728. https://doi.org/10.1016/j.envpol.2020.114728

Li J, Xing X, Li J, Shi M, Lin A, Xu C, Zheng J, Li R (2018) Preparation of thiol-functionalized activated carbon from sewage sludge with coal blending for heavy metal removal from contaminated water. Environ Pollut 234:677–683. https://doi.org/10.1016/j.envpol.2017.11.102

Lu F, Astruc D (2020) Nanocatalysts and other NMs for water remediation from organic pollutants. Coord Chem Rev 408:213180. https://doi.org/10.1016/j.ccr.2020.213180

Macías-García A, García-Sanz-Calcedo J, Carrasco-Amador JP, Segura-Cruz R (2019) Adsorption of paracetamol in hospital wastewater through activated carbon filters. Sustainability 11(9):2672. https://doi.org/10.3390/su11092672

Madhura L, Singh S, Kanchi S, Sabela M, Bisetty K, Inamuddin (2019) Nanotechnology-based water quality management for wastewater treatment. Environ Chem Lett 17:65–121. https://doi.org/10.1007/s10311-018-0778-8

Madima N, Mishra S, Inamuddin I, Mishra A (2020) Carbon-based NMs for remediation of organic and inorganic pollutants from wastewater. A review. Environ Chem Lett 18:1169–1191. https://doi.org/10.1007/s10311-020-01001-0

Maduraiveeran G, Jin W (2021) Carbon NMs: synthesis, properties and applications in electrochemical sensors and energy conversion systems. Mater Sci Eng B 272:115341. https://doi.org/10.1016/j.mseb.2021.115341

Mashkoor F, Nasar A, Inamuddin (2020) Carbon nanotube-based adsorbents for the removal of dyes from waters: a review. Environ Chem Lett 18:605–629. https://doi.org/10.1007/s10311-020-00970-6

Mousavi SM, Hashemi SA, Amani AM, Esmaeili H, Ghasemi Y, Babapoor A, Mojoudi F, Arjomand O (2018) Pb (II) removal from synthetic wastewater using Kombucha Scoby and graphene oxide/Fe_3O_4. Phys Chem Res 6(4):759–771. https://doi.org/10.22036/PCR.2018.133392.1490

Nimibofa A, Newton EA, Cyprain AY, Donbebe W (2018) Fullerenes: synthesis and applications. J Mater Sci 7:22–33. https://doi.org/10.5539/jmsr.v7n3p22

Obotey Ezugbe E, Rathilal S (2020) Membrane technologies in wastewater treatment: a review. Membranes 10(5):89. https://doi.org/10.3390/membranes10050089

Ouni L, Ramazani A, Taghavi Fardood S (2019) An overview of carbon nanotubes role in heavy metals removal from wastewater. Front Chem Sci Eng 13:274–295. https://doi.org/10.1007/s11705-018-1765-0

Park C M, Wang D, Su C (2018) Recent developments in engineered NMs for water treatment and environmental remediation. In: Handbook of NMs for industrial applications, pp 849–882. https://doi.org/10.1016/B978-0-12-813351-4.00048-1

Saffarimiandoab F, Gul BY, Tasdemir RS, Ilter SE, Unal S, Tunaboylu B, Menceloglu YZ, Koyuncu İ (2021) A review on membrane fouling: membrane modification. Desalin Water Treat 216:47–70. https://doi.org/10.5004/dwt.2021.26815

Sarkar B, Mandal S, Tsang YF, Kumar P, Kim K-H, Ok YS (2018) Designer carbon nanotubes for contaminant removal in water and wastewater: a critical review. Sci Total Environ 612:561–581. https://doi.org/10.1016/j.scitotenv.2017.08.132

Sharma RK, Solanki K, Dixit R, Sharma S, Dutta S (2021) Nanoengineered iron oxide-based sorbents for separation of various water pollutants: current status, opportunities and future outlook. Environ Sci Water Res Technol 7(5):818–860. https://doi.org/10.1039/D1EW00108F

Sweetman MJ, May S, Mebberson N, Pendleton P, Vasilev K, Plush SE, Hayball JD (2017) Activated carbon, carbon nanotubes and graphene: materials and composites for advanced water purification. C 3(2):18. https://doi.org/10.3390/c3020018

Tabish TA, Memon FA, Gomez DE, Horsell DW, Zhang S (2018) A facile synthesis of porous graphene for efficient water and wastewater treatment. Sci Rep 8(1):1817. https://doi.org/10.1038/s41598-018-19978-8

Thines R, Mubarak N, Nizamuddin S, Sahu J, Abdullah E, Ganesan P (2017) Application potential of carbon NMs in water and wastewater treatment: a review. J Taiwan Inst Chem Eng 72:116–133. https://doi.org/10.1016/j.jtice.2017.01.018

Thirunavukkarasu A, Nithya R, Sivashankar R (2020) A review on the role of NMs in the removal of organic pollutants from wastewater. Rev Environ Sci Bio Technol 19:751–778. https://doi.org/10.1007/s11157-020-09548-8

Thongam DD, Chaturvedi H (2021) Advances in NMs for heterogeneous photocatalysis. Nano Express 2(1):012005. https://doi.org/10.1088/2632-959X/abeb8d

Tong Y, McNamara PJ, Mayer BK (2019) Adsorption of organic micropollutants onto biochar: a review of relevant kinetics, mechanisms and equilibrium. Environ Sci Water Res Technol 5(5):821–838. https://doi.org/10.1039/C8EW00938D

Wang S, Sun H, Ang H-M, Tadé MO (2013) Adsorptive remediation of environmental pollutants using novel graphene-based NMs. Chem Eng J 226:336–347. https://doi.org/10.1016/j.cej.2013.04.070

Wu Y, Pang H, Liu Y, Wang X, Yu S, Fu D, Chen J, Wang X (2019) Environmental remediation of heavy metal ions by novel-NMs: a review. Environ Pollut 246:608–620. https://doi.org/10.1016/j.envpol.2018.12.076

Xu L, Wang J (2017) The application of graphene-based materials for the removal of heavy metals and radionuclides from water and wastewater. Crit Rev Environ Sci Technol 47(12):1042–1105. https://doi.org/10.1080/10643389.2017.1342514

Yang X, Zhou T, Ren B, Hursthouse A, Zhang Y (2018) Removal of Mn (II) by sodium alginate/graphene oxide composite double-network hydrogel beads from aqueous solutions. Sci Rep 8(1):10717. https://doi.org/10.1038/s41598-018-29133-y

Zazouli MA, Yousefi Z, Yazdani Cherati J, Tabarinia H, Tabarinia F, Akbari Adergani B (2014) Evaluation of L-Cysteine functionalized single-walled carbon nanotubes on mercury removal from aqueous solutions. J Maz Univ Med Sci 24(111):10–21. http://jmums.mazums.ac.ir/article-1-3452-en.html

Zhang X, Huang Q, Deng F, Huang H, Wan Q, Liu M, Wei Y (2017) Mussel-inspired fabrication of functional materials and their environmental applications: progress and prospects. Appl Mater Today 7:222–238. https://doi.org/10.1016/j.apmt.2017.04.001

Zhang C-Z, Chen B, Bai Y, Xie J (2018) A new functionalized reduced graphene oxide adsorbent for removing heavy metal ions in water via coordination and ion exchange. Sep Sci Technol 53(18):2896–2905. https://doi.org/10.1080/01496395.2018.1497655

Zhang YJ, Huang GX, Winter LR, Chen JJ, Tian L, Mei SC, Zhang Z, Chen F, Guo ZY, Ji R, You YZ (2022) Simultaneous nanocatalytic surface activation of pollutants and oxidants for highly efficient water decontamination. Nat Commun 13(1):3005. https://doi.org/10.1038/s41467-022-30560-9

Zheng S, Hao L, Zhang L, Wang K, Zheng W, Wang X, Zhou X, Li W, Zhang L (2018) Tea polyphenols functionalized and reduced graphene oxide-ZnO composites for selective Pb^{2+} removal and enhanced antibacterial activity. J Biomed Nanotechnol 14(7):1263–1276. https://doi.org/10.1166/jbn.2018.2584

Zhu Y, Zhu R, Xi Y, Zhu J, Zhu G, He H (2019) Strategies for enhancing the heterogeneous Fenton catalytic reactivity: a review. Appl Catal B 15(255):117739. https://doi.org/10.1016/j.apcatb.2019.05.041

Chapter 6
Photocatalytic Application of Nanomaterials in Wastewater Research

Ankita Yadav, Anjali Kajal, Tanishika Narang, and Geeta Dhania

Abstract Urbanization and technical advancements have given rise to large wastewater-generating industries like biotechnology, petrochemical, agriculture, and pharmaceuticals. This effluent is difficult to treat because of its diverse composition, which includes organic compounds, colors, pharmaceutical waste, residual contaminants, and hazardous substances. Conventional wastewater treatment methods, including ion exchange, adsorbents, reverse osmosis, and electrostatic precipitation, have the disadvantages of being expensive, having a small recycling capacity, and being inefficient. Photocatalytic techniques have shown promising effect in treating wastewater, especially when contaminated with hazardous dyes. These dyes are highly dangerous to the environment and public health. Researchers have created environmental friendly techniques, like photocatalytic degradation of metal-based nanomaterials (NMs), to effectively and efficiently break down these harmful pollutants. With their strong reactivity, functionalization, and size-dependent qualities, straight band gap, high optical absorption coefficient, low cost, and low toxicity, NMs such as metal nanoparticles (NPs), oxides, carbon compounds, zeolite, and filtration membranes are perfect for treating wastewater. The efficacy of photocatalytic NPs can be increased by coupling these NPs with appropriate species, which may help to avert further water shortages. Research in this area can help address the limitations of photocatalytic NPs in wastewater treatment. The several photocatalytic NM metal-based and metal-free that have been investigated for the treatment and purification of water and wastewater are discussed in this review. This review opens up new directions for the study of water treatment by offering insightful information on how solar energy is transformed for photocatalytic reduction of hazardous metal ions and degradation of dye-contaminated wastewater.

Keywords Wastewater · Photocatalytic · Nanomaterials · Nanoparticles · Industries · Dye removal

---

A. Yadav · A. Kajal · T. Narang · G. Dhania (✉)
Department of Environmental Science, Maharshi Dayanand University, Rohtak, India
e-mail: geetadhania@rediffmail.com

N. Agarwal et al. (eds.), *Nanomaterials in Wastewater Research*, Advances in Wastewater Research, https://doi.org/10.1007/978-981-96-3908-3_6

6.1 Introduction

All living beings depend on water, although just 3% of it is usable due to the majority of it being frozen in polar ice caps and glaciers (Baker et al. 2016). A healthy economy and the advancement of civilization depend on clean water. However, huge wastewater has been generated as a result of industrialization, urbanization, population growth, and agricultural activity, tainting and polluting the water (Pendergast and Hoek 2011). Water shortage is a significant environmental problem, and developing countries' fast population expansion has increased demand for clean water. With only 80% of contaminated water being treated before being recycled back into the environment, the issue is growing increasingly concerning. The fact that 1.8 billion people depend on tainted water that is rife with diseases and germs emphasizes how urgently sustainable water management is needed (Khan and Malik 2019).

To remove organic contaminants and heavy metals from wastewater, a variety of treatment procedures have been developed. These techniques include filtration, flocculation, coagulation, biological treatment, absorption/adsorption, sedimentation, physical and chemical treatment, and membrane processes. These techniques do have certain drawbacks like poor adsorption capacity, expensive regeneration, and column fouling (Fu and Wang 2011; Lieu et al. 2016). Nanotechnology has become a cutting-edge solution for wastewater treatment and water scarcity. It provides multipurpose and incredibly efficient solutions that enable the economic exploitation of non-traditional water sources to expand the available water supply (Bhat et al. 2022). The special qualities of nanostructured materials may offer long-term fixes for issues with wastewater treatment. Additionally, nanotechnology minimizes labor, time, and investment requirements in the wastewater treatment business while solving environmental challenges. Nanotechnology-enabled water and wastewater treatment provides a possible answer to current problems while also expanding the water supply (Qu et al. 2013). Photocatalysis is a highly efficient and affordable process that transforms photonic energy into chemical energy and is being explored as an alternative to chemically based AOPs. In this procedure, photocatalysts like sunshine, UV light, and visible light are employed. Photocatalysis has the potential to be an efficient, non-selective, and successful method for treating wastewater (Jorfi et al. 2016). This process can help mineralize complicated pollutants into less complex materials like water, carbon dioxide, and inorganic ions while also eliminating a variety of pollutants (Mishra and Sundaram 2022). Wastewater treatment techniques are frequently limited by the high expenses involved in these procedures, the requirement for sophisticated equipment, and the absorption of UV light. These techniques exacerbate the energy dilemma by requiring additional chemicals as oxidants and energy (Sewnet et al. 2022). Furthermore, these techniques frequently produce hazardous waste products and fail to entirely remove contaminants. The desire to overcome these obstacles and enhance water remediation techniques is what is driving the development and advancement of nanotechnology in wastewater treatment (Ahmad et al. 2021). This chapter emphasizes how important photocatalysts are for clearing organic contaminants and lowering harmful concentrations of heavy metal ions from aquatic ecosystems. It addresses the benefits of nanoscale photocatalysts over conventional ones, describes the process and factors impacting degradation, and goes over different synthesis methods. Future developments and relevant viewpoints are also covered.

6.2 NMs/Photocatalyst

NMs that have at least one dimension smaller than around 100 nm are referred to be nanoscale materials. The diameter of a human hair is 100,000 times larger than a nanometer, which is one-millionth of a millimeter. Because special optical, magnetic, electrical, and other properties appear at this scale; NMs are of great interest (Alagarasi 2013). NPs are minuscule particles with distinct adsorptive and catalytic properties, measuring between 1 and 100 nm. These NPs can be distinguished from two-dimensional NMs and one-dimensional NMs having two dimensions larger than the nanoscale. NPs are divided into a number of classes, such as metal NPs, ceramic NPs, polymeric NPs, and fullerenes. These NPs possess distinct chemical and physical characteristics, as well as a large surface area despite their small size.

NMs are small, very effective particles that enhance chemical stability, nonlinear optical performance, thermal conductivity, and catalytic reactivity. They also have significant surface area-to-volume ratios (Agarwal et al. 2017). Both organic and inorganic NPs are possible; semiconductor and noble metal NPs are examples of inorganic NPs, while carbons NPs (fullerenes) are examples of organic NPs (Kalpana and Rajeswari 2017).

6.2.1 Metal and Metal Oxide-Based NMs/Photocatalyst

Metallic NPs have enormous potential for various forms of environmental cleanup because of their adaptability. One, two, or three metals and their oxides make up a variety of classes of NMs. Through physical and chemical techniques, the characteristics of metal-based NMs, such as shape control, stability, and monodispersing has been investigated. It is feasible to use these beneficial properties for water treatments such as chemical disinfection, membrane separation, adsorption, and photodegradation. With their 0D nanostructures, TiO_2 (Titanium dioxide), (Ag) silver, and (ZnO) zinc oxide are the most well-studied metals and metal-oxides in relation to energy and environmental concerns (Ahmad et al. 2020a). Because of their potent antibacterial, antifungal, and antiviral qualities, these materials are used as water disinfectants and in water treatment procedures, self-cleaning surfaces, and water filtration systems (Shao et al. 2017). Moreover, in a single wastewater treatment process, these oxides and their hybrids can simultaneously degrade organic molecules and kill various species (Deshmukh et al. 2019).

6.2.2 Carbon-Based NMs/Photocatalyst

NPs made of carbon due to their intriguing characteristics have excellent stability (chemical and mechanical), cheaper, and integrated operation. Most carbon-based NMs have high specific surface area, area functionalization, and chemical stability because of their distinctive structural characteristics. A wide range of carbon-based NMs have demonstrated significant promise for water filtration, including graphene oxide (Choi et al. 2013), SWCNT (Gao et al. 2014), and MWCNT (Azelee et al. 2017). These nanotubes are being utilized more often as long-lasting holes in water treatment membranes (Baby et al. 2019).

6.2.3 Organic NMs/Photocatalyst

In hybrid coordination networks, which are three-dimensional linked structures with substantial pore volume ratios that facilitate effective diffusion paths for guest species, the development of metal-organic frameworks (MOFs) has advanced significantly (Nakata and Fujishima 2012). Water that has been polluted can be separated and purified using MOF. Numerous techniques have been utilized to alter the MOF, either before to or following synthesis, in order to overcome its constraints, which include restricted activity and structural stability (Yu et al. 2020).

6.2.4 Hybrid NMs/Photocatalyst

Applications for hybrid NMs in polymer nanocomposites, health, and the environment are numerous. To make a new composite material, they can be mixed with other materials (Barsan and Brett 2016). Improved removal efficiency and adsorption capacity have been demonstrated by hybrid NMs (Mamah et al. 2020). In order to degrade organic dyes and industrial effluents, Gade et al. produced layer-type perovskite materials using ion exchange and studied those using spectroscopic techniques. These substances provide encouraging adsorption outcomes (Gade et al. 2018).

6.3 Conventional Method for Wastewater Treatment

Various physical, chemical, and biological treatment processes are applied to industrial effluents before they are released into a body of water (Fig. 6.1). Examples of traditional treatment methods include membrane separation, chemical precipitation, oxidation/reduction, and biological therapy. Many methods based on adsorption, physiochemical treatment, and electrochemical treatments have been used by researchers for many years to remove heavy metals from industrial effluents (Razzak et al. 2022).

6.3.1 Electrochemical Processes

An electrochemical process is a chemical reaction in which an electrical current pass through an aqueous metal-carrying solution when a voltage is applied between a cathode plate and an insoluble anode. This process precipitates heavy metal hydroxides in an acidic or neutralized electrolyte (Tran et al. 2017).

6.3.2 Electro Flotation and Electrocoagulation

A simple procedure called electro flotation causes particles floating in a fluid to rise above by encouraging them to stick to microscopic H_2 and O_2 bubbles that form on the flotation cell's cathode and anode. One important factor in determining a chemical species' capacity to float is the collector-to-metal ion ratio. Ion flotation requires that the collector be at least stoichiometric and have foamy characteristics (Mota et al. 2015).

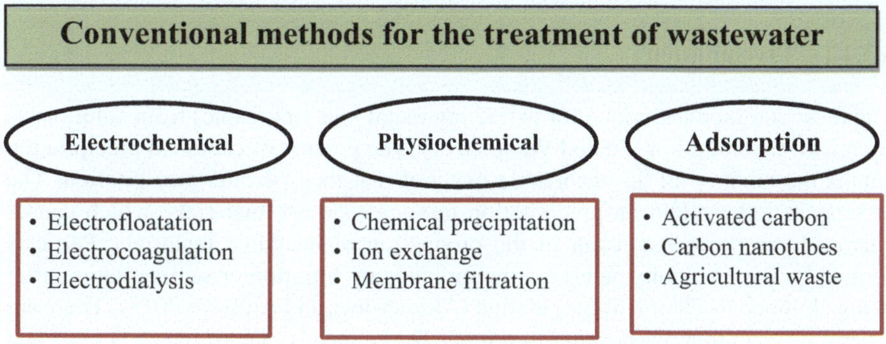

Fig. 6.1 Conventional methods for wastewater treatment (Razzak et al. 2022)

When treating water and wastewater, electrocoagulation is a popular technique for efficiently removing impurities such bacteria, organic materials, heavy metals, hardness, and turbidity (Zhu et al. 2005). An electrical current is used in the electrochemical process of electrocoagulation to extract metals from solutions. An electrocoagulation apparatus can be used to remove tannins, colors, suspended particles, and dissolved metals. The contaminants in wastewater are kept in solution by an electrical charge. Ions and other charged particles become destabilized and precipitate when neutralized by ions generated by the electrocoagulation system that have the opposite electrical charge (Bazrafshan et al. 2015).

6.3.3 Electrodialysis

A lot of water is needed for several manufacturing processes. Reducing, recycling, recovering materials, and repurposing (4Rs) the wastewater generated therefore offers a significant financial advantage. Electrodialysis (ED) is an effective treatment technique for industrial wastewater. By employing charged ion exchange membranes, the membrane technique known as electrodialysis can extract metal ions that are charged from a liquid solution (Cifuentes et al. 2009). By using ionized materials such as sodium chloride, this technique creates sodium and chloride ions in solution. Ions carry the current stream when terminals are immersed in a salt solution. Cations are particles that have a positive charge and are drawn to the negative cathode. The positive anode is approached by negatively charged anions (Oncel et al. 2013). Anion channels block anion but allow cations, while cation channels allow anions but block cations. Electrodialysis uses layers or channels containing certain anions and cations between cathodes (Samiey et al. 2014).

6.3.4 The Physiochemical Processes

6.3.4.1 Precipitation

The most common technique for extracting metal ions (inorganic) from solutions is chemical -precipitation (Fu and Wang 2011). The process of chemical precipitation entails the creation of an unsolvable material that then permits precipitation. The process includes adding a precipitation mixture to waste materials, which causes chemical reactions that result in the production of insoluble materials. Particles formed by the insoluble material are eliminated by filtration or sedimentation after being clumped by chemical coagulation (Akinterinwa and Adibayo 2018). There are several ways to treat wastewater to remove heavy metals, one of which is chemical precipitation.

6.3.4.2 Ion Exchange

In addition to lowering sludge volume, ion exchange also offers other advantages such selective repossession and achieving strict proclamation specifications (Zewail and Yousef 2015). Copper was extracted from wastewater by using INDION 225H high-density ion exchange resin (Thakare and Jana 2015). Nickel was extracted from wastewater by using expanded bed adsorption and a high-density ion exchange resin (Shaidan et al. 2012). Large volumes of wastewater were simultaneously treated and heavy metals were effectively removed by the resin. About 97% of the nickel in the wastewater was effectively recovered by the researchers. For the treatment of wastewater, this high-density ion exchange resin provides hydrodynamic benefits.

6.3.4.3 Coagulation and Flocculation

The method for treating wastewater that is most frequently employed is coagulation. The neutralization of the particle charges is referred to as coagulation. According to a study heavy metal solids can be removed from solutions using this technique (Verma et al. 2012). Whether water treatment instruments are used or not, it is a very useful way of controlling heavy metal pollution. This method destabilizes the charges on the particles by highly charged tiny molecules introduce into the wastewater. Additionally, this method is usually combined with flocculation, which is process by which colloidal particles separate from the suspension and combine to create deposits (Wang et al. 2005).

6.3.4.4 Membrane Filtration

Procedures for this method are categorized based on kind of membrane that is utilized. Rapid separation, no-phase transitions, energy-saving and eco-friendly are just a few of the benefits that filtration offers over other traditional procedures (Zhu et al. 2014).

6.3.4.5 Reverse Osmosis

In reverse osmosis a solvent is forced from the area of increased concentration to the area of low concentration, a physical filtration process which removes unwanted molecules from water by forcing it through a permeability barrier that have specific properties. This process requires an intense pressure than the osmotic pressure (Kanamarlapudi et al. 2018). It can only be used to remove modest amounts of heavy metals so reverse osmosis is not thought to be a viable process to remove heavy metals from industrial wastewater (Ranitha et al. 2016).

6.3.5 *Adsorption Process*

This method is an inexpensive process. Adsorption offers a number of benefits over traditional techniques, including less biological, cheap cost, increased efficacy, metal's recovery and regeneration of the adsorbent. Adsorbents used for heavy metal removal include activated carbon, zeolites, manganese-oxides and waste products of agriculture (Sardella et al. 2015).

6.3.5.1 Activated Carbon

This can be made from any material that has a high carbon amount, such as wastes from industries. Many agricultural wastes are used to make activated carbon, such as lignin, coconut buttons, grape and olive stones, moso and bamboo (Renu and Singh 2017). A group of workers produced activated carbon for wastewater treatment by using residual coconut buttons to remove Pb, Hg, and Cu (Anirudhan and Sreekumari 2011).

6.3.5.2 Carbon Nano-Tubes

In the past ten years, carbon nanotubes have drawn a lot of attention due to their exceptional mechanical and electrical capabilities, stability, and exceptional adsorbent qualities. Kosa et al. used 8-hydroxyquinoline to modify multi-walled carbon nano-tubes with the purpose of eliminating zinc, cadmium, lead and copper. Over 80% of the cadmium was found to be removed efficiently (Kosa et al. 2012).

6.4 Role of NMs in Wastewater Treatment

Numerous NMs have been shown to play pivotal roles in wastewater cleanup applications (Ahmad et al. 2021). Wastewater treatment systems are essential to the growth and development of the economy and the preservation of public health. As a result; creating and implementing cutting-edge technologies that treat water with high efficiency and low energy usage is essential (Sadegh et al. 2017). Heavy metal ions and azo dyes are the most prevalent hazardous substances in wastewater that cause specific issues (Islam et al. 2017). Wastewater can contain a variety of common and uncommon chemicals, including suspended solids, heavy metals, oil, grease, gases, and complex organic compounds including plant matter, excreta, and biological microbes like coliform, viruses, and protozoan cysts. Green NPs can be used by engineers to create highly efficient and reasonably priced environmentally friendly wastewater treatment systems. Substances created via nanotechnology that have a diameter of less than 100 nm are known as NPs. Green NPs are made of natural components that are safe for people and the environment (Neeti et al. 2023). Green NPs can be used by engineers to create highly efficient, reasonably priced wastewater treatment systems that are also ecologically benign. These new features could be very beneficial to the medical and technological sectors. Safe for both the environment and people, green NPs are composed of natural ingredients. Combining green chemistry and nanotechnology is the result of producing green NPs. Treatment systems for wastewater and water frequently use metal oxide NPs (Neeti et al. 2023). Water contaminants such as heavy metals, inorganic, organic, and dangerous microorganisms were removed from the water throughout the treatment process by the strong reactivity and good adsorption properties of NPs. The common applications of NPs/NMs in environmental remediation include membranes, absorbents, photocatalysts, and chemical disinfectants.

6.4.1 Adsorbents

Adsorption's great efficiency, affordability, and simplicity of use make it one of the finest options for treating wastewater. The limitation is that contaminants can only move to the adsorbent, which must be produced frequently (Muruganandham et al. 2014). The most important part of an adsorption system is a well-designed adsorbent (Kumari et al. 2019). Among the benefits of nano-adsorbents are their strong adsorption capacity, affordability, non-toxicity to the environment, simplicity of separation, resilience, and reusability (Wang et al. 2020). When compared to larger-sized particles, nanosized particles are typically more efficient (Sarma et al. 2019). Adsorption typically occurs through the following mechanisms: migration into the adsorbent's pores, diffusion to the surface, and monolayer accumulation of the adsorbate. Adsorption mechanism depends on electrochemical potential, ion-exchange process, and surface characteristics (Baby et al. 2019). For instance, charged pollutants have a tendency to adsorb on adsorbents having an oppositely charged surface through electrostatic attraction. Higher surface area and mesoporous architectures in adsorbents result in higher adsorption kinetics and capacities (Bao et al. 2013). An essential component of contaminant adsorption efficiency is metal and metal-oxide NPs. Many active sites, low-cost, strong mechanical qualities, and environmental sustainability are among the characteristics of NPs that make them appropriate for this use (Ahmad et al. 2021).

6.4.2 Disinfectants

Disinfectants are the chemicals that are employed in wastewater treatment to inactivate and eliminate germs on inert surfaces. Disinfection frequently results in the corrosion of cell walls, disruption of protoplasm, permeability of cells, or enzyme activity. These disruptions prevent bacteria from multiplying and ultimately lead to their death (Shannon et al. 2008). Disinfection, however, also has to contend with the hazardous disinfectant By-Products (DBP) of traditional chemical disinfectants, such as ozone, chlorine, and chloramines, which can combine with different types of wastewater components to generate DBP. Because NMs act without causing DBP, they can eliminate this problem. Ag, TiO_2, and Zinc Oxide nano-particles are the most widely used inorganic antibacterial NMs (Ahmad et al. 2021).

6.4.3 Nano-Composite Membranes

Due to the effectiveness of membrane separation methods in eliminating contaminants, they have appeared as a new technique in the field of treating wastewater. The separation properties and transport methods across the membrane are mostly determined by the membrane's structure. In addition to having a less environmental impact and being simpler to install, operate, and scale up, this technique has been shown to be more fruitful in terms of removal efficiency (Dickhout et al. 2017). Membrane fouling, on the other hand, is the primary issue with membrane technology. Because fouling affects both the water permeability performance and the rejection of organics and salt, it is the primary factor limiting the production of membrane-thin films. When material accumulates on solid surfaces unintentionally, fouling occurs, which compromises function (Xu et al. 2008). Through the alteration of properties of membrane surface, significant efforts have been made for developing and improving membrane filtration (Igunnu and Chen 2014). Electrostatic, sieving, and absorption processes are generally necessary for membrane separation. The hydrophilic/hydrophobic behavior of the solute and membrane is consistent with these mechanisms. To achieve synergistic effects on the process of filtration, the researchers have therefore designed membranes that incorporate NM. A wide variety of membranes with varying levels of rejection, structural integrity, and fouling liability are made possible by the introduction of functional NPs. Different-sized NPs have been investigated as modifiers to increase the hydrophilicity of the polymeric membrane and thus upgrade selectivity, strength and permeability (Jafarinejad 2017).

6.4.4 Photocatalysts

Photocatalysis is an important method for removing environmental contaminants. NMs play a key role in using light to treat water and waste water. Photocatalytic degeneration is an advanced oxidation process that uses ultra-violet light as a source of energy (Khin et al. 2012). Irradiation with UV light generates the highly reactive hydroxyl radical (•OH) to degrade persistent chemicals in wastewater (Muruganandham et al. 2014). Photocatalysis offers numerous benefits over other AOPs, such as Fenton and catalytic reactions of Photon-Fenton, and H_2O_2. It can run under ambient conditions, using inexpensive and harmless photocatalysts as well as atmospheric air as an oxidant (Fadlalla et al. 2019). It has been observed that soluble photocatalyst particles perform with great efficiency (Argurio et al. 2018). In comparison to photocatalyst particles immobilized in or on a membrane, they have a greater active surface area, which offers a good interaction b/w the pollutants and photocatalysts. Additionally, the solubilized particles (photocatalysts) are doped with additional alien ions, such as non-metal and transitional metal doping, to improve the photocatalytic performance (Lv et al. 2019).

6.5 Materials Used in Photocatalyst Synthesis for Wastewater Treatment

6.5.1 Metal Oxides (MO)

- TiO_2: Because of its oxidizing, robust, affordable, biocompatible, non-toxic, and photocatalytic qualities, it is well recognized for its significance as a semiconductor photocatalyst in the treatment of wastewater (Zhang et al. 2016).
- ZnO: Excellent chemical stability, low cost, non-toxicity, and high catalytic efficiency.
- MgO: Because of its stable, cost-effective, large energy band gap, chemical, mechanical, optical, and electrical characteristics, and lack of toxicity, it is an excellent photocatalyst.
- CuO: 2.17 eV band gap P-type semiconductor with strong photocatalytic properties.
- WO_3: Absorbs sunlight with a strong wavelength of < 480 nm.
- Nb_2O_5: Good chemical stability, strong photoactivity under favorable particle size conditions, and semiconductivity similar to TiO_2.
- $BiVO_4$, also known as monoclinic scheelite bismuth vanadate, is a very innovative and promising visible light-driven photocatalyst.
- Bi_2MoO_6 (BMO): Has a band gap energy of 2.3–2.7 eV, is non-toxic, and is inexpensive.
- SnO_2: Size, greater oxidation potential, resistance to corrosion, size, large 3.7 eV band gap, chemical motionlessness, significant photochemical steadiness, visual qualities, & affordable price.
- CeO_2: This material can function as a superb photocatalyst when mixed with other materials to create heterojunctions.
- Metal oxide quantum dots (MOQDs): At the moment, micromolecules and nanomolecules bigger than 10 nm are unable to replicate the special properties of MOQDs like their physicochemical and photo-luminescent properties (Ahuja et al. 2023).

6.5.2 Iron-Based Minerals in Photocatalysis

The low cost and environmental friendliness of iron-based materials make them striking candidates intended for use in photocatalysis.

- $FeTiO_3$: A stable and efficient catalyst, $FeTiO_3$ is an antiferromagnetic semiconductor with band gap energy of 2.4–2.9 eV.
- Fe_3O_4: An inexpensive, environmentally friendly iron-based substance with a high capacity to generate hydroxyl radicals in a water matrix and strong magnetic characteristics.
- $FeNi_3$: $FeNi_3$ readily reacts with oxygen and loses its magnetic- properties, although having a strong chemical activity. Magnetic NPs having a core-shell configuration are one type of prevention strategy.
- $BiFeO_3$: Because of its tiny band gap energy, the coexistence of ferromagnetic and ferroelectric characteristics, chemical inertness, and non-toxicity, perovskite bismuth ferrite is an intriguing visible light photocatalyst (Ahuja et al. 2023).

6.5.3 Metal Sulfides and Photocatalysts

Because of their distinctive layered structure, metal sulfides and selenides offer more active sites while reducing electron mobility.

- MoS_2 is a promising photocatalyst because of its laminar, graphene-like structure, adjustable band gap, and high photocatalytic activity, despite its low quantum yield and tendency toward aggregation and restacking, which decreases surface area. MoS_2 can form good heterojunction composites, such as MoS_2/WO_3 and MoS_2/C_3N_4, which show great photocatalytic potential.
- Chalcogenide semiconductors, such as $CuInS_2$, $AgIn_5S_8$, $ZnIn_2S_4$, and Stannum indium sulfide, show promise for use in heterogeneous photocatalysis.
- The photocatalyst semiconductor CdS is driven by visible light and has poor activity, photo corrosion, and photodissociation. CdS/ZnO hybrids can circumvent these limitations by selectively absorbing UV radiation.
- $CuFeS_2$

  In an acidic medium, $CuFeS_2$ accelerates the short-term breakdown of H_2O and H_2O_2 to create non-selective •OH. Nevertheless, the process as a whole and the development of hotspots are impacted by its low specific surface area (Ahuja et al. 2023).

6.5.4 *Carbon-Based Substances for Photocatalytic Applications*

- Due to their high optical activity, large surface area, and excellent electrical conductivity, carbon-based substances such as carbon nanotubes (CNT), carbon quantum dots (CQDs), polymer materials, reduced graphene oxide (rGO), and activated carbon.
- Activated carbon (AC) are perfect for photocatalytic applications. AC enhances textile dye absorptivity and photocatalytic activity by supporting transition metal NPs.
- rGO, a low Fermi energy zero band gap material, has an exceptional chemical stability and a distinctive two-dimensional structure.
- CQDs improve semiconductors' optical and electrical characteristics, and their green synthesis from natural carbon sources is gaining attention. Because of its attractive band structure, ease of synthesis, and stability, g-C_3N_4 is a widely used material in photocatalysis (Ahuja et al. 2023).

6.6 Photocatalytic Process for Wastewater Treatment

The electrochemical process of photolysis of water at a semiconductor electrode was described by Fujishima and Honda. The method is described as a sequential progression of (AOPs), i.e., advanced oxidation processes, which mitigate some of its disadvantages, including high price, fractional mineralization, and high OH radicle requirements. Since light energy is needed for the photocatalyst to become active during the photocatalysis process, this is interesting because light or photon sources can be used to control the reaction (Ong et al. 2018). The study of substances that, when used as a catalyst, raise or speed up a chemical reaction rate while leaving the catalyst unaltered after the reaction, and their synthesis, modification, and mechanism is known as catalysis. Compared to a typical reaction, the catalyst requires less activation energy, hence the reaction will proceed more quickly. The energy required for a reaction to happen is known as the activation energy or threshold energy. The catalyst allows for a different, lower-activation-energy pathway to be used in place of the typical reaction pathway, which speeds up the reaction. Nonetheless, the outcome and general thermodynamics are the same (Sinar Mashuri et al. 2020). The mechanism of photocatalyric reaction is described in Fig. 6.2.

Photocatalysis exhibits great promise for recovering green technologies and environmental practices. It encourages scientists to improve this technology by anticipating the greatest kind of reactors and photocatalysts. It is being used in numerous applications (Wu et al. 2017), including air pollution treatment (Bansal et al. 2018), environmental pollutants, and coating technology (Mera et al. 2018). Photocatalysis, in theory, needs light in order to activate the photocatalyst and start the process. Ohtani states that the photocatalyst's preferred crystal facets determine whether a

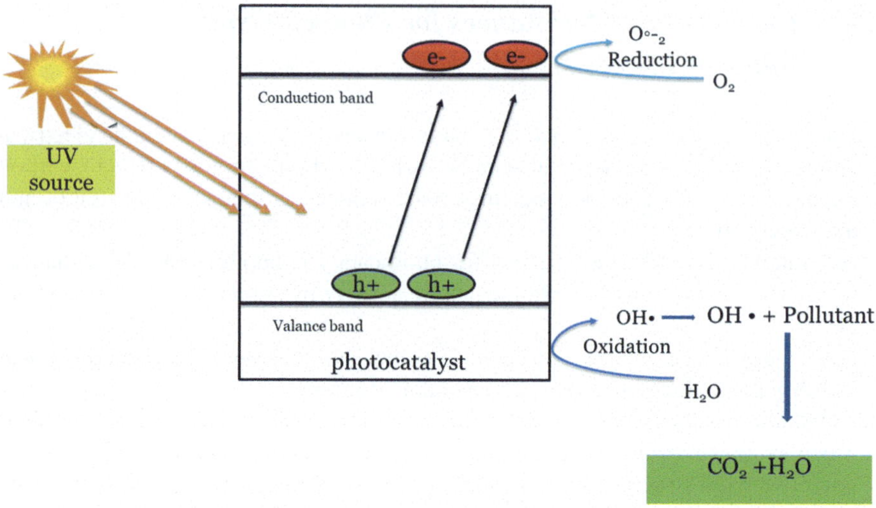

Fig. 6.2 Mechanism of photocatalytic reaction processes (Sinar Mashuri et al. 2020)

reaction is photocatalytic or catalytic (Ohtani 2010). As a result, while the catalytic process requires active sites to take place, the intensity of the irradiation light will influence the reaction's kinetics, producing electron-hole pairs that cause the reaction to happen (Ibrahim et al. 2020). The mass of active spots and the reaction kinetics coexist in catalysis. Catalytic reactions are often restricted to spontaneous reactions when the Gibbs free energy is more than zero (Farid et al. 2018). However, the ability of the material to absorb light and convert it into chemical energy is what enables both spontaneous and non-spontaneous reactions when using photocatalysts.

6.6.1 Photocatalysis-Induced Reduction of Heavy Metal Ions to a Less Dangerous Form

The potential for resolving environmental issues has drawn attention to the photocatalytic destruction of heavy metal ions, such as Cr (VI) and Cr (III). Because of its compassion, Cr (III) is less versatile than Cr (VI), which is an oxidation state with several negative effects. Reverse osmosis, precipitation, adsorption processes, photocatalysis, and ion transfer are some of the techniques used to remove Cr. Nevertheless, these approaches are too resource-intensive and inappropriate for smaller businesses. Due of its promise, TiO_2, a semiconductor with better stability, intriguing chemical properties, and a low cost, has drawn a lot of interest. Electron-hole pairs are produced during photolysis and have the ability to engage in redox reactions or recombine. Under visible light irradiation, the concentration of Cr (VI) rapidly decreased when the high-energy ball milling treatment and sol-gel method were used to fabricate multi-doped NSC-TiO_2. As a sustainable water treatment technique and a greener substitute for metal-doped TiO_2 the treatment of contaminated water by non-metal-doped TiO_2 exhibits significant promise.

6.6.2 Organic Pollutant Removal

Degradation of organic pollutants occurs when hydroxyl free radicals react with the pollutant's organic group, changing it into non-hazardous forms. Atomic hydrocarbons, phenols, cleaners, antibiotics, and their byproducts are released into water ways by organic pollutants that come from petrochemical facilities, urban sewage systems, pharmaceutical companies, and agro-industries. These pollutants are persistent. Before being disposed of, these pollutants need to be cleaned up and any dangerous compounds eliminated.

6.6.3 Pesticides Removal

In agricultural areas, pesticides are common and dangerous organic pollutants that are employed to eradicate pests and safeguard crops. On the other hand, overuse can activate harmful substances in crops and farms, leading to illness. To reduce the toxicity of pesticides, nanocomposites, nanofiltration, and nano-based catalysts are employed. In agricultural settings, carbamate insecticides are frequently employed. TiO_2, a special-structured photocatalyst, can aid in the breakdown of pesticides. Under particular circumstances, including catalyst, temperature, pH, and UV light, TiO_2 degrades most readily. Two common organophosphorus pesticides are monocrotophos and chlorpyrifos. Using the sol-gel method, pest controllers were combined and placed within a glass jacket together with a TiO_2 nano-photocatalyst.

6.6.4 Dye Wastewater Degradation Using Photocatalysis

Water contamination by fertilizers, organic dyes, industrial chemicals, and home wastes raises environmental issues. Every year, the textile, paper, cosmetic, and leather sectors generate about one million tons of organic dyes. When mixed with natural water, this non-biodegradable, colored, toxic, and carcinogenic effluent can lead to serious illnesses. Sludge generation and pollutant elimination are two disadvantages of traditional technologies like coagulation and adsorption. Photocatalysts that have demonstrated efficacy and promise in the treatment of dye wastewater include ZnO, TiO_2, SiO_2, and CeO_2. Several photocatalysts used in recent experiments to remove pollutants are shown in Table 6.1.

Table 6.1 Various photocatalysts used along with their degradation capacity for specific pollutant

Photocatalyst/NM	Source of light	Pollutant (percentage removal)	References
AgxO/FeOx/ZnO	Suntest solar simulator	Tetracycline 85%	Yu et al. (2019)
$CoFe_2O_4$/mpg-C_3N_4 (Sono-catalyst)	Ultrasonic bath apparatus (WUC-D10H, 40 kHz, 665 W	Methylene blue (MB) 93%	Hassani et al. (2018)
CNTs/TiO_2/AgNPs/ surfactant nanocomposites		Methylene blue dye (MB) 100%	Azzam et al. (2019)
Z-scheme CdS/g C_3N_4	Tungsten lamp (100 W)—90 min	Methyl orange (MO) 85% above	Pourshirband and Nezamzadeh-Ejhieh (2021)
Polyethylene TiO_2	Low-intensity solar simulated light	Methylene blue dye (MB) 98%	Rtimi et al. (2015)
WO_3/TiO_2		Pesticide (Malathion) 63%	Ramos-Delgado et al. (2013)
Hybrid acrylic/TiO_2 films	Sunlight (low intensity)	Escherichia coli (inactivation of bacteria) 100% in 240 min	Bonnefond et al. (2015)
Nanoscale cobalt aluminate		Navy blue textile dye 67	Jabbar et al. (2019)
Cerium oxide (CeO_2)		Phenol approx. 93%	Ahmad et al. (2020b)
ZnO/activated carbon (AC)	8 by 8 W UV	COD = 100%	Suresh et al. (2015)
N-doped TiO_2 (on Polystyrene spheres)	LED lamp	Escherichia coli inactivation = 93%	Sacco et al. 2018)

6.6.5 Factors Affecting Photocatalytic Performance in Wastewater Treatment

6.6.5.1 pH

The chemistry, oxidation efficiency, and fate of organic molecules are all impacted by the pH of the water or media. It regulates the isoelectric point and surface charge of photocatalysts, with the zero-point charge being essential for researching the efficiency of photocatalytic oxidation. The rate of oxidation, hydroxyl radical assault, and conduction band electron reduction all affect the mechanism. Adsorption capabilities require pH adjustment. Anionic dyes in acidic conditions, such as textile effluent, boost photodegradation efficiency. Research has indicated that all catalysts perform best at pH 3.0 for degradation.

6.6.5.2 Time and Temperature

As time increases, the removal of pollutants also increases but only up to a certain extent. The organic reaction under UV light is highly effective at a temperature range of 20–80 °C. The majority of photocatalytic processes takes place at normal temperature; however, because of the lower desorption rate below 0 °C, the activation energy increases. However, reactant adsorption may be restricted by temperature rises above 80 °C. Using $FeTiO_3$ as a photocatalyst, Garcia-Munoz et al., investigated the temp-dependent hospital waste degradation. The study discovered that there was increased elimination of TOC and COD from 76% at 30 °C to 82% at 40 °C and 83% at 50 °C, as well as mild oxidation at 30 °C and entire oxidation at 50 °C (García-Muñoz et al. 2017).

6.6.5.3 Concentration of Pollutant

Reactions involving photocatalysis have pseudo first order kinetics and are represented by the Langmuir–Hinshelwood model. The amount of pollution present has an impact on how long it takes for mineralization to occur. The degrading performance of photocatalysis is impacted by a rise in organic molecules, which results in adsorption on the surface of the catalyst and decreased photocatalytic activity. The amount of organic pollutants that are sorbed on the surface of the photocatalyst and the production of OH radicals with the molecules of contaminants determines the degradation percentage. The efficiency of photodegradation increases as the initial concentration of pollutants decreases due to the increased adsorption of photocatalysts. Pollutants covering the photocatalytic surface reduce photons and electrons, which lower photo degradation beyond the concentration, limit (Kiriakidou et al. 1999).

6.6.5.4 Light Intensity Effect

As light intensity rises, there is a noticeable energy loss during degradation, which raises expenses. High energy loss is the result of uncontrolled light intensity those with energy below the critical photon energy are unable to ignite pairs of electrons and holes (Yang and Liu 2007). More energy waste can result from larger light intensities than from the breakdown of organic contaminants. In order to minimize energy loss due to electron-hole recombination, it is imperative to develop desirable light intensities (Dillert et al. 2013).

6.6.5.5 Dose of Catalyst

A photocatalyst's increased dose provide more active sites for reactions and light absorption, which increases the amount of •OH produced and the amount of photoactivity. However, because of turbidity and particle-particle interactions, higher dosages result in less effectiveness (Coleman et al. 2007). After treating textile wastewater for 300 min at various catalyst doses, Souza et al. discovered that 0.2 g/L produced the best degradation (Souza et al. 2016). In another investigation into the impact of MgO dosage on photocatalytic COD degradation, a rise in COD removal up to 100% was observed at 0.8 g/L of catalyst dosage and a fall at 1.4 g/L. The significance of catalyst dose in photocatalytic degradation is emphasized in both investigations. High doses can decrease the photo degradation rate for pollutants.

6.6.6 Limitation of Photocatalysts

Although photocatalysis has many benefits, there are obstacles because there aren't enough appropriate materials. Certain materials are unstable in aqueous solutions and have a narrow absorbance range, such as semiconductors with sulfides, oxides, or nitrides. The drawbacks of metal oxides include increased recombination rates, less recovery, and less adsorption (Mohamed et al. 2020). Although TiO_2-based elements offer advantages, they perform poorly and have poor charge transportation, little oxygen generation, and a high recombination rate (Lum et al. 2020). ZnO is unstable in both acidic and alkaline environments, and it has a lower adsorption potential. Byproducts are one of the difficulties faced by microorganisms in fuel generation; hence, newer reactors and modified microbes are required. Although they are very effective, weak contacts and reduced charge transport are problems for nano-photocatalysts. Catalysts with minimal or no metal content are required for industrial uses.

6.7 Conclusion and Future Perspective

The increasing urbanization, population expansion, and climate change are driving up the demand for clean water globally. For the purposes of agriculture, industry, human consumption, and ecosystem management, purified water is essential. The disadvantages of traditional technologies include high energy consumption, limited removal efficiency, excessive sludge production, and secondary pollution development. Researchers are working on low-cost, low energy, and environmentally friendly water treatment methods, such as photocatalysts. Compared to conventional techniques, photocatalysis has benefits such faster response times, greater catalytic efficiencies, and the absence of secondary hazardous products. For the photodegradation of dyes in wastewater, many photocatalysts have been produced; nano-photocatalysts are more effective. Improved dye degradation has been demonstrated using doped photocatalysts, semiconductor and spinel composites, magnetic photocatalysts, and green photocatalysts built of metal sulfide nanostructures. UV light has the highest energy and lowest wavelength of any light source, making it the most effective for photocatalytic oxidation—a technique that effectively removes chemicals that are resistant to removal from wastewater. For its widespread practical application and the possible utilization of solar light, more research is required.

References

Agarwal H, Kumar SV, Rajeshkumar S (2017) A review on green synthesis of zinc oxide NPs–an eco-friendly approach. Resour Effic Technol 3(4):406–413

Ahmad NA, Goh PS, Zulhairun AK, Ismail AF (2020a) Antifouling property of oppositely charged titania nanosheet assembled on thin film composite reverse osmosis membrane for highly concentrated oily saline water treatment. Membranes 10(9):237

Ahmad T, Iqbal J, Bustam MA, Zulfiqar M, Muhammad N, Al Hajeri BM et al (2020b) Phytosynthesis of cerium oxide NPs and investigation of their photocatalytic potential for degradation of phenol under visible light. J Mol Struct 1217:128292

Ahmad NA, Goh, PS, Zulhairun AK, Wong TW, Ismail AF (2021) The role of functional NMs for wastewater remediation. In: Functional hybrid nanomaterials for environmental remediation. The Royal Society of Chemistry, pp 1–28

Ahuja T, Brighu U, Saxena K (2023) Recent advances in photocatalytic materials and their applications for treatment of wastewater: a review. Journal of Water Process Engineering 53:103759

Akinterinwa A, Adibayo I (2018) Chemical precipitation approach to the removal of heavy metals from wastewater for discharge into sanitary sewerage bioremediation of water view project environment view project. Intl J Chem 1(2):167–171

Alagarasi A (2013) Chapter-introduction to NMs. Indian Institute of Technology Madras, pp 1–24

Anirudhan TS, Sreekumari SS (2011) Adsorptive removal of heavy metal ions from industrial effluents using activated carbon derived from waste coconut buttons. J Environ Sci 23(12):1989–1998

Argurio P, Fontananova E, Molinari R, Drioli E (2018) Photocatalytic membranes in photocatalytic membrane reactors. Processes 6(9):162

Azelee IW, Goh PS, Lau WJ, Ismail AF, Rezaei-DashtArzhandi M, Wong KC, Subramaniam MN (2017) Enhanced desalination of polyamide thin film nanocomposite incorporated with acid treated multiwalled carbon nanotube-titania nanotube hybrid. Desalination 409:163–170

Azzam EM, Fathy NA, El-Khouly SM, Sami RM (2019) Enhancement the photocatalytic degradation of methylene blue dye using fabricated CNTs/TiO_2/AgNPs/Surfactant nanocomposites. J Water Process Eng 28:311–321

Baby R, Saifullah B, Hussein MZ (2019) Carbon NMs for the treatment of heavy metal-contaminated water and environmental remediation. Nanoscale Res Lett 14(1):341

Baker BH, Aldridge CA, Omer AR (2016) Water: availability and use. Mississippi State University Extension

Bansal P, Verma A, Talwar S (2018) Detoxification of real pharmaceutical wastewater by integrating photocatalysis and photo-Fenton in fixed-mode. Chem Eng J 349:838–848

Bao M, Zhu G, Wang L, Wang M, Gao C (2013) Preparation of monodispersed spherical mesoporous nanosilica–polyamide thin film composite reverse osmosis membranes via interfacial polymerization. Desalination 309:261–266

Barsan MM, Brett CM (2016) Recent advances in layer-by-layer strategies for biosensors incorporating metal NPs. TrAC Trends Anal Chem 79:286–296

Bazrafshan E, Mohammadi L, Ansari-Moghaddam A, Mahvi AH (2015) Heavy metals removal from aqueous environments by electrocoagulation process–a systematic review. J Environ Health Sci Eng 13:1–16

Bhat SA, Sher F, Hameed M, Bashir O, Kumar R, Vo DVN et al (2022) Sustainable nanotechnology based wastewater treatment strategies: achievements, challenges and future perspectives. Chemosphere 288:132606

Bonnefond A, González E, Asua JM, Leiza JR, Kiwi J, Pulgarin C, Rtimi S (2015) New evidence for hybrid acrylic/TiO_2 films inducing bacterial inactivation under low intensity simulated sunlight. Colloids Surf B 135:1–7

Choi W, Choi J, Bang J, Lee JH (2013) Layer-by-layer assembly of graphene oxide nanosheets on polyamide membranes for durable reverse-osmosis applications. ACS Appl Mater Interfaces 5(23):12510–12519

Cifuentes L, García I, Arriagada P, Casas JM (2009) The use of electrodialysis for metal separation and water recovery from $CuSO_4$–H_2SO_4–Fe solutions. Sep Purif Technol 68(1):105–108

Coleman HM, Vimonses V, Leslie G, Amal R (2007) Degradation of 1, 4-dioxane in water using TiO_2 based photocatalytic and H_2O_2/UV processes. J Hazard Mater 146(3):496–501

Da Mota IDO, de Castro JA, de Góes Casqueira R, de Oliveira Junior AG (2015) Study of electroflotation method for treatment of wastewater from washing soil contaminated by heavy metals. J Market Res 4(2):109–113

Deshmukh SP, Patil SM, Mullani SB, Delekar SD (2019) Silver NPs as an effective disinfectant: a review. Mater Sci Eng C 97:954–965

Dickhout JM, Moreno J, Biesheuvel PM, Boels L, Lammertink RGH, De Vos WM (2017) Produced water treatment by membranes: a review from a colloidal perspective. J Colloid Interface Sci 487:523–534

Dillert R, Engel A, Große J, Lindner P, Bahnemann DW (2013) Light intensity dependence of the kinetics of the photocatalytic oxidation of nitrogen (ii) oxide at the surface of TiO_2. Phys Chem Chem Phys 15(48):20876–20886

El Jabbar Y, ElHafdi M, Benchikhi M, El Ouatib R, Er-Rakho L, Essadki A (2019) Photocatalytic degradation of navy blue textile dye by nanoscale cobalt aluminate prepared by polymeric precursor method. Environ Nanotechnol Monit Manag 12:100259

Fadlalla MI, Senthil Kumar P, Selvam V, Ganesh Babu S (2019) Recent advances in NMs for wastewater treatment. In: Advanced nanostructured materials for environmental remediation, pp 21–58

Farid MAA, Hassan MA, Taufiq-Yap YH, Ibrahim ML, Hasan MY, Ali AAM et al (2018) Kinetic and thermodynamic of heterogeneously K_3PO_4/AC-catalysed transesterification via pseudo-first order mechanism and Eyring-Polanyi equation. Fuel 232:653–658

Fu F, Wang Q (2011) Removal of heavy metal ions from wastewaters: a review. J Environ Manage 92(3):407–418

Gade R, Ahemed J, Yanapu KL, Abate SY, Tao YT, Pola S (2018) Photodegradation of organic dyes and industrial wastewater in the presence of layer-type perovskite materials under visible light irradiation. J Environ Chem Eng 6(4):4504–4513

Gao SJ, Shi Z, Zhang WB, Zhang F, Jin J (2014) Photoinduced superwetting single-walled carbon nanotube/TiO_2 ultrathin network films for ultrafast separation of oil-in-water emulsions. ACS Nano 8(6):6344–6352

García-Muñoz P, Pliego G, Zazo JA, Munoz M, de Pedro ZM, Bahamonde A, Casas JA (2017) Treatment of hospital wastewater through the CWPO-photoassisted process catalyzed by ilmenite. J Environ Chem Eng 5(5):4337–4343

Hassani A, Eghbali P, Metin Ö (2018) Sonocatalytic removal of methylene blue from water solution by cobalt ferrite/mesoporous graphitic carbon nitride ($CoFe_2O_4$/mpg-C_3N_4) nanocomposites: response surface methodology approach. Environ Sci Pollut Res 25:32140–32155

Ibrahim ML, Khalil NNANA, Islam A, Rashid U, Ibrahim SF, Mashuri SIS, Taufiq-Yap YH (2020) Preparation of Na_2O supported CNTs nanocatalyst for efficient biodiesel production from waste-oil. Energy Convers Manage 205:112445

Igunnu ET, Chen GZ (2014) Produced water treatment technologies. Int J Low Carbon Technol 9(3):157–177

Islam MS, San Choi W, Nam B, Yoon C, Lee HJ (2017) Needle-like iron oxide@ $CaCO_3$ adsorbents for ultrafast removal of anionic and cationic heavy metal ions. Chem Eng J 307:208–219

Jafarinejad S (2017) A comprehensive study on the application of reverse osmosis (RO) technology for the petroleum industry wastewater treatment. J Water Environ Nanotechnol 2(4):243–264

Jorfi S, Barzegar G, Ahmadi M, Soltani RDC, Takdastan A, Saeedi R, Abtahi M (2016) Enhanced coagulation-photocatalytic treatment of Acid red 73 dye and real textile wastewater using UVA/synthesized MgO NPs. J Environ Manage 177:111–118

Kalpana VN, Rajeswari VD (2017) Biosynthesis of metal and metal oxide NPs for food packaging and preservation: a green expertise. In: Food biosynthesis, pp 293–316. Academic Press

Kanamarlapudi SLRK, Chintalpudi VK, Muddada S (2018) Application of biosorption for removal of heavy metals from wastewater. Biosorption 18(69):70–116

Khan ST, Malik A (2019) Engineered NMs for water decontamination and purification: from lab to products. J Hazard Mater 363:295–308

Khin MM, Nair AS, Babu VJ, Murugan R, Ramakrishna S (2012) A review on NMs for environmental remediation. Energy Environ Sci 5(8):8075–8109

Kiriakidou F, Kondarides DI, Verykios XE (1999) The effect of operational parameters and TiO_2-doping on the photocatalytic degradation of azo-dyes. Catal Today 54(1):119–130

Kosa SA, Al-Zhrani G, Salam MA (2012) Removal of heavy metals from aqueous solutions by multi-walled carbon nanotubes modified with 8-hydroxyquinoline. Chem Eng J 181:159–168

Kumari P, Alam M, Siddiqi WA (2019) Usage of NPs as adsorbents for waste water treatment: an emerging trend. Sustain Mater Technol 22:e00128

Lieu M, Smith GP, Giles PA, Ziparo F, Maughan BJ, Démoclès J et al (2016) The XXL Survey-IV. Mass-temperature relation of the bright cluster sample. Astron Astrophys 592:A4

Lum PT, Foo KY, Zakaria NA, Palaniandy P (2020) Ash based nanocomposites for photocatalytic degradation of textile dye pollutants: a review. Mater Chem Phys 241:122405

Lv C, Lan X, Wang L, Yu Q, Zhang M, Sun H, Shi J (2019) Alkaline-earth-metal-doped TiO_2 for enhanced photodegradation and H_2 evolution: insights into the mechanisms. Catal Sci Technol 9(21):6124–6135

Mamah SC, Goh PS, Ismail AF, Amin MAM, Ahmad NA, Suzaimi ND, Raji YO (2020) Facile preparation of palygorskite/chitin nanofibers hybrids NM with remarkable adsorption capacity. Mater Sci Eng B 262:114725

Mera AC, Martínez-de la Cruz A, Pérez-Tijerina E, Meléndrez MF, Valdés H (2018) Nanostructured BiOI for air pollution control: Microwave-assisted synthesis, characterization and photocatalytic activity toward NO transformation under visible light irradiation. Mater Sci Semicond Process 88:20–27

Mishra S, Sundaram B (2022) Efficacy of NPs as photocatalyst in leachate treatment. Nanotechnol Environ Eng 7(1):173–192

Mohamed AS, Abukkhadra MR, Abdallah EA, El-Sherbeeny AM, Mahmoud RK (2020) The photocatalytic performance of silica fume based Co_3O_4/MCM-41 green nanocomposite for instantaneous degradation of Omethoate pesticide under visible light. J Photochem Photobiol A 392:112434

Muruganandham M, Suri RPS, Jafari S, Sillanpää M, Lee GJ, Wu JJ, Swaminathan M (2014) Recent developments in homogeneous advanced oxidation processes for water and wastewater treatment. Int J Photoenergy 2014:1–12

Nakata K, Fujishima A (2012) TiO_2 photocatalysis: design and applications. J Photochem Photobiol C 13(3):169–189

Neeti K, Singh R, Ahmad S (2023) The role of green NMs as effective adsorbents and applications in wastewater treatment. Mater Today Proc 77:269–276

Ohtani B (2010) Photocatalysis A to Z—what we know and what we do not know in a scientific sense. J Photochem Photobiol C 11(4):157–178

Oncel MS, Muhcu A, Demirbas E, Kobya M (2013) A comparative study of chemical precipitation and electrocoagulation for treatment of coal acid drainage wastewater. J Environ Chem Eng 1(4):989–995

Ong CB, Ng LY, Mohammad AW (2018) A review of ZnO NPs as solar photocatalysts: synthesis, mechanisms and applications. Renew Sustain Energy Rev 81:536–551

Pendergast MM, Hoek EM (2011) A review of water treatment membrane nanotechnologies. Energy Environ Sci 4(6):1946–1971

Pourshirband N, Nezamzadeh-Ejhieh A (2021) An efficient Z-scheme CdS/g-C_3N_4 nano catalyst in methyl orange photodegradation: focus on the scavenging agent and mechanism. J Mol Liq 335:116543

Qu X, Alvarez PJ, Li Q (2013) Applications of nanotechnology in water and wastewater treatment. Water Res 47(12):3931–3946

Ramos-Delgado NA, Hinojosa-Reyes L, Guzman-Mar IL, Gracia-Pinilla MA, Hernández-Ramírez A (2013) Synthesis by sol–gel of WO_3/TiO_2 for solar photocatalytic degradation of malathion pesticide. Catal Today 209:35–40

Ranitha M, Nurlidia M, Muhammad Rashid S, Yoshimitsu U (2016) Bioremoval of lead in industrial wastewater by microalgae. J Eng Sci Technol 11:43–49

Razzak SA, Faruque MO, Alsheikh Z, Alsheikhmohamad L, Alkuroud D, Alfayez A et al (2022) A comprehensive review on conventional and biological-driven heavy metals removal from industrial wastewater. Environ Adv 7:100168

Renu AM, Singh K (2017) Methodologies for removal of heavy metal ions from wastewater: an overview. Interdiscip Environ Rev 18(2):124–142

Rtimi S, Pulgarin C, Sanjines R, Kiwi J (2015) Kinetics and mechanism for transparent polyethylene-TiO_2 films mediated self-cleaning leading to MB dye discoloration under sunlight irradiation. Appl Catal B 162:236–244

Sacco O, Vaiano V, Rizzo L, Sannino D (2018) Photocatalytic activity of a visible light active structured photocatalyst developed for municipal wastewater treatment. J Clean Prod 175:38–49

Sadegh H, Ali GA, Gupta VK, Makhlouf ASH, Shahryari-Ghoshekandi R, Nadagouda MN et al (2017) The role of NMs as effective adsorbents and their applications in wastewater treatment. J Nanostructure Chem 7:1–14

Samiey B, Cheng CH, Wu J (2014) Organic-inorganic hybrid polymers as adsorbents for removal of heavy metal ions from solutions: a review. Materials 7(2):673–726

Sardella F, Gimenez M, Navas C, Morandi C, Deiana C, Sapag K (2015) Conversion of viticultural industry wastes into activated carbons for removal of lead and cadmium. J Environ Chem Eng 3(1):253–260

Sarma GK, Sen Gupta S, Bhattacharyya KG (2019) NMs as versatile adsorbents for heavy metal ions in water: a review. Environ Sci Pollut Res 26:6245–6278

Sewnet A, Abebe M, Asaithambi P, Alemayehu E (2022) Visible-light-driven g-C_3N_4/TiO_2 based heterojunction nanocomposites for photocatalytic degradation of organic dyes in wastewater: a review. Air Soil Water Res 15:11786221221117266

Shaidan NH, Eldemerdash U, Awad S (2012) Removal of Ni (II) ions from aqueous solutions using fixed-bed ion exchange column technique. J Taiwan Inst Chem Eng 43(1):40–45

Shannon MA, Bohn PW, Elimelech M, Georgiadis JG, Marinas BJ, Mayes AM (2008) Science and technology for water purification in the coming decades. Nature 452(7185):301–310

Shao F, Xu C, Ji W, Dong H, Sun Q, Yu L, Dong L (2017) Layer-by-layer self-assembly TiO_2 and graphene oxide on polyamide reverse osmosis membranes with improved membrane durability. Desalination 423:21–29

Sinar Mashuri SI, Ibrahim ML, Kasim MF, Mastuli MS, Rashid U, Abdullah AH et al (2020) Photocatalysis for organic wastewater treatment: from the basis to current challenges for society. Catalysts 10(11):1260

Souza RP, Freitas TK, Domingues FS, Pezoti O, Ambrosio E, Ferrari-Lima AM, Garcia JC (2016) Photocatalytic activity of TiO_2, ZnO and Nb_2O_5 applied to degradation of textile wastewater. J Photochem Photobiol A 329:9–17

Suresh P, Vijaya JJ, Kennedy LJ (2015) Synergy effect in the photocatalytic degradation of textile dyeing waste water by using microwave combustion synthesized zinc oxide supported activated carbon. React Kinet Mech Catal 114:767–780

Thakare YN, Jana AK (2015) Performance of high density ion exchange resin (INDION225H) for removal of Cu (II) from waste water. J Environ Chem Eng 3(2):1393–1398

Tran TK, Chiu KF, Lin CY, Leu HJ (2017) Electrochemical treatment of wastewater: selectivity of the heavy metals removal process. Int J Hydrogen Energy 42(45):27741–27748

Verma AK, Dash RR, Bhunia P (2012) A review on chemical coagulation/flocculation technologies for removal of colour from textile wastewaters. J Environ Manage 93(1):154–168

Wang LK, Vaccari DA, Li Y, Shammas NK (2005) Chemical precipitation. Physicochemical treatment processes. Humana Press, Totowa, NJ, pp 141–197

Wang L, Shi C, Pan L, Zhang X, Zou JJ (2020) Rational design, synthesis, adsorption principles and applications of metal oxide adsorbents: a review. Nanoscale 12(8):4790–4815

Wu B, Lu S, Xu W, Cui S, Li J, Han P (2017) Study on corrosion resistance and photocatalysis of cobalt superhydrophobic coating on aluminum substrate. Surf Coat Technol 330:42–52

Xu P, Drewes JE, Heil D (2008) Beneficial use of co-produced water through membrane treatment: technical-economic assessment. Desalination 225(1–3):139–155

Yang L, Liu Z (2007) Study on light intensity in the process of photocatalytic degradation of indoor gaseous formaldehyde for saving energy. Energy Convers Manage 48(3):882–889

Yu J, Kiwi J, Zivkovic I, Rønnow HM, Wang T, Rtimi S (2019) Quantification of the local magnetized nanotube domains accelerating the photocatalytic removal of the emerging pollutant tetracycline. Appl Catal B 248:450–458

Yu J, Zhang Y, Chen J, Cui L, Jing W (2020) Solvothermal-induced assembly of 2D–2D rGO-TiO_2 nanocomposite for the construction of nanochannel membrane. J Membr Sci 600:117870

Zewail TM, Yousef NS (2015) Kinetic study of heavy metal ions removal by ion exchange in batch conical air spouted bed. Alex Eng J 54(1):83–90

Zhang AY, Wang WK, Pei DN, Yu HQ (2016) Degradation of refractory pollutants under solar light irradiation by a robust and self-protected ZnO/CdS/TiO_2 hybrid photocatalyst. Water Res 92:78–86

Zhu B, Clifford DA, Chellam S (2005) Comparison of electrocoagulation and chemical coagulation pretreatment for enhanced virus removal using microfiltration membranes. Water Res 39(13):3098–3108

Zhu WP, Sun SP, Gao J, Fu FJ, Chung TS (2014) Dual-layer polybenzimidazole/polyethersulfone (PBI/PES) nanofiltration (NF) hollow fiber membranes for heavy metals removal from wastewater. J Membr Sci 456:117–127

Chapter 7
Sustainable Nanomaterials in Wastewater Remediation

Amrit Krishna Mitra and Saikat Nayak

Abstract A serious risk to the wellbeing of terrestrial and aquatic ecosystems is wastewater that contains harmful materials. It is essential to develop toxin-free technologies for wastewater treatment in order to support the sustainable advancement of global civilization. The field of green nanotechnology aims to achieve sustainability in a variety of applications. It also possesses a major influence on the development of greener, cleaner technologies that have substantial positive effects on human health and the environment. Owing to their unique features, nanomaterials (NMs) are currently the subject of extensive research for wastewater treatment. Various types of NMs are utilized for wastewater treatment, depending on the kind of contaminants and the required level of treatment efficiency. Novel and emerging NMs are evolving as the field of green nanotechnology continues to grow. For example, covalent organic frameworks, single-walled and multiple-walled carbon nanotubes (SWCNTs and MWCNTs), metal, and metal oxide-based nanoparticles (NPs) are among the NMs which can be utilized to remediate wastewater. The applications of graphene-based NPs, their oxides (GO), and reduced graphene oxide (rGO) are also extremely promising in wastewater treatment. Higher adsorption capabilities for organic contaminants are demonstrated by ferric oxide. Furthermore, magnetic nanopowder confers a low adsorption capacity for phenols. With excellent adsorption, pyrrolidone-reduced graphene oxide (PVP-RGO)-based NPs have been employed as adsorbents for the removal of inorganic target contaminant copper. Again, it has been demonstrated that zero-valent metal and metal oxide NPs—particularly iron oxides—are more efficient at recovering heavy metals from wastewater than conventional carbon nanotubes (CNTs). This chapter provides a thorough analysis of current developments in common NMs used for sustainable wastewater treatment. This study also evaluates in detail the appropriate uses, potential, and existing constraints of green nanotechnology in wastewater treatment. The novelty of the chapter lies in its review of contemporary applications of different kinds of NMs and

A. K. Mitra (✉)
Department of Chemistry, Government General Degree College, Singur, West Bengal, India
e-mail: ambrosia12june@gmail.com

S. Nayak
Department of Chemistry, Rani Rashmoni Green University, Tarakeswar, West Bengal, India

N. Agarwal et al. (eds.), *Nanomaterials in Wastewater Research*, Advances in Wastewater Research, https://doi.org/10.1007/978-981-96-3908-3_7

related research studies. Such an approach will be helpful to acquire insights into technological breakthroughs, applications, and future problems of nanotechnology deployment for wastewater treatment.

Keywords Wastewater · Nanocrystals · Nanomaterials · Membranes · Nanofibrils · Adsorbents · Absorbents · Disinfectants · Flocculants

7.1 Introduction

Water is a fundamental resource vital to the survival of various forms of life on Earth, making it one of our most valuable and essential resources for human development (Choudhary et al. 2020). However, the quality of water is deteriorating due to numerous human activities such as mining, deforestation, and agriculture, coupled with increasing demands for cleaner and safer water (Somsesta et al. 2020; Choudhury 2020). Industries including food processing, paper and pulp production, textiles, and pharmaceuticals are major contributors to water pollution (Alharbi 2018; de Lima et al. 2018). Access to clean drinking water is a global concern, exacerbated by inefficient wastewater management practices that are chemically complex and environmentally hazardous (Li et al. 2019; Elgarahy et al. 2019a).

Industrial wastewater often contains dyes and heavy metals, posing significant risks and necessitating effective treatment processes. Traditional methods like flocculation, filtration, and adsorption are commonly used due to their effectiveness, versatility, and affordability (Asghar and Mushtaq 2023; Sayyed et al. 2021). However, NPs have emerged as a promising solution for wastewater treatment due to their superior purification capabilities, biodegradability, and non-toxic nature (Soliman and Moustafa 2020; Elgarahy et al. 2019b). Water pollution encompasses a broad spectrum of contaminants, including pharmaceutical residues (e.g., paracetamol, tramadol, ciprofloxacin), herbicides, pesticides, phenols, and various heavy metals such as arsenic, aluminum, cadmium, lead, as well as dyes (Afolabi et al. 2020; Romero et al. 2020).

Water treatment methods are categorized into mechanical, thermal, chemical, physical, and biological processes. Among these, electrocoagulation is favored for its efficiency, reduced chemical usage, and minimal sludge production (Wang et al. 2020; Elgarahy et al. 2021). The exploration of green synthesized NMs for water remediation has been documented (Singh et al. 2021). This review paper delves into various green synthesis methods for NPs and their applications, including nanofiltration, nano-adsorbents, nanocatalysts, and sensor-based NMs. NMs, with their exceptional adsorbent properties and beneficial physicochemical characteristics, have proven effective in removing a wide range of pollutants. Nanotechnology offers innovative solutions to the challenges of water scarcity and wastewater treatment. Nanotechnology-based wastewater treatment methods present significant improvements over traditional approaches, addressing environmental issues while reducing labor, time, and costs (Bhat et al. 2022). Reviews of recent studies highlight the effectiveness of newly developed NMs in wastewater treatment, focusing on four main categories: nanotechnology combined with biological techniques, nano-adsorbents, nanocatalysts, and nanomembranes (Anjum et al. 2019).

7.2 Sustainable NMs by Design

Utilizing the unique chemical and physical properties of engineered NMs offers immense potential for advancing technologies that address global sustainability challenges. These include applications in high-performance solar cells, lightweight lithium-ion batteries, water treatment, environmental remediation, and agricultural productivity. To analyze these findings, two key concepts are employed: "ecosystem" and "value chain" (Babbitt and Moore 2018).

7.2.1 Ecosystem

The concept of an "innovation ecosystem" is widely referenced in the literature on innovation systems. This concept draws inspiration from the ecological term "ecosystem" (Jackson 2011). Malsch expanded this definition to include three interconnected communities of actors responsible for regulating nanotechnology (Malsch 2014). This approach allows diverse stakeholders (business, government, and others) to analyze decision-making processes regarding integrated risk management and sustainable NM manufacturing at a group level (Malsch 2014). The SUNDS decision support tool is being developed specifically for this context (Besley et al. 2008).

7.2.2 Value Chain

The term "value chain," coined by Porter (1985), refers to how a business adds value to acquired resources by transforming and selling them through core operations in subsystems, focusing on inbound logistics, operations, outbound logistics, marketing, and sales services. These subsystems are supported by auxiliary activities like infrastructure development, procurement, human resources management, and technological development (Yusof et al. 2018). Unlike the Nano futures plan, this value chain emphasizes business rather than technology (Malsch et al. 2015).

The European Green Deal represents a major legislative effort aimed at creating a sustainable, circular, inclusive, zero-pollution, and climate-neutral economy. It proposes a new industrial strategy to minimize adverse impacts on human health and the environment. The Safe-by-Design (SbD) concept involves identifying risks to human health and the environment to reduce uncertainties and potential hazards.

This approach ensures the material's safety throughout its entire life cycle, from research and development to production, recycling, and disposal (Babbitt and Moore 2018).

7.3 NMs Green Synthesis

Green synthesis is increasingly recognized as an effective approach for developing innovative NMs that can revolutionize large-scale production while ensuring safety for both human health and the environment (Duan et al. 2015). Green chemistry principles guide these methods, emphasizing the creation of eco-friendly chemical processes and products that minimize or eliminate harmful elements (Fig. 7.1). This approach aligns with environmentally friendly nanotechnology, which aims to support sustainable health and mitigate technological risks (Elgarahy et al. 2021).

The primary objectives of green nanotechnology are to develop novel NMs and produce products that can be applied across various fields, such as wastewater treatment (Ran et al. 2008). The greener synthesis of NMs establishes standards for creating safer, more stable, cleaner, and more sustainable NMs and nanoproducts (Wei Guo 2011). Green NMs are synthesized using natural resources like microorganisms and plants. Compared to chemical or microbiological methods, green NP synthesis is simpler, less expensive, more efficient, and less harmful to the environment. Adjusting reaction conditions, such as temperature and pH levels, can further refine the properties of these NMs.

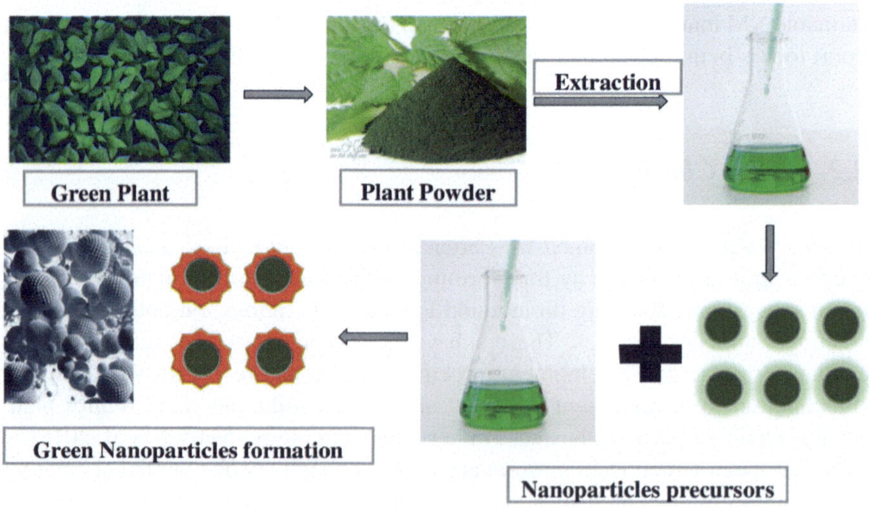

Fig. 7.1 Green synthesis of NMs using plant extracts (Elgarahy et al. 2021)

While green nanotechnology faces challenges, particularly in terms of lacking regulations and understanding the new risks it may present, the pursuit of sustainable and green solutions remains steadfast. Various techniques have been employed to synthesize a range of multifunctional, green-based NMs with distinct shapes, structures, scales, and crystallinities, exploring their potential applications (Gawande et al. 2013).

7.4 Green Synthesis Methods

NPs can be synthesized using natural reducing agents derived from biological microbes like bacteria and fungi, or from plant-based sources such as polysaccharides and other botanical extracts.

7.4.1 *Biological Synthesis*

This environmentally friendly method uses microorganisms like algae, bacteria, fungi, and actinomycetes to reduce metal ions, forming metallic NPs. The reduction can occur either within the cell or externally in the surrounding environment (Hasan 2015).

7.4.2 *Plant Extract-Mediated Synthesis*

Metal NPs can be synthesized using extracts from various plant species, sometimes involving live plants (Vijayaraghavan and Ashokkumar 2017; Kalia et al. 2020). The process typically involves mixing the plant extract with a metal salt solution, heating, and continuous stirring, sometimes assisted by microwave or ultrasonic radiation. Chemical compounds such as alkaloids, tannins, terpenoids, and coenzymes in the plant extract act as reducing agents (Khandel et al. 2018).

7.4.3 *Chemical Synthesis*

Various techniques are employed to create NPs, including enzymatic treatment, polyelectrolyte complexation, acid hydrolysis, ultrasonic methods, chemo-mechanical therapy, and nanoprecipitation through micro-emulsion methods and ionotropic gelation (Elgarahy et al. 2021).

7.4.3.1 Polyelectrolyte Complexation

This method involves the interaction of oppositely charged polymers or their salts to form polyelectrolyte complexes (PECs). An example is the creation of chitosan/streptokinase NPs (Baharifar et al. 2020).

7.4.3.2 Enzymatic Treatment

Enzymes like pullulanase break down highly branched amylopectin. For instance, experiments with pullulanase have shown that starch NPs, produced in a gelatinized maize starch solution, vary in size from 10 to 15 nm and have irregular morphologies (Lin et al. 2020).

7.4.3.3 Ultrasonic Synthesis

Ultrasonic piezoelectric conversion, operating at high voltage and frequencies of 20 kHz, is used to create NPs. Studies have shown how ultrasound pre-treatment affects the starch content, water content, pore size, and oil absorption in potato slices, as well as the ratio of immobilized to bound water (Zhang et al. 2021).

7.4.3.4 Acid Hydrolysis

This straightforward method is commonly used to produce semi-crystalline starch NPs and nanocrystals (Kumari et al. 2020).

7.4.3.5 Chemo-Mechanical Treatment

This technique combines microfluidization with TEMPO (2,2,6,6-tetramethylpiperidine-1-oxyl radical) oxidation to create nanofibers. For example, TEMPO-oxidized cellulose nanofibers derived from bagasse were used as sustainable raw materials for new nanocomposites, which were then applied in water treatment to remove dyes (Elgarahy et al. 2021).

7.4.3.6 Nanoprecipitation

Nanoprecipitation is a simple, reproducible single-step method for producing synthetic polymer NPs. For instance, starch NPs can be prepared by precipitating amorphous starch in ethanol (Safari and Sadeghi 2017; Zhang et al. 2020). Techniques like ionotropic gelation and micro-emulsion are used to produce NPs, where the size distribution can be refined to below 100 nm. In micro-emulsion, NPs form in the aqueous core of reverse micelles and are cross-linked by an agent (Elwakeel et al. 2021; Moradi et al. 2020). Excess surfactants can be removed by calcium chloride precipitation, and organic solvents are eliminated through low-pressure evaporation (Dang and Lai 2024).

7.5 Role of NMs

7.5.1 Carbon-Based NMs

7.5.1.1 Carbon Nanotubes (CNTs)

Carbon-based NMs, particularly CNTs, have garnered significant attention in wastewater treatment due to the large volume of wastewater generated and the abundant carbon resources available in nature. Since their discovery in 1991, CNTs have been valued for their exceptional mechanical, thermal, chemical, and electrical properties, making them ideal for various applications, including energy storage, medical devices, nanocomposites, micro-electronics, adsorption, and contaminant treatment. They are particularly effective in environmental clean-up, pollutant transformation, and catalysis. Their high adsorption capacity and efficiency make CNTs especially useful for removing contaminants like lead, benzene, copper, zinc, cadmium, and dyes from water bodies (Khan et al. 2021).

CNTs are classified into single-walled CNTs (SWCNTs) and multi-walled CNT (MWCNTs). Compared to other carbon-based materials, CNTs offer a shorter equilibrium time, lower membrane fouling rates, and higher adsorption capacity. Their solid mesoporous structure, high aspect ratio, large-surface-area-to-volume ratio, and fibrous shape make CNT-based or organic membranes highly effective in various applications (Kurwadkar et al. 2019).

The ion exchange capacity of CNTs and their ability to absorb waste varies depending on the type of oxidation applied by different functional groups. For instance, the adsorption capacity for cadmium ions (Cd^{2+}) can be significantly enhanced by treating CNTs with H_2O_2 and HNO_3, increasing the adsorption capacity to 2.6 and 5.1 mg/g, respectively, at comparable equilibrium concentrations. Although $KMnO_4$-oxidized CNTs have a smaller functional group surface area than HNO_3-oxidized CNTs, they exhibit higher Cd^{2+} adsorption capacity, possibly due to the adsorption of residual MnO_2 particles. However, further research is needed to fully understand Cd^{2+} adsorption by $KMnO_4$-oxidized CNTs (Li et al. 2007).

By integrating the magnetic properties of iron oxide with CNTs' adsorption capabilities, CNTs can effectively remove chromium. Additionally, combining CNTs with various metal materials allows for modifications to their mechanical, electrical, and optical properties. It was demonstrated that graphene-coated CNTs with polarized edges could act as electron-donating π-donors, where electron donors may adhere to the denser adsorbent's imperfect areas. Furthermore, an additional n-π

electron-donor-acceptor (EDA) reaction may occur, where electrons from amino or hydroxyl groups interact with the electron-deficient areas of CNTs, acting as π-electron acceptors (Kurwadkar et al. 2019).

Membrane technology represents another avenue where CNTs are utilized in wastewater treatment. CNTs are extensively researched and applied as adsorbents in membrane technologies for filtration and water purification. CNTs composite adsorbents have successfully reduced fluoride levels in groundwater. Currently, 6% of the CNTs produced worldwide are used in wastewater treatment. While CNTs have shown no toxicity to algae in natural water, the risk to aquatic life increases at higher temperatures at the point of release (Ahmed et al. 2022; Nasrollahzadeh et al. 2021).

7.5.1.2 Graphene and Graphene Oxide-Based NMs

Graphene, similar to natural graphite, is a single-layer carbon-based structure composed of just one carbon atom. This carbon-layered material possesses remarkable properties, making it an effective choice for various applications. Graphene is highly valued for its large surface area, diverse functional groups, cost-effectiveness, sensitivity, high selectivity, and exceptional permeability to a wide range of chemical compounds. These characteristics, combined with its simple structure, make graphene a highly sought-after material for pollutant removal in wastewater treatment (Yin et al. 2020).

Graphene oxide (GO) is produced by chemically oxidizing layers of graphite, resulting in a two-dimensional material. GO's ability to mask oxidative defects improves its membrane properties, and its chemical inertness adds to the durability of GO membranes. Graphene sieves can effectively remove ions from water, and researchers have achieved a 97% reduction of sodium chloride, indicating that graphene oxide can be used to separate ions from saline water. Additionally, graphene has demonstrated the capability to remove various heavy metals, including lead, cadmium, zinc, copper, and cobalt. While it may have lower adsorption selectivity, graphene exhibits strong adsorption capabilities for these metals. The carboxyl and hydroxyl functional groups on GO play a crucial role in attracting and adsorbing heavy metals from wastewater (Ahmad et al. 2020).

Graphene nanotube (GNT) composites, a combination of graphene and carbon nanotubes, are also effective in removing contaminants like Methyl Blue (MB). Due to its exceptional mechanical strength, light weight, and flexibility, GO is a valuable tool for removing harmful heavy metals from wastewater, protecting aquatic ecosystems. Additionally, dropped 3D graphene has shown a 90% removal efficiency of ibuprofen from wastewater within just 180 min (Ponnusamy et al. 2021). Graphene has also been proven effective in removing antibiotics from wastewater, making it a powerful material for addressing water pollution (Ahmed et al. 2022).

7.5.1.3 Carbon Dots (C-Dots)

Carbon dots (C-dots) are an emerging type of carbon NM that has gained significant attention in recent years. Among various non-biomedical applications, C-dots hold great potential in fields like photocatalysis and optical devices. These NMs can be synthesized through various methods, including hydrothermal reactions, electrochemical oxidation, microwave irradiation, laser ablation, ultrasonic tin, and reflux techniques.

C-dots exhibit remarkable semiconductor properties, photoluminescent characteristics, and strong antibacterial capabilities. These attributes make them suitable for detecting heavy metal ions, removing organic and inorganic pollutants, and catalytically breaking down wastewater contaminants (Ramar et al. 2018). Due to their ability to emit light across different wavelengths, C-dots can function as photo-Fenton catalysts. They have been shown to catalyze the degradation of hydrogen peroxide (H_2O_2) in photocatalytic systems. Studies have explored the potential of C-dots to enhance the photo-Fenton reaction by accelerating H_2O_2 decomposition to produce free hydroxyl radicals, which in turn help eliminate Methyl Blue. The results indicate that C-dots with a broader light absorption range exhibit enhanced catalytic activity in photo-Fenton reactions. This makes the C-dots photo-Fenton reaction highly promising for wastewater treatment, as it can operate effectively across a wide pH range, from acidic to basic conditions.

The abundance of functional groups and various polar moieties on the surface of C-dots make them excellent adsorbents for removing toxic substances from wastewater. In one example, a nanocomposite adsorbent was created by combining magnetic materials ($MnFe_2O_4$) with polyethyleneimine-functionalized C-dots (Huang et al. 2019). This composite showcases the versatility and potential of C-dots in environmental remediation (Ahmed et al. 2022).

7.6 Approaches of NMs in Wastewater Treatment

7.6.1 Adsorption of Dyes

Dyes are organic compounds widely used in industries such as textiles, paper, plastics, and food. When these dyes are released into water bodies through industrial effluents, they can cause significant environmental issues due to their toxicity and potential to pollute water sources. Adsorbents are materials that can attract and hold dye molecules on their surface through physical or chemical interactions. Common adsorbents for dye removal include activated carbon, silica gel, zeolites, and various clay minerals. These materials offer a large surface area and possess functional groups that can form bonds with dye molecules, effectively removing them from water. Adsorption processes are commonly used in wastewater treatment facilities to reduce the environmental impact of dye-laden effluents (O'Connell et al. 2008).

7.6.2 Adsorption of Heavy Metals

Heavy metals such as lead, cadmium, mercury, and chromium are toxic pollutants frequently found in industrial wastewater. Exposure to these heavy metals can lead to severe health problems in humans and disrupt aquatic ecosystems. Adsorbents play a vital role in removing heavy metals from water through adsorption processes. Materials like activated carbon, zeolites, clay minerals, and modified polymers have been extensively studied and applied as adsorbents for heavy metal removal. For example, in 2016, Sirviö et al. developed bisphosphonate-functionalized cellulose nanocrystals (CNCs) for the removal of vanadium ions. The electrostatic interaction between cationic vanadium species and anionic acid groups, along with the complexation of vanadium ions with the bisphosphonate-functionalized CNCs, proved highly specific at very low pH levels (2–3). The maximum adsorption capacity (qmax) of vanadium ions was found to be 1.98 mmol/g. (Sirviö et al. 2016).These adsorbents possess specific surface properties and functional groups that enhance the binding of heavy metal ions, effectively reducing their concentration in water to safe levels.

7.6.3 Adsorption of Other Water Contaminants

In addition to dyes and heavy metals, water often contains various other contaminants such as organic pollutants, pesticides, pharmaceuticals, and emerging contaminants like microplastics and perfluorinated compounds. Adsorbents are versatile tools for removing these contaminants due to their high surface area and selective adsorption capabilities. Different types of adsorbents are used depending on the nature of the contaminants. For example, activated carbon is commonly employed to remove organic pollutants due to its high adsorption capacity and chemical stability. In 2014, Chen et al. developed β-cyclodextrin-modified CNCs@Fe_3O_4@SiO_2 superparamagnetic nanorods for the adsorption of pharmaceutical compounds, such as procaine hydrochloride and imipramine hydrochloride (Chen et al. 2014). The hydrophobic cavity of the β-cyclodextrin-modified nanorods demonstrated the ability to adsorb pharmaceutical residues. Emerging contaminants like microplastics may require specialized adsorbents that can be separated from an aqueous solution using a magnetic field. Recent studies, such as one by Herrera-Morales et al. in 2017, have explored the use of polyethylene glycol (PEG) modified carbon nanocomposites for the adsorption of pharmaceutical compounds like acetaminophen, N,N-diethyl-meta-toluamide, and sulfamethoxazole (Herrera-Morales et al. 2017). Porous polymers or modified zeolites have also been designed to effectively capture smaller particles. Overall, adsorption processes provide a cost-effective and efficient method for treating water contaminated with a broad range of pollutants (Mohammed et al. 2018).

7.7 Nanocatalysts

Nanocatalysts are materials engineered at the nanoscale to accelerate chemical reactions, leveraging their high surface area and specialized catalytic sites. The distinct characteristics of NMs—such as their high surface-to-volume ratio, quantum confinement effects, and enhanced surface reactivity—make them highly effective for various chemical transformations (Tiwari et al. 2008). Nanocatalysts encompass a broad spectrum, including metal NPs, nanotubes, nanocomposites, and bimetallic NPs and are utilized in fields such as environmental remediation, energy conversion, pharmaceutical synthesis, and petrochemical industries (Hsien et al. 2001). Commonly used NMs for catalysis include metal NPs (e.g., gold, silver, palladium), metal oxides (e.g., titanium dioxide, iron oxide), carbon-based materials (e.g., carbon nanotubes, graphene), and zeolites. For instance, titanium dioxide (TiO_2) NPs are noted for their photocatalytic properties and are frequently employed in water purification (Prieto et al. 2005; Nowack 2008). Iron catalysts are valued for their recyclability due to their ferromagnetic properties. Sherin et al. utilized Terminalia bellerica kernel extract to synthesize silver NPs, which demonstrated catalytic reduction of organic pollutants such as methyl orange and 4-nitrophenol (Sherin et al. 2020). Zero-valent iron is effective for reducing both organic and inorganic contaminants, with reaction rates 25–30 times faster than granular iron. Photocatalytic methods, which do not require costly oxidizing agents compared to other advanced oxidation techniques, are increasingly used for toxin removal from wastewater. Nanocatalysts offer significant advantages over traditional catalysts, including higher catalytic activity, selectivity, stability, and reduced reaction times and energy consumption (Kalidindi and Jagirdar 2012).

7.8 Nanomembranes

Nanomembranes are ultra-thin films or sheets with nanoscale thicknesses, ranging from a few nanometers to several hundred nanometers. These membranes are engineered with specific pore sizes, surface properties, and molecular selectivity to precisely control the transport of molecules and ions. Nanomembranes are used in a variety of applications, including water purification, gas separation, drug delivery, and sensing (Ulbricht 2006). In water purification, nanomembranes effectively remove contaminants such as bacteria, viruses, heavy metals, and organic pollutants through filtration or separation processes. Common materials for nanomembrane fabrication include polymers (e.g., polyamide, polysulfone), ceramics (e.g., alumina, titania), and composites incorporating NMs like carbon nanotubes or graphene oxide (Jie et al. 2015; Kazemimoghadam 2010). Inorganic membranes often feature asymmetric, dense, porous ceramic supports made from materials like silica, zirconia, and alumina. Ceramic membranes aim to develop ultra-filtration membranes with water permeability similar to nanofiltration and reverse osmosis membranes. Zeolite ceramic membranes are widely studied for reverse osmosis desalination in wastewater treatment. Additionally, carbon nanotubes are important in synthesizing polymer composite membranes, which exhibit high flexibility, large aspect ratios, and low mass density, enhancing their performance. Recent studies have highlighted hybrid nanostructured membranes with thin film nanocomposite properties, showing potential for antimicrobial activity and effective separation. Nanomembranes offer numerous advantages, including high flux rates, excellent anti-fouling properties, mechanical strength, proton conductivity, oxidative and thermal stability, and tunable selectivity, making them promising for next-generation membrane technologies (Lou et al. 2016; Danquah and Sharadwata Pan 2018).

7.9 Metal and Metal Oxide NPs

Zero-valent metals such as Fe^0, Sn^0, Ti^0, Al^0, Zn^0, Mg^0, Ni^0, and Pb^0 are among the most abundant elements in the Earth's crust. These metals are highly reactive and have magnetic properties, making them effective for reducing heavy metals in wastewater treatment (Rambabu et al. 2021). Nano-sized metal oxides (NMOs) like zinc oxide, titanium oxide, magnesium oxide, aluminum oxide, and iron oxide are also used as adsorbents for pollutant removal. The reduction in size from micrometers to nanometers increases the stability of metal oxides due to reduced surface energy. Magnetic composite adsorbents based on NMOs facilitate easy separation and reduce operational costs in wastewater treatment. Different forms of NMOs exhibit varying adsorption capacities for adsorbates at the same temperature. For example, ferric oxide (Fe_2O_3) and ferric oxyhydroxide (FeOOH) have adsorption capacities of 26.6 mg/g and 149.25 mg/g, respectively, at 25 °C. NMOs are extensively studied for their potential to expel contaminants from wastewater due to their high adsorption capabilities and rapid energy transfer. However, the toxicity and cytotoxicity of metal oxide NPs remain a concern, and eco-friendly production methods and performance improvements are on-going challenges (Agrawal and Sahu 2006).

7.9.1 Zinc Oxides

Zinc oxide NPs (ZnO-NPs) are known for their ability to inhibit microbial growth, exhibit high bioactivity, and facilitate bacterial elimination. ZnO also enhances photocatalytic activity, making it effective in dye removal under UV irradiation. The removal efficiency of dyes increases with radiation intensity, though higher pH levels can hinder this process. ZnO nanoflowers synthesized using sea buckthorn fruit achieve 100% degradation efficiency within a third of the original photocatalyst's degradation time (Taju et al. 2014; Rupa et al. 2019a). Somu and Paul produced ZnO-NPs using casein as a reducing and capping agent, which imparts a negative charge to the NPs, enhancing metal cation adsorption. ZnO-NPs also display antimicrobial properties and effective dye degradation (Somu and Paul 2018).

7.9.2 Titanium Oxides

Titanium dioxide (TiO_2) NPs are known for their high stability and toxicity, making them effective in wastewater treatment through biological processes (Sethy et al. 2020). TiO_2 NPs, with diameters of 18 nm, demonstrate efficient removal of chemical oxygen demand (COD) and lead (Pb^{2+}) through photocatalytic degradation. TiO_2 has garnered attention since 1972, when Fujishima and Honda demonstrated its photocatalytic capabilities. TiO_2 is valued for its high photocatalytic activity, affordability, and stability under UV and biological conditions. It has a wide bandgap (3.2 eV), requiring UV light for excitation and charge separation, which leads to the generation of reactive oxygen species (ROS) capable of degrading contaminants quickly (Fig. 7.2). TiO_2 NPs are effective across a broad range of contaminants, including dyes, pesticides, arsenic, cyanide, and polycyclic aromatic hydrocarbons. Recent studies have focused on TiO_2 magnetic NPs synthesized in a spinning disk reactor, allowing for efficient recovery by magnetic trapping, making the process suitable for industrial applications (Lu et al. 2016).

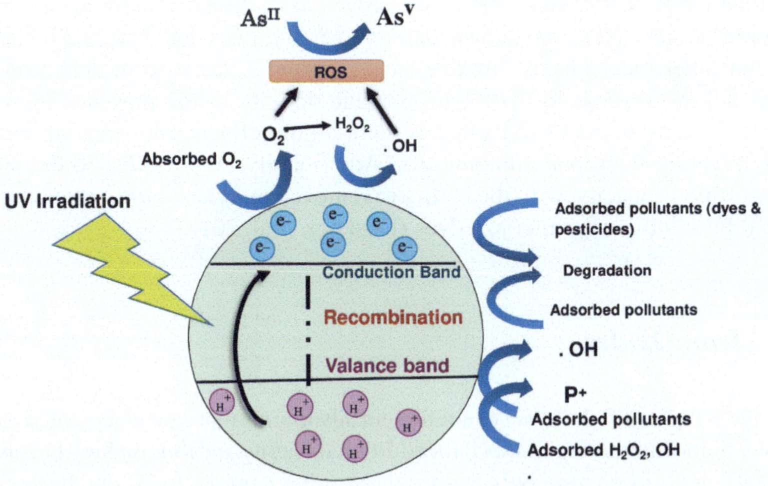

Fig. 7.2 Microbial synthesis of TiO_2 NPs

7.9.3 Magnesium Oxides

Phosphorus is an essential component in agriculture and industry, but its excessive presence in water systems can have detrimental effects on ecosystems. Therefore, the removal of phosphorus from water is crucial. Research has shown that magnesium oxide NPs (MgO-NPs) are effective in removing phosphorus from water. Additionally, MgO-NPs are capable of removing heavy metals such as lead and chromium. Studies indicate that MgO-NPs can effectively reduce Pb^{2+} to Pb^0 and Cr^{3+} to Cr^{3+}, with higher adsorption activity observed at elevated pH levels. The degradation efficiency of MgO-NPs positively correlates with increasing pH (Mohamadi et al. 2021). Furthermore, MgO-NPs have shown promise in treating wastewater by reducing petrochemicals like toluene, benzene, xylene, and ethylbenzene through reduction processes. Biosynthesized MgO-NPs have also been demonstrated to decolorize wastewater and reduce levels of total suspended solids (TSS), chemical oxygen demand (COD), and total dissolved solids (TDS), while exhibiting antimicrobial activity. Green-synthesized MgO-NPs have been effective in reducing bacterial activity on cotton fabrics, thereby improving water systems.

7.9.4 Aluminum Oxides

Aluminum oxide NPs (Al_2O_3-NPs) have proven to be effective in impacting microbial communities. NPs are known to damage microbes by disrupting bacterial DNA and other mechanisms. Studies have reported a decrease in dehydrogenase activity and an increase in hydrolytic enzyme activity in the presence of Al_2O_3-NPs. The toxicity of Al_2O_3-NPs makes them highly effective for treating wastewater by targeting microbial communities. Additionally, Al_2O_3-NPs exhibit significant adsorption capacity, with the ability to remove methylene blue, indicating their potential for eliminating other azo dyes (Doskocz et al. 2018).

7.9.5 Iron Oxides

Iron oxide NPs (Fe_2O_3-NPs) serve as efficient adsorbents that can be separated using a magnetic field, eliminating the need for additional energy and thus making the process energy-efficient. Iron oxide NPs, with sizes ranging from 21 to 70 nm, have a wide range of applications due to their cost-effectiveness. Fe_2O_3-NPs, synthesized from iron salts and NaOH to produce maghemite (γ-Fe_2O_3) and magnetite (Fe_3O_4), are particularly effective in removing toxic mercury (Hg) from water, thereby protecting environmental safety. Functionalizing iron oxide with glycolipid biosurfactants like rhamnolipid, which is extracted from *Pseudomonas aeruginosa*, has been shown to reduce toxicity and enhance the biodegradation capacity of chemical dyes. Fe_2O_3-NPs have also been studied for their ability to adsorb COD and ammonia nitrogen (NH_3-N). The findings suggest that iron oxide NPs are not only eco-friendly but also highly efficient and reusable, making them a promising option for water treatment applications (Vélez et al. 2016).

7.10 Antimicrobial NMs

Current disinfection methods are effective at removing pathogens from drinking water (Kidd et al. 2020). Recent research has addressed gaps in water treatment, particularly in balancing disinfection and the formation of disinfection by-products (Mazhar et al. 2020). In the industry, chlorine, ozone, and chloramines are commonly used as disinfectants in water treatment. However, these substances can react with other compounds to produce harmful disinfection by-products, many of which are carcinogenic.

TiO_2 is a widely used NP that is highly effective in inactivating microbes in drinking water, surface water, and other water sources. TiO_2 exhibits antibacterial properties due to the production of reactive oxygen species. Solar disinfection using titanium dioxide (SDO TiO_2) is particularly significant because it enhances TiO_2's ability to absorb visible light and UV-A rays, improving its antimicrobial effectiveness. Like TiO_2, silver NPs also demonstrate strong antimicrobial properties. Silver NPs are extensively used in both medical applications and external treatments. They act as antifungal agents and are commonly employed in water decontamination. Silver NPs deactivate essential enzymes and -SH groups by ionizing silver, which disrupts cell membranes and inhibits DNA replication. The toxicity of silver depends on the rate at which silver ions are released (Fig. 7.3). In developing countries, silver NMs are used to protect against bacterial contamination (Qu et al. 2013).

Similarly, zinc oxide (ZnO) exhibits excellent antibacterial properties and is used in products like paint, sunscreen, and coatings due to its high UV absorption capacity (Rupa et al. 2019b). However, ZnO NMs can dissolve zinc species in water. Due to their permeable or insoluble characteristics, ZnO NMs present some limitations in water treatment applications. Table 7.1 displays the antimicrobial mechanisms of various NMs.

7.11 NMs for Heavy Metal Recovery

The development of nuclear energy has led to significant environmental challenges, including air pollution and the release of heavy metals like cadmium (Cd) and uranium (U) into the environment (Dong et al. 2018). To ensure the sustainable growth of nuclear energy while minimizing its environmental impact, it is crucial to recover these heavy metals from wastewater.

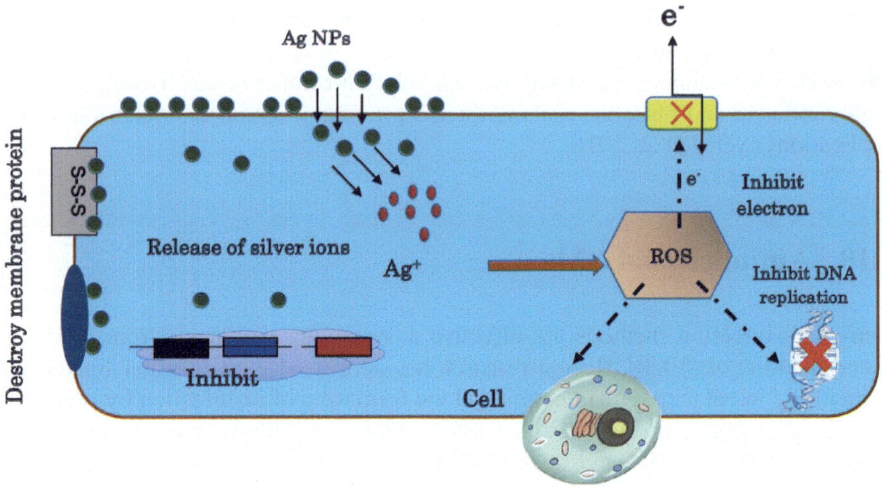

Fig. 7.3 Schematic diagram mechanisms of silver NPs (Bhat et al. 2022)

Table 7.1 NM's antimicrobial mechanism (Qu et al. 2013)

NM	Antimicrobial mechanisms
Nano-ZnO	Release of zinc ions, production of H_2O_2, membrane damage
Graphene-based NMs	Membrane damage, oxidation stress
Nano-MgO	Membrane damage
Nano-Ag	Release of silver ions, protein damage, suppression of DNA replication
Nano-Ce_2O_4	Membrane damage
nC_{60}	ROS-independent oxidation
Fullerol and aminofullerene	Production of ROS
Nano-TiO_2	Production of ROS
Carbon nanotubes	Membrane damage, oxidation stress

Traditional methods for heavy metal recovery include ion exchange, solid–liquid extraction, biological enrichment, and reductive precipitation. However, these techniques are often costly, limiting their widespread practical use.

Metal-free graphitic carbon nitride (g-C_3N_4) has emerged as a promising material for heavy metal recovery due to its low cost, environmental friendliness, high chemical and thermal stability, and effectiveness as a heavy metal scavenger (Hao et al. 2018, 2019).

Industrial wastewater often contains heavy metals like mercury (Hg), cadmium (Cd), and uranium (U), which are typically found in acidic conditions, necessitating adsorbents with high stability. Advanced covalent organic frameworks (COFs), which are linked by strong covalent bonds, offer high stability across a range of aqueous environments. For instance, Huang et al. (2017) synthesized a stable TAPB-BMTTPA-COF by incorporating methyl sulfide units onto the phenyl groups' edges and sulfur-containing functionalities into the COF structure. The active sulfur sites effectively capture Hg^{2+} ions (Huang et al. 2017).

In recent years, COFs have shown promise as adsorbents for heavy metal recovery, demonstrating high adsorption efficiency for various metal ions, including soft Lewis acids (Ag^{+}, Pb^{2+}), hard Lewis acids (Fe^{3+}, Cr^{3+}), and borderline Lewis acids (Cu^{2+}, Ni^{2+}), with removal efficiencies exceeding 85% (Zhou et al. 2018).

7.12 Safety, Toxicity, and Environmental Impact of NMs

The application of nanotechnology in water treatment raises significant concerns regarding safety and environmental impact. Due to the limited understanding of NMs' behavior, their large-scale use in wastewater treatment has been delayed.

Environmental Impact: One of the main concerns is the potential danger and harmful effects that NMs may have on human health and the environment. NPs can enter the bloodstream through the skin, inhalation, or ingestion, leading to damage to membranes and other biological effects. More comprehensive studies are needed to prevent contamination in water and ensure a safe environmental impact (Ali 2012).

Toxicity: The toxicity of NMs is a critical concern that cannot be ignored, although it remains poorly understood and highly uncertain. The toxic nature of NMs is primarily due to their strong chemical reactivity, particularly in metal oxides and carbon nanotubes. A study by Salehpour et al. examined the ecotoxicological and biodegradation effects of pure polyvinyl alcohol (PVA) films and cellulose nanofibrils (CNFs) in municipal solid waste remediation. The results showed that PVA films degraded more quickly than CNFs and did not negatively impact the growth and germination of vegetable species such as spinach and cress (Salehpour et al. 2018).

Before NMs can be widely used in water treatment, it is essential to identify and address key challenges, including safety assessment, large-scale manufacturing facilities, safe waste disposal, and energy efficiency (Table 7.2). Scientists must invest time and effort to bridge this knowledge gap, ensuring that the use of NMs in water treatment has positive impacts on human health and the environment.

7.13 Sustainable Nanotechnological Approaches in Wastewater Treatment

As the global demand for clean water rises due to heavy industrialization, urbanization, and other factors, the availability of fresh water is decreasing while wastewater volumes are increasing. This situation has heightened the need for effective

Table 7.2 Self-toxic effect of NMs (Bhat et al. 2022)

NM	Self-toxic effect observation	References
Au/Carbon Composite	Lower-range non-toxic	Jia et al. (2005)
Carbon, Al_2O_3 composites, and metal	Concentration and time-dependent	Ajdary et al. (2018)
Ag/Carbon composite	Dose and time-dependent but toxic	Karlsson et al. (2008)
CdSe quantum dots	Acute toxicity observed	Ray et al. (2009)
Single and multi-walled carbon nanotubes	Increase in toxicity at concentrations above 15 μg/cm	Rajkumar et al. (2016)
Hexadecylcetyltrimethyl ammonium bromide (CTAB)	Extremely poisonous in certain concentrations	Takahashi et al. (2006)
Fe_2O_3 and composite carbon nanotubes	Toxic effect and damage to DNA at the lowest concentrations	Karlsson et al. (2008)
Carbon Composite/TiO_2/Al_2O_3	More toxic at the level of microns	Hubbs et al. (2011)

and sustainable water purification methods, particularly in developing and industrialized nations. Nanotechnology offers significant potential for detecting contaminants in wastewater and enhancing treatment processes. Traditional decontamination methods like ozonation and chlorination consume large amounts of chemical agents and can produce harmful by-products, often referred to as disinfection by-products. Nanotechnological advancements, therefore, offer a promising pathway for sustainable water treatment solutions (Iravani 2021).

7.13.1 Sustainable Groundwater Treatment Using Nanotechnology

In industrial settings, groundwater is often contaminated with solvents like benzene and organic pollutants such as ethylene. Conventional treatment methods focus on removing contaminants, destroying pollutants, and returning purified water to the environment. NMs, with their large surface areas, offer advantages due to their high reactivity, which can create more sustainable treatment options. Iron-based NMs, such as zero-valent iron NPs and Carbon-Iron nanocomposites, are particularly effective in cleaning up pollution (Mackenzie et al. 2012). These nanoscale iron particles can immediately remediate a range of organic compounds in water, including ethanes, benzenes, and chlorinated biphenyls. They are also used to remediate dense non-aqueous phase liquid contaminants in aquifers. Despite some limitations, certain NMs can effectively remove arsenic from groundwater. Additionally, cellulosic NMs are used to remove contaminants due to their unique properties, such as low cost and renewability. For example, nano-fibrillated cellulose poly aerogel has proven effective in removing heavy metal ions like Pb^{2+}, Cd^{2+}, Ni^{2+}, and Zn^{2+}, with adsorption capacities of 165, 135, 117, and 138 mg/g, respectively. Carbon nanotubes, metal, and metal oxide NPs are other promising NMs for destroying bacterial cell membranes without producing toxic by-products, offering a significant advantage over traditional disinfection methods like chlorination and ozonation (Elliott and Zhang 2001).

7.13.2 Sustainable Wastewater Treatment Using Nanotechnology

NMs possess unique chemical and physical properties, making them suitable for various applications, including wastewater treatment. Their high specific surface area and chemical reactivity make them particularly effective in this field. Industrial sources, including factories and laboratories, discharge wastewater containing hazardous organic compounds like herbicides, pesticides, azo dyes, and phenols. These contaminants must be removed before the wastewater is released into the environment to prevent serious harm to living organisms. Conventional treatment methods, such as biological processes, combustion, adsorption, activated carbon, and chemical oxidation, are costly and often fail to fully degrade organic compounds, leading to secondary contamination. NM-based photocatalysis, on the other hand, offers a more sustainable solution for water purification due to its environmentally friendly nature and complete mineralization of hazardous organic chemicals.

Photocatalysis involves a light-induced reaction enhanced by the presence of a catalyst. Semiconductor NPs and nanocomposites such as TiO_2, ZnO, ZnS, Si-Ti, Zn-CeO_2, and CdS-TiO_2 are highly effective for the photocatalytic degradation of organic contaminants. The main advantage of nanosized semiconductors is the quantum size effect, which increases the bandgap energy as the particle size decreases. When photons with high bandgap energy strike the surface of these nanosized semiconductors, valence electrons are excited and move to the conduction band, where they undergo charge separation and oxidize organics at the semiconductor surface. These methods are low-cost, reusable, eco-friendly, and capable of complete degradation, making them more sustainable than conventional water treatment methods.

Polyethylene glycol (PEG)-modified cellulose nanocrystal composites have shown promise in absorbing pharmaceutical compounds from aqueous solutions, effectively removing hydrophobic drugs from water (Norrahim et al. 2021). Carbon nanotube-based nano-adsorbents are also active in dye removal due to their cylindrical hollow structure, large surface area, and hydrophobic walls. However, the high production cost of carbon nanotubes limits their application in large-scale municipal wastewater treatment. The hydrophobicity, high reactivity, and reusability of graphene-based materials further enhance their potential for water treatment applications.

With the increasing demand for clean water driven by urbanization, scientists and industries are focusing on treating saline water sustainably. Biofouling is a critical challenge in surface and seawater treatment, and reverse osmosis is the primary method used to address it. Adsorption-based techniques, including the use of carbon nanotubes and graphene oxide membranes, are proving effective in treating saline water (Tom 2021).

7.14 Contemporary and Emerging Applications in Wastewater Treatment

Sustainable NPs and nanostructured materials have become increasingly valuable in water purification due to their cost-effectiveness and environmental benefits. Nanotechnology plays a crucial role in addressing the challenges of wastewater treatment, enabling the recycling of wastewater and the removal of contaminants to achieve purified water. The potential applications and functionalities of NMs in this field are explored below (Table 7.3). Traditional remediation techniques often suffer from drawbacks such as high costs, limited effectiveness, and time-consuming processes. In contrast, sustainable NMs offer solutions to these challenges, providing promising avenues for wastewater purification.

Adsorption is a widely used method for removing organic and inorganic contaminants in wastewater treatment. Compared to conventional adsorbents, nano-adsorbents provide advantages such as a higher specific surface area, shorter intra-particle diffusion distances, and enhanced surface chemistry. Carbon nanotubes, for example, have a superior adsorption capacity and can adsorb various polar organic compounds. This efficiency is due to interactions between carbon nanotubes and contaminants through π-π interactions, covalent bonding, and electrostatic forces (Yang and Xing 2010). Research has demonstrated that carbon nanotubes are more effective than activated carbon in removing heavy metals like Cu^{2+}, Pb^{2+}, Cd^{2+}, and Zn^{2+} (Li et al. 2003; Lu et al. 2006).

Graphene-based NMs have also shown efficacy as adsorbents for treating wastewater containing pharmaceuticals. Graphene oxide, in particular, is effective in removing heavy metals due to its high colloidal stability, large surface area, and ease of recovery. Additionally, nanocatalysts exhibit photocatalytic activity under UV light, with low toxicity to humans. Metal and metal oxide NPs such as MgO, TiO_2, ZnO, Al_2O_3, and Fe_2O_3 possess antifungal and antimicrobial properties, making them useful in wastewater remediation. These materials not only sense and monitor contaminants but also treat wastewater to produce clean water.

Looking ahead, it is crucial to continue research in this area to overcome existing challenges and develop even more sustainable approaches for using NMs in wastewater remediation.

Table 7.3 Current and potential applications of nanotechnology in water and wastewater treatment (Qu et al. 2013)

Applications	Representative NMs	Desirable NM properties	Enabled technologies
Membranes and membrane processes	Nano-zeolites	Molecular sieve, hydrophilicity	High permeability thin film nanocomposite membranes
	Nano-Ag	Strong and wide-spectrum antimicrobial activity, low toxicity to humans	Anti-biofouling membranes
	Carbon nanotubes	Antimicrobial activity (unaligned carbon nanotubes)	Anti-biofouling membranes
		Small diameter, atomic smoothness of inner surface, tunable opening chemistry, high mechanical and chemical stability	Aligned carbon nanotube membranes
	Aquaporin Nano-TiO_2	High permeability and selectivity Photocatalytic activity in UV and possibly visible light range, low human toxicity, high stability, low cost	Aquaporin membranes Photocatalytic reactors, solar disinfection systems
	Nano-magnetic	Tunable surface chemistry, superparamagnetic	Forward osmosis
Adsorption	Carbon nanotubes	High specific surface area, highly accessible adsorption sites, diverse contaminant-CNT interactions, tunable surface chemistry easy reuse	Contaminant preconcentration/ detection of recalcitrant contaminates
	Nanoscale metal oxide	High specific surface area, short intraparticle diffusion distance, more adsorption sites, compressible without significant surface area reduction, easy reuse, some are superparamagnetic	Adsorption media filters, slurry reactors
	Nanofibers with core-shell structure	Tailored shell surface chemistry for selective adsorption, reactive core for degradation, short internal diffusion distance Molecular sieve, hydrophilicity	Reactive nano-adsorbents
Disinfection and microbial control	Nano-Ag	Strong and wide-spectrum antimicrobial activity, low toxicity to humans, ease of use	POU water disinfection, anti-biofouling surface

(continued)

Table 7.3 (continued)

Applications	Representative NMs	Desirable NM properties	Enabled technologies
	Carbon nanotubes	Antimicrobial activity, fiber shape, conductivity	POU water disinfection, anti-biofouling surface
	Nano-TiO_2	Photocatalytic ROS generation, high chemical stability, low human toxicity and cost	POU of full scale disinfection and decontamination
Sensing and monitoring	Carbon nanotubes	Large surface area, high mechanical strength and chemical stability, excellent electronic properties	Electrochemical detection, sample preconcentration
	Quantum dots	Broad absorption spectrum, narrow, bright and stable emission which scales with the particle size and chemical component	Optical detection
	Dye-doped silica NPs	High sensitivity and stability, rich silica chemistry for easy conjugation	Optical detection
	Magnetic NPs	Tunable surface chemistry, superparamagnetism	Sample preconcentration and purification
	Noble metal NPs	Enhanced localized surface plasmon resonances, high conductivity	Optical and electrochemical detection

7.15 Major Challenges, Knowledge Gaps, and Issues with NMs in Water Treatment

Nanotechnology has made significant advancements in wastewater purification, offering cost-effective solutions for contaminant removal. However, several key challenges must be addressed before NMs can be widely applied in practical water treatment processes. Despite the numerous benefits, the potential impact of NPs on human health and ecosystems remains a critical concern due to their toxic nature (Soni et al. 2020).

The small size of NMs presents a significant challenge in their recovery and separation after use in wastewater treatment processes. Additionally, nanofibers, while possessing unique properties, face issues such as fouling and shrinkage, which can lead to poor mechanical performance and structural deterioration after treating wastewater. Although strategies like using magnetic fields to recover carbon nanotubes from water after wastewater treatment have shown promise, the lack of comprehensive knowledge and understanding of nanotechnology poses challenges in recovering NMs. To commercialize nanotechnology effectively, it is essential to address these knowledge gaps and gain a systematic understanding of in vitro toxicity (Asghar and Mushtaq 2023; Mohammed et al. 2018).

7.16 Future Perspectives

Research has shown that sustainable NMs hold great promise for wastewater treatment and water purification. Various nanotechnologies, including nanocatalysts, nano-adsorbents, antimicrobial agents, and molecularly imprinted polymers, have proven effective in removing impurities from water. Magnetic NMs, which can attract magnetic and metallic particles in water, and membrane filters with filtering and transport capabilities, are among the novel functionalities of NMs currently under investigation.

As a result, there is growing demand from researchers, industry, and policymakers to enhance the properties of NMs. This includes focusing on the development of nanostructures with versatile dimensions and optimizing their production methods to ensure environmental sustainability (Anjum et al. 2019).

Looking ahead, the innovation of multifunctional NMs, especially those combined with nanocomposites (such as iron-based carbon nanotubes) and multimetallic futuristic nanocomposites with antimicrobial, adsorptive, and photocatalytic properties, will be crucial for water purification and wastewater remediation.

Future research should prioritize the following areas:

1. Developing novel and innovative synthetic methods to reduce production costs and enhance catalytic efficiency.
2. Minimizing the use of magnetic NMs and employing mild, water-soluble, low-cost reductants for pollutant degradation.
3. Utilizing biowaste and natural materials to support the synthesis of NMs for water treatment applications.
4. Focusing on the use of safe and cost-effective methods in wastewater treatment to protect human health and the environment (Nasrollahzadeh et al. 2021).

7.17 Conclusion

Sustainable wastewater treatment processes are essential for human health, environmental protection, and ensuring the availability of clean water. Currently, nanotechnology offers a remarkable and effective approach to eliminating chemical and biological pollutants and degrading hazardous contaminants in developing water treatment technologies. NMs, with their novel and distinctive characteristics, are highly efficient at removing harmful organic compounds, heavy metal ions, and microorganisms from wastewater.

NMs have been successfully developed and investigated for their roles in wastewater remediation, including nano-adsorbents, carbon nanotubes, graphene, graphene oxide, photocatalysts (ZnO, TiO_2, CdS), electrocatalysts (Pt, Pb), nanomembranes, and metal and metal oxide NPs (MgO, Fe_2O_3, Al_2O_3, etc.). These nano-adsorbents and zero-valent metals show great potential for removing heavy metals like Cr, Hg, Zn, Ni, Pb, Vb, and As from wastewater. NMs are also promising for treating groundwater, wastewater, and saline water in a sustainable manner.

However, the use of NMs presents challenges and key considerations due to the current lack of comprehensive knowledge. Therefore, the future development of low-cost, highly effective, and more sustainable NMs is essential. The field of wastewater remediation using sustainable NMs is of great interest to both scientists and industry, and it is anticipated that sustainable NMs will play a significant role in environmental and wastewater remediation in the near future.

Acknowledgements Amrit Krishna Mitra would like to acknowledge the financial assistance provided by the Department of Science & Technology and Biotechnology, Government of West Bengal, India (Memo No. 860(Sanc.)/STBT-11012(25)/5/2019-ST SEC dated 03/11/2023). The authors would like to acknowledge Prof. Ashutosh Ghosh, Vice-Chancellor of Rani Rashmoni Green University, for his guidance and encouragement throughout the process of preparing this manuscript.

References

Afolabi IC, Popoola SI, Bello OS (2020) Modeling pseudo-second-order kinetics of orange peel-paracetamol adsorption process using artificial neural network. Chemom Intell Lab Syst 203:104053. https://doi.org/10.1016/j.chemolab.2020.104053

Agrawal A, Sahu KK (2006) Kinetic and isotherm studies of cadmium adsorption on manganese nodule residue. J Hazard Mater 137(2):915–924. https://doi.org/10.1016/j.jhazmat.2006.03.039

Ahmad SZN, Salleh WNW, Ismail AF, Yusof N, Yusop MZM, Aziz F (2020) Adsorptive removal of heavy metal ions using graphene-based NMs: toxicity, roles of functional groups and mechanisms. Chemosphere 248:126008. https://doi.org/10.1016/j.chemosphere.2020.126008

Ahmed SF, Mofijur M, Ahmed B, Mehnaz T, Mehejabin F, Maliat D, Shafiullah GM (2022) NMs as a sustainable choice for treating wastewater. Environ Res 214:113807. https://www.sciencedirect.com/science/article/pii/S0013935122011343

Ajdary M, Moosavi MA, Rahmati M, Falahati M, Mahboubi M, Mandegary A, Varma RS et al (2018) Health concerns of various NPs: a review of their in vitro and in vivo toxicity. Nms 8(9):634. https://doi.org/10.3390/nano8090634

Alharbi OM (2018) Sorption, kinetic, thermodynamics and artificial neural network modelling of phenol and 3-amino-phenol in water on composite iron nano-adsorbent. J Mol Liq 260:261–269. https://doi.org/10.1016/j.molliq.2018.03.104

Ali I (2012) New generation adsorbents for water treatment. Chem Rev 112(10):5073–5091. https://doi.org/10.1021/cr300133d

Anjum M, Miandad R, Waqas M, Gehany F, Barakat MA (2019) Remediation of wastewater using various nano-materials. Arab J Chem 12(8):4897–4919. https://doi.org/10.1016/j.arabjc.2016.10.004

Asghar F, Mushtaq A (2023) The future of NM in wastewater treatment: a review. Int J Chem Biochem Sci 23:150–157. https://www.iscientific.org/wp-content/uploads/2023/05/17-IJCBS-23-23-23.pdf

Babbitt CW, Moore EA (2018) Sustainable NMs by design. Nat Nanotechnol 13(8):621–623. https://doi.org/10.1038/s41565-018-0235-7

Baharifar H, Khoobi M, Bidgoli SA, Amani A (2020) Preparation of PEG-grafted chitosan/streptokinase NPs to improve biological half-life and reduce immunogenicity of the enzyme. Int J Biol Macromol 143:181–189. https://doi.org/10.1016/j.ijbiomac.2019.11.157

Besley JC, Kramer VL, Priest SH (2008) Expert opinion on nanotechnology: risks, benefits, and regulation. Journal of NP Research 10:549–558. https://doi.org/10.1007/s11051-007-9323-6

Bhat SA, Sher F, Hameed M, Bashir O, Kumar R, Vo DVN, Lima EC et al (2022) Sustainable nanotechnology based wastewater treatment strategies: achievements, challenges and future perspectives. Chemosphere 288:132606. https://doi.org/10.1016/j.chemosphere.2021.132606

Chen L, Berry RM, Tam KC (2014) Synthesis of β-cyclodextrin-modified cellulose nanocrystals (CNCs)@ Fe_3O_4@SiO_2 superparamagnetic nanorods. ACS Sustain Chem Eng 2(4):951–958. https://doi.org/10.1021/sc400540f

Choudhary M, Peter CN, Shukla SK, Govender PP, Joshi GM, Wang R (2020) Environmental issues: a challenge for wastewater treatment. In: Green materials for wastewater treatment, pp 1–12. https://doi.org/10.1007/978-3-030-17724-9_1

Choudhury AR (2020) Synthesis of a novel gellan-pullulan nanogel and its application in adsorption of cationic dye from aqueous medium. Carbohyd Polym 227:115291. https://doi.org/10.1016/j.carbpol.2019.115291

Dang HNT, Lai NH (2024) U.S. Patent No. 11,857,931. U.S. Patent and Trademark Office, Washington, DC. https://patents.google.com/patent/US11857931B2/en

Danquah MK, Sharadwata Pan AB (2018) Engineered NMs for wastewater. In: Fundamentals of NPs: classifications, synthesis methods, properties and characterization, p 129. https://doi.org/10.1016/B978-0-323-51255-8.00006-9

de Lima CRM, Gomes DN, de Morais Filho JR, Pereira MR, Fonseca JLC (2018) Anionic and cationic drug sorption on interpolyelectrolyte complexes. Colloids Surf B 170:210–218. https://doi.org/10.1016/j.colsurfb.2018.05.071

Dong A, Dai T, Ren M, Zhao X, Zhao S, Yuan Y, Wang N et al (2018) Functionalization and fabrication of soluble polymers of intrinsic microporosity for CO_2 transformation and uranium extraction. Eng Sci 5(3):56–65. https://doi.org/10.30919/es8d613

Doskocz N, Affek KA, Załęska-Radziwiłł M (2018) Effect of aluminium oxide NPs on the enzymatic activity on microorganisms of activated sludge. In: 10th conference on interdisciplinary problems in environmental protection and engineering EKO-DOK 2018. EDP Sciences. https://doi.org/10.1051/e3sconf/20184400033

Duan H, Wang D, Li Y (2015) Green chemistry for NP synthesis. Chem Soc Rev 44(16):5778–5792. https://doi.org/10.1039/C4CS00363B

Elgarahy AM, Elwakeel KZ, Elshoubaky GA, Mohammad SH (2019a) Untapped sepia shell–based composite for the sorption of cationic and anionic dyes. Water Air Soil Pollut 230:1–23. https://doi.org/10.1007/s11270-019-4247-1

Elgarahy AM, Elwakeel KZ, Elshoubaky GA, Mohammad SH (2019b) Microwave-accelerated sorption of cationic dyes onto green marine algal biomass. Environ Sci Pollut Res 26:22704–22722. https://doi.org/10.1007/s11356-019-05417-2

Elgarahy AM, Elwakeel KZ, Akhdhar A, Hamza MF (2021) Recent advances in greenly synthesized nanoengineered materials for water/wastewater remediation: an overview. Nanotechnol Environ Eng 6:1–24. https://doi.org/10.1007/s41204-021-00104-5

Elliott DW, Zhang WX (2001) Field assessment of nanoscale bimetallic particles for groundwater treatment. Environ Sci Technol 35(24):4922–4926. https://doi.org/10.1021/es0108584

Elwakeel KZ, Elgarahy AM, Guibal E (2021) A biogenic tunable sorbent produced from upcycling of aquatic biota-based materials functionalized with methylene blue dye for the removal of chromium (VI) ions. J Environ Chem Eng 9(2):104767. https://doi.org/10.1016/j.jece.2020.104767

Gawande MB, Branco PS, Varma RS (2013) Nano-magnetite (Fe_3O_4) as a support for recyclable catalysts in the development of sustainable methodologies. Chem Soc Rev 42(8):3371–3393. https://doi.org/10.1039/C3CS35480F

Hao X, Chen R, Liu Q, Liu J, Zhang H, Yu J, Wang J (2018) A novel U (vi)-imprinted graphitic carbon nitride composite for the selective and efficient removal of U (vi) from simulated seawater. Inorg Chem Front 5(9):2218–2226. https://doi.org/10.1039/C8QI00522B

Hao X, Zhang H, Liu Q, Liu J, Chen R, Yu J, Wang J (2019) In situ anchoring of pyrrhotite on graphitic carbon nitride nanosheet for efficient immobilization of uranium. Chem Eur J 25(2):590–597. https://doi.org/10.1002/chem.201804289

Hasan S (2015) A review on NPs: their synthesis and types. Res J Recent Sci 2277:2502. https://www.researchgate.net/profile/Saba-Hasan/publication/273203342_A_Review_on_NPs_Their_Synthesis_and_Types/links/54fb49ff0cf270426d0dcb29/A-Review-on-NPs-Their-Synthesis-and-Types.pdf?utm_medium=email&utm_source=transaction

Herrera-Morales J, Morales K, Ramos D, Ortiz-Quiles EO, Lopez-Encarnacion JM, Nicolau E (2017) Examining the use of nanocellulose composites for the sorption of contaminants of emerging concern: an experimental and computational study. ACS Omega 2(11):7714–7722. https://doi.org/10.1021/acsomega.7b01053

Hsien YH, Chang CF, Chen YH, Cheng S (2001) Photodegradation of aromatic pollutants in water over TiO_2 supported on molecular sieves. Appl Catal B Environ 31(4):241–249. https://www.sciencedirect.com/science/article/pii/S0926337300002836

Huang N, Zhai L, Xu H, Jiang D (2017) Stable covalent organic frameworks for exceptional mercury removal from aqueous solutions. J Am Chem Soc 139(6):2428–2434. https://doi.org/10.1021/jacs.6b12328

Huang S, Jiang S, Pang H, Wen T, Asiri AM, Alamry KA, Wang S et al (2019) Dual functional nanocomposites of magnetic $MnFe_2O_4$ and fluorescent carbon dots for efficient U (VI) removal. Chem Eng J 368:941–950. https://doi.org/10.1016/j.cej.2019.03.015

Hubbs AF, Mercer RR, Benkovic SA, Harkema J, Sriram K, Schwegler-Berry D, Sargent LM et al (2011) Nanotoxicology—a pathologist's perspective. Toxicol Pathol 39(2):301–324. https://doi.org/10.1177/0192623310390705

Iravani S (2021) NMs and nanotechnology for water treatment: recent advances. Inorg Nano Met Chem 51(12):1615–1645. https://doi.org/10.1080/24701556.2020.1852253

Jackson, DJ (2011) What is an innovation ecosystem? National Science Foundation, Arlington, VA. http://urenio.org/wp-content/uploads/2011/05/What-is-an-Innovation-Ecosystem.pdf

Jia G, Wang H, Yan L, Wang X, Pei R, Yan T, Guo X et al (2005) Cytotoxicity of carbon NMs: single-wall nanotube, multi-wall nanotube, and fullerene. Environ Sci Technol 39(5):1378–1383. https://doi.org/10.1021/es0487291

Jie G, Kongyin Z, Xinxin Z, Zhijiang C, Min C, Tian C, Junfu W (2015) Preparation and characterization of carboxyl multi-walled carbon nanotubes/calcium alginate composite hydrogel nano-filtration membrane. Mater Lett 157:112–115. https://www.sciencedirect.com/science/article/pii/S0167577X15008022

Kalia A, Manchanda P, Bhardwaj S, Singh G (2020) Biosynthesized silver NPs from aqueous extracts of sweet lime fruit and callus tissues possess variable antioxidant and antimicrobial potentials. Inorg Nano Met Chem 50(11):1053–1062. https://doi.org/10.1080/24701556.2020.1735420

Kalidindi SB, Jagirdar BR (2012) Nanocatalysis and prospects of green chemistry. Chemsuschem 5(1):65–75. https://doi.org/10.1002/cssc.201100377

Karlsson HL, Cronholm P, Gustafsson J, Moller L (2008) Copper oxide NPs are highly toxic: a comparison between metal oxide NPs and carbon nanotubes. Chem Res Toxicol 21(9):1726–1732. https://doi.org/10.1021/tx800064j

Kazemimoghadam M (2010) New nanopore zeolite membranes for water treatment. Desalination 251(1–3):176–180. https://www.sciencedirect.com/science/article/pii/S0011916409013241

Khan FSA, Mubarak NM, Tan YH, Khalid M, Karri RR, Walvekar R, Mazari SA et al (2021) A comprehensive review on magnetic carbon nanotubes and carbon nanotube-based buckypaper for removal of heavy metals and dyes. J Hazard Mater 413:125375. https://doi.org/10.1016/j.jhazmat.2021.125375

Khandel P, Yadaw RK, Soni DK, Kanwar L, Shahi SK (2018) Biogenesis of metal NPs and their pharmacological applications: present status and application prospects. J Nanostructure Chem 8:217–254. https://doi.org/10.1007/s40097-018-0267-4

Kidd J, Westerhoff P, Maynard AD (2020) Public perceptions for the use of NMs for in-home drinking water purification devices. NanoImpact 18:100220. https://www.sciencedirect.com/science/article/pii/S2452074820300148

Kumari S, Yadav BS, Yadav RB (2020) Synthesis and modification approaches for starch NPs for their emerging food industrial applications: a review. Food Res Int 128:108765. https://doi.org/10.1016/j.foodres.2019.108765

Kurwadkar S, Hoang TV, Malwade K, Kanel SR, Harper WF, Struckhoff G (2019) Application of carbon nanotubes for removal of emerging contaminants of concern in engineered water and wastewater treatment systems. Nanotechnol Environ Eng 4:1–16. https://doi.org/10.1007/s41204-019-0059-1

Li YH, Zhao YM, Hu WB, Ahmad I, Zhu YQ, Peng XJ, Luan ZK (2007) Carbon nanotubes-the promising adsorbent in wastewater treatment. J Phys Conf Ser 61(1):698. https://doi.org/10.1088/1742-6596/61/1/140

Li Y, Bai P, Yan Y, Yan W, Shi W, Xu R (2019) Removal of Zn^{2+}, Pb^{2+}, Cd^{2+}, and Cu^{2+} from aqueous solution by synthetic clinoptilolite. Microporous Mesoporous Mater 273:203–211. https://doi.org/10.1016/j.micromeso.2018.07.010

Li YH, Ding J, Luan Z, Di Z, Zhu Y, Xu C, Wei B et al (2003) Competitive adsorption of Pb^{2+}, Cu^{2+} and Cd^{2+} ions from aqueous solutions by multiwalled carbon nanotubes. Carbon 41(14):2787–2792. https://www.sciencedirect.com/science/article/pii/S0008622303003920

Lin X, Sun S, Wang B, Zheng B, Guo Z (2020) Structural and physicochemical properties of lotus seed starch NPs prepared using ultrasonic-assisted enzymatic hydrolysis. Ultrason Sonochem 68:105199. https://doi.org/10.1016/j.ultsonch.2020.105199

Lou J, Liu Y, Wang Z, Zhao D, Song C, Wu J, Deng T (2016) Bioinspired multifunctional paper-based rGO composites for solar-driven clean water generation. ACS Appl Mater Interfaces 8(23):14628–14636. https://doi.org/10.1021/acsami.6b04606

Lu C, Chiu H, Liu C (2006) Removal of zinc (II) from aqueous solution by purified carbon nanotubes: kinetics and equilibrium studies. Ind Eng Chem Res 45(8):2850–2855. https://doi.org/10.1021/ie051206h

Lu H, Wang J, Stoller M, Wang T, Bao Y, Hao H (2016) An overview of NMs for water and wastewater treatment. Adv Mater Sci Eng 2016(1):4964828. https://doi.org/10.1155/2016/4964828

Mackenzie K, Bleyl S, Georgi A, Kopinke FD (2012) Carbo-iron–an Fe/AC composite—as alternative to nano-iron for groundwater treatment. Water Res 46(12):3817–3826. https://www.sciencedirect.com/science/article/pii/S0043135412002552

Malsch I (2014) Nano-education from a European perspective: Nano-training for non-R&D jobs. Nanotechnol Rev 3(2):211–221. https://doi.org/10.1515/ntrev-2013-0039

Malsch I, Subramanian V, Semenzin E, Hristozov D, Marcomini A (2015) Supporting decision-making for sustainable nanotechnology. Environ Syst Decis 35(1):54–75. https://doi.org/10.1007/s10669-015-9539-4

Mazhar MA, Khan NA, Ahmed S, Khan AH, Hussain A, Changani F, Vambol V (2020) Chlorination disinfection by-products in municipal drinking water—a review. J Clean Prod 273:123159. https://www.sciencedirect.com/science/article/pii/S0959652620332042

Mohamadi L, Bazrafshan E, Rahdar A, Labuto G, Kamali AR (2021) Nanostructured MgO-enhanced catalytic ozonation of petrochemical wastewater. Boletín de la sociedad española de cerámica y vidrio 60(6):391–400. https://www.sciencedirect.com/science/article/pii/S0366317520300650

Mohammed N, Grishkewich N, Tam KC (2018) Cellulose NMs: promising sustainable NMs for application in water/wastewater treatment processes. Environ Sci Nano 5(3):623–658. https://doi.org/10.1039/C7EN01029J

Moradi S, Shayesteh K, Behbudi G (2020) Preparation and characterization of biodegradable ligninsulfonate NPs using the microemulsion method to enhance the acetylation efficiency of ligninsulfonate. Int J Biol Macromol 160:632–641. https://doi.org/10.1016/j.ijbiomac.2020.05.157
Nasrollahzadeh M, Sajjadi M, Iravani S, Varma RS (2021) Carbon-based sustainable NMs for water treatment: state-of-art and future perspectives. Chemosphere 263:128005. https://doi.org/10.1016/j.chemosphere.2020.128005
Norrrahim MNF, Kasim NAM, Knight VF, Misenan MSM, Janudin N, Shah NAA, Yunus WMZW et al (2021) Nanocellulose: a bioadsorbent for chemical contaminant remediation. RSC Adv 11(13):7347–7368. https://doi.org/10.1039/D0RA08005E
Nowack B (2008) Pollution prevention and treatment using nanotechnology. Nanotechnology 2:1–15. https://medien.umbreitkatalog.de/pdfzentrale/978/352/731/Leseprobe_l_9783527317356.pdf
O'Connell DW, Birkinshaw C, O'Dwyer TF (2008) Heavy metal adsorbents prepared from the modification of cellulose: A review. Bioresour Technol 99(15):6709–6724. https://www.sciencedirect.com/science/article/pii/S0960852408000710
Ponnusamy G, Farzaneh H, Tong Y, Lawler J, Liu Z, Saththasivam J (2021) Enhanced catalytic ozonation of ibuprofen using a 3D structured catalyst with MnO_2 nanosheets on carbon microfibers. Sci Rep 11(1):6342. https://doi.org/10.1038/s41598-021-85651-2
Prieto O, Fermoso J, Nuñez Y, Del Valle JL, Irusta R (2005) Decolouration of textile dyes in wastewaters by photocatalysis with TiO_2. Solar Energy 79(4):376–383. https://www.sciencedirect.com/science/article/pii/S0038092X05001453
Qu X, Alvarez PJ, Li Q (2013) Applications of nanotechnology in water and wastewater treatment. Water Res 47(12):3931–3946. https://doi.org/10.1016/j.watres.2012.09.058
Rajkumar KS, Kanipandian N, Thirumurugan RJAN (2016) Toxicity assessment on haemotology, biochemical and histopathological alterations of silver NPs-exposed freshwater fish Labeo rohita. Appl Nanosci 6:19–29. https://doi.org/10.1007/s13204-015-0417-7
Ramar V, Moothattu S, Balasubramanian K (2018) Metal free, sunlight and white light based photocatalysis using carbon quantum dots from Citrus grandis: a green way to remove pollution. Solar Energy 169:120–127. https://doi.org/10.1016/j.solener.2018.04.040
Rambabu K, Bharath G, Banat F, Show PL (2021) Green synthesis of zinc oxide NPs using Phoenix dactylifera waste as bioreductant for effective dye degradation and antibacterial performance in wastewater treatment. J Hazard Mater 402:123560. https://www.sciencedirect.com/science/article/pii/S0304389420315466
Ran N, Zhao L, Chen Z, Tao J (2008) Recent applications of biocatalysis in developing green chemistry for chemical synthesis at the industrial scale. Green Chem 10(4):361–372. https://doi.org/10.1039/B716045C
Ray PC, Yu H, Fu PP (2009) Toxicity and environmental risks of NMs: challenges and future needs. J Environ Sci Health C 27(1):1–35. https://doi.org/10.1080/10590500802708267
Romero V, Fernandes SP, Kovář P, Pšenička M, Kolenko YV, Salonen LM, Espiña B (2020) Efficient adsorption of endocrine-disrupting pesticides from water with a reusable magnetic covalent organic framework. Microporous Mesoporous Mater 307:110523. https://doi.org/10.1016/j.micromeso.2020.110523
Rupa EJ, Kaliraj L, Abid S, Yang DC, Jung SK (2019a) Synthesis of a zinc oxide nanoflower photocatalyst from sea buckthorn fruit for degradation of industrial dyes in wastewater treatment. Nms 9(12):1692. https://doi.org/10.3390/nano9121692
Rupa EJ, Anandapadmanaban G, Chokkalingam M, Li JF, Markus J, Soshnikova V, Yang DC (2019) Cationic and anionic dye degradation activity of Zinc oxide NPs from Hippophae rhamnoides leaves as potential water treatment resource. Optik 181:1091–1098. https://www.sciencedirect.com/science/article/pii/S0030402618320862
Safari J, Sadeghi M (2017) Nanostarch: a novel and green catalyst for synthesis of 2-aminothiazoles. Monatshefte Für Chemie-Chemical Monthly 148:745–749. https://doi.org/10.1007/s00706-016-1805-8

Salehpour S, Jonoobi M, Ahmadzadeh M, Siracusa V, Rafieian F, Oksman K (2018) Biodegradation and ecotoxicological impact of cellulose nanocomposites in municipal solid waste composting. Int J Biol Macromol 111:264–270. https://www.sciencedirect.com/science/article/pii/S0141813017344410

Sayyed AJ, Pinjari DV, Sonawane SH, Bhanvase BA, Sheikh J, Sillanpää M (2021) Cellulose-based NMs for water and wastewater treatments: a review. J Environ Chem Eng 9(6). https://www.sciencedirect.com/science/article/pii/S2213343721016031

Sethy NK, Arif Z, Mishra PK, Kumar P (2020) Green synthesis of TiO_2 NPs from Syzygium cumini extract for photo-catalytic removal of lead (Pb) in explosive industrial wastewater. Green Process Synth 9(1):171–181. https://doi.org/10.1515/gps-2020-0018

Sherin L, Sohail A, Mustafa M, Jabeen R, Ul-Hamid A (2020) Facile green synthesis of silver NPs using Terminalia bellerica kernel extract for catalytic reduction of anthropogenic water pollutants. Colloid Interface Sci Commun 37:100276. https://doi.org/10.1016/j.colcom.2020.100276

Singh NB, Susan BH, Guin M (2021) Applications of green synthesized NMs in water remediation. Curr Pharm Biotechnol 22(6):733–761. https://doi.org/10.2174/1389201021666201027160029

Sirviö JA, Hasa T, Leiviskä T, Liimatainen H, Hormi O (2016) Bisphosphonate nanocellulose in the removal of vanadium (V) from water. Cellulose 23:689–697. https://doi.org/10.1007/s10570-015-0819-4

Soliman NK, Moustafa AF (2020) Industrial solid waste for heavy metals adsorption features and challenges; a review. J Market Res 9(5):10235–10253. https://doi.org/10.1016/j.jmrt.2020.07.045

Somsesta N, Sricharoenchaikul V, Aht-Ong D (2020) Adsorption removal of methylene blue onto activated carbon/cellulose biocomposite films: equilibrium and kinetic studies. Mater Chem Phys 240:122221. https://doi.org/10.1016/j.matchemphys.2019.122221

Somu P, Paul S (2018) Casein based biogenic-synthesized zinc oxide NPs simultaneously decontaminate heavy metals, dyes, and pathogenic microbes: a rational strategy for wastewater treatment. J Chem Technol Biotechnol 93(10):2962–2976. https://doi.org/10.1002/jctb.5655

Soni R, Pal AK, Tripathi P, Lal JA, Kesari K, Tripathi V (2020) An overview of nanoscale materials on the removal of wastewater contaminants. Appl Water Sci 10(8):189. https://doi.org/10.1007/s13201-020-01275-3

Taju G, Majeed SA, Nambi KSN, Hameed AS (2014) In vitro assay for the toxicity of silver NPs using heart and gill cell lines of Catla catla and gill cell line of Labeo rohita. Comp Biochem Physiol Part C Toxicol Pharmacol 161:41–52. https://www.sciencedirect.com/science/article/pii/S153204561400009X

Takahashi H, Niidome Y, Niidome T, Kaneko K, Kawasaki H, Yamada S (2006) Modification of gold nanorods using phosphatidylcholine to reduce cytotoxicity. Langmuir 22(1):2–5. https://doi.org/10.1021/la0520029

Tiwari DK, Behari J, Prasenjit Sen PS (2008) Application of NPs in waste water treatment. http://www.idosi.org/wasj/wasj3(3)/11.pdf

Tom AP (2021) Nanotechnology for sustainable water treatment—a review. Mater Today Proc 10:16–78. https://doi.org/10.1016/j.matpr.2021.05.629

Ulbricht M (2006) Advanced functional polymer membranes. Polymer 47(7):2217–2262. https://www.sciencedirect.com/science/article/pii/S0032386106001303

Vélez E, Campillo GE, Morales G, Hincapié C, Osorio J, Arnache O, Jaramillo F et al (2016) Mercury removal in wastewater by iron oxide NPs. J Phys Conf Ser 687(1):012050. https://doi.org/10.1088/1742-6596/687/1/012050

Vijayaraghavan K, Ashokkumar T (2017) Plant-mediated biosynthesis of metallic NPs: a review of literature, factors affecting synthesis, characterization techniques and applications. J Environ Chem Eng 5(5):4866–4883. https://doi.org/10.1016/j.jece.2017.09.026

Wang X, Chen A, Chen B, Wang L (2020) Adsorption of phenol and bisphenol A on river sediments: effects of particle size, humic acid, pH and temperature. Ecotoxicol Environ Saf 204:111093. https://doi.org/10.1016/j.ecoenv.2020.111093

Wei Guo K (2011) Green nanotechnology of trends in future energy. Recent Pat Nanotechnol 5(2):76–88. https://doi.org/10.2174/187221011795909198
Yang K, Xing B (2010) Adsorption of organic compounds by carbon NMs in aqueous phase: polanyi theory and its application. Chem Rev 110(10):5989–6008. https://doi.org/10.1021/cr100059s
Yin Z, Cui C, Chen H, Duoni, Yu X, Qian W (2020) The application of carbon nanotube/graphene-based NMs in wastewater treatment. Small 16(15):1902301. https://doi.org/10.1002/smll.201902301
Yusof MN, Ismail R, Ismail SA, Bakar AHA (2018) Knowledge retention for sustaining competitive advantage in construction industry: a conceptual. Malays Constr Res J (MCRJ), 10. https://www.researchgate.net/profile/Arasinah-Kamis/publication/327351329_Construction_Management_Graduates_Roles_And_Tasks_Portfolio_Analysing_Using_Rasch_Measurement_Model_SCOPUS/links/5c2c287d299bf12be3a72c14/Construction-Management-Graduates-Roles-And-Tasks-Portfolio-Analysing-Using-Rasch-Measurement-Model-SCOPUS.pdf#page=21
Zhang R, Chu F, Hu Y, Hu H, Hu Y, Liu H, Wang H et al (2020) Preparation of photo-crosslinking starch colloidal particles. Starch-Stärke 72(5–6):1900175. https://doi.org/10.1002/star.201900175
Zhang J, Yu P, Fan L, Sun Y (2021) Effects of ultrasound treatment on the starch properties and oil absorption of potato chips. Ultrason Sonochem 70:105347. https://doi.org/10.1016/j.ultsonch.2020.105347
Zhou Z, Zhong W, Cui K, Zhuang Z, Li L, Li L, Yu Y (2018) A covalent organic framework bearing thioether pendant arms for selective detection and recovery of Au from ultra-low concentration aqueous solution. Chem Commun 54(71):9977–9980. https://doi.org/10.1039/C8CC05369C

Chapter 8
Nano-Bioremediation of Industrial Wastewater: Applications, Limitations, and Future Perspectives

Ganesh Swain and Tapas Das

Abstract Recent times have witnessed a rapid rise in the utilization of organic and inorganic composites due to increasing human activity, creating a large amount of wastewater. A lot of research work has been accomplished and investigated various physical and chemical methods to mitigate the impact of hazardous pollutants on ecosystems. However, each of these methods is associated with several limitations including high costs, limited effectiveness, the production of toxic by-products, and the requirement for large capital expenditure. Hence, an environmentally sustainable, efficient, and low-cost method will be preferred. Integrated approaches are more beneficial in reducing the limitations of a single technology and work as the best alternative to traditional remediation techniques. One novel technique that has recently been employed is nano-bioremediation. Nano-bioremediation is a hybrid technology consisting of both nanotechnology and bioremediation. The remediation process involved in this technique is comparatively more effective, less time-consuming, and eco-friendly. This chapter presents an overview of nanotechnology and various nano-based materials for the removal of both organic and inorganic pollutants from wastewater. Additionally, an extensive analysis of the nano-bioremediation method, process, and usage has been summarized.

Keywords Bioremediation · NP · Industrial wastewater · Carbon nanotubes · Toxicity

G. Swain (✉)
Department of Chemical and Biochemical Engineering, IIT, Patna, Bihar, India
e-mail: ganeshswain24@iitp.ac.in

T. Das
Department of Chemical Engineering, NIT, Rourkela, Odisha, India

N. Agarwal et al. (eds.), *Nanomaterials in Wastewater Research*, Advances in Wastewater Research, https://doi.org/10.1007/978-981-96-3908-3_8

8.1 Introduction

In the last decade, environmental pollution has developed into a critical issue, characterized by harmful substances, i.e., pesticides, petroleum derivatives, dyes, and heavy metals (Swain et al. 2021; Khan et al. 2023). The growth of industrial and agricultural sectors has led to the release of various pollutants, such as synthetic polymers, and noxious compounds into the environment (Ghasemi et al. 2022). The major reasons for water pollution are the unrestricted deployment of hazardous compounds in various consumer products and the discharge of industrial effluents into water bodies. These pollutants exhibit either biodegradable or non-biodegradable properties, posing significant threats to living animals as they enter the food chain system and accumulate through biomagnification. Due to their elemental composition, these toxic chemicals persist in the environment for prolonged periods, necessitating global concern and prompting extensive research efforts to develop effective remediation strategies for polluted sites (Pal et al. 2021).

Physical and chemical methods utilized for wastewater remediation were found to be inefficient both economically and environmentally due to their extensive reliability on chemicals, costly equipment, and the generation of secondary by-products (Eniola et al. 2023; Tanveer et al. 2023). On the other hand, biological remediation is preferred as a more trustworthy alternative in terms of flexibility, cost-effectiveness, and eco-friendliness (Lin et al. 2023; Liu et al. 2023). However, bioremediation encounters challenges, such as more time-consuming, limited substrate degradability, and being greatly affected by environmental factors (Muhammad et al. 2024; Banerjee et al. 2022). Furthermore, its utilization becomes limited where sites are heavily contaminated with extremely toxic and hazardous pollutants.

Over the last decades, nanotechnology has been widely applied in various sectors, including pharmaceuticals, textiles, electronics, optics, cosmetics, sports, etc. (Fig. 8.1) (Din et al. 2024). Nanotechnology shows a promising approach in the direction of wastewater treatment. Recently, sustainable remediation has gained significant importance, focusing on reducing contaminant concentrations to risk-based levels and mitigating additional environmental impacts (Shabbir et al. 2024). Recent developments in this field induce the integration of multiple technologies into single systems, aiming to provide comprehensive solutions for economical and time-efficient methods. Hence, there is a growing preference for a synergistic approach merging nanomaterials (NMs) with bioremediation, known as nano-bioremediation. Nano-bioremediation refers to an environmentally sustainable and economically feasible remediation technique, leveraging NMs to enhance the degradation efficacy of microbial cells (Salama et al. 2021).

Nano-based technology has been preferred as a suitable alternative to current water treatment methods. Nano-based technologies emerged as efficient and versatile processes for effective water treatment, eliminating the need for extensive infrastructure. These technologies can also be integrated with other water treatment methods, such as adsorption, membrane-based separation, coagulation, and photocatalysis (Din et al. 2024; Mahmoud et al. 2021). Furthermore, a sudden rise

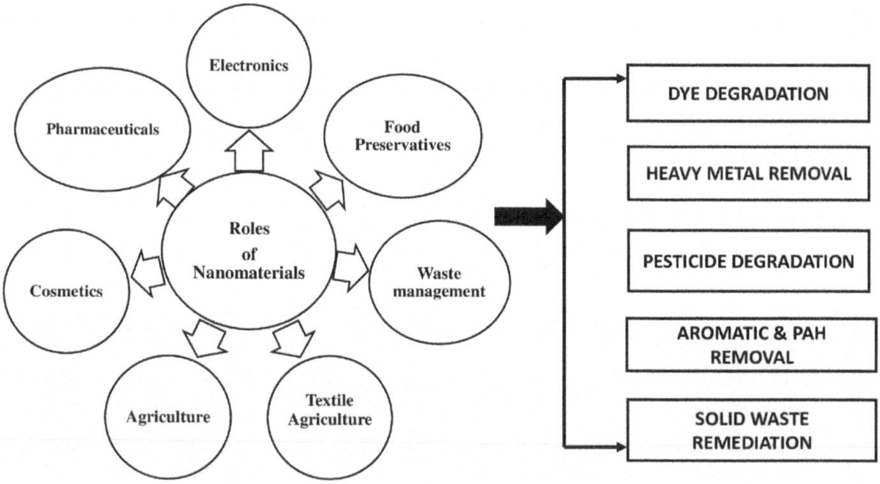

Fig. 8.1 Utilization of NMs for various applications

of interest in developing materials using nanotechnology principles in conjunction with engineering techniques has emerged. These advanced methods aim to address water quality issues through the application of nanotubes, nanocatalysts, nano-adsorbents, nanostructured catalytic membranes, and micro-molecules (Zhao et al. 2022). Each remediation method has its own advantages and limitations, making the integration of various techniques a viable solution for addressing remediation challenges. One such method that has garnered significant attention in recent years is nano-bioremediation. This approach combines the benefits of nanotechnology with the advantages of bioremediation, offering a promising strategy for effective environmental remediation.

The objective of this chapter is to investigate recent advancements in nano-bioremediation, with a specific focus on strategically eliminating various pollutants. Additionally, the various materials employed for nano-bioremediation of wastewater and the compatibility of nanoparticles (NPs) with microbial growth were analyzed. Furthermore, the chapter addresses the bottlenecks and possible future trends in nano-bioremediation of wastewater.

8.2 Classification of NMs Used for Wastewater Treatment

A wide range of NMs has been effectively utilized in the bioremediation of wastewater. These materials can be classified broadly into metallic NP, carbon-based NPs, nanocomposites, and nano-zeolites.

8.2.1 Metallic NPs

NPs have been utilized to remediate persistent organic compounds, demonstrating rapid reactions through microbial activities. In this direction, zero-valent iron has shown promising results in the removal of heavy metals due to its high reactivity. Besides, various NPs are synthesized from biological sources, such as algae, bacteria, plant extracts, and fungi. Various biologically produced polyphenolic substances including aldehydes, phenolics, flavonoids, alkaloids, etc., become useful to synthesize different NPs. However, the production of biogenic substances depends upon the quantity and quality of the natural chemicals. Various literature works have been carried out to investigate the key mechanism and working principle of the green synthesis of NPs (i.e., silver, copper, gold, zinc oxide, etc.) for wastewater treatment. Metallic NPs are composed of metals, which are reduced to nanoscale dimensions. These NPs often undergo self-oxidation to form metal oxide NPs that exhibit better reactivity and efficacy compared to their metallic complements. Various metal oxides, such as zinc oxide, nickel oxide, Fe_2O_3, and CuO, have been utilized along with microbes and enzymes in the bioremediation of biodegradable and non-biodegradable pollutants (Din et al. 2024; He et al. 2019). For instance, ZnO NPs derived from *Phoenix dactylifera* species have effective dye degradation ability and also possess antipathogenic effects (Rambabu et al. 2021). In a recent study, nano-bimetallic iron and copper NPs effectively adsorbed 82, 96, and 98% of color, COD, and BOD, respectively, from real textile wastewater (Mahmoud et al. 2021).

Jain et al. (Jain et al. 2014) synthesized quasi-spherical, symmetrical, polydisperse zinc oxide NPs, sized 80–120 nm, which were well distributed. The surface-modified nanocatalysts showed superior photocatalytic activity and minimized the recombinant phenomena. Additionally, the synthesized cost-effective and environment friendly catalyst was able to degrade more than 90% of the dye within 30 min of batch operation. The application of fungal cellular (like *Cylindrocladium floridanum*) derived NPs (i.e., silver NPs) has a high potential for the treatment of phenolic wastewater. These biogenic NPs catalyze the conversion of one phenolic compound (4-NP) to another (4-aminophenol) in the presence of sodium borohydride. Silver NPs catalyze this reduction homogeneously, while fungal-matrixed gold nanocomposites catalyze it heterogeneously, allowing for the recovery and recycling of the AuNPs catalyst (Singh et al. 2023a). Additionally, a highly stable silver nanoparticle (AgNPs) was synthesized using an aqueous extract of *Caulerpa racemosa*, with the phytoconstituents responsible for capping, stabilization, and preventing aggregation, thus ensuring high stability. These AgNPs demonstrated excellent catalytic activity for the degradation of methylene blue (Viswanathan et al. 2024). Furthermore, gold NPs (size of 6.07 nm) synthesized from the protein extract of *Pycnoporus sanguineus* exhibited a superior degradation ability. The research work revealed that the nanocatalyst was able to adsorb 100% of 4-nitroaniline (12.5 μmol) within 6 min of time period (Salunke et al. 2016).

The coupled utilization of NPs and microbes holds significant promise for remediating herbicide-contaminated soil. Among these, Fe_3O_4 NPs are the most studied and have proven effective in such environments. Research has shown that iron NPs degrade the herbicide 2,4-D by reducing it to several phenolic compounds like 2,4-dichlorophenol, chlorophenols, and phenol (Zhou et al. 2024). This strong degradation capability facilitated a higher surface area availability and excellent chemical reactivity of Fe_3O_4 NPs, which facilitated the remarkable removal of 2,4-D. A higher dichlorination was observed at a low pH media due to the increase of a number of proton radicals, which ultimately enhanced the phenolic degradation. Fe_3O_4 NPs, containing both Fe^{2+} and Fe^{3+}, act as electron donors, reducing chlorinated organic compounds. However, very little research work was reported on the effects of Fe_3O_4 nanocatalysts in contaminated soil samples. Fe_3O_4 nanocatalysts enhance the microbial population and its enzymatic activity in the soil, promoting the reductive dechlorination of 2,4-D (Tao et al. 2023). Similarly, Gomes et al. (Gomes et al. 2015) used iron NPs for remediating polychlorinated biphenyls (PCB)-contaminated soils, finding that acidic conditions supported maximum PCB degradation.

8.2.2 Carbon-Based NPs

Carbon-based NPs are comprised of sp^2 hybridized carbon atoms, exhibiting multidimensional structures. Various carbon-based NPs are carbon nanotubes, fullerene, carbon-based quantum dots, graphene, nanodiamonds, etc. Notably, carbon nanotubes possess remarkable adsorption capacity for pollutant removal, facilitated by various interactions, such as electrostatic, hydrogen bonding, hydrophobic, π–π stacking, and van der Waals interactions (Manimegalai et al. 2023; Cheng et al. 2019). Carbon nanotubes have revealed a synergistic effect on microbial growth enhancement and facilitation of pollutant degradation (Ma et al. 2024). Other carbon NMs, including carbon nanotubes and graphene oxide NPs, have been employed with microbial agents to degrade various pollutants, such as aromatic compounds (benzene, toluene, ethylbenzene, xylene (BTEX), etc.), azo dyes, and inorganic pollutants (Zhao et al. 2022; Sargin et al. 2020). Additionally, graphene oxides have been used as matrix supports for the immobilization of bacteria, protecting them from adverse conditions and augmenting their efficiency in combating various organic pollutants (Zhong et al. 2020).

8.2.3 Nanocomposites

Nanocomposites are formed by combining various metallic NPs and nano supports that offer excellent flexibility along with a higher surface area and adsorption capacity (Surendra et al. 2020). Their superior adsorption efficacy makes them well suited for removing pollutants. Moreover, these nanocomposites serve as a shield for microorganisms against high substrate concentration, acidic or alkaline pH, and temperature.

For instance, the nanocomposite based on biomass-derived biochar showed significant removal capacity for various pollutants via surface adsorption mechanisms (Yu et al. 2023). Furthermore, bacterial cells immobilized onto nanocomposite exhibited greater removal performance over free cell methods due to lesser inhibition effect and higher metabolic activity (Chauhan et al. 2023). Goethite-chitosan nanocomposites enhanced the water and soil bioremediation by stimulating the growth and efficacy of native microbial strains (Partovinia et al. 2023).

Composites represent solid materials with multiple phases, one of which has a nanoscale dimension (less than 100 nm). These materials combine various metallic NPs and nano supports, offering increased adsorption sites and capacity. Moreover, nanocomposites provide a protective layer over the microbial cell, which provides shelter as well as a strong barrier toward extreme environmental conditions. Their high adsorption efficacy makes them suitable for removing polyaromatic hydrocarbons (PAHs). For instance, green tea residual-based magnetic nanocomposite effectively removed various petroleum contaminants, such as benzo fluoranthene and chrysene, through physical adsorption via van der Waals force of interaction (Chauhan et al. 2023).

Furthermore, nanocomposites have been observed to enhance microbial growth and stimulate microbial activity in petroleum degradation (Hosny et al. 2024). Nanocomposite nanofibrous webs, serving as immobilizing supports, have shown significant potential in crude oil remediation. Compared to free cell methods, the immobilized system exhibited enhanced bacterial growth, metabolism, enzyme production, and favorable biochemical reactions (Swain et al. 2021). Another example is goethite-chitosan nanocomposites, which serve as a nutrient source for scaling up soil and water bioremediation. These composites stimulate the growth and reactivity of native microbial strains, thereby enhancing the bioremediation process (Chauhan et al. 2023).

8.2.4 Bio-Based NPs

Conventional methods (physical or chemical processes) for the synthesis of NPs have significant drawbacks including environmental impact and ill effects on human well-being (Puri et al. 2024). These traditional approaches usually involve the application of hazardous chemicals and high temperatures. Hence, further research was carried out for developing NPs through environmentally friendly processes. One such emerging approach is green nanotechnology, an interdisciplinary field focused on minimizing environmental harm. Green NMs are safe and eco-friendly, which are widely adopted across various industries (Rai et al. 2023). These materials include natural substances and their derivatives, such as plant extracts, wines, amino acids, and glucose (Patani et al. 2023). Microorganisms like algae, bacteria, and yeasts are also employed in the synthesis of green NPs.

Nanoparticle-microbial enzyme conjugates have higher efficacy in degrading carcinogenic azo dyes in a very short period. In this direction, a conventional nanoparticle, i.e., titanium dioxide was immobilized with fungal laccase, was utilized for the treatment of textile wastewater. Conversely, the degradation of phenol red was achieved using *Catharanthus roseus*-mediated zinc oxide NPs, with efficient dye degradation accomplished within 8 h. Methyl orange dye, widely used in various industries, poses challenges due to its resistance to degradation and mutagenic and carcinogenic properties. Copper NPs synthesized using *Centella asiatica* L. leaf extract act as catalysts for the photocatalytic degradation of methyl orange (Raina et al. 2020). Furthermore, carbon nanotubes (CNTs) show promise in wastewater remediation due to superior thermo-chemical stability, outstanding affinity, and adsorption characteristics for the removal of various organic and inorganic contaminants.

The green synthesis method involves utilizing plant extracts, considered the most reliable means of producing environmentally friendly and biocompatible NPs (Joshi et al. 2023). This approach offers numerous advantages, such as easiness of availability and non-toxic, and cost-effective raw materials for nanoparticle synthesis. Various plant parts including fruits, flowers, leaves, and roots are combined with biomolecules like amino acids and carbohydrates (Bukhari et al. 2021). Alternatively, employing microorganisms for green synthesis offers several benefits, particularly in terms of metal ion reduction under controlled bacterial growth (Rasheed et al. 2023). The growing preference for bacterial nanoparticle synthesis is attributed to the eco-friendly nature of the microorganisms. Another green pathway involves biosynthesis mediated by fungi which is capable of producing a greater quantity of NP due to the presence of diverse intracellular enzymes (Rasool et al. 2023). Lastly, nanoparticle manufacturing utilizing algae, disregarding leaf or root structure, allows for the production of various metal types using microalgae (Singh et al. 2023b). Nevertheless, the drawback of employing algae lies in the requirement to wash seaweeds and dry them under sunlight, potentially leading to the loss of several components. The schematic representation describing the synthesis mechanism of NPs via the biogenetic process is depicted in Fig. 8.2.

A summary of various NPs with their synthesis process and efficacy for wastewater treatment are summarized in Table 8.1.

8.3 Immobilization of Microorganisms on NPs for Wastewater Treatment

However, low biodegradation rates, sludge generation, and substrate inhibition hinder the efficacy of free cells in bioremediation (Narayanan et al. 2023). Additionally, immobilized cell technology offers long-term stabilization for the removal of toxic pollutants. Various materials including alginate, chitosan, cellulose, starch, and polyvinyl alcohol have been utilized for microbial immobilization. However,

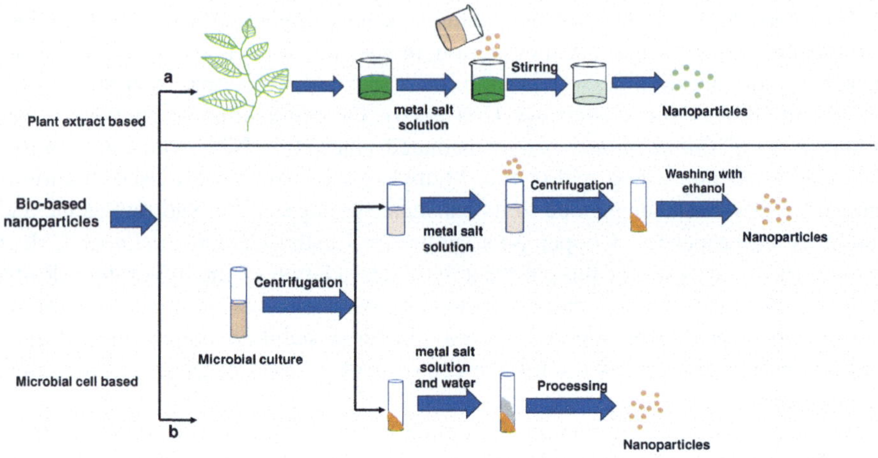

Fig. 8.2 Synthesis of bio-based NPs via **a** plant extract-based and **b** microbial cell-based

Table 8.1 A brief description of NMs preparation and removal efficiency for wastewater treatment

Composition of NMs	Classification	Synthesis method	Pollutant	Removal efficiency (%)	References
Fe-Pd@C	Metallic	Mechanochemical	Phenol	95	He et al. (2019)
Co-doped ZnO	Metallic	Co-precipitation (chemical)	Methylene blue dye	100	Din et al. (2024)
ZnO	Nanocomposites	Bio (plant)-based	Brilliant flavine dye	89	Rasool et al. (2023)
Ag	Nanocomposites	Bio (plant)-based	Azo dyes	53.7–93.2	Rasheed et al. (2023)
Silica NP	Non-metallic	Sol-gel	Methylene blue dye	98	Joshi et al. (2023)
Ag-doped indium oxide ceramic	Non-metallic	Wet-chemical	Crystal violet dye	87.2	Shabbir et al. (2024)
$ZnFe_2O_4$	Nanocomposites	Probe sonication	Azo dyes	88–98	Surendra et al. (2020)

the major drawbacks of the immobilized cell technique include limited substrate transport and mechanical fragility. An alternative method involves the use of NPs as immobilization carriers.

Fungi emerged as an eminent solution in bioremediation for wastewater treatment, utilizing fungal biomass, enzymes, and NPs (Li et al. 2023; He et al. 2017). Magnetic NPs draw significant attention as they have a high surface area and offer a porous matrix for the entrapment of microbes (Fouda et al. 2021). Fe_3O_4 NPs have demonstrated high efficiency in degrading various organic compounds, i.e., volatile organic compounds, azo dyes, phenols, and polycyclic aromatic hydrocarbons (Bestawy et al. 2020; Bhattu and Singh 2023). Biogenic magnetite NPs exhibit a higher adsorption capacity for metal cations compared to commercially available magnetite, making them potential candidates for microbial magnetite formation in toxic metal cation removal from water (Hassan et al. 2020). Iron oxide NPs possess significant potential as wastewater remediation agents due to their strong adsorption capacity, chemical inertness, and superparamagnetic properties. Fe_3O_4 magnetic NPs have been found to enhance the production and activity of degrading enzymes, as well as membrane permeability, making them widely studied immobilization supports in bioremediation (Azeez et al. 2022).

It is widely recognized that various microorganisms (*Pseudomonas* sp., *Burkholderia pickettii*, *Klebsiella pneumoniae*, and *Comamonas* sp.) possess the capability to transform herbicides into less toxic compounds (Magnoli et al. 2020; Modi et al. 2023). Recently, peroxidase enzyme immobilized onto Fe_3O_4 magnetic NPs demonstrated enhanced removal ability and reusability of the enzyme (stable for up to 100 cycles) pH and thermal stability (Bilal et al. 2022). Table 8.2 summarizes the application of NP combined with microorganisms for various pollutants degradation from wastewater. Scanning electron microscope (SEM) images confirmed the complete distribution of bacterial cells on the surface of α-Fe_2O_3 NPs. The immobilized bacterial cells exhibited high tolerance to a wide range of atrazine concentrations up to 300 mg L^{-1} at a pH range of 4.0–9.0 and temperature of 20–45 °C. Moreover, the immobilized cells efficiently degraded more than 90% of atrazine, an herbicide, within 20 days of process time (Khatoon and Rai 2018). Recently, a new technique was adopted for the immobilization of peroxidase enzyme onto Fe_3O_4 NPs followed by co-precipitation with glutaraldehyde. The obtained composite enhanced the thermal stability, resistivity against extreme pH, and reusability cycle of the matrix. The specified enzyme, such as peroxidase, is a well-known biocatalyst usually used for dye wastewater treatment due to its various suitable properties. The NPs-immobilized peroxidase effectively removes artificial coloring agents of various compositions in a laboratory-scale bioreactor, showing the robustness and high effectiveness of the system.

The immobilization method enhances enzyme stability, making them suitable for real applications, notably in the remediation of pollutants. A prominent example is the remediation of environmental contaminants using immobilized laccase on NP. Laccase is a multicopper oxidase capable of catalyzing various aromatic phenol and derivative substances containing amine groups. Some non-substrate dyes can also be eliminated from the contaminated liquid solution by using the specified enzyme (laccase) in the presence of small molecular redox mediators. For instance, a magnetic nanocatalyst of Fe_3O_4/SiO_2 (< 30 nm) was formulated to adhere to laccase for the decontamination of phenolic as well as dye waste effluent (Patani et al. 2023). Similarly, various dyes including Procion Red MX-5B and azophloxine were efficiently removed by immobilized laccase. Additionally, immobilized *P. sanguineus* laccase on copper tetra-amino phthalocyanine (CuTAPc)-Fe_3O_4 NP exhibited enhanced

Table 8.2 A summary of nano-bioremediation techniques for various pollutant removal

NMs	Microbial species	Types of pollutant	Concentration	Removal efficiency (%)	References
CuO	*Fusarium oxysporum* OSF18	Color (textile wastewater)	–	90	Li et al. (2023)
Fe_3O_4-biochar	*R. capsulatus*	COD, NH^{4+}, Total phosphorus (Synthetic wastewater)	877, 29.7, and 3.6 mg/L	83, 61, 63	He et al. (2017)
α-Fe_2O_3 and MnO	*Aspergillus carbonarius* D-1	COD (textile and tannery wastewater)	203 mg/L	92.1	Fouda et al. (2021)
Fe_3O_4	*Enterobacter cloacae and Pseudomonas otitidis*	Total hydrocarbon	400 mg/L	85	Bestawy et al. (2020)
ZnO	*Bacillus* sp., *Enterobacter* sp., *Staphylococcus* sp., *Brevibacterium* sp., *Klebsiella* sp.	Methylene blue	10	95.5	Modi et al. (2023)
Fe_3O_4	*Proteus mirabilis*, *Pseudomonas batumici*, and *Providencia vermicola*	COD (dairy effluent)	500–700 mg/L	63.9–67.9	Salama et al. (2021)

COD: Chemical oxygen demand

thermal and storage stability. Further, it has the ability to remove more than 80% of initial dye even after multiple uses.

This was due to the increase of resistivity of the enzyme at a higher dye concentration and other adverse parameters, which results in higher reusability with more degradation efficiency in a single and multiple batch operation. Generally, the membrane of the biocells is less susceptible to the toxin and they tend to rupture when exposed to higher concentrations. This leads to the generation of reactive radicals including O_2^-, H_2O_2, and OH^-, which cause stress and death of the cells. To overcome this situation, coupling of nanoremediation with membrane separation method might be useful. In this direction, the immobilization of laccase on TiO_2-blended polymer membranes was utilized for wastewater treatment effectively (Awad et al. 2021). The thermodynamic and chemical strength grows sharply in addition to pH tolerability after immobilization, which provides more efficiency compared to the suspension system. Similarly, a noble catalyst made of immobilized *P. chrysosporium* impregnated onto nitrogen-doped TiO_2 was utilized to demineralize the contaminants present in a landfill leachate at a very low BOD/COD ratio (Ghasemi et al. 2013).

8.4 Response and Reactivity of Microorganisms During Nano-Bioremediation

The effectiveness of nano-bioremediation depends exclusively on the interaction between NP and microbes. This interaction can be described by analyzing various aspects like adaptability, microbial growth, and toxic effects with NPs.

8.4.1 Role of NP in Microbial Growth

The NPs coupled with microorganisms have excellent removal ability of pollutants highlighting their impact on microbial growth and metabolic activity (Narayanan et al. 2023). The zero-valent iron and iron-copper NPs positively influenced bacterial growth, increasing removal efficiency by 53.84% and 84.61%, respectively (Chauhan et al. 2023). Additionally, microbial growth exhaustively depends upon the concentration of NPs and environmental factors. A remarkable rise in cell growth and biosurfactant was observed even at lower concentrations of iron NPs. However, a declining trend of microbial growth as well as surfactant production was observed at higher concentrations. This could be attributed to the potential toxicity of high NP concentrations to microorganisms resulting in inhibition of microbial growth.

NPs play a pivotal role in enhancing microbial reactivity due to their smaller dimensional structure, and high compatibility in a minimum space providing a more surface-to-volume ratio leading to increased collisions and accelerated microbiological reaction rates. Moreover, NPs exhibit quantum effects by interfering in chemical reactions and reducing the required activation energy, transforming long-chain aromatic derivatives into less harmful substances (Zheng et al. 2023). Studies have reported that NPs, in conjunction with microbial communities, stimulate microbe activity at appropriate concentrations. However, water repellant nature of the hydrophobic matrix has limitation to accommodate high microbial cells inside it to degrade the pollutants present in aqueous media. To address this challenge, the surface tension can be minimized by utilizing specific biosurfactant-generating microbes, which can alter hydrocarbon solubility (Rai et al. 2023).

However, biosurfactant extraction and synthesis are limited by metabolic activity and source utilization pathways. For instance, biofunctionalized cellulose nanocrystals were utilized as emulsifiers, reducing diesel oil droplet size. Surface functionalization of NPs can upgrade oil emulsification in water, utilizing agents like chitosan biopolymer and amphiphilic polypeptide polymer to enhance hydrophobicity and reduce interfacial tension (Gopi et al. 2018). Additionally, certain NPs, such as metal oxide NPs, induce biosurfactant production in microorganisms by stimulating gene expression. Iron NPs, acting as micronutrients, influence microbial biosurfactant synthesis potential, suggesting their use to improve biosurfactant production. Studies using Fe-silica NPs and ZnO_2 NPs have shown increased biosurfactant production in various microbial strains (Sohni et al. 2018). However, the concentration of NPs must be optimized to avoid toxicity, which can inhibit both microbial growth as well as surfactant generation. Therefore, determining the ideal NPs concentration is crucial for enhancing biosurfactant production.

8.4.2 Compatibility and Resistivity of Microorganisms for NPs

NPs play a vital role in enhancing the metabolic activity of microorganisms due to their small size and unique structure. The increased specific surface area offers more collisions, which accelerates microbiological reaction rates (Zheng et al. 2023). Additionally, NPs exhibit quantum effects by reducing the minimum activation energy requirement for the chemical reactions, which ultimately transforms aliphatic hydrocarbons into less harmful ones (Ramezani Zheng et al. 2023; Marcon et al. 2021). However, the lesser biocompatibility (that means less adhesion between the microbial site and the solid surface) adversely affects the remediation in soil and aquatic environments (Ambaye et al. 2022). Furthermore, NPs (such as metal oxide NPs) can induce the generation of biosurfactants. In this direction, the synthesis of biosurfactants in *Bacillus amyloliquefaciens* can be increased by the addition of supplementary nutrients (i.e., iron). However, a higher NP concentration can restrict microbial growth and biosurfactant production due to potential toxicity.

8.4.3 Toxic Effect of NMs on Microorganisms

The toxicity of NPs in microbes includes cell membrane rupture, reduction of bacterial performance, and damage to DNA. The adverse effects of NP toxicity on microbial cells depend largely on factors, such as the dimensional configuration (size, types) of NPs, and several external factors influencing the interaction of microbial cells with NPs. Some NPs (like gold) exhibit minimal toxic effects. In the meantime, Ag NPs possess antibiotic effects on various microorganisms (Pal et al. 2021). In a separate study, it was noted that ZnO NPs slightly decreased the expression of the catabolic gene catechol 2,3 dioxygenase, while Ag NPs caused a significant reduction (80%) (El-Sayed et al. 2019). Furthermore, Ag NPs were found to negatively impact bacterial species involved in petroleum degradation by inhibiting their proliferation. Consequently, NPs demonstrating moderate toxicity are often utilized in bioremediation experiments.

Nanoparticle size also plays a critical role, with smaller-sized NPs possessing higher microbicidal potential compared to those larger than 10 nm, which exhibit reduced penetration. Moreover, numerous studies in bioremediation have empirically demonstrated that toxicity is directly proportional to NP concentration. Higher doses can be harmful to metabolic kinetics and the life cycle of microorganisms. It was reported that the iron NPs have a high phenolic compound degradation ability, which exhibits detrimental effects on the growth of bacteria even at a low concentration of 0.5 g/L (Banerjee et al. 2022).

The lethal concentration of NPs compromises membrane permeability by denaturing transport proteins and lipopolysaccharides, facilitating NP ingress into cells. Once inside, NPs react with hydrogen peroxide to generate reactive oxygen species, leading to increased oxidative stress within microbial cells (Pal et al. 2021). Gram-positive bacterial species are particularly susceptible to NPs due to charge interactions with negatively charged cell wall components, potentially leading to nutrient deprivation and cell death. To mitigate these effects, strategies, such as polymer coating, chelating agent capping, doping, or tethering of antioxidant molecules onto NP surfaces, are employed. These strategies impart a negative charge, reducing interaction with cell surfaces, ion dissolution, and reactive oxygen species synthesis. Coating ZVI with polyelectrolyte significantly reduces toxicity at exposure levels of 0.1–0.5 g/L by reducing adhesion to bacterial cells (Daniel et al. 2023). Additionally, inert organic surface coat polymers like guar gum and sodium carboxymethyl cellulose not only reduce toxicity but also enhance NP mobility, stability, and reactivity. Some NPs (such as gold NPs) exhibit negligible toxicity levels, whereas silver NPs possess biocidal effects on various microorganisms. For instance, it was observed that ZnO NPs led to a slight decrease in the expression of the catabolic gene catechol 2,3 dioxygenase, whereas silver NPs caused a significant reduction of 80% (Hu et al. 2021). Therefore, NPs having moderate toxicity effects are often employed for wastewater treatment. NPs with smaller sizes (less than 10 nm) have antibacterial properties compared to larger-sized NPs, which have lower penetrative ability (Singh et al. 2023b). These interactions affect microbial cells by blocking membranous pores, thereby promoting cell death (Hu et al. 2021). Hence, various techniques, such as polymer coating and entrapping with a chelating agent, have been utilized to mitigate such adverse effects (Daniel et al. 2023).

8.5 Integration of Nanobiotechnology with Other Techniques

8.5.1 Photocatalysts as Microbial Agents

Photocatalysis, which accelerates photoreactions using suitable catalysts, has emerged as an efficient technique for wastewater remediation. The artificial nano-photocatalysts offer several merits including more active surface area, high surface adsorption activity, and specific surface area. Previously, most of the research works were based on two well-known catalysts (i.e., ZnO and TiO_2) for pollutants removal. However, other NMs, such as Fe_2O_3, MgO, CdS, ZnS, CdTe, and CdSe, also hold potential as effective photocatalysts (Surendra et al. 2020; Yu et al. 2023).

Sahu et al. (Sahu et al. 2021) reported the fabrication of ZnO photocatalysts using a unique approach (sense and shoot) could be able to remove microbial contaminants from wastewater. Moreover, various studies found that the efficiency of photocatalysts can be enhanced by modifying their materials. For instance, previous research work reported that depositing Ag on TiO_2-coated films increased antimicrobial activity against *E. coli* by 6.9 times compared to TiO_2 alone under visible light. This enhanced activity is attributed to the larger surface area, providing more active sites for high activity of biocells (Ghasemi et al. 2022). However, the presence of extracellular polymeric substances (EPS) in bacterial cells can resist the antimicrobial action of photocatalysts by controlling their bactericidal activity. It has been suggested that removing EPS from biocells could improve the efficacy of photocatalysts during wastewater treatment. Photocatalysis-based disinfection holds great promise for water treatment, but commercialization has faced significant challenges. Hence, it is very much essential to look out for other suitable options to increase the compatibility of microbial cells with photocatalysts for higher removal efficiency.

8.5.2 Microbial Fuel Cells

The utilization of microbial fuel cell (MFC) technology for bioremediation has garnered attention due to its dual benefits including wastewater decontamination as well as electricity generation from various pollutants (Ghasemi et al. 2013). The generated energy by MFCs has the potential to reduce the electricity demand of conventional treatment technologies. Simultaneously, microbes can degrade complex organic pollutants (azo dyes, aromatic compounds, etc.) and various simpler carbon sources (such as glucose, sucrose, and starch), into water and carbon dioxide. However, the commercialization of MFC technology faces challenges due to its lower efficiency compared to other fuel cell-based technologies. To make this technology commercially viable, it is necessary to either reduce membrane maintenance costs or use more affordable membranes and cathode catalysts. Additionally, selecting appropriate materials is crucial to improve MFC efficiency. This could involve modifying the electrode surface to enhance MFC conductivity, as the cathode surface requires high catalytic activity for reactive oxygen species. Enhancing certain properties with the use of NPs could be a viable option to address these challenges.

The performance of MFCs is currently being enhanced by using an economical nanocomposite matrix (i.e., nanostructured carbon). These electrodes offer mechanical stability, excellent electrical conductivity, higher electrochemical catalytic activity, and increased surface area. Ghasemi et al. investigated the use of CNTs as electrodes and platinum as the catalyst in MFCs with varying concentrations of COD at 100, 500, 1000, and 2000 mg/l (Ghasemi et al. 2013). The highest power generation was achieved with CNT/Pt electrodes across all COD concentrations, followed by the individual use of CNT and Pt. These findings indicate that the catalytic activity of Pt was enhanced by CNTs due to the increased surface area of the electrode. The superior performance of CNT electrodes is attributed to their unique structural properties and high conductivity. Consequently, CNT/Pt can replace conventional cathodes in MFCs. Additionally, CNTs can be coated with polymers, such as polypyrrole and polyaniline to form nanocomposites that enhance microbial attachment, leveraging the negatively charged CNTs.

8.5.3 Algal Membrane-Based Bioreactor with NPs

Membranes are commonly made from polyvinylidene fluoride (PVDF), polysulfone (PSF), and polyethersulfone (PES) due to their unique mechanical strength and chemical stability (Sahu et al. 2021). However, membrane fouling poses a significant challenge in membrane technology, primarily due to hydrophobic interactions between microbial cells and membrane components. The above bottleneck can be minimized by several new techniques (like surface coating, plasma treatment, etc.), which ultimately enhance the hydrophilicity and help to reduce the fouling of the membrane. For example, combining CNTs and TiO_2 with PSF hollow fiber membranes enhances surface hydrophilicity and antifouling properties (Awad et al. 2021). It was previously reported that coating reverse osmosis (RO) membranes with TiO_2 improves the polyvinyl layer, reducing fouling through a self-cleaning mechanism (Tawalbeh et al. 2023). Additionally, TiO_2 photocatalysts, with their hydrophilic character and high surface area, can be used for air pollution remediation, thereby decreasing hydrophobicity and fouling when incorporated into membranes. In another study, the use of a PVDF membrane impregnated with TiO_2 was found to be more resilient and effective for wastewater treatment (Sahu et al. 2021). Further, the membrane bioreactor was able to remove almost 75% of the initial phosphorus and nitrogen content in wastewater while exhibiting minimal fouling.

8.6 Limitations and Future Perspectives of Nano-Bioremediation

Understanding the application of nano-bioremediation for petroleum hydrocarbons presents a multifaceted challenge. While nanotechnology holds promise for addressing environmental issues, its implementation faces significant hurdles that necessitate careful consideration. One such challenge is the potential persistence of NPs in the environment and their accumulation in organisms, posing risks of toxicity. Moreover, the field of nano-bioremediation is still in its early stages, requiring standardization for the use of NMs in bioremediation practices. The production cost of NPs and their integration with microbial agents adds complexity, hindering scalability for large-scale implementation. Real-world application of nano-bioremediation faces soil complexity, with NP interacting intricately with soil components like humic acids and clay particles. Characterization and analysis of NPs within soil are challenging, often conducted with soil suspensions rather than intact soil.

Organic contaminants' mobility in soil is mitigated by organic matter sorption, affecting petroleum hydrocarbon migration in subsurface environments. Nanoparticle introduction in soil can alter alkalinity, salinity of soil structure, health, nutrient composition, microbial diversity, and plant growth. Understanding the interaction between contaminants, NPs, and microorganisms is vital for assessing nanotreatment effects on soil biodiversity and long-term impacts. NMs' introduction in soil matrices can shift soil microbial composition toward species involved in contaminant remediation, potentially reducing soil heterogeneity.

Future prospects in nano-bioremediation involve developing less toxic NPs with lower environmental footprints and standardized protocols to enhance safety and efficacy. Collaboration among researchers, industry, and governmental bodies is imperative for advancing nano-bioremediation, addressing concerns like cost and scalability. Integration of nano-bioremediation with genetic engineering and artificial intelligence holds promise for more efficient environmental remediation solutions. The utilization of nanotechnology for wastewater remediation presents a promising approach for addressing several environmental issues. However, nano-bioremediation poses some bottlenecks and challenges, which must be overcome for real-field implementation. The accumulation of persistent NPs in the environment becomes a major shortcoming of this process. Since nano-bioremediation research is still in its early stages, standardization of NM usage is necessary. The high production cost of NPs and their integration with microbes pose challenges for scalability and large-scale applications. Employing nano-bioremediation in real environmental conditions is further complicated by the composition of wastewater, where NPs interact with aqueous phases and components. Additionally, introducing NP into water can alter pH levels, nutrient availability, and microbial dynamics. Understanding the interaction between contaminants, NPs, and microbial species is crucial for assessing the long-term effects of wastewater treatment. Despite these challenges, nano-bioremediation offers numerous prospects. Ongoing research aims to develop less toxic NP with lower environmental impacts. However, standardized protocols for NMs usage in bioremediation are currently lacking, necessitating further research for improved safety and efficacy.

8.7 Conclusions

Nanotechnology stands poised to transform various industries, including pollution control, by enhancing existing technologies. By developing innovative products to replace current production processes, nanotechnologies contribute significantly to improving performance, potentially leading to environmental and cost savings. Additionally, they offer opportunities to reduce material consumption and promote the development of more sustainable production processes. While NMs show promise in water treatment, concerns have been raised about their potential human and environmental toxicity. However, nanotechnology offers effective alternatives to address pollution-related issues, such as heavy metal contamination, chemical pollutants, and oil emissions, without causing harm to the environment. Certain plants, fungi, and bacteria, known as "super accumulators," have been identified for their ability to collect high metal concentrations, and NMs derived from these sources can be utilized to extract pollutants from the atmosphere and contaminated sites. In highly polluted environments, NPs derived from natural sources play a valuable role in detoxifying and remediating soil, water, and other ecosystems. Through modification and enhancement, nanotechnology has the potential to improve the quality and duration of bioremediation processes. The interdisciplinary nature of nanotechnology, coupled with its historical success and future potential, underscores its growing importance in addressing environmental challenges. Xenobiotics pose a persistent threat to water, animal, and human health worldwide, necessitating urgent remediation efforts. While nanoparticle-assisted microbial remediation offers significant potential for removing harmful chemicals. However, the commercialization of nanotechnology applications remains a significant challenge. Developing nano-bio formulations for commercial use at contaminated sites is essential. While laboratory-scale studies have been informative, further research for real application is essential to fully comprehend how NP and microbial inoculants can effectively contribute to wastewater treatment. Collaboration between researchers and government entities is crucial for the development of long-term, feasible, and cost-effective nano-biotechniques for industrial application. Continued modifications and adaptations in nanobiotechnology will enhance the efficacy and longevity of future remediation processes.

References

Ambaye TG, Vaccari M, Prasad S, Rtimi S (2022) Recent progress and challenges on the removal of per-and poly-fluoroalkyl substances (PFAS) from contaminated soil and water. Environ Sci Pollut Res 29:58405–58428. https://doi.org/10.1007/s11356-022-21513-2

Awad ES, Sabirova TM, Tretyakova NA, Alsalhy QF, Figoli A, Salih IK (2021) A mini-review of enhancing ultrafiltration membranes (UF) for wastewater treatment: performance and stability. Chemeng 5:34. https://doi.org/10.3390/chemengineering5030034

Azeez RA, Hasan MM, Alawi NM, Nguyen HM (2022) A review study of immobilized microbial-NP: techniques and biotechnology applications. Iraqi J Oil Gas Res 2:87–101. https://doi.org/10.55699/ijogr.2022.0202.1030

Banerjee M, Mukherjee S, Mukherjee S (2022) Facets of nanoparticle-microbe interactions and their roles in nanobioremediation of environmental pollutants: biochemical, molecular, and technological perspectives. In: Nano-bioremediation: fundamentals and applications, pp 111–145. Elsevier. https://doi.org/10.1016/B978-0-12-823962-9.00038-6

Bestawy EE, El-Shatby BF, Eltaweil AS (2020) Integration between bacterial consortium and magnetite (Fe_3O_4) NP for the treatment of oily industrial wastewater. World J Microbiol Biotechnol 36:1–16. https://doi.org/10.1007/s11274-020-02915-1

Bhattu M, Singh J (2023) Recent advances in NMs based sustainable approaches for mitigation of emerging organic pollutants. Chemosphere 321:138072. https://doi.org/10.1016/j.chemosphere.2023.138072

Bilal M, Rashid EU, Zdarta J, dos Santos JC, Fernandes PC, Cheng H, Jesionowski T (2022) Engineering magnetic nanobiocatalytic systems with multipurpose functionalities for biocatalysis, biotechnology and bioprocess applications. Sustain Chem Pharm 30:100866. https://doi.org/10.1016/j.scp.2022.100866

Bukhari A, Ijaz I, Gilani E, Nazir A, Zain H, Saeed R, Alarfaji SS, Hussain S, Aftab R, Naseer Y (2021) Green synthesis of metal and metal oxide NP using different plants' parts for antimicrobial activity and anticancer activity: a review article. Coatings 11:1374. https://doi.org/10.3390/coatings11111374

Chauhan P, Imam A, Kanaujia PK, Suman SK (2023) Nano-bioremediation: an eco-friendly and effective step towards petroleum hydrocarbon removal from environment. Environ Res 231:116224. https://doi.org/10.1016/j.envres.2023.116224

Cheng J, Sun Z, Yu Y, Li X, Li T (2019) Effects of modified carbon black NP on plant-microbe remediation of petroleum and heavy metal co-contaminated soils. Int J Phytoremediation 21:634–642. https://doi.org/10.1080/15226514.2018.1556581

Daniel AI, Keyster M, Klein A (2023) Biogenic zinc oxide NP: a viable agricultural tool to control plant pathogenic fungi and its potential effects on soil and plants. Sci Total Environ 897:165483. https://doi.org/10.1016/j.scitotenv.2023.165483

Din MI, Khalid R, Hussain Z, Gul S, Mujahid A (2024) Synthesis and characterization of cobalt doped zinc oxide NP and their application for catalytic reduction of methylene blue dye. Desalin Water Treat 317:100002. https://doi.org/10.1016/j.dwt.2024.100002

El-Sayed WS, Elbahloul Y, Saad ME, Hanafy AM, Hegazi AH, ElShafei GM, Elbadry M (2019) Impact of NP on transcriptional regulation of catabolic genes of petroleum hydrocarbon-degrading bacteria in contaminated soil microcosms. J Basic Microbiol 59:166–180. https://doi.org/10.1002/jobm.201800186

Eniola JO, Sizirici B, Fseha Y, Shaheen JF, Aboulella AM (2023) Application of conventional and emerging low-cost adsorbents as sustainable materials for removal of contaminants from water. Environ Sci Pollut Res 30:88245–88271. https://doi.org/10.1007/s11356-023-28399-8

Fouda A, Hassan SED, Abdel-Rahman MA, Farag MM, Shehal-deen A, Mohamed AA, Alsharif SM, Saied E, Moghanim SA, Azab MS (2021) Catalytic degradation of wastewater from the textile and tannery industries by green synthesized hematite (α-Fe_2O_3) and magnesium oxide (MgO) NP. Curr Res Biotechnol 3:29–41. https://doi.org/10.1016/j.crbiot.2021.01.004

Ghasemi M, Ismail M, Kamarudin SK, Saeedfar K, Daud WRW, Hassan SH, Heng LY, Alam J, Oh SE (2013) Carbon nanotube as an alternative cathode support and catalyst for microbial fuel cells. Appl Energy 102:1050–1056. https://doi.org/10.1016/j.apenergy.2012.06.003

Ghasemi R, Merdan G, Tore GY (2022) Emerging biotechnologies for treatment of antibiotic residues from pharmaceutical waste waters for sustainable environment a case study under visible light with NFC-doped titania. In: Waste management, pp 218–244. CRC Press. https://doi.org/10.1201/9780429341106-12

Gomes HI, Dias-Ferreira C, Ottosen LM, Ribeiro AB (2015) Electroremediation of PCB contaminated soil combined with iron NP: effect of the soil type. Chemosphere 131:157–163. https://doi.org/10.1016/j.chemosphere.2015.03.007

Gopi S, Amalraj A, Sukumaran NP, Haponiuk JT, Thomas S (2018) Biopolymers and their composites for drug delivery: a brief review. Macromol Symp 380:1800114. https://doi.org/10.1002/masy.201800114

Hassan M, Naidu R, Du J, Liu Y, Qi F (2020) Critical review of magnetic biosorbents: their preparation, application, and regeneration for wastewater treatment. Sci Total Environ 702:134893. https://doi.org/10.1016/j.scitotenv.2019.134893

He S, Zhong L, Duan J, Feng Y, Yang B, Yang L (2017) Bioremediation of wastewater by iron oxide-biochar nanocomposites loaded with photosynthetic bacteria. Front Microbiol 8:823. https://doi.org/10.3389/fmicb.2017.00823

He D, Niu H, He S, Mao L, Cai Y, Liang Y (2019) Strengthened Fenton degradation of phenol catalyzed by core/shell Fe–Pd@ C nanocomposites derived from mechanochemically synthesized Fe-Metal organic frameworks. Water Res 162:151–160. https://doi.org/10.1016/j.watres.2019.06.058

Hosny M, Mubarak MF, El-Sheshtawy HS, Hosny R (2024) Break oily water emulsion during petroleum enhancing production processes using green approach for the synthesis of SnCuO@FeO nanocomposite from microorganisms. Sci Rep 14:8406. https://doi.org/10.1038/s41598-024-56495-3

Hu C, Yang Y, Lin Y, Wang L, Ma R, Zhang Y, Feng X, Wu J, Chen L, Shao L (2021) GO-based antibacterial composites: application and design strategies. Adv Drug Deliv Rev 178:113967. https://doi.org/10.1016/j.addr.2021.113967

Jain N, Bhargava A, Panwar J (2014) Enhanced photocatalytic degradation of methylene blue using biologically synthesized protein-capped ZnO NP. Chem Eng J 243:549–555. https://doi.org/10.1016/j.cej.2013.11.085

Joshi U, Roy R, Bhosale P, Nikam S, Kamble T, Satsangi PG (2023) Sustainable economic production of silica NP from rice husks for adsorptive removal of anionic and cationic dyes. Biores Technol Rep 24:101685. https://doi.org/10.1016/j.biteb.2023.101685

Khan MF, Hof C, Niemcová P, Murphy CD (2023) Recent advances in fungal xenobiotic metabolism: enzymes and applications. World J Microbiol Biotechnol 39:296. https://doi.org/10.1007/s11274-023-03737-7

Khatoon H, Rai JPN (2018) Augmentation of Atrazine biodegradation by two Bacilli immobilized on α-Fe_2O_3 magnetic NP. Sci Rep 8:17831. https://doi.org/10.1038/s41598-018-36296-1

Li OM, Darwesh H, Matter IA (2023) Nano-bioremediation of textile industry wastewater using immobilized CuO-NPs myco-synthesized by a novel Cu-resistant Fusarium oxysporum OSF18. Environ Sci Pollut Res 30:16694–16706. https://doi.org/10.1007/s11356-022-23360-7

Lin J, Ye W, Xie M, Seo DH, Luo J, Wan Y, Van der Bruggen B (2023) Environmental impacts and remediation of dye-containing wastewater. Nat Rev Earth Environ 4:785–803. https://doi.org/10.1038/s43017-023-00489-8

Liu N, Sun Z, Zhang H, Klausen LH, Moonhee R, Kang S (2023) Emerging high-ammonia-nitrogen wastewater remediation by biological treatment and photocatalysis techniques. Sci Total Environ 875:162603. https://doi.org/10.1016/j.scitotenv.2023.162603

Ma W, Zhang X, Han H, Shi X, Kong Q, Yu T, Zhao F (2024) Overview of enhancing biological treatment of coal chemical wastewater: new strategies and future directions. J Environ Sci 135:506–520. https://doi.org/10.1016/j.jes.2022.11.008

Magnoli K, Carranza CS, Aluffi ME, Magnoli CE, Barberis CL (2020) Herbicides based on 2, 4-D: its behavior in agricultural environments and microbial biodegradation aspects: a review. Environ Sci Pollut Res 27:38501–38512. https://doi.org/10.1007/s11356-020-10370-6

Mahmoud AS, Mostafa MK, Peters RW (2021) A prototype of textile wastewater treatment using coagulation and adsorption by Fe/Cu NP: techno-economic and scaling-up studies. Nanomater Nanotechnol 11:18479804211041180. https://doi.org/10.1177/18479804211041181

Manimegalai S, Vickram S, Deena SR, Rohini K, Thanigaivel S, Manikandan S, Govarthanan M (2023) Carbon-based NM intervention and efficient removal of various contaminants from effluents–a review. Chemosphere 312:137319. https://doi.org/10.1016/j.chemosphere.2022.137319

Marcon L, Oliveras J, Puntes VF (2021) In situ nanoremediation of soils and groundwaters from the nanoparticle's standpoint: a review. Sci Total Environ 791:148324. https://doi.org/10.1016/j.scitotenv.2021.148324

Modi S, Yadav VK, Amari A, Osman H, Igwegbe CA, Fulekar MH (2023) Nanobioremediation: a bacterial consortium-zinc oxide nanoparticle-based approach for the removal of methylene blue dye from wastewater. Environ Sci Pollut Res 30:72641–72651. https://doi.org/10.1007/s11356-023-27507-y

Muhammad M, Batool S, Hivare V, Li WJ, Waheed A, Sinha D (2024) Bioremediation techniques—classification, principles, advantages, limitations, and prospects. In: Microbiome-assisted bioremediation, pp 1–23. Academic Press. https://doi.org/10.1016/B978-0-443-21911-5.00003-9

Narayanan M, Ali SS, El-Sheekh M (2023) A comprehensive review on the potential of microbial enzymes in multipollutant bioremediation: mechanisms, challenges, and future prospects. J Environ Manage 334:117532. https://doi.org/10.1016/j.jenvman.2023.117532

Pal A, Goswami R, Roy DN (2021) A critical assessment on biochemical and molecular mechanisms of toxicity developed by emerging NMs on important microbes. Environ Nanotechnol Mon Manage 16:100485. https://doi.org/10.1016/j.enmm.2021.100485

Partovinia A, Khanpour-Alikelayeh E, Talebi A, Kermanian H (2023) Improving mass transfer rates in microbial cell immobilization system for environmental applications: synergistic interaction of cells on crude oil biodegradation. J Environ Manag 326:116729. https://doi.org/10.1016/j.jenvman.2022.116729

Patani A, Patel A, Prajapati D, Khare N, Singh S (2023) Various agriculture crop plant-based bioactive compounds and their use in NM synthesis and applications. In: NMs from agricultural and horticultural products, pp 223–241. Springer Nature Singapore, Singapore. https://doi.org/10.1007/978-981-99-3435-5_12

Puri A, Mohite P, Maitra S, Subramaniyan V, Kumarasamy V, Uti DE, Atangwho IJ (2024) From nature to nanotechnology: The interplay of traditional medicine, green chemistry, and biogenic metallic phytoNP in modern healthcare innovation and sustainability. Biomed Pharmacother 170:116083. https://doi.org/10.1016/j.biopha.2023.116083

Rai RS, Bajpai V, Khan MI, Elboughdiri N, Shanableh A, Luque R (2023) An eco-friendly approach on green synthesis, bio-engineering applications, and future outlook of ZnO NM: a critical review. Environ Res 221:114807. https://doi.org/10.1016/j.envres.2022.114807

Raina S, Roy A, Bharadvaja N (2020) Degradation of dyes using biologically synthesized silver and copper NP. Environ Nanotechnol Mon Manage 13:100278. https://doi.org/10.1016/j.enmm.2019.100278

Rambabu K, Bharath G, Banat F, Show PL (2021) Green synthesis of zinc oxide NP using *Phoenix dactylifera* waste as bioreductant for effective dye degradation and antibacterial performance in wastewater treatment. J Hazard Mater 402:123560. https://doi.org/10.1016/j.jhazmat.2020.123560

Ramezani Zheng Z, Tian S, Feng Y, Zhao S, Li X, Wang S, He Z (2023) Recent advances of photocatalytic coupling technologies for wastewater treatment. Chinese J Catal 54:88–136. https://doi.org/10.1016/S1872-2067(23)64536-X

Rasheed A, Hussain S, Mushtaq W, Zubair M, Siddique K, Attia K, Khan N, Fiaz S, Chen Y (2023) Application of silver NP synthesized through varying biogenic and chemical methods for wastewater treatment and health aspects. Environ Sci Pollut Res, 1–18. https://doi.org/10.1007/s11356-022-24761-4

Rasool A, Kiran S, Gulzar T, Abrar S, Ghaffar A, Shahid M, Nosheen S, Naz S (2023) Biogenic synthesis and characterization of ZnO NP for degradation of synthetic dyes: a sustainable environmental cleaner approach. J Clean Prod 398:136616. https://doi.org/10.1016/j.jclepro.2023.136616

Sahu JN, Karri RR, Zabed HM, Shams S, Qi X (2021) Current perspectives and future prospects of nano-biotechnology in wastewater treatment. Sep Purif Rev 50:139–158. https://doi.org/10.1080/15422119.2019.1630430

Salama AM, Abedin RMA, Elwakeel KZ (2021) Influences of greenly synthesized iron oxide NP on the bioremediation of dairy effluent using selected microbial isolates. Int J Environ Sci Technol 19(8):1–12. https://doi.org/10.1007/s13762-021-03625-3

Salunke BK, Sawant SS, Lee SI, Kim BS (2016) Microorganisms as efficient biosystem for the synthesis of metal NP: current scenario and future possibilities. World J Microbiol Biotechnol 32:1–16. https://doi.org/10.1007/s11274-016-2044-1

Sargin I, Baran T, Arslan G (2020) Environmental remediation by chitosan-carbon nanotube supported palladium NP: Conversion of toxic nitroarenes into aromatic amines, degradation of dye pollutants and green synthesis of biaryls. Sep Purif Technol 247:116987. https://doi.org/10.1016/j.seppur.2020.116987

Shabbir MN, Alabada R, Aadil M, Ahmad Z, Alothman AA, Almuhous NA, Khan WA, Mohamed RM (2024) Synthesis, structural characterizations, and photocatalytic application of silver-doped indium oxide ceramic NP. Ceram Int 50:4404–4414. https://doi.org/10.1016/j.ceramint.2023.11.143

Singh KR, Natarajan A, Pandey SS (2023a) Bioinspired multifunctional silver NP for optical sensing applications: a sustainable approach. ACS Appl Bio Mater 6:4549–4571. https://doi.org/10.1021/acsabm.3c00669

Singh D, Verma SK, Singh V, Shyam P (2023b) Green functional NMs: synthesis and application. In: Modern nanotechnology: volume 2: green synthesis, sustainable energy and impacts. Springer Nature Switzerland, Cham, pp 45–65. https://doi.org/10.1007/978-3-031-31104-8_3

Sohni S, Nidaullah H, Gul K, Ahmad I, Omar AM (2018) Nanotechnology for safe and sustainable environment: realm of wonders, NMs and their fascinating attributes-development and prospective applications of. Nanosci Nanotechnol 2:37–117. https://doi.org/10.2174/9781681086453118020004

Surendra BS, Shekhar TS, Veerabhadraswamy M, Nagaswarupa HP, Prashantha SC, Geethanjali GC, Likitha C (2020) Probe sonication synthesis of $ZnFe_2O_4$ NPs for the photocatalytic degradation of dyes and effect of treated wastewater on growth of plants. Chem Phys Lett 745:137286. https://doi.org/10.1016/j.cplett.2020.137286

Swain G, Sonwani RK, Giri BS, Singh RS, Jaiswal RP, Rai BN (2021) A study of external mass transfer effect on biodegradation of phenol using low-density polyethylene immobilized *Bacillus flexus* GS1 IIT (BHU) in a packed bed bioreactor. Water Environ J 35:285–294. https://doi.org/10.1111/wej.12626

Tanveer R, Yasar A, Nizami AS, Tabinda AB (2023) Integration of physical and advanced oxidation processes for treatment and reuse of textile dye-bath effluents with minimum area footprint. J Clean Prod 383:135366. https://doi.org/10.1016/j.jclepro.2022.135366

Tao Z, Zhou Q, Zheng T, Mo F, Ouyang S (2023) Iron oxide NP in the soil environment: adsorption, transformation, and environmental risk. J Hazard Mater 459:132107. https://doi.org/10.1016/j.jhazmat.2023.132107

Tawalbeh M, Qalyoubi L, Al-Othman A, Qasim M, Shirazi M (2023) Insights on the development of enhanced antifouling reverse osmosis membranes: Industrial applications and challenges. Desalination 553:116460. https://doi.org/10.1016/j.desal.2023.116460

Viswanathan S, Palaniyandi T, Shanmugam R, Karunakaran S, Pandi M, Wahab MRA, Baskar G, Rajendran BK, Shivaji A, Moovendhan M (2024) Synthesis, characterization, cytotoxicity, and antimicrobial studies of green synthesized silver NP using red seaweed Champia parvula. Biomass Convers Biorefin 14:7387–7400. https://doi.org/10.1007/s13399-023-03775-z

Yu S, Zhou J, Ren Y, Yang Z, Zhong M, Feng X, Su B, Lei Z (2023) Excellent adsorptive-photocatalytic performance of zinc oxide and biomass derived N, O-contained biochar nanocomposites for dyes and antibiotic removal. Chem Eng J 451:138959. https://doi.org/10.1016/j.cej.2022.138959

Zhao HQ, Hou N, Wang YR, Li WQ, Liu Q, Lu P, Mu Y (2022) Carbon nanotubes mediated chemical and biological decolorization of azo dye: understanding the structure-activity relationship. Environ Res 210:112897. https://doi.org/10.1016/j.envres.2022.112897

Zheng T, Ouyang S, Zhou Q (2023) Synthesis, characterization, safety design, and application of NPs@ BC for contaminated soil remediation and sustainable agriculture. Biochar 5:5. https://doi.org/10.1007/s42773-022-00198-3

Zhong Y, Mahmud S, He Z, Yang Y, Zhang Z, Guo F, Chen Z, Xiong Z, Zhao Y (2020) Graphene oxide modified membrane for highly efficient wastewater treatment by dynamic combination of nanofiltration and catalysis. J Hazard Mater 397:122774. https://doi.org/10.1016/j.jhazmat.2020.122774

Zhou H, Lv L, Ye M, Baig SA, Luo Y, Chen J, Hu S, Zhang H, Wang J (2024) Improvement strategy of citrate and biochar assisted nano-palladium/iron composite for effective dechlorination of 2, 4-dichlorophenol. Environ Sci Pollut Res 31(23):1–14. https://doi.org/10.1007/s11356-024-33475-8

Chapter 9
Harnessing Nanomembrane Technology for Advanced Wastewater Remediation

Satish Kumar, Mukesh Bharti, and Mihir Kumar Purkait

Abstract The contamination of water is a global concern that is affecting millions of people, causing different diseases, such as diarrhea, typhoid fever, and hepatitis. A potential remedy for these problems is provided by nanomembranes. Traditional water restoration techniques have some drawbacks, viz. exorbitant prices, detrimental effects on the ecosystem, and an incapacity to get rid of recently discovered pollutants. However, nanomembranes provide benefits over conventional techniques, including decreased environmental impact, cost-effectiveness, and flexibility to change contaminants. Nowadays, several membranes are utilized to purify water incorporating 2D materials, such as graphene, MOFs, zeolite, TMDCs, MXenes, C_3N_4, and hBN. This chapter provides insightful information on developing the next generation of membranes for effective, scalable, and affordable water filtration for readers interested in this exciting study area. Furthermore, a detailed analysis of nanomembrane synthesis utilizing various nanomaterials (NMs), such as metal oxides, metallic materials, and metal-organic frameworks which has further enhanced the membrane performance by improving mechanical strength, adsorption capacities, and separation efficacy. Moreover, the production processes for these membranes are conducted, emphasizing both their benefits and drawbacks are also discussed. Nevertheless, there are drawbacks to employing nanomembranes in water filtration, including scaling, fouling, regeneration, and possible negative environmental consequences. Furthermore, the chapter undertakes a comprehensive examination that covers the background, technical applications, scalable, difficulties, constraints, possible applications, and reusability of these NMs.

Keywords Wastewater · Contaminants · Water remediation · Nanomembranes · NMs · Environment

S. Kumar · M. Bharti · M. K. Purkait (✉)
Department of Chemical Engineering, Indian Institute of Technology Guwahati, Guwahati, Assam, India
e-mail: mihir@iitg.ac.in

N. Agarwal et al. (eds.), *Nanomaterials in Wastewater Research*, Advances in Wastewater Research, https://doi.org/10.1007/978-981-96-3908-3_9

9.1 Introduction

Access to clean water is a vital need for the continuation of life and is widely acknowledged as a basic right of every person. Unfortunately, the widespread occurrence of water pollution has escalated into a global issue, endangering the well-being and safety of millions of people globally. Contaminated water has the potential to transmit several illnesses, including typhoid, hepatitis A, dysentery, and cholera. The World Health Organization (WHO) reports that around 3.4 million individuals, mostly children under the age of five, succumb annually to illnesses resulting from contaminated water and inadequate sanitation. In the past, traditional methods like filtration, sedimentation, chemical treatment, disinfection, and flocculation, using procedures, such as ozonation, UV radiation, and chlorination, have been used to remove contaminants from water sources. Nevertheless, these technologies pose notable obstacles, including the difficulties of efficiently eradicating certain contaminants, the substantial costs associated with their implementation, and the detrimental environmental consequences (Ahmed et al. 2021).

Conventional water remediation technologies have a tremendous problem in effectively removing new pollutants, such as pesticides, pharmaceuticals, and endocrine disruptors. These chemicals can potentially cause detrimental health consequences in both people and animals. These pollutants are often present in minute amounts, rendering traditional procedures ineffective. As a result, they may remain in water and gradually build up, creating long-term dangers to both human health and the environment. Another disadvantage of traditional approaches is the significant expense. Chemical treatment techniques, such as coagulation and flocculation, need substantial quantities of equipment, energy, and chemicals, which may be resource-intensive and expensive. In addition, filtration and sedimentation systems involve substantial expenses for both maintenance and operation, which might be a hindrance in certain situations. Conventional water treatment techniques might have harmful effects on the environment (Bharti et al. 2023).

Chemical treatment methods may emit hazardous by-products that adversely affect aquatic habitats and animals. The process of sedimentation and filtration may produce a substantial quantity of trash, which presents difficulties in terms of efficient handling and disposal (Puri et al. 2021). Nanomembranes provide an effective remedy to the challenges linked to conventional water filtering methods. Nanomembranes are ultra-thin membranes, typically with a thickness of less than 100 nm that possess the ability to selectively allow the passage of some substances while blocking others. These membranes may potentially remove toxins from water dependent on criteria, such as size, charge, and other properties. Nanomembranes provide several benefits in comparison with traditional approaches, including the capacity to eliminate new pollutants, financial efficiency, and less ecological footprint. Nanomembranes have a high level of effectiveness in the removal of new pollutants, such as medicines and pesticides. Nanomembranes have very tiny pore sizes that enable them to efficiently segregate even the tiniest particles, such as viruses and bacteria. Consequently, they are very efficient in eradicating recently discovered pollutants, which are often found in minimal quantities and difficult to eradicate by traditional means (Deepti et al. 2023).

Nanomembranes provide economic advantages in terms of operating expenses. Nanomembranes need lower energy consumption and a reduced amount of equipment, as opposed to chemical treatment procedures. Furthermore, they have little maintenance costs, which makes them a pragmatic choice for water filtration in environments with restricted resources. Nanomembranes have a reduced ecological footprint in comparison with traditional approaches. They generate a smaller quantity of trash and use a diminished amount of chemicals, hence reducing their influence on aquatic environments and animals. Furthermore, their reduced energy use results in a smaller carbon impact. Despite the many potential benefits of nanomembranes, they nevertheless face several obstacles that need attention and resolution.

The implementation of nanomembranes in water remediation is associated with certain disadvantages (Kumar et al. 2023). Accumulation of contaminants on the membrane surface gives rise to fouling and scaling, which significantly reduce the efficiency of the membrane over time. Membrane fouling occurs when suspended particles, usually bigger than 0.01 microns, are present, whereas scaling occurs when dissolved compounds, particularly salts, exceed their solubility limit. Fouling and scaling may occur due to several reasons, such as the feed water quality, the kind of membrane used, and the operating circumstances. To maintain the long-term efficiency of nanomembranes, it is crucial to address the issues of fouling and scaling. Fouling is the process in which impurities build up on the membrane's surface, leading to a progressive decline in effectiveness.

Scaling is the process of mineral deposits building up on the surface of a membrane, which leads to a reduction in water flow and an increase in pressure. Fouling and scaling may occur due to several reasons, such as the feed water quality, the kind of membrane used, and the operating circumstances. To reduce fouling and scaling, many methods may be used, including adding antifouling coatings, membrane cleaning and maintenance, and optimizing operational parameters. However, these techniques may be expensive and prolonged, and their efficacy relies on the particular application and membrane type (Deepti et al. 2023; Sharma et al. 2023a).

Another drawback of nanomembranes is their lack of cost efficiency. While nanomembranes have minimal operating costs, their production costs may be exorbitant, rendering them unavailable in some situations. The high cost of nanomembranes mostly stems from the costly raw materials, such as ceramics and polymers, as well as the intricate synthesis process required to produce membranes with nanoscale properties (Purkait et al. 2018). Scientists and researchers continuously endeavor to decrease the cost of nanomembranes by developing more cost-effective production techniques and investigating alternative materials. An example of this is the investigation into graphene use and other two-dimensional materials as substitute membrane materials, which might provide benefits in terms of performance, scalability, and economic viability (Mondal and Purkait 2019). Another consideration to be mindful of concerning nanomembranes is their possible environmental impact.

While nanomembranes are more environmentally friendly than traditional techniques, concerns exist over the potential emission of nanoparticles (NPs) into nature during the production and disposal of these nanomembranes. Additional research is necessary to have a thorough comprehension of the potential ramifications of NMs on aquatic ecosystems and species since their impacts are presently not well established (Aiswarya et al. 2022; Sharma et al. 2023b). To effectively use nanomembranes in practical applications, it is necessary to overcome certain challenges and barriers, although nanomembranes have shown potential in controlled laboratory environments, their practical efficacy and long-term resilience are yet uncertain. Furthermore, the widespread use of nanomembranes may be hindered by regulatory and legal constraints that must be resolved.

Although nanomembranes provide significant benefits over traditional water filtration techniques, there are still unsolved concerns that need attention. The issues included in this context are scalability and fouling, economic, eco-friendly, and practical implementation. It is crucial to effectively use micron-sized membranes in water purification to achieve the goal of providing clean water to everyone to address these challenges. This review chapter provides a thorough overview of the use of several innovative nano-sized materials in the process of water cleanup. The materials consist of metal-organic frameworks (MOFs), carbon-based nanostructured materials, metal oxide, nano-adsorbents, metallic NMs, and nanomembranes. The chapter delves into the concepts and procedures associated with nanomembrane adsorption and filtration. Furthermore, this chapter offers a thorough analysis of the chronology, technical execution, difficulties, limitations, potential for reutilization, and future possibilities of these nanomembranes.

9.2 Nanomembranes Chronology

The evolution of nanomembranes chronicles the advancements and discoveries in the field of nanotechnology, specifically focusing on the expansion of thin, flexible coverings known as nanomembranes. The advent of the fabrication of very thin materials in the latter part of the twentieth century initiated the voyage of nanomembranes, owing to the groundbreaking efforts of researchers. At first, the emphasis was on understanding the fundamental NM properties and their potential applications. In the 1990s and early 2000s, significant progress was made in fabrication processes. Technological advancements like as molecular layer deposition and self-assembly methods have enabled precise control over the thickness and content of membranes (Sharma et al. 2023a).

A breakthrough occurred in the historical timeline with the introduction of graphene, which is a two-dimensional lattice composed of a single sheet of carbon atoms. This occurrence sparked a revolution in the field of nanomembrane research. Graphene's exceptional mechanical, electrical, and thermal properties have opened up possibilities for a diverse array of uses, such as in electronics, sensors, and biomedical devices. Transition metal dichalcogenides (TMDs) and hexagonal boron nitride (hBN) have become viable options for nanomembranes, alongside graphene, thereby expanding the selection of materials (Dhara et al. 2024). Since the mid-2000s, there has been a substantial surge in research focused on nanomembranes. The rise in activity may mostly be ascribed to progress in nanofabrication methods, computer modeling, and multidisciplinary cooperation. Researchers have explored several applications of nanomembranes, including flexible electronics, energy storage, filtration, and biological implants. Moreover, efforts have been made to address challenges, such as the ability to manage bigger workloads, provide stability, and seamlessly integrate with current technologies, therefore enabling the implementation of the project. Advancements in the creation and investigation of nanomembranes are anticipated to persist and evolve in the future. Potential future developments might include integrating nanomembranes with emerging technologies including Internet of Things (IoT) sensors, wearable devices, and advanced medical implants. Furthermore, the incorporation of various fields, such as materials science, physics, chemistry, and engineering, will be crucial in pushing forward cutting-edge nanomembrane research, allowing for the investigation of new possibilities for groundbreaking applications in various domains of science and technology.

9.3 Methods for Producing Nanomembranes

Various methodologies are used to create nanomembranes, which include a range of procedures that may be classified as either bottom-up or top-down methods. Top-down procedures include the creation of membranes by decreasing the size of bigger structures to nanoscale levels. This includes procedures like lithography, in which intricate designs are engraved into a substrate to generate nanoscale characteristics. In contrast, bottom-up techniques construct nanomembranes via the assembly of tiny components to form bigger structures. Illustrative instances include self-assembly methodologies, whereby molecules or NP autonomously organize themselves into structured arrangements, and molecular beam epitaxial, a process that incrementally develops thin films at the atomic level. Researchers may achieve precise customization of nanomembrane characteristics by integrating both top-down and bottom-up methodologies, resulting in enhanced adaptability for multi-faceted applications (Sharma et al. 2022). Figure 9.1 depicts the sequential procedure for creating nanomembranes, highlighting the complex approach from choosing materials to assembling them. It provides a visual guide for advanced membrane synthesis.

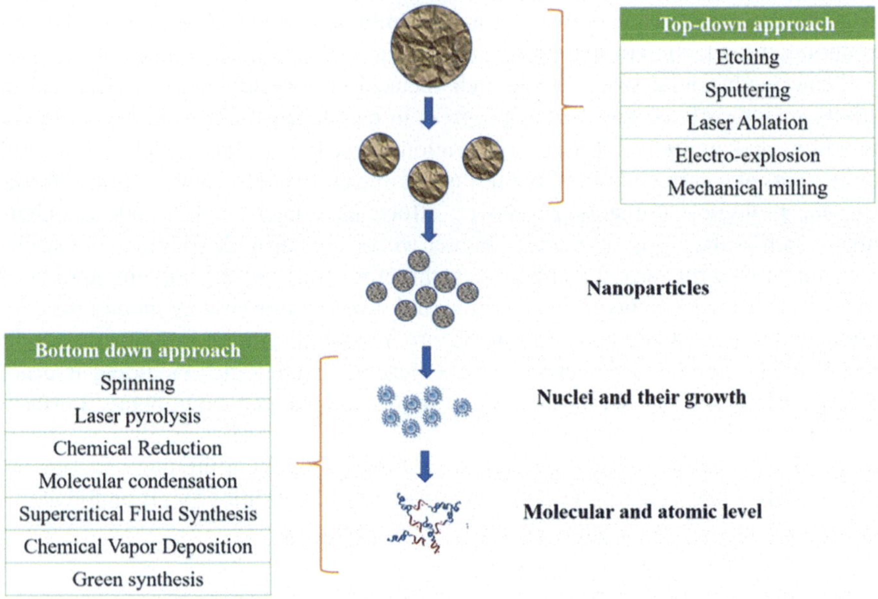

Fig. 9.1 Strategy of preparation of nanomembranes

9.3.1 Anodization

There are two techniques available for anodizing pre-anodized aluminum oxide. These states are stable, rather than undergoing any significant changes. Steady-state anodization may be categorized into two types: moderate and severe. Upon comparing the two processes, it is evident that the hard-anodizing reaction is more intense, but the development of the porous anodic aluminum oxide layer takes place more rapidly. These features result in the generation of a substantial amount of thermal energy, necessitating the need for additional cooling equipment. Two kinds of anodization are unstable. Periodic and pulsed anodization are terms used to describe these particular techniques. The periodic anodizing process is superior to other techniques in regulating the porosity structure within the anodic aluminum oxide layer. Anodic aluminum oxide nanomembranes have distinctive topologies that render them very appropriate for a diverse array of applications, mostly due to their visual attributes (Kim et al. 2022). Applying a coating to the metal surface is an essential step in the anodizing process. Applying voltage to the anodic layer modifies its properties, including the thickness of the protective layer and the size of the pores. Fluctuations in voltage and current influence the surface of aluminum. Electric polishing uses a combination of low and high voltages. When there is a rise in voltage and a drop in current, a permeable layer is formed. When there is a low current and high voltage, a coating of aluminum oxide is created. Anodic aluminum oxide patterns are gradually created when metal is anodized in acid solutions under the influence of an electric field (García-Torres et al. 2023).

9.3.2 Lithography

Lithography is widely used in several sectors. Currently, the objective is to achieve attributes on a smaller scale with exceptional accuracy. To improve the results in the production of nanomembranes, it is crucial to further investigate lithography methods. There are two distinct lithography techniques: maskless and masked lithography (Cigané et al. 2021). Masked lithography is a precise process that uses a mask to imprint patterns onto a substrate material. Maskless lithography is a technique that creates different designs by writing them one after another, without the need for a mask. Two instances of masked techniques include photolithography, with a minimum feature size of 2–3 μm, and nanoimprint lithography, with a minimal feature size of 6–40 nm. Electron beam lithography is an alternative technique that eliminates the requirement for masks. According to Soh et al., it can achieve a minimum feature size of 5 nm (Soh et al. 2001).

9.3.3 Micromachining

Micromachining encompasses a variety of techniques that may be used on nano- and microscale membranes. Nanoimprint lithography is a micromachining technique that has gained significant attention in the area of nanomembrane development. Laser micromachining is a technique used to do precise and intricate machining using lasers. Laser micromachining is a very accurate method that uses a laser beam to trace a predetermined path to create a certain design. Laser technology enables the production of diverse nanostructures on an assortment of materials, including metals (e.g., Ni, Ag), semiconductors (e.g., MoS_2, gallium nitride, and silicon), organics (e.g., polyimide foil and protein), and polymers (including polyimide foil and protein). This micromachining technique is specifically developed to create membranes with very fine dimensional tolerances and the utmost degree of precision (Elman et al. 2010).

9.3.4 Chemical Vapor Deposition (CVD)

Solid surfaces are formed through CVD, which involves a chemical reaction taking place in the vapor phase. At higher temperatures, the precursor undergoes breakdown, resulting in the production of gaseous atoms. Subsequently, these atoms adhere to

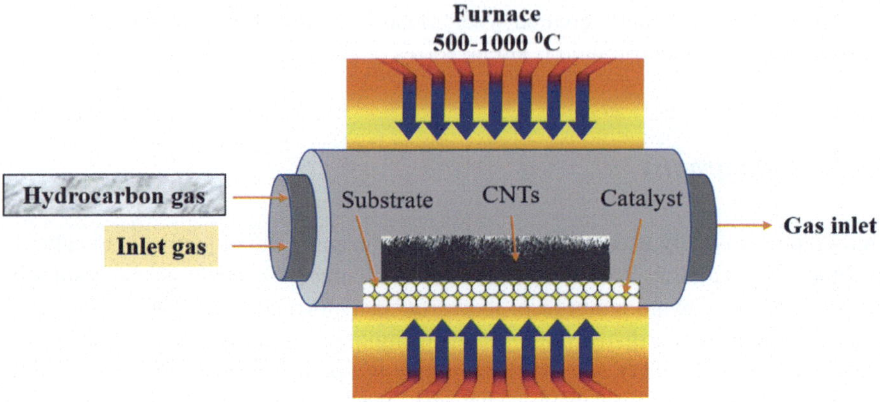

Fig. 9.2 Chemical vapor deposition (CVD)

the substrate and are applied by the CVD process, as seen in Fig. 9.2. CVD is a technique used to manufacture various materials, including oxides, such as iron, tin, and titanium, as well as polymers like polyaniline, polythiophene, and other nanoporous materials. This process is beneficial for the production of semiconductors, namely metal alloys like platinum-cobalt and other materials that include nanopores (Cigané et al. 2021).

9.3.5 Sequential Layer Deposition

The layer sequential deposition process entails consecutive molecule adsorption with opposing charges onto a surface, leading to the creation of nanomembranes. The substrate is immersed in a cationic polyelectrolyte-containing solution. A monolayer of molecules with a thickness of around 1 nm absorbs the polyelectrolyte (Wang et al. 2022). Subsequently, the wafer undergoes a process of purification and dehydration. Furthermore, a solution comprising polymer particles with a reduced concentration and negative charges is used to immerse the substrate that has been coated with polish. A new monolayer is created on top of the previous one. Subsequently, the wafer undergoes an additional process of cleansing and dehydration. To get the accurate number of layers, it is important to iterate the process using either the same materials or other ones. Multilayers are formed by the step-by-step application of separate layers. The thickness varies between 5 nm and over 500 nm (Barbhuiya et al. 2022).

9.3.6 Sol-Gel Process

Several kinds of nanomembranes may be developed through the sol-gel technique. The wet chemical process involves four main stages for formation: The first step involves creating a homogeneous solution containing either salt or metal. In the second step, it is important to enhance the concentration of the homogeneous solution. The gel is formed by condensation reactions during the third step of synthesis. The last step entails complete desiccation of the gel. The quality of the nanomembrane is contingent upon the chemical composition of the solutions, the centrifugation procedure, the temperature, and the technique of cooling. The sol-gel process enables the creation of structures that demonstrate exceptional magnetic, optical, electrical, thermal, and mechanical characteristics. This technique may be used to fabricate nanomembranes from different materials, such as polymers and metal oxide (Kumar et al. 2022).

9.3.7 3D Printing Methods

The field of membrane research is seeing tremendous growth in the use of 3D-printed nanomembranes. Two-photon polymerization is a technique used to create membranes with the utmost precision and resolution (De Pasquale 2021). 3D printing, also known as additive manufacturing (AM), is a process that makes use of a variety of materials to produce intricate and varied geometric shapes. Layer-by-layer manufacturing, or 3D printing, is the technique that joins materials. According to the rules and in compliance with global ISO/ASTM standards, the seven primary categories of additive manufacturing processes are vat photopolymerization, material extrusion, directed energy deposition (DED), material jetting, binder jetting, powder bed fusion, and sheet lamination. The field of membrane science is now undergoing a cutting-edge and growing area of study centered on 3D-printed nanomembranes. Two-photon polymerization method is used to fabricate membranes with the highest level of intricacy. The resolution is around 100 nm. This approach relies on the phenomenon of dual-photon absorption occurring inside a transparent resin (Wang et al. 2021). 3D printing may use a wide range of materials. Polymers are often used in 3D printing because of their inherent malleability, cost efficiency, compatibility with living organisms, and desired mechanical properties. Ceramics and hybrid materials, such as composites, may be used in 3D printing to enhance both bioactivity and mechanical stability. However, it is essential to recognize that there are some disadvantages associated with this strategy. Unfortunately, the current limitations of 3D printing, namely its low resolution, hinder the manufacturing of nanomembranes.

9.4 Nanomembranes and Their Applications in Water Remediation

Nanomembranes are categorized into inorganic, organic, and hybrid variations, as well as synthetic biomembranes, depending on the materials used in their production (Fig. 9.3).

9.4.1 Organic Nanomembranes

Organic nanomembranes are a unique kind of self-supporting nanomembrane since they are made entirely of one or more organic components (Dhara et al. 2023). Diverse organic compounds and their amalgamations provide potential for future use. However, it is essential to recognize that not every organic component is suitable for the creation of nanomembranes. Some chemical components, due to being in a gaseous or liquid state, cannot create membrane-like structures. Organic materials are carbon compounds that do not consist of pure carbon in any of its many forms. They also contain essential carbon compounds like cyanide, carbon oxides, carbonates, and carbides. Many macromolecular and polymeric structures, such as synthetic lipids, polysaccharides, synthetic polymers, proteins, DNA, and RNA-based membranes, may be used to construct freestanding nanomembranes (Dhara et al. 2024).

Macromolecular nanomembranes have potential advantages; however, they are very susceptible to temperature variations, and their beneficial characteristics are only maintained within a limited temperature range. Moreover, they are vulnerable

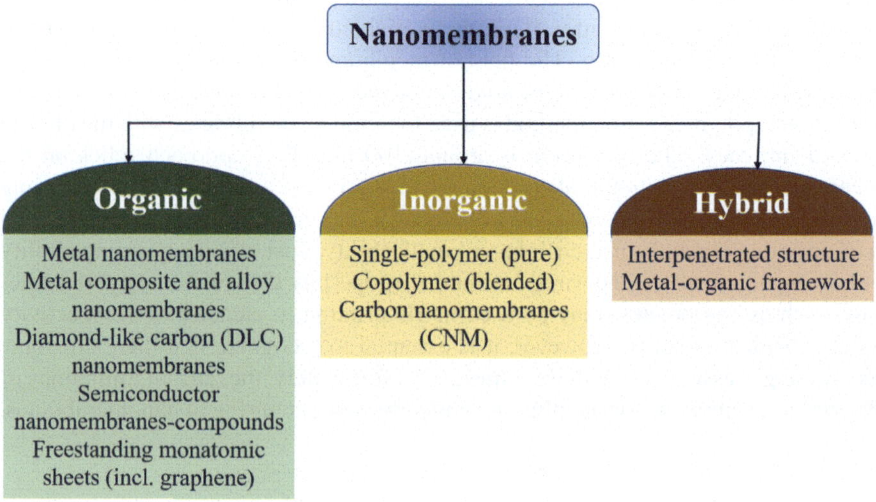

Fig. 9.3 Classification of nanomembranes

to attack and disintegration by organic solvents and may be endangered as a result of heightened humidity. Furthermore, the mechanical properties of these membranes worsen as their thickness decreases, and they often exhibit a low Young's modulus. When subjected to continuous pressure, the majority of macromolecular nanomembranes experience contraction and long-lasting deformation. Pyroxylin sometimes referred to as nitrocellulose or collodion is considered to be the first instance of an organic nanomembrane. Three types of organic nanomembranes may be distinguished: copolymer-based mixed nanomembranes (CNM), pure organic nanomembranes (made of a single polymer), and pure organic nanomembranes (composed of several polymers) (Gaikar et al. 2023).

9.4.2 Inorganic Nanomembranes

Aside from traditional inorganic carbon and organic NP, transition metal NMs and other NMs have been suggested for water remediation. Specifically designed to collect and remove 2,4,6-trichlorophenol (TCP) from wastewater, the application makes use of magnetic halloysite nanotube (MHNT) composites reinforced with molecularly imprinted polymers (MIPs). This technique has promise for the creation of financially viable goods. By using magnetic halloysite nanotube particles (MHNTs) as the foundation material, the researchers synthesized a magnetic molecularly imprinted polymer (MMIP). This MMIP was designed with the particular purpose of selectively identifying and differentiating the existence of (MHNTs) and (TCP) which were produced using a high-temperature process using ferric triacetylacetonate soaked in 1-methyl-2-pyrrolidone. The procedure included the formation of chemical bonds between carboxylic acid-functionalized halloysite nanotubes (HNTs−COOH) and magnetic NP.

According to their study article, García-Torres et al. developed a method for creating flexible electronic hybrid materials with nanostructured surfaces (García-Torres et al. 2023). This was achieved by using two-dimensional nanomembranes that are perforated and self-sustaining, serving as a substrate for precisely organized one-dimensional nanostructures made of metal. The manufacturing technique involves the deposition of consecutive layers of perforated poly(3,4-ethylenedioxythiophene) and poly (lactic acid) (PLA). Afterward, the aforesaid layers are combined with copper metallic nanowires (NWs) by the technique of electrodeposition. Subsequently, a coating of silver is applied onto the topmost layer of PLA including nanoscale pores via a transmetallation process. This method combines the adaptability and plasticity of very thin, pliable polymeric nanomembranes with the exceptional electrical characteristics of metals, making it exceptionally well suited for incorporation into bio-integrated electronic systems. This study showcased the ability of the nanomembranes surface chemistry to accurately identify and measure the presence of H_2O_2. The display included a wide concentration range, ranging from 0.35 to 10 mM. The detection limit was 7 μm, and the sensitivity was 120 $\mu A\ cm^{-2}\ mM^{-1}$. The hybrid nanomembranes showed both pliability and flexibility. They demonstrated a high degree of selectivity toward H_2O_2, as well as strong durability and reliable consistency. The use of XPS, EDX, XRD, SEM, EIS, and CV investigations performed the verification of these features.

9.4.3 Hybrid Nanomembranes

Recently, the usage of inorganic nanomembranes has been increasing due to their potential uses in water treatment. These membranes often include inorganic materials, such as ceramics, metal oxides, and metals. Graphene oxide is highly esteemed as a very promising inorganic nanomembrane for water filtration. This material has exceptional mechanical and thermal characteristics, together with a significant surface area, enabling effective adsorption of pollutants. Other inorganic NMs that have been researched for water remediation include zeolite, mesoporous silica, and MOFs. MOFs are porous substances having modifiable characteristics that may be used for the adsorption and degradation of organic contaminants. In addition, inorganic nanomembranes have been used to efficiently eliminate heavy metals from water (Haldar et al. 2020). Shayesteh and his colleagues (2016) provide a detailed account of the production and assessment of titania-Y-alumina multilayer nanomembranes (Shayesteh et al. 2016). A recent research examined the rate at which a multilayer membrane was rejected in both basic and acidic environments. The slip-casting technique, which included the use of alpha-alumina tubes, was used to create the nanomembrane support. The support was layered with gamma-alumina as a lower coat and titania as an upper coat in a sequential application. Afterward, an evaluation was performed to ascertain the water permeability and flux of the membrane. The research also assessed the nanomembranes capacity to repel bacteria and other ions in simulated wastewater with varying pH values. An investigation was conducted to assess the efficacy of a nanomembrane in the treatment of simulated wastewater. Applying pressure within the range of 1–10 bar resulted in a decrease in the permeability of water across the nanomembrane. Nevertheless, the permeability ultimately reached a stable state at increased pressures, resulting in an increased rate of water flow. Rejection tests were performed on a simulated wastewater model including ions. The results indicated that the nanomembrane generated using the support provided by the slip-casting technique was capable of partially obstructing the movement of ions and successfully isolating all bacteria. It was shown that altering the pH level led to an increased rate of ion rejection. Silicon-on-insulator (SOI) is a commonly used composite material in the semiconductor device fabrication sector. The configuration comprises a slender crystalline Si layer, referred to as the template layer, which is isolated from a larger wafer by a SiO_2 film. This approach was pioneered around 15 years ago and rapidly received substantial acclaim in the industry for its efficient and dependable production of a thin Si template layer. Single-crystal silicon-on-insulator (SOI) offers several benefits compared to bulk silicon crystal, especially in low-power circuit applications used in portable electronic devices. The incorporation of a thin Si layer on top of an oxide has greatly enhanced the efficacy of semiconductor devices. Titanium dioxide (TiO_2) is a coating material that has advantageous qualities like non-toxicity, stability, affordability, and photocatalytic capabilities, as highlighted by several researchers (Duarah et al. 2023). Recently, the performance of CA/PEG membranes was enhanced by introducing different concentrations of TiO_2. These improved membranes were then used in reverse osmosis (RO) and membrane distillation (MD) processes. In 2018, a study conducted by Shafiq and his colleagues, it was shown that membranes made of CA/PEG with TiO_2, which had different weight percentages (5–25 wt.%), exhibited desalination efficiencies of 80, 90, 95.4, 85, and 80%, respectively. These statistics suggest that the use of a 15 wt. % loading was the most efficient in achieving the maximum amount of desalination. By applying a layer of TiO_2 to the membranes and exposing them to UV light, their resistance to lipids was improved and they gained the capacity to self-clean (Shafiq et al. 2018). The abundance of TiO_2 resulted in the obstruction of the membrane's pores, hence diminishing its functioning. Emami et al. discovered that membranes coated with TiO_2 NP exhibited remarkable self-cleaning properties when exposed to UV radiation (Emami et al. 2018). In 2001, Kwak and colleagues found that a TFC membrane, which had been augmented with TiO_2 and consisted of inorganic and organic hybrids, had superior resistance to fouling in reverse osmosis (RO) than a pure PA membrane (Kwak et al. 2001).

In 2015, Safarpour and colleagues developed a reverse osmosis membrane called thin film nanocomposite (TFN-RO) utilizing interfacial polymerization. The membrane was then coated with reduced graphene oxide/TiO_2. The modified membrane exhibited enhanced resistance to lipid absorption and fouling properties in comparison with the unaltered membrane (Safarpour et al. 2015). Further, in 2017, Ren and his colleagues fabricated a membrane consisting of electrospun nanofibers coated with a layer of TiO_2. The membrane demonstrated a significant permeate flow rate of 73.4 L per square meter per hour and an outstanding salt rejection rate of 99.99% (Ren et al. 2017). Zinc oxide (ZnO) NP are increasingly being used as an additive because they are cost-effective and possess exceptional stability in terms of physical, chemical, mechanical, and thermal properties. They also have a large surface area, can be functionalized on the surface, and exhibit remarkable antibacterial and anti-corrosive properties. Studies have shown that the use of ZnO in mixed membranes improves their ability to prevent the absorption of lipids, resulting in enhanced permeability and resistance to fouling (Sharifpour et al. 2018). To enhance the antibacterial characteristics of a cellulose acetate (CA) membrane for reverse osmosis (RO), ZnO NP were integrated into the membrane by the electrospinning method.

9.5 Utilization of Nanomembranes

The unique features and adaptability of nanomembranes make them very promising for a wide range of water treatment applications. They may be used for duties, such as desalination, eliminating heavy metals and contaminants, and improving

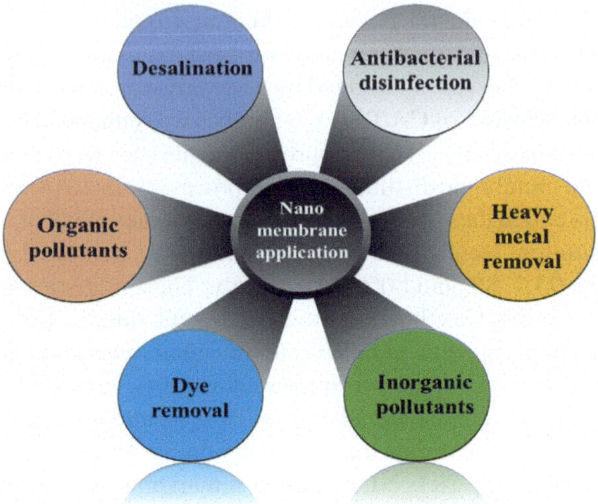

Fig. 9.4 Application of nanomembranes

water purification operations. Nanomembranes with precise nanopores are used in desalination to effectively separate salt from water molecules, providing a sustainable approach to address the issue of freshwater shortage. In addition, nanomembranes can selectively separate heavy metals and pollutants, thus enhancing the quality of water in both industrial and environmental contexts. Moreover, their elevated surface area-to-volume ratio amplifies their ability to adsorb organic chemicals and bacteria, so guaranteeing the provision of clean drinking water. In summary, nanomembranes provide a hopeful solution for effectively and sustainably tackling various water treatment obstacles. Figure 9.4 depicts the utilization of nanomembranes in water treatment processes.

9.5.1 Removal of Natural Organic and Inorganic Pollutants

It is important to eliminate both inorganic and organic impurities from water since they may alter the inherent taste, hue, and aroma of the water. The existence of both non-polar and polar functional groups in these pollutants leads to the fouling of the membrane. Therefore, it is crucial to efficiently and effectively remove these contaminants. Esfahani et al. published a study on a new membrane made of a nanocomposite material including gold NP (AuNPs) and polysulfone (PSF) (Esfahani et al. 2019). The membrane is often known as CT-GNPs/PSF. This membrane has inherent self-cleaning properties and effectively removes humic acid (HA), which is a kind of organic fouling agent. The hydrophilic functional groups and negative surface charges of the nanocomposite membrane resulted in increased permeability and breakdown capacity toward HA. The nanocomposite membrane was created exhibited remarkable permeability (160 $L/m^2/h$), the greatest rate of blocking hazardous substances (87% for HA), the most efficient rate of flux recovery (95%), and improved resistance to fouling when compared to the original PSF membranes. Thin film nanocomposite (TFN) membranes exhibit high efficiency in water filtration at the nanoscale owing to their very small pore diameters, ranging from 1 to 200 nm. Purkait and Singh have shown remarkable water permeability, robust resistance to different pH levels, and amazing longevity. In their study, Celik et al. examined the effectiveness of polyether sulfone/MWCNT (C/P) membranes in removing TOC from water. The formation of the membranes occurred by the establishment of hydrogen bonds between the sulfonic and carboxylic groups of polyether sulfone (PES) and functionalized multi-walled carbon nanotubes (MWNTs), respectively (Celik et al. 2011). The research found that increasing the amount of MWNTs (up to 2%) improved the porosity, hydrophilicity, and water flow capacity of the nanomembrane. The results suggest that carbon-based nanomembranes are more effective than PES membranes without any additional coatings or changes.

A recent study conducted by researchers examined the use of a versatile and self-sustaining nanomembrane made of sodium titanate nanobelts (Na-TNB) as a material capable of efficiently attracting and eliminating complicated radioactive Sr(II), Cs(I) ions, as well as oil spills from water. The ongoing capture of radioactive cations Sr(II) and Cs(I) inside the nanomembrane by ion exchange throughout the process of water treatment. When the nanomembrane was subjected to a strong acid solution, it experienced a phase transition to rutile, which allowed it to efficiently eliminate oil. Furthermore, the nanomembrane was restored by an alkaline treatment of rutile, allowing for the process to be repeated (Wen et al. 2016). The activated carbon NP in the mixed matrix nanomembranes demonstrate remarkable effectiveness in the nanofiltration process for eliminating sulfate and copper ions from water (Singh et al. 2021). The study discovered that incorporating 0.5 weight percent of activated carbon NP into polyether sulfone (PES) yielded the most effective outcome in eliminating ions from wastewater. Li and colleagues performed a study on a new composite membrane that was designed to remove trichloroethene from water via dichlorination (Li et al. 2019). The researchers used Fe–Pd bimetallic NP to form a coating on the acrylic acid microgels. These coated microgels were then put onto the pores of a poly (vinylidene fluoride) membrane. The presence of nanoscale surface pores on the composite membrane obstructed the contamination of Fe-Pd NPs by limiting the penetration of colloids and bovine serum albumin into the membrane's inner region. The composite membrane demonstrated efficient performance in terms of swift dichlorination, quick isolation, and excellent separation of products. Customized nanomembranes have significant promise in nanofiltration operations.

Ren et al. examined the production of nanofiber membranes by using an electrospinning technique with polyvinylidene fluoride (PVDF) that included immobilized NZVI and GO (Ren et al. 2019). The objective of these membranes was to efficiently remove Cd(II) and trichloroethylene (TCE) from wastewater. The nanomembranes exhibited a synergistic effect resulting from the incorporation of their constituent materials, namely graphene oxide (GO), nanoscale zero-valent iron (NZVI), and polyvinylidene fluoride (PVDF). The presence of GO enhanced the hydrophilic properties of the nanomembrane, resulting in an augmentation of the filtration flow. Furthermore, GO improved the capacity to immobilize nanoscale zero-valent iron (NZVI) on the nanomembrane. The efficacy of NZVI in eliminating Cd(II) was much higher, reaching about 100% elimination by an adsorption mechanism. In contrast, when a nanofiber membrane was used to remove TCE, there was a slightly higher flow rate of around 317 LMH, and the clearance rate attained was about 82%. This illustrates the effectiveness of combining GO-NZVI with PVDF.

9.5.2 Desalination

Nanocomposite membranes have shown enhanced performance characteristics, including higher water permeability and enhanced salt rejection, as well as strong resistance to fouling. Researchers developed durable and strong nanomembranes by combining multi-walled carbon nanotubes (MWNT) with polymeric membranes. These nanomembranes were very effective in removing salts and humic acid from water. According to research, using a 15 mg/g MWNT (multi-walled carbon nanotube) and a 10% PA (polyamide) casting solution led to the creation of a strong and durable membrane that produced outstanding outcomes (Shawky et al. 2011). This combination resulted in a substantial increase in the extraction of NaCl, with a quantity of 3.17 times more than before. Additionally, the amount of HA extracted was 1.67 times greater than before. Some researchers fabricated desalination TFN by constructing a composite material consisting of polymer-coated metal (M), metal oxide (MOx), and multi-walled carbon nanotubes (MWNTs). Comparative research was undertaken to examine the performance of several M/MOx (Ag, Al_2O_3, Fe_2O_3, and TiO_2) loaded MWNT coatings on a thin polymeric film of polydopamine (PDA) for nanofiltration. The goal was to evaluate their effectiveness in filtering solutions that include both monovalent and divalent salts. The findings demonstrated that M/MOx/MWNT/PDA nanofiltration membranes had improved performance in comparison with thin film composite membranes. A separate research conducted an experiment aimed at enhancing the effectiveness of a cellulose acetate/triacetate (CTA) membrane by including a hydrophilic and extensively porous copper-based MOF (Cu-BTC MOF, btc = benzene 1, 3, 5-tricarboxylate) (Haldar et al. 2020).

Zirehpour et al. demonstrated that this improvement had a significant beneficial effect on forward osmosis desalination. The incorporation of MOF NPs enhanced the overall porosity, connectedness among pores, and hydrophilicity of the nanomembrane, leading to a water permeability of 72% and heightened selectivity. Scientists have produced mixed matrix nanomembranes by blending GO-polysulfone (GO-PSF) with varying amounts of graphene oxide, ranging from around 0.25 to 2.0 wt.% (Zirehpour et al. 2016). In 2020, Camacho et al. demonstrated the use of nanomembranes in a direct contact membrane distillation process (Camacho et al. 2020). The membrane with a greater concentration of GO at around 2.0 wt. % had the highest permeate flow rate. In contrast, the membrane with a lower concentration of GO, around 1.0 wt.%, exhibited improved salt rejection and porosity. The results suggest that there is a limit to how much graphene oxide may be effectively used in nanomembranes.

9.5.3 Antibacterial Disinfections

Polymer nanofibers have potential in several domains, including material reinforcement, energy retention, filtration, biological investigation, and water decontamination. The electrospun nano-fibrous membranes have a high level of conductivity for drug adsorption and water permeability due to their nanostructure, which promotes the wound-healing effect (Singh et al. 2021). Furthermore, the scientific community has shown considerable enthusiasm about the antibacterial characteristics of graphene and graphene oxide (GO) nanocomposites for water cleanup. Membranes encounter a substantial hindrance in their use for filtration as a result of biofouling, which not only limits their utility but also results in subsequent contamination. Furthermore, the microorganisms interact with membranes, leading to a decrease in their resistance and the efficiency of water flow. Hence, it is essential to include antibacterial properties in the fabrication of membranes in membrane technology. Scientists have developed and examined electrospun nano-fibrous membranes that possess antibacterial characteristics.

According to Baite et al., the membranes consisted of a combination of chitosan (CS), graphene oxide (GO), and poly (vinyl alcohol) (PVA) (Baite et al. 2022). The nanomembrane demonstrated strong antibacterial efficacy against gram-positive S. bacteria. Staphylococcus aureus and gram-negative Escherichia coli. Coli. Li et al. devised a novel method to create support sheets using a combination of guanidine polymer (PHGC-polyhexamethylene guanidine hydrochloride) and polyethylene glycol (PEG) functionalized GO (Li et al. 2015). The sheets were used as an antibacterial substance known as GO-PEG-PHGC. The results showed that the novel material had enhanced antibacterial efficacy in comparison with native GO, GO-PEG, and GO-PHGC alone. The enhanced dispersion of the antibacterial agent (GO-PEG-PHGC) is due to the presence of PEG. The enhanced dispersion of the NM resulted in a more robust interaction with the bacterial membrane, leading to heightened cellular damage. Furthermore, the research conducted by Zhou et al. revealed that the attachment of AgNPs to PDA-functionalized GO sheets resulted in significant antibacterial effects. The application of PDA coating provided stability to the nanocomposite and enabled the chemisorption of AgNPs, resulting in the formation of reduction sites. Hence, nanocomposites comprising of polymer/inorganic membranes and NP demonstrate remarkable efficiency in the process of nanofiltration owing to their hydrophilic characteristics, porosity, and improved mechanical robustness (Zhou et al. 2013).

Zarpelon et al. created a unique kind of nanomembranes that are thin and consist of silver NPs (Ag NPs) that are joined together using glutaraldehyde (Zarpelon et al. 2016). The nanomembranes were purposefully engineered for purifying industrial effluent. The nanomembranes successfully eradicated over 93% of the total coliform bacteria, with a special focus on E. The presence of coli bacteria was detected in the wastewater sample. In a recent work, He et al. used AMIB (1-aminoethyl-3-methylimidazolium bromide) to fabricate a polyamide thin film known as PIP-TMC-AMIB, which exhibits remarkable efficacy in salt removal and antibiotic extraction from water by nanofiltration antibacterial membrane (He et al. 2020). The nanomembrane demonstrates exceptional water permeability and effectively filters several medicines, such as ciprofloxacin, tetracycline, oxytetracycline, levofloxacin, and erythrocin, with rejection rates over 90%. Furthermore, it efficiently segregates combinations of antibiotics and sodium chloride, while exhibiting improved antibacterial characteristics.

9.5.4 Applications for Removing Dye

Nanomembranes demonstrate exceptional efficacy in eliminating colors from wastewater, while simultaneously achieving an optimal equilibrium between cost-effectiveness and energy efficiency. The main process for removing dye is via the physical adsorption or electrostatic interaction between dye molecules and functional groups on the surface of the nanocomposite membranes. In their study, Lv et al. developed robust nano-fibrous membranes by using highly charged polyether sulfone (PES) material (Lv et al. 2019). These membranes were designed to effectively remove anionic pigments and microbes from wastewater. PMETAC, a polymer including quaternary ammonium salt, exhibited exceptional recyclability and potent antibacterial properties when evaluated against Staphylococcus aureus and E. coli. Furthermore, it demonstrated an impressive ability to absorb anionic CR dyes. Li et al. conducted research and found a polypropylene nanomembrane (PDA/ZIF-67@PP) that demonstrated self-cleaning capabilities.

Zhou et al. performed a new work on microporous PVDF membranes. The membranes were improved by using halloysite nanotubes and zwitterionic poly(sulfobetaine methyl methacrylate) (PSBMA/HNTs/PVDF) (Zhou et al. 2021). This alteration was done to improve the membranes' ability to absorb heavy metal ions and soluble dyes from wastewater. Recent research has shown that including HNTs (halloysite nanotubes), PDA (polydopamine), and PSBMA (poly (sulfobetaine methacrylate)) into wastewater filters increases the availability of active sites, hence improving the adsorption of soluble dyes and metal ions. The PVDF system with increased properties exhibited remarkable resistance to fouling when exposed to viscous crude oil and soybean oil, thereby highlighting the nanomembranes high effectiveness. The researchers conducted a second study where they created adsorbent membrane capsules utilizing hierarchical $Mg(OH)_2$ nanostructures, with an average diameter of 1.1 mm. The objective of these capsules is to eradicate the alkaline hue seen in textile industry effluent. The cost-effective electro-spraying method was used to fabricate membrane capsules containing a hierarchical $Mg(OH)_2$ nanostructure, which served as an exceptionally effective adsorptive filler. Magnesium hydroxide ($Mg(OH)_2$) exhibits two distinct forms—cylindrical and flower-like. Both of these nanostructures have shown remarkable effectiveness in the removal of basic dyes. Table 9.1 compiles the data from prior studies on the use of various nanomembranes in the context of wastewater treatment.

Table 9.1 Applications of different types of nanomembranes in water treatment

Type of nanomembrane	Size (nm)	Water treatment application	Removal capacity/ efficiency	References
Silica KCC-1, oleic acid, and chitosan	300–400	Pb(II) removal	168 mg/g	Zarei et al. (2019)
Sawdust carbon with EDTA-modified Fe_3O_4	638	Removal of cadmium	63.3 mg/g	Kataria and Garg (2018)
Poly(vinyl imidazole) grafted xanthan gum/montmorillonite		E. coli Malachite green (MG) and S. aureus, dye removal	(S. aureus) 26.0 ± 0.25 (E. coli) 24.0 ± 0.81 (MG adsorption) 909.1 mg/g	Abu et al. (2021)
Ag–Cu_2O/glucose rGO	< 200 (Cu_2O) 60 (Ag)	Methyl orange	90	Sharma et al. (2018)
ZnO–Ag	–	Antimicrobial activity against E. coli and S. aureus Photocatalytic activity against phenol	97.2% (photocatalytic activity) Inhibition zone: 15 ± 0.21 (E. coli) 18 ± 0.24 (S. aureus)	Zarei et al. 2019)
Al-MOF/Fe_3O_4/PDA@Ag	54.31	Ciprofloxacin, methyl orange, norfloxacin	(Ciprofloxacin) 81.7% (Methyl orange) 98.2% (Norfloxacin) 80.1%	Wang et al. (2020)
MOF-based material (Sn(II)-Benzene-1,4-dicarboxylic acid (BDC))	17	Eosin yellow (EY), Eriochrome Black T (EBT), and Congo Red (CR) removal	(EY) 208.3 mg/ g (EBT) 125 mg/g (CR) 95.2 mg/g	Ghosh and Das (2020)
BUC-21/titanate nanotube		Cr (VI) reduction and Cr (III) removal	(Cr(VI) reduction) 100%, (Cr(III) removal) 97.6%	Wang et al. (2019)

9.6 Challenges with Nanomembranes-Enhanced Water Remediation

The structural stability of nanomembranes is very important. However, some nanomembranes are now considered unsuitable for bioengineering because they lack sufficient mechanical strength and show signs of biological incompatibility. Moreover, it has negative repercussions on individuals. The manufacturing technique and the amount of porosity may modify the mechanical characteristics of nanomembranes. Nanomembranes are categorized into three groups according to their chemical composition. The three categorizations are inorganic, organic, and hybrid. Arranging the openings of inorganic nanomembranes improves their characteristics and expands their range of applications. Conversely, organic nanomembranes are very cost-effective and may be used in live organisms without any complications. Although the construction process of hybrid nanomembranes may provide challenges, they possess several advantageous properties. The mechanical strength of nanomembranes falls exponentially as their porosity rises. The study extensively examined several techniques for fabricating nanomembranes, including anodization, 3D printing, lithography, micromachining, chemical vapor deposition, and layer-by-layer deposition. Nevertheless, lithography is linked to substantial costs due to the necessary equipment. Micromachining is likewise a costly technique due to similar factors. Chemical vapor deposition outperforms other technologies in terms of mass manufacturing. 3D printing is a recently developed technique for fabricating nanomembranes, albeit it is limited by its poor resolution. The aforementioned techniques are not suited for the production of nanomembranes. Advancements in 3D printing technology have made it more possible to produce nanomembranes. Ongoing research is continually exploring various production strategies to create nanomembranes with extraordinary mechanical resilience and improved quality. Nanomembrane technologies in bioengineering, including applications in skin and bone regeneration and medication transport, encounter several challenges and opportunities within this field. The correlation between the mechanical robustness and biocompatibility of nanomembranes will be influenced by factors, such as the incorporation of novel concepts, distinctive amalgamations of materials, the potential for large-scale production, and the sustainability of innovative methodologies.

9.7 Conclusion

Nanomembranes may be categorized according to their distinct features and production processes as a nanomembrane's mechanical qualities can be influenced by its composition, procedure, or porosity. One distinction pertains to the manufacturing procedure of nanomembranes. Furthermore, considerations including porosity, strength, permeability, chemical stability, and thermal stability should be considered while selecting the best technique for producing nanomembranes. Before undertaking expensive research, it is essential to possess a thorough comprehension of the fundamental concepts and origins that have played a role in the development of nanomembranes in the last ten years. Considering this, we have conducted a thorough analysis and categorization of several artificial nanomembranes. Fabrication techniques include a variety of methods, such as sol-gel processing, chemical vapor deposition, layer-by-layer deposition, lithography, anodization, and micromachining. Continual progress in manufacturing techniques seeks to create nanomembranes that have improved quality and greater mechanical durability. The use of nanomembrane technology in bioengineering, including tissue engineering and drug delivery systems, faces both problems and opportunities. The relationship between the strength and compatibility with living organisms of a nanomembrane depends on technological progress, innovative material choices, the ability to increase manufacturing methods, and the durability of the materials over time. This review study contributes to the existing research on nanomembranes in the constantly progressing field of bioengineering.

References

Abu MH, Goda ES, Abdallah HM, Esmail A, Gamal H, Ro K (2021) Innovative bactericidal adsorbents containing modified xanthan gum/montmorillonite nanocomposites for wastewater treatment. Int J Biol Macromol 167:1113–1125. https://doi.org/10.1016/j.ijbiomac.2020.11.065

Ahmed SF, Mofijur M, Nuzhat S, Chowdhury AT, Rafa N, Uddin MA, Inayat A, Mahlia TMI, Ong HC, Chia WY, Show PL (2021) Recent developments in physical, biological, chemical, and hybrid treatment techniques for removing emerging contaminants from wastewater. J Hazard Mater 416:125912. https://doi.org/10.1016/j.jhazmat.2021.125912

Aiswarya V, Das S, Kumar S, Datta A (2022) Preparation and characterization of magnetized GO nanoparticle enhanced microencapsulated phase change material for thermal energy storage application, Indian. Chem Eng 64:519–529. https://doi.org/10.1080/00194506.2022.2066577

Baite TN, Mandal B, Purkait MK (2022) Antioxidant-incorporated poly(vinyl alcohol) coating: preparation, characterization, and influence on ripening of green bananas. ACS Omega 7:42320–42330. https://doi.org/10.1021/acsomega.2c05271

Barbhuiya NH, Misra U, Singh SP (2022) Biocatalytic membranes for combating the challenges of membrane fouling and micropollutants in water purification: a review. Chemosphere 286:131757. https://doi.org/10.1016/j.chemosphere.2021.131757

Bharti M, Das PP, Purkait MK (2023) A review on the treatment of water and wastewater by electrocoagulation process: advances and emerging applications. J Environ Chem Eng 11:111558. https://doi.org/10.1016/j.jece.2023.111558

Camacho LM, Pinion TA, Olatunji SO (2020) Behavior of mixed-matrix graphene oxide—polysulfone membranes in the process of direct contact membrane distillation. Sep Purif Technol 240:116645. https://doi.org/10.1016/j.seppur.2020.116645

Celik E, Park H, Choi H, Choi H (2011) Carbon nanotube blended polyethersulfone membranes for fouling control in water treatment. Water Res 45:274–282. https://doi.org/10.1016/j.watres.2010.07.060

Ciganė U, Palevičius A, Janušas G (2021) Review of nanomembranes: materials, fabrications and applications in tissue engineering (bone and skin) and drug delivery systems. J Mater Sci 56:13479–13498. https://doi.org/10.1007/s10853-021-06164-x

De Pasquale G (2021) Additive manufacturing of micro-electro-mechanical systems (Mems). Micromachines 12:1374. https://doi.org/10.3390/mi12111374

Deepti A, Dhara A, Purkait MK (2023) Membrane and disinfection technologies for industrial wastewater treatment. In Shah MP (ed) Industrial wastewater reuse: applications, prospects and challenges. Springer Nature Singapore, Singapore, pp 89–112. https://doi.org/10.1007/978-981-99-2489-9_5

Dhara S, Shekhar Samanta N, Das PP, Uppaluri RVS, Purkait MK (2023) Ravenna grass-extracted alkaline lignin-based polysulfone mixed matrix membrane (MMM) for aqueous Cr(VI) removal. ACS Appl Polym Mater 5:6399–6411. https://doi.org/10.1021/acsapm.3c00999

Dhara S, Sontakke AD, Shekhar Samanta N, Venkata Sai Uppaluri R, Purkait MK (2024) Biowaste-alkaline lignin and GO integrated polysulfone ultrafiltration membrane fabrication for Pb^{2+} and Eosin Y dye removal. Sep Purif Technol 34:126894. https://doi.org/10.1016/j.seppur.2024.126894

Duarah P, Das PP, Purkait MK (2023) Technological advancement in the synthesis and application of nanocatalysts. In Shah MP (ed) Advanced application of nanotechnology to industrial wastewater. Springer Nature Singapore, Singapore, pp 191–214. https://doi.org/10.1007/978-981-99-3292-4_10

Elman NM, Daniel K, Jalali-Yazdi F, Cima MJ (2010) Super permeable nano-channel membranes defined with laser interferometric lithography. Microfluid Nanofluidics 8:557–563. https://doi.org/10.1007/s10404-009-0537-z

Emami F, Shekarriz S, Shariatinia Z, Moridi Mahdieh Z (2018) Self-cleaning properties of nylon 6 fabrics treated with corona and TiO_2 NP under both ultraviolet and daylight irradiations. Fibers Polym 19:1014–1023. https://doi.org/10.1007/s12221-018-1025-4

Esfahani MR, Koutahzadeh N, Esfahani AR, Firouzjaei MD, Anderson B, Peck L (2019) A novel gold nanocomposite membrane with enhanced permeation, rejection and self-cleaning ability. J Memb Sci 573:309–319. https://doi.org/10.1016/j.memsci.2018.11.061

Gaikar P, Sangale S, Wadhawa G (2023) Chapter 16: the Langmuir-Blodgett method for metal oxide nanostructures. In: Mane R, Jadhav V, Al-Enizi A (eds) Solution methods for metal oxide nanostructures. Elsevier, pp 369–392. https://doi.org/10.1016/B978-0-12-824353-4.00017-8

García-Torres J, Lázaro C, Sylla D, Lanzalaco S, Ginebra MP, Alemán C (2023) Combining 2D organic and 1D inorganic nanoblocks to develop free-standing hybrid nanomembranes for conformable biosensors. J Nanostructure Chem 13:507–517. https://doi.org/10.1007/s40097-022-00482-5

Ghosh A, Das G (2020) Green synthesis of Sn(II)-BDC MOF: preferential and efficient adsorption of anionic dyes. Microporous Mesoporous Mater 297:110039. https://doi.org/10.1016/j.micromeso.2020.110039

Haldar D, Duarah P, Purkait MK (2020) Chemosphere MOFs for the treatment of arsenic, fluoride and iron contaminated drinking water: a review. Chemosphere 251:126388. https://doi.org/10.1016/j.chemosphere.2020.126388

He B, Peng H, Chen Y, Zhao Q (2020) High performance polyamide nanofiltration membranes enabled by surface modification of imidazolium ionic liquid. J Memb Sci 608:118202. https://doi.org/10.1016/j.memsci.2020.118202

Kataria N, Garg VK (2018) Green synthesis of Fe_3O_4 NP loaded sawdust carbon for cadmium (II) removal from water: regeneration and mechanism. Chemosphere 208:818–828. https://doi.org/10.1016/j.chemosphere.2018.06.022

Kim J, Gong J, Lin W, Lal S, Su X, Vincent D, Cho SJ, Zhou J, Min S, Kim D, Ma Z (2022) Low contact-resistivity and high-uniformity Ni/Au ohmic contacts on Si nanomembranes grafted to Si substrates via low-temperature rapid thermal annealing. Mater Sci Semicond Process 151:106988. https://doi.org/10.1016/j.mssp.2022.106988

Kumar S, Bhawna, Sharma R, Gupta A, Dubey KK, Khan AM, Singhal R, Kumar R, Bharti A, Singh P, Kant R, Kumar V (2022) TiO_2 based Photocatalysis membranes: an efficient strategy for pharmaceutical mineralization. Sci Total Environ 845:157221. https://doi.org/10.1016/j.scitotenv.2022.157221

Kumar S, Mondal P, Purkait MK (2023) Hazardous effects of heavy metals from industrial wastewaters and their remediation through green technology. In: Mathuriya AS, Pandit S, Singh NK (eds) Green technologies for industrial waste remediation. Springer Nature Switzerland, Cham, pp 29–55. https://doi.org/10.1007/978-3-031-46858-2_3

Kwak S-Y, Kim SH, Kim SS (2001) Hybrid organic/inorganic reverse osmosis (RO) membrane for preparation and characterization of TiO_2 nanoparticle self-assembled aromatic polyamide membrane. Environ. Sci. Technol. 35:2388–2394

Li P, Sun S, Dong A, Hao Y, Shi S, Sun Z, Gao G, Chen Y (2015) Developing of a novel antibacterial agent by functionalization of graphene oxide with guanidine polymer with enhanced antibacterial activity. Appl Surf Sci 355:446–452. https://doi.org/10.1016/j.apsusc.2015.07.148

Li B, Chen X, Li K, Zhang C, He Y, Du R, Wang J, Chen L (2019) Coupling membrane and Fe–Pd bimetallic NP for trichloroethene removing from water. J Ind Eng Chem 78:198–209. https://doi.org/10.1016/j.jiec.2019.06.013

Lv C, Chen S, Xie Y, Wei Z, Chen L, Bao J, He C, Zhao W, Sun S, Zhao C (2019) Positively-charged polyethersulfone nanofibrous membranes for bacteria and anionic dyes removal. J Colloid Interface Sci 556:492–502. https://doi.org/10.1016/j.jcis.2019.08.062

Mondal P, Purkait MK (2019) Preparation and characterization of novel green synthesized iron–aluminum nanocomposite and studying its efficiency in fluoride removal. Chemosphere 235:391–402. https://doi.org/10.1016/j.chemosphere.2019.06.189

Puri N, Gupta A, Mishra A (2021) Recent advances on nano-adsorbents and nanomembranes for the remediation of water. J Clean Prod 322:129051. https://doi.org/10.1016/j.jclepro.2021.129051

Purkait MK, Sinha MK, Mondal P, Singh R (2018) Chapter 1—introduction to membranes. In Purkait MK, Sinha MK, Mondal P, RBT-IS, Singh T (eds) Stimuli responsive polymeric membranes—smart polymeric membranes. Elsevier, pp 1–37. https://doi.org/10.1016/B978-0-12-813961-5.00001-2

Ren LF, Xia F, Chen V, Shao J, Chen R, He Y (2017) TiO2-FTCS modified superhydrophobic PVDF electrospun nanofibrous membrane for desalination by direct contact membrane distillation. Desalination 423:1–11. https://doi.org/10.1016/j.desal.2017.09.004

Ren J, Woo YC, Yao M, Lim S, Tijing LD, Shon HK (2019) Nanoscale zero-valent iron (nZVI) immobilization onto graphene oxide (GO)-incorporated electrospun polyvinylidene fluoride (PVDF) nanofiber membrane for groundwater remediation via gravity-driven membrane filtration. Sci Total Environ 688:787–796. https://doi.org/10.1016/j.scitotenv.2019.05.393

Safarpour M, Khataee A, Vatanpour V (2015) Thin film nanocomposite reverse osmosis membrane modified by reduced graphene oxide/TiO_2 with improved desalination performance. J. Memb. Sci. 489:43–54. https://doi.org/10.1016/j.memsci.2015.04.010

Shafiq M, Sabir A, Islam A, Khan SM, Gull N, Hussain SN, Butt MTZ (2018) Cellulaose acetate based thin film nanocomposite reverse osmosis membrane incorporated with TiO_2 NP for improved performance. Carbohydr Polym 186:367–376. https://doi.org/10.1016/j.carbpol.2018.01.070

Sharifpour E, Ghaedi M, Nasiri Azad F, Dashtian K, Hadadi H, Purkait MK (2018) Zinc oxide nanorod-loaded activated carbon for ultrasound-assisted adsorption of safranin O: central composite design and genetic algorithm optimization. Appl Organomet Chem 32:e4099. https://doi.org/10.1002/aoc.4099

Sharma K, Maiti K, Kim NH, Hui D, Lee JH (2018) Green synthesis of glucose-reduced graphene oxide supported Ag-Cu_2O nanocomposites for the enhanced visible-light photocatalytic activity. Compos Part B Eng 138:35–44. https://doi.org/10.1016/j.compositesb.2017.11.021

Sharma M, Das PP, Chakraborty A, Purkait MK (2022) Clean energy from salinity gradients using pressure retarded osmosis and reverse electrodialysis: a review. Sustain Energy Technol Assess 49:101687. https://doi.org/10.1016/j.seta.2021.101687

Sharma M, Mondal P, Sontakke AD, Chakraborty A, Purkait MK (2023a) High performance graphene-oxide doped cellulose acetate based ion exchange membrane for environmental remediation applications. Int J Environ Anal Chem 103:7751–7772. https://doi.org/10.1080/03067319.2021.1975276

Sharma M, Das PP, Kumar S, Purkait MK (2023b) Polyurethane foams as packing and insulating materials. In: Polyurethanes: preparation, properties, and applications volume 3: emerging applications, pp 4–83. https://doi.org/10.1021/bk-2023-1454.ch004

Shawky HA, Chae SR, Lin S, Wiesner MR (2011) Synthesis and characterization of a carbon nanotube/polymer nanocomposite membrane for water treatment. Desalination 272:46–50. https://doi.org/10.1016/j.desal.2010.12.051

Shayesteh M, Samimi A, Shafiee Afarani M, Khorram M (2016) Synthesis of titania–γ-alumina multilayer nanomembranes on performance-improved alumina supports for wastewater treatment. Desalin Water Treat 57:9115–9122. https://doi.org/10.1080/19443994.2015.1030703

Singh R, Volli V, Lohani L, Purkait MK (2021) Polymeric ultrafiltration membranes modified with fly ash based carbon nanotubes for thermal stability and protein separation. Case Stud Chem Environ Eng 4:100155. https://doi.org/10.1016/j.cscee.2021.100155

Soh HT, Guarini KW, Quate CF (2001) Introduction to scanning probe lithography. In: Soh HT, Guarini KW, Quate CF (eds) Scanning probe lithography. Springer US, Boston, MA, pp 1–22. https://doi.org/10.1007/978-1-4757-3331-0_1

Wang X, Liu W, Fu H, Yi X-H, Wang P, Zhao C, Wang C-C, Zheng W (2019) Simultaneous Cr(VI) reduction and Cr(III) removal of bifunctional MOF/titanate nanotube composites. Environ Pollut 249:502–511. https://doi.org/10.1016/j.envpol.2019.03.096

Wang Y, He L, Li Y, Jing L, Wang J, Li X (2020) Ag NPs supported on the magnetic Al-MOF/PDA as nanocatalyst for the removal of organic pollutants in water. J Alloys Compd 828:154340. https://doi.org/10.1016/j.jallcom.2020.154340

Wang C, Park MJ, Seo DH, Phuntsho S, Gonzales RR, Matsuyama H, Drioli E, Shon HK (2022) Inkjet printed polyelectrolyte multilayer membrane using a polyketone support for organic solvent nanofiltration. J Memb Sci 642:119943. https://doi.org/10.1016/j.memsci.2021.119943

Wang N, Thameem Dheen S, Fuh JYH, Senthil Kumar A (2021) A review of multi-functional ceramic NP in 3D printed bone tissue engineering. Bioprinting 23:e00146. https://doi.org/10.1016/j.bprint.2021.e00146

Wen T, Zhao Z, Shen C, Li J, Tan X, Zeb A, Wang X, Xu A-W (2016) Multifunctional flexible free-standing titanate nanobelt membranes as efficient sorbents for the removal of radioactive 90Sr2+ and $137Cs^+$ ions and oils. Sci Rep 6:20920. https://doi.org/10.1038/srep20920

Zarei F, Marjani A, Soltani R (2019) Novel and green nanocomposite-based adsorbents from functionalised mesoporous KCC-1 and chitosan-oleic acid for adsorption of Pb(II). Eur Polym J 119:400–409. https://doi.org/10.1016/j.eurpolymj.2019.07.043

Zarpelon F, Galiotto D, Aguzzoli C, Carli LN, Figueroa CA, Baumvol IJR, Machado G, Crespo JDS, Giovanela M (2016) Removal of coliform bacteria from industrial wastewaters using polyelectrolytes/silver NP self-assembled thin films. J Environ Chem Eng 4:137–146. https://doi.org/10.1016/j.jece.2015.11.013

Zhou H, Liu Y, Chi W, Yu C, Yu Y (2013) Preparation and antibacterial properties of Ag@polydopamine/graphene oxide sheet nanocomposite. Appl Surf Sci 282:181–185. https://doi.org/10.1016/j.apsusc.2013.05.099

Zhou L, Xiao G, He Y, Wu J, Shi H, Zhong F, Yin X, Li Z, Chen J (2021) Multifunctional filtration membrane with anti-viscous-oils-fouling capacity and selective dyes adsorption ability for complex wastewater remediation. J Hazard Mater 413:125379. https://doi.org/10.1016/j.jhazmat.2021.125379

Zirehpour A, Rahimpour A, Khoshhal S, Firouzjaei MD, Ghoreyshi AA (2016) The impact of MOF feasibility to improve the desalination performance and antifouling properties of FO membranes. RSC Adv 6:70174–70185. https://doi.org/10.1039/c6ra14591d

Chapter 10
Waste Water Remediation and Nanoadsorbents

Prabhat Kumar Patel, Hrishikesh Kumar Singh, and Asit Singh

Abstract Water supplies are becoming more and more polluted every day for a variety of reasons, such as the unregulated release of contaminants from point and indirect streams. Consequently, we must immediately begin to cultivate cost-effective and competent water purification methods. This chapter seeks to assess various methods for water treatment, their fundamental ideologies, costs, and adaptability for the eradication of wastewater contaminants. Among these know-hows, the adsorption process stands out as a cost-effective and environmentally friendly option due to its universality and inexpensiveness. Such a chapter delves into various adsorbents such as activated carbon, chitosan, zeolite, clays, activated alumina, limestone, silica gel, etc., detailing their capabilities for mitigation of contaminants. It also discusses the lacunas of using traditional sorbents in wastewater treatment processes. At present, nanotechnology, as this century's foremost exciting and underexplored scientific branch, offers a promising alternative to traditional adsorbents due to their unique properties. It holds the potential to address contemporary challenges, particularly in wastewater treatment. This chapter underscores the significance of adsorption and nanoparticles (NPs) in advancing wastewater treatment, utilizing unconventional water resources. Additionally, it goes over the idea of sorption and the variety of nanosorbents that are employed to treat sewage and the contaminants that may be removed by these materials, with a focus on regenerative options.

Keywords Wastewater · Nanosorbent · Sustainability · Adsorption · Dyes · Contaminants

P. K. Patel (✉) · H. K. Singh · A. Singh
Department of Civil Engineering, Institute of Engineering and Technology Lucknow, Lucknow, Uttar Pradesh, India
e-mail: pprabhat1991@gmail.com

N. Agarwal et al. (eds.), *Nanomaterials in Wastewater Research*, Advances in Wastewater Research, https://doi.org/10.1007/978-981-96-3908-3_10

10.1 Introduction

The existence of all living forms on Earth depends on water. The need for potable water is on the rise due to industrialization and better living standards (Kumari et al. 2019). We rely on clean water in our homes, but it is also essential for many commercial and industrial processes. As a result of rising water consumption and the limited availability of fresh water sources, people have begun turning to non-traditional water sources, such as rainwater harvesting, polluted fresh water, wastewater, brackish water, and saltwater. However, these newly acquired resources cannot be employed immediately (Kumari et al. 2019). It is necessary to treat water before employing it for domestic, commercial, or farming uses. Under these conditions, treating wastewater has paramount significance. The treatment of wastewater has grown into its own industry in recent decades, and this development has contributed to environmental contamination in several ways. Before the modern era, the range of water treatment methods was extremely restricted. However, as human knowledge keeps growing as well as technologies keep evolving, experts have devised a variety of effective and environmentally favorable techniques for treating polluted water (Gupta et al. 2011). The mitigation of toxic contaminants including heavy metal ions, organic and inorganic contaminants, microplastics, dyes, etc. from discarded liquid systems is an interesting as well as challenging problem. In the precise field that targets the removal of toxic pollutants from spent supplies, several processes such as membrane separation, ion exchange resins, electrochemical reduction, nanomaterials (NMs), adsorption process, chemical precipitation, coagulation and flocculation, sedimentation filtration, reverse osmosis, flotation, electrodialysis, solvent extraction, bio removal, photocatalysis, phyto remediation, and ultrafiltration have been reported for their efficacious performance. In addition, the issue of sustainable and economically viable technology for heavy metal removal is one of the most challenging problems for adequate resolution. This is due to pertinent issues such as environmental friendliness, affordability, permanence, and quicker and easier research-ensuring platforms (Patel et al. 2023a). Further, as per adopted technologies, the heavy metal contaminated wastewater treatment technologies can be primarily divided into three categories. These are (a) chemical precipitation, electrochemical reduction, and electrolysis that involve chemical reactions for metal removal; (b) biomass adsorption, metal extraction, ion exchange resins, reverse osmosis, evaporation, enrichment, and procedures that target heavy metal removal through a phase change and through the principle of sorption, and (c) biosorption, bio flocculation, artificial wetlands, and other biological processes that involve biological system-based adsorption and enrichment mechanisms for heavy metal removal (Patel et al. 2023b). Figure 10.1 provides such conventional wastewater treatment methods for heavy metal-contaminated wastewater effluents.

When it comes to cleaning and decontaminating water, NPs have risen to prominence as an intriguing solution in the last decade. The development of NPs with organized topologies and assets has been the subject of much research and development. NPs are almost identical to their bulk counterparts but on a smaller size.

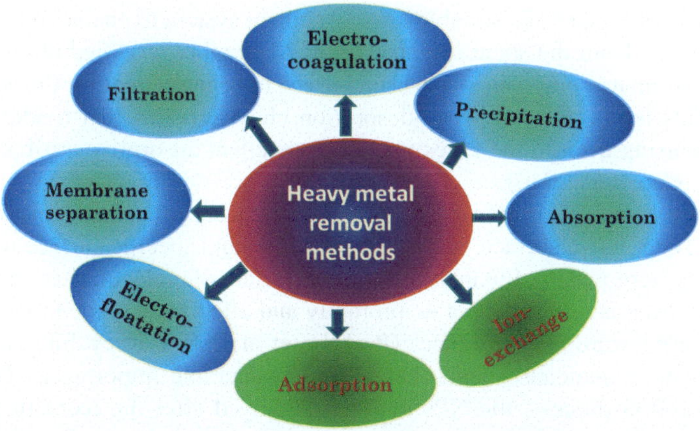

Fig. 10.1 Pictorial representations of various wastewater treatment methods

Typically, NPs have a size of less than 100 nm in one spatial dimension. Their minimized surface defects and very high specific surface area provide them with remarkable physical, optical, electrical, magnetic, and biological attributes (Kumari et al. 2019). Because of all these qualities, they are great adsorbents and can remove a wide variety of contaminants. Highly effective and inexpensive alternatives that depend less on massive infrastructures and more on technology are modular and multipurpose procedures driven by nanotechnology for water treatment (Qu et al. 2013a). In addition to solving the problems that persist with traditional techniques of wastewater treatment, the advent of nanotechnology and NPs has opened the door to the possibility of making full use of hitherto untapped unorthodox water sources. NPs have a significant influence on human health and the environment, as well as any possible negative consequences of therapeutic procedures.

Ion exchange and chelating resin-based physisorption/chemisorption have been proven to be very efficient in extracting heavy metals such as chromium, copper, zinc, iron, and lead. These materials also exhibited very good recovery of heavy metals from aqueous systems being even characterized with extremely low solution concentrations. Thereby, they have been inferred to be among the few competent technologies that cater towards sustainable research commercialization. In addition, compared to synthetic cheating resins, biopolymers such as chitosan, fruit, and vegetable wastes have been proposed to be competitive due to their ease of preparation, cheaper cost, and excellent separation capabilities (Anwar et al. 2010; Nasernejad et al. 2005). Chemisorption utilizing chelating resins has been more promising than physisorption. This is due to the physisorption functionality based on adsorbate attraction through Vander Waal's attraction forces, and chemical adsorption functionality based on the chemical bonding between dominating metal ions present in the adsorbate and various chelating functional groups in the resins. As a result, chemisorption enhances heavy metal removal and reuse of regenerated spent adsorbent. Thereby, the technology minimizes waste production and improves metal desorption-based recovery through the deployment of a suitable low-cost eluent system to enable the extraction of heavy metals from the spent chelated resin. Also, other modified physisorbents, such as Na Alginate composite, have been proven to be inefficient for the mentioned purpose. This is due to their weak desorption characteristics that render them to be non-promising even in the cases involving excellent adsorptive attributes of the adsorbents (Cao et al. 2010).

NPs with a strong adsorption affinity can emanate from either inorganic or organic resources. Their active surface, substantial porosity, and specific surface area make them ideal for removing pollutants with a broad variety of molecular sizes (Santhosh et al. 2016). Nanoadsorbents operate promptly and effectively while preventing the release of any hazardous substances (Pacheco et al. 2006). Recycling is an additional attribute of nanoadsorbents that holds considerable importance. Therefore, after the sorption process, the NPs may be employed after the recreating option, which involves the separation of the sorbate from the NPs for reuse. In 2007, studies were conducted to determine the efficacy of carbon NPs in eradicating hydrocarbons from water (Yang and Xing 2007).

10.2 Types of Contaminants Present in Wastewater

Based on their type and radix, toxins of water and wastewater are classed as inorganic, biological, or organic pollutants. Due to their bioaccumulation, heavy metals pose one of the greatest threats to all the pollutants found in wastewater and water. When the quantity of a substance in an organism increases to a level higher than its environmental concentration, this phenomenon is called bioaccumulation. When elements are absorbed in and stored by biological systems, they do not need to be broken down or expelled until later (Verma and Dwivedi 2013). Metals with specific densities more than 5 g cm^{-3} and atomic weights ranging from 63.5 to 200.6 are often referred to as "heavy metals" (Wang and Shi 2001). Although these metals play an essential role in several metabolic and biological reactions in living things at low quantities, they become harmful when their concentrations are beyond certain thresholds. The amount of harmful contaminants in wastewater systems throughout the globe has grown due to the extensive use of chemicals. Economic stability, environmental quality, and human health all decline as a result. Therefore, it is necessary to improve the efficacy of remedial technologies in order to raise water quality (Beyersmann and Hartwig 2008). One of the most significant classes of water contaminants is organic materials. An indicator of its presence is the total organic carbon (TOC) level. Components both dissolved and particulate are included in the TOC. As a result, changes in suspended solids, which were discernible in rivers, can affect TOC. A large variety of waste products and effluents include organic carbon, which is derived from organic matter and found in freshwater environments. Consequently, TOC is seen as a helpful indicator of contamination level. Aquatic organisms and their life cycles are major contributors to organic contaminants in wastewater, water, and ordinary run-off and debris. The different organic material compositions in water cause it to have a bad taste and smell and prevent it from being treated for industrial and household uses (El-sayed 2020).

10.3 Types of NanoAdsorbents

The treatment of wastewater employs a wide range of nanoadsorbents. It has been assessed that they are exceptionally efficient at eliminating contaminants from water. Although it would be preferable to be allowed to compare capacities directly, this may not always be feasible due to factors including NP shape and size, operating circumstances (such as pH, climate, and reaction period), and experimentation form (such as batch or column runs). A variety of NM frameworks, including particles, filaments, and quantum dots, have been fabricated and implemented in diverse purposes (Sadegh et al. 2017). A study investigated the efficacy of treating water and wastewater using synthetic nanosorbent as highly efficient materials with distinctive characteristics (Wang 2012). It also employed nanosorbents to eradicate dyes and heavy metals from wastewater and water, and various investigations have looked at the results of employing such materials for wastewater and water treatment.

The removal of pollutants from water and wastewater has been the subject of numerous advancements and improvements. Consequently, much of the current research is focused on one or more of the following types of nanosorbents:

- Carbon-based nanosorbents
- Polymeric nanosorbents
- Zeolites
- Metal-based nanosorbents.

10.3.1 Carbon-Based Nanosorbents

The most significant and distinguishing qualities of any nanoadsorbent have been its enormous surface area and porosity. Activated carbon, zeolites, pillared clays, polymers, metal–organic frameworks, and many more of these materials are already accessible (Sen Gupta and Bhattacharyya 2012). On the other hand, carbon-based adsorbents such as carbon nanotubes (CNTs), activated carbon, porous carbon, graphene, and fullerenes are very effective and have excellent thermal stability (Wang et al. 2012).

10.3.1.1 CNTs

One form of carbon is the allotrope known as carbon nanotubes. In its most basic form, CNTs are a rolled-up, cylindrical structure. Two main varieties predominate: (a) Single-Walled CNTs (SWCNTs), which consist of one sheet of graphene rolled into a tube, and (b) Multi-Walled CNTs (MWCNTs), which are composed of many sheets of graphene wrapped into a tube. Among many carbon-based sorbents; CNTs are one of the nanosorbents that have received the most investigation. A literature review demonstrates that CNTs are very efficient when it comes to the adsorption of organic and inorganic contaminants that are found in water (Gupta and Saleh 2013). In comparison to activated carbon powder, the CNTs have an exceptionally high sorption capacity due to their extremely high surface-to-volume ratio of active sites (Pan et al. 2008). A CNT's outer surface is its active surface for adsorption (Yang and Xing 2010). According to a research CNTs in water form dispersed formations that reduce surface area (Pan and Xing 2008). However, organic molecules may be adsorbed to a great extent in the spaces within the intermediate channels.

10.3.1.2 Activated Carbon

Activated carbon does not work effectively with larger organic compounds, even though its surface area is similar to that of CNTs. Many large organic compounds, such as the majority of antibiotics (Chowdhury and Balasubramanian 2014) and medicines (Ji et al. 2009), are unable to pass through their micropores. Thus, CNTs are better at adsorbing such compounds because of their bigger pores and more accessible sorption sites. For organic chemical adsorption in water, researchers employed carbon NPs (Pan and Xing 2008). Results have discovered that activated carbon does not have a strong attraction for organic molecules with low molecular weight. Conversely, CNTs exhibit a wide variety of pollutant–CNT interactions, which allows them to effectively adsorb the majority of these organic molecules (Qu et al. 2013b).

10.3.1.3 Graphene

Although adsorbents from the graphene family are efficient against many water contaminants, they have not been studied to the same extent as CNTs and activated carbon (Novoselov et al. 2000). Adsorbents such as graphene oxide (GO) and reduced GO, have been described by many authors to be capable of removing contaminants such as synthetic dyes (Chowdhury and Balasubramanian 2014) and heavy metals (Huang et al. 2011). The utilization of GO as a replacement for CNTs and as an appropriate material for the treatment of wastewater has been observed in a variety of cases. The fabrication cost of GO is somewhat cheaper. Unlike other materials, the inner walls of CNTs cannot be absorbed by water (Sitko et al. 2013). Fullerenes have also been investigated for their adsorption capabilities on organic contaminants in water, but at a far lesser scale than other carbon allotropes.

10.3.2 Polymeric Nanosorbents

The large surface area, adjustable chemical composition, excellent structural stiffness, distribution of pore sizes, and feasible synthesis under specific circumstances of polymeric sorbents have contributed to their evaluation as an alternative to traditional sorbents (Zare et al. 2018). These sorbents may be divided into four categories depending on their arrangement: dendrimer-based, CNT/polymer, metal and metal oxide/polymer, and graphene/polymer (Bushra 2018). Dendrimers, which consist of molecules that are repeatedly divided, serve as an ideal illustration of the effectiveness of polymeric nanoadsorbents in eliminating both inorganic and organic contaminants. Adsorption of heavy metals can be facilitated via the external sections, whereas adsorption of organic pollutants can be achieved via their inner hydrophobic segments. Using dendrimers in conjunction with an ultrafiltration approach, a group of co-workers was able to remove copper from the water, which eventually culminated in the elimination of entire copper cations (Diallo et al. 2005). Changes in pH have a big effect on how the nanoadsorbent materials respond. Due to the intricate multistage synthesis process involved, the production of commercial dendrimers is limited to a handful of companies in recent years. According to a study dendrimers were ineffective in extracting anionic substances like dyes. To improve their performance, researchers synthesized a combination product called chitosan-dendrimer nanostructure. With an impressive dye removal effectiveness of over 99%, these materials are not only non-toxic but also biocompatible and biodegradable (Sadeghi-Kiakhani et al. 2013).

10.3.3 Zeolites

The aluminosilicate crystal has an exterior morphology that is saturated with liquid molecules and cations via its many electrostatic pores. Ion exchange and cyclic dehydration are made possible by the large range of motion. While most zeolites are on the microscale, it is possible to fabricate nanozeolites with sizes ranging from 10 to 500 nm. Nanoscale zeolite outperforms conventional zeolite in terms of its ability to adsorb, contact area (400–1000 $m^2\ g^{-1}$), and density. Therefore, nanozeolite is very effective and has a strong affinity for contaminants. Since 1980, the combination of zeolite and silver has been recognized. Silver ions might be incorporated into the zeolite framework due to its porous shape. A study investigated and contrasted the antimicrobial actions of several nanosilver-containing materials, including zeolites (Egger et al. 2009). They found that silver assaults and hinders microbial development. When metals react with solutions, a small amount of silver cations are released from their surfaces (Nagy et al. 2011). A chemical with antibacterial capabilities was manufactured by multiple companies using zeolites that contain silver ions. A study employed nanozeolites generated through laser-induced disintegration of zeolite at the microscale to remediate water and effluents (Tiwari 2008).

When applied as nanosorbents, both CNTs and nanometals are remarkably efficient at removing inorganic contaminants such as Pb, As, and Cd. Nanometals and zeolites are useful because they are inexpensive and can be utilized with current water treatment methods. This is because they can be used as permanent sorbents in a variety of granules and spheres. Although CNTs are costly to produce, they need additional operational equipment like membrane filtration facilities to ensure that no NPs drop to the bottom of the liquid. When it comes to removing micro contaminants, one major advantage of CNT is its strong ability to adsorb polar organic compounds. Comparatively, CNTs are exceptionally economical in comparison to wide-ranging sorbents like activated carbon. The study of how CNTs may be deployed in very small amounts for extremely specialized sorption methods will be the focus of future developments. Polymeric nanoadsorbents are still making progress in the market, even though nanometals and CNTs are already available for a variety of purposes. Polymeric nanoadsorbents have been very innovative chemicals that allow for the simultaneous elimination of organic and inorganic contaminants. Rising fabrication expenses and a lack of feasible techniques to improve polymeric dendrimers are the main obstacles to the widespread adoption of these innovative nanosorbent materials. Because of their genesis and characteristics, chemicals such as zeolites, nanometals, and CNTs have been classified as innocuous substances concerning environmental toxicity. The size and form of nanosorbents, as well as chemical stabilizers and surface modifications, all have an impact on their hazards. However; it is necessary to reassess the ecotoxicity of recognized NMs; with new alterations.

10.3.4 Nanometals and Nanometal Oxides

In contrast to activated carbon, nanometal oxides exhibit enhanced adsorption capabilities, rendering them viable substitutes for the elimination of heavy metals. Moreover, nanometal oxides may be thwarted without sacrificing surface area and have a relatively small intraparticle diffusion distance because of their large surface area. Nanomagnetite is one of many superparamagnetic nanometal oxides that may be recovered and separated using a low-gradient magnetic force. Adsorptive medium filters and slurry reactors may be treated with nanoscale metal oxides (Qu et al. 2013b). The use of nano iron hydroxide, an adsorbent with a huge surface area and great abrasion resistance, to successfully remove As from water or wastewater is another example of nanometal oxides (Pan et al. 2008). For simple incorporation into current treatment protocols, nanometal oxides and nanometals may be powdered or compacted into porous pellets (Gehrke et al. 2015).

10.4 Pollutants and Their Removal Using nanoadsorbents

Several studies have attempted to separate particulate materials from the treated water using membranes or other methods, but doing so increases the overall expense. Thus, research in the future will concentrate on expanding our understanding and enhancing the use of materials for nanoadsorbents. Nowadays, there are various kinds of water contaminants present in the water and wastewater streams. According to their nature, they may generally be grouped into three types: organic, inorganic, and biological. Adsorption utilizing nanosorbents may be an effective measure to eradicate biological pollutants alongside inorganic and organic pollutants.

10.4.1 Inorganic Pollutants

The eradication of inorganic contaminants from wastewater using adsorption technology is a proven reliable method. A cursory examination of the current corpus of literature indicates that the predominant approach for eliminating inorganic ions is through the utilization of NPs composed of oxides of titanium, iron, and aluminum. Researchers have demonstrated those iron oxide NPs are an excellent adsorbent for facilitating the effective elimination of inorganic species due to their wide surface area and simplicity to fabricate (Patel et al. 2024a). Heavy metal contaminants, such as arsenate, cadmium, chromium, copper, etc., have been removed using NPs of different metals and metal oxides.

10.4.1.1 Arsenic (As)

Arsenic is perhaps the most hazardous of all metal contaminants because it may pollute subsurface water. As a result of As pollution in subsurface water, people all over the world are consuming water that is tainted with As (Brammer and Ravenscroft 2009). According to research conducted in Bangladesh, after screening hundreds of urines, nail, hair, and water samples; researchers identified that about 900 communities had higher concentrations of As in subsurface water than the allowed permitted level. Numerous methods and approaches have been investigated and tested by researchers to successfully extract As from groundwater. One of the most efficient methods for this aim is adsorption. As(III) was removed by cobalt and manganese ferrite NPs synthesized by Martinez-Vargas et al. (2017). The Freundlich adsorption isotherm provided the best corresponding with the sorption data for both NPs.

10.4.1.2 Chromium (Cr)

Cr has been utilized extensively in the dye and mining sectors. There have been two oxidation states of Cr in water: Cr (III), which is rather innocuous, and Cr (VI), which is very portable in surroundings and malignant to living species (Depault et al. 2006). Therefore, techniques for its efficient, widespread elimination from water must be developed. Remarkable results were obtained when Cr (VI) was adsorbed onto rGO-Co_3O_4 NPs (fabricated from reduced cobalt oxide and GO) (Al Nafiey et al. 2017). A single-stage creation of rGO-CO_3O_4 NPs was carried out under moderate circumstances. When compared to other adsorbents based on graphene, the adsorption capacity (208.8 mg g^{-1}) was determined to be quite efficient. The researchers developed a new environmentally friendly method for extracting Cr (VI) from water by synthesizing zero-valent iron NPs (NZVI) employing plant extracts (Fazlzadeh et al. 2017). The next step was to conduct a batch experiment for Cr (VI) sorption after the NPs characterization. Subsequent improvements in contact time resulted in the complete elimination of contaminants, confirming that the adsorption capacity was at least 90%. Acidic environments exhibited the highest rate of sorption. According to the research on Cr (VI) sorption, the process adhered to the pseudo-second-order kinetic model. Copper NPs have been developed and were utilized in water purification processes (Gupta et al. 2016). The most effective way to achieve maximal Cr (VI) sorption on these NPs was to optimize the batch trials by adjusting variables such as pH, time, metal concentration, adsorbent dosage, etc. A uniform distribution of sorption energy over the adsorbent and the formation of a monolayer of Cr (VI) ions were proposed using the isotherm model. Cr was isolated from water by synthesizing activated carbon from old rubber tires (Gupta et al. 2013). The process has been demonstrated to be pH dependent, and equilibrium was reached after 1 h.

10.4.1.3 Cadmium (Cd)

Cd is a heavy metal that is very dangerous and does not break down. It builds up in the body and does devastating harm (Keshavarz et al. 2013). As soon as it enters the body, it stays there and accumulates throughout the years. It demineralizes bones and damages kidneys. A lot of people are trying to figure out how to effectively remove Cd ions from effluent to deal with the issue of Cd pollution in water. To reiterate, adsorption has the potential to be one of the most effective strategies for the elimination of cadmium from water streams. Adsorption of Cd (II) metal ions was achieved employing nanoscale zerovalent iron particles supported on reduced graphene oxide (NZVI/rGOs), which were developed using the plasma conversion process. These NPs have an exceptionally high uptake capacity of 425.72 mg g^{-1}. It only took 50 min for the sorption to finish. Also, it was determined that the adsorption process was endothermic and occurred spontaneously (Li et al. 2016). AzVIN was synthesized to remove Cd (II) adopting a response surface approach and the results demonstrated that at pH 7, after 60 min of contact, the % removal was determined to be 79.68% (Savasari et al. 2015). γ-Al_2O_3-NPs were synthesized and used in a variety of methods, including XRD, SEM, TGA, and IR, to characterize them (Tabesh et al. 2018). Afterward, the effectiveness of removing Cd (II) from γ-Al_2O_3-NPs was investigated by changing variables such as pH, contact duration, metal concentration, and sorbent dose. The Freundlich isotherm model was used to describe the adsorption process. Pb^{2+} had a sorption capacity of 47.08 mg g^{-1}, whereas Cd^{2+} had a sorption capacity of 17.22 mg g^{-1}.

10.4.1.4 Lead (Pb)

Heavy metal application has soared with the industrial chemical production industry's meteoric rise. Lead is among the most harmful heavy metals. Lead ions are chronically detrimental and may induce major blood vessel disorders (Chen et al. 2011). A concentration of 0.010 mg L^{-1} of lead has been regarded safe for human consumption (Luo et al. 2014). Previously, Cu pipes had been utilized for piping water in residences and welded with Pb. Water might quickly become polluted with lead if the conduits or fittings contain Pb. Consequently, galvanized or plastic pipework has become appreciated. A research was conducted in which magnetite Fe_3O_4/Chitosan NPs (Fe_3O_4/CSNPs) had been fabricated and utilized to efficiently eliminate heavy metals (Fan et al. 2017). The uptake capacity and rate of Pb (II) for Fe_3O_4/CSNPs are both improved over pure Fe_3O_4 because of the significant binding property of chitosan. The maximal uptake capacity was determined to be 79.29 mg g^{-1}.

To extract lead (II) ions from effluents, a study employed sulfonated NPs. Since they have an extra active site known as a "sulfo-group," these NPs have been proven as highly effective sorbents (Chen et al. 2017). Adsorption has been shown to reach 99% after 24 h of contact. The Freundlich isotherm approach was superseded by the Langmuir isotherm model in the adsorption process. The reaction happened very quickly, which fits with the idea of pseudo-second-order dynamics. After sorption, spent sorbent was readily desorbed due to its superparamagnetic ability. The efficacy and process of MgO-NPs to remove Pb(II) from water-based solutions was studied recently (Xiong et al. 2015). Batch procedures have been employed to investigate sorption, and it became apparent that the sorption data followed pseudo-second-order. The Langmuir isotherm model was followed by the sorption process. The optimum uptake capacity was achieved as 2614 mg m^{-1}. Their exceptional removal efficacy and ease of fabrication serve as an ideal sorbent for treating heavy metal-contaminated wastewater.

10.4.2 Organic Pollutants

Adsorption techniques have been employed to eliminate a wide variety of organic contaminants from wastewater (Patel et al. 2023c). These pollutants include different types of dyes, insecticides, and petroleum products. Many investigators are devoting their efforts to the study of sorption to create efficient adsorbents that may remove organic contaminants from water.

10.4.2.1 Dye

Industries such as textile, rubber, paper, printing, painting, etc., use a large quantity of dyeing compounds for their production processes, which is a significant contributor to dye pollution in water bodies. The current situation indicates that there is a growing need to use dyes, particularly azo dyes, to make objects more visually appealing. This is leading to a rise in the concentration of dyes in the water bodies where these effluents are disposed of. Among azo dyes, eriochrome black T (EBT) has been employed in many industrial applications. Consequently, a wide variety of colors may be eliminated from wastewater using an adsorbent that is efficient towards azo dyes. It is possible for a reactive azo dye to have one or more azo groups, which function as chromophores within the structure of the molecule (Ahmad and Hameed 2010). There are a number of studies that have been published in the scientific literature that demonstrate the use of photocatalytic treatment for the purpose of decontaminating colors in wastewater. In a comparative evaluation of its employment for photocatalytic degradation of organic dyes, researchers synthesized ZnO nanoparticles using various distinct ways (Saravanan et al. 2013a). Similarly, nanocatalysts containing CeO_2/V_2O_5 or CeO_2/CuO may be used to remediate methylene blue in sewage utilizing visible light (Saravanan et al. 2013b). The use of ultrasound-assisted Auramine-O (AO) contaminated wastewater by Asfaram et al. (2015) demonstrated the outstanding sorption potential of ZnS:Cu-NP-AC, with a maximum capacity of 183.15 mg g^{-1} for AO.

10.4.2.2 Pesticides

Household operations, forestry, farming, and other industrial processes (such as food, cosmetics, and textile production) are the biggest contributors to pesticide pollution (Ahmaruzzaman and Gupta 2011). Nevertheless, research has also demonstrated that pesticide contamination can also be detected in the atmosphere. As they float through the air, dust particles pick up pesticides from forests, farms, and other home activities. These pollutants eventually end up in our water, sediment, and soil sources. Numerous researchers are currently engaged in the development of strategies to mitigate pesticide contamination. Toxic pesticides like malathion and lindane are widely used and recommended in India (Gupta et al. 2002). Employing NPs for pesticide adsorption has shown to be the most successful and promising approach.

To effectively remove permethrin insecticide from water, Moradi et al. synthesized chitosan-zinc oxide NPs (CS-ZnONPs) composite beads (Moradi Dehaghi et al. 2014). To use NPs for adsorption, batch tests were conducted. The parameter optimization process yielded a maximum removal rate of 99% for the tested pesticide concentrations, adsorbent dosage, contact duration, and pH. A NaOH solution with a concentration of 0.1 M may be used to perform a complete regeneration of the adsorbent, after which it can be used once again for adsorption applications. From both an economic and ecological perspective, this renders these NPs crucial. Furthermore, endosulfan and chlorpyrifos seriously harm amphibians as well. 10% of male frogs grown in atrazine-contaminated water turned into females. Genetically male frogs phenotypically developed ovaries within their testes. They also acquired the propensity to produce viable eggs and mate with other males. Herbicide spraying in weedy fish nurseries gradually lowers the level of protection that juvenile fish need to hide from predators, which lowers the reproductive capacity of aquatic life forms.

After activating the alumina, a column of NPs was passed across the solution comprising pesticides, and a coating of gold and silver was applied. For a relatively longer period, this particular tool was employed for the goal of considerably reducing the extent of water pollution. When used in combination with the gold and silver solution for wastewater treatment, alumina is a more effective adsorbent for removing pesticides than when used alone.

10.4.2.3 Other Organic Contaminants

Volatile organic compounds, polyaromatic hydrocarbons, phenol, toluene, aromatic amines, humic acid, and other substances are among the additional organic pollutants. The ecosystem is negatively impacted by these organic contaminants. The brain and nervous system are affected right away by volatile organic chemicals like toluene. Continuous interaction with these pollutants can damage the lungs, liver, and kidneys. The primary application of aromatic compounds in industry is for the synthesis of chemicals. These polyaromatic hydrocarbons are also harmful because they affect the kidneys, liver, and eyes, as well as induce nausea, vomiting, cataracts, and other health problems. Due to the extreme toxicity of these chemicals, elimination is required.

10.4.3 Biological Pollutants

To determine the extent to which activated carbon treated with Ag and CuO NPs eliminated T4 bacteria, (Shimabuku et al. 2017) conducted a perpetual flow study in residence filters. The NP-based filter was subjected to a water flow rate of 450 mL min^{-1} throughout the experiment. Then, the filtered water was examined for bacteriophage and discharge of Ag and CuO. An extremely high molecular load has been demonstrated by the porous framework that included Ag and CuO for the bacteriophage. It has been determined that the quantity of Ag and CuO discharged into the water was exceptionally low, below the limits allowed for ingesting water. This research found that activated carbon treated with Ag and CuO NPs may be employed as a filter to eliminate viruses from potable water. Copper (Cu), silver (Ag), and iron (Fe) NPs were synthesized in an eco-friendly way utilizing extracts from black and green tea leaves (Asghar et al. 2018). Another group synthesized magnetic Fe_3O_4

NPs using both top-down (mechanical ball milling) and bottom-up (chemical reduction process) techniques. The sorption capability of the synthesized NPs for both gram-negative *Escherichia coli* and gram-positive *Staphylococcus aureus* bacteria has been demonstrated to be surface charge sensitive. For both bacteria, the sorption process on Fe_3O_4 NPs resembled the Langmuir isotherm model, with the pseudo-second-order kinetic model providing an excellent fit. Therefore, these NPs may be utilized as an inexpensive adsorbent alternative for the straightforward and quick elimination of hazardous germs and microorganisms from water.

10.5 Factors Affecting the Adsorption Process

The efficacy of liquid phase sorption, and hence the proper functioning of the water treatment facilities, is dependent on various characteristics. The sorption performance can be affected by physicochemical variables; the sorbent's features, and its functional groups. Additionally, the molecular characteristics, and reactivity of the adsorbate should be considered (Darabdhara et al. 2017). The amount of time needed to eliminate the impurity in a batch process could be affected by how quickly the water-based mixture is blended. A solid sample with known mass multiplied by a liquid phase with known composition allows the concentration to alter consistently until homeostasis is attained. The duration of this process, which can be significantly decreased through stirring, is ascertained through initial kinetic measurements. The quantity adsorbed can be determined by employing the initial and equilibrium compositions, as well as the material quantities (solid mass and liquid volume). The adsorbent, pollutant, and adsorption techniques used in the experiment all have an impact on the sorption rate. Generally, minimizing resistance to mass transfer by raising the rate will accelerate the biosorption removal rate of adsorbed contaminants; however, this may come at the expense of the biosorbent's material properties (Patel et al. 2022). Industrial water treatment sewage is multicomponent, unlike most laboratory research. There are plenty of harmful substances that can be found in industrial wastewater. These include acetic acid, ammonium sulfate, caustic soda, formic acid, hydrochloric acid, hydrogen peroxide, hydrosulfates, a leveling agent, organic reagent, organic solvent, oxalic acid, soap, softener, sulfuric acid, polyethylene emulsion, PV acetate, and wetting agent. Sewage contains many different types of pollutants, including pesticides, heavy metals, pharmaceutical residues, colors, and colloidal particles. All of these may have an effect on adsorption removal by competing for binding sites or interfering with other processes. The target contaminant's biosorption clearance efficiency tends to decrease as the concentrations of competing pollutants increase (Patel et al. 2024b). Adsorption process design needs to account for the impacts of all of these variables. The development of dye removal technologies on an industrial scale will be substantially facilitated by optimizing such circumstances. Data from studies conducted in the preceding years demonstrate that the variables (temperature, pH of the solution, adsorbent volume/particle size, and initial dye concentration) have the most impact. Overarching patterns have been identified from the acquired data, considering the variables' influences.

10.5.1 The Effect of Initial Dye Concentration

Due to its unintended impact on the efficacy of dye degradation by altering the abundance of binding sites on the sorbent surface, the initial concentration of the dye can serve as one of the key variables affecting the sorption process. Typically, as the original pigment dosage increases, the proportion of dye degradation reduces. In this instance, the ability of the sorbent improves in tandem with the initial concentration. This has occurred because large initial dye concentrations generate an important force for mass transfer. The adsorption process is propelled by the initial solute concentration, which facilitates mass transfer and diffusion from the solution (containing an increased quantity of dye) to the sorbent's free surface (Patel et al. 2024b).

10.5.2 The Effect of Solution pH

Several articles state that the solution's pH has been an extremely important variable in almost all sorption methods. Both the sorbent''s performance and the process's efficacy has been affected by this aspect. The pH can affect the solution chemistry, reactions of functional groups, the effect of coexisting ions, and the zeta potential of the sorbent. Aqueous medium pH may affect adsorbent characteristics, sorption procedure, and dye molecular dissociation. The pH of a solution may change the sorbent and the dye's structure chemically. Also, it can alter the ionization of ions and surface charge (Patel et al. 2024a).

10.5.3 The Effect of Adsorbent Dosage

One of the most significant parameters that have an impact on the adsorption procedure is the quantity of adsorbent. The dose of the adsorbent is an essential factor since it establishes the adsorbent capability for a certain initiating concentration (Wang et al. 2012). When the starting concentration remains constant, Kroeker's rule states that the specific adsorbed volume will decrease as the adsorbent mass increases. Dye extraction effectiveness and efficacy are strongly associated with rising adsorbent dosage. More active adsorption sites and surface area become accessible for contaminant removal when the dose of the adsorbent is increased, but the concentrations of the contaminants remain constant (Patel et al. 2023b).

The effectiveness of pollutant removal (E%) rises with increasing biomass concentration (adsorbent quantity), however, there is no direct relationship between biomass and pollutant removal (qe). This is because as the quantity of the biosorbent goes up, the shape of the sorption isotherm changes. The reduction in specific adsorbed quantity is most likely owing to the fact that particular surfaces or surface groups may not be covered in the higher concentrated solutions. It has been shown in multiple investigations that after efficiency approaches a saturation point, it gradually falls, sometimes by a modest amount. This phenomenon can be elucidated by the fact that maximal adsorption occurs at an identified adsorbent dose, after which the count of ions bound to the adsorbent and unbound ions remains uniform, regardless of the additional inclusion of the sorbent (Patel et al. 2024a).

10.5.4 The Effect of Adsorbent Particle Size

While particle size is not frequently examined in biosorption investigation, it can significantly impact diverse chemical reactions and sorption (Patel et al. 2024a). Particles of smaller sizes have a larger specific surface area. Factors influencing SSA include particle size, material composition, and opacity. Pollutant uptake and efficacy are enhanced when employing batch procedures with reduced particle sizes, which leads to an increased amount of active sites available for interaction. Since lighter particles float and so cannot establish interaction with the solution, the sorption capacity may be decreased if the diameter of the particles is too fragile. Biosorption makes it difficult to separate these small particles from water.

10.5.5 The Effect of Solution Temperature

Since changes in temperature may change an endothermic reaction into an exothermic one, or vice versa, temperature is a further vital physico-chemical element that impacts the treatment process. In addition, it may significantly alter the sorption rate, either increasing or decreasing it. Various sorbents and pollutants are susceptible to temperature's effect on sorption effectiveness. While it generally improves the biosorption of adsorbed contaminants by raising the adsorbate's kinetic energy and surface activity, it also has the potential to harm the biosorbent's chemical composition.

The sorption process (efficacy) diminishes as temperature increases (exotherm). One possible explanation is that when the temperature increases, the sorptive forces between the adsorbate and the sorbent's active sites weaken, leading to a drop in the dye elimination rate. Because the dye ions become more mobile when heat is introduced to the environment, exothermic sorption is often used for diffusion regulation.

The endotherm states that the sorption process (efficacy) improves as temperature rises. This is because the sorbent surface becomes more activated at higher temperatures, rendering more active sites available. One possible explanation for the observed rise in sorption capability levels as a function of temperature is that the substantial dye ions become more mobile at higher temperatures (Patel et al. 2023d).

In general, if the rate of sorption increases with increasing temperature, it might mean that the process is endothermic, while it is exothermic at reduced temperatures. Estimations of thermodynamic coefficients are performed once the influence of initial temperature on sorption has been carried out. It is widely recognized that the sorption processes are significantly influenced by their operating temperature, which is regulated by thermal variables such as the standard enthalpy change (DH0, J mol^{-1}), the standard entropy change (DS0, J mol^{-1}), and the standard free Gibbs energy change (DG0, J mol^{-1}) of the sorption processes. The Gibbs–Helmholtz equation is used to determine these values. The values of DG, DH, and DS are dependent on the final and initial states of the sorption system since enthalpy, DG, and DS are all state functions. Since these thermal quantities have extended properties, it is important to devote careful consideration to the quantity of material that Gibbs free energy, enthalpy, and entropy correlate with.

When the temperature increases during dye molecule sorption, the entropy (DS) and enthalpy (DH) values might go up or down, respectively. Prior to sorption, molecules have free movement in all three dimensions; however, once adsorbed, their mobility is constrained to the surface, and their disarray diminishes; as a consequence, the entropy falls, suggesting that the process is exothermic. This could also be because dyes were more easily dissolved at higher temperatures, but interactions between the sorbate and the sorbent weakened, which led to less sorption. Endothermic processes are characterized by an increase in both entropy and enthalpy.

10.5.6 Activation of Solid Sorbent, Surface Modification

To enhance the efficacy and uptake capacity, a variety of chemical as well as physical surface modification techniques may be adopted. The most prevalent physical alteration processes are freezing, crushing, boiling/heating, and drying. Typically, these surface modification methods result in the destruction of the biomass's cell membrane, which liberates cellular components that could potentially facilitate the absorption of contaminants. Most physical methods for alteration are easy to use and rarely expensive, but they fail to succeed compared to chemical methods. Polymerization, functional group modification, and pretreatment are some of the chemical modification strategies that are currently under investigation. Washing is the most popular chemical treatment since it is easy to use and produces good results. Washing biomass with acid, alkali, or detergent, or crosslinking it with organic solvents is two of the most typical chemical preparatory procedures. After undergoing basic processing steps like cutting or grinding, some sorbents generate biosorbent particles that are both stable and effective. To get particles with appropriate mechanical characteristics, it is often necessary to embed the sorbent in an artificial polymer matrix or transplant it onto a natural carrier substance like silica. A variety of approaches to modulating biomass adsorbents to enhance distinct facets of biosorption have been expounded upon by numerous researchers.

10.6 Regeneration of Adsorbent

An essential factor in determining and justifying the cost-effectiveness of any utilized sorption medium is its ability to regenerate ability. The versatility and utility of any sorbent in water treatment methods depend substantially on how easily it can be reproduced. Because of the obvious economic and ecological benefits, any adsorbent that can be readily recycled will perpetually take precedence. Researchers synthesized Fe@Au bimetallic NPs with Graphene Oxide (Fe@Au-ATPGO) and utilized them to catalyze the decomposition of 2,4-nitrophenol (Gupta et al. 2014). He demonstrated that the nanoparticles may be recycled at least ten times without a discernible drop in stable conversion effectiveness and can be revived in fifteen seconds employing merely a magnetic field.

Fe/Cu NP revival after As (III) and As (V) adsorption was also investigated (Babaee et al. 2018). For recuperation, they utilized NaOH solutions of 0.10 and 0.01 mol L^{-1}. Researchers found that sorption was more effective in acidic media while desorption was much easier in basic ones. Using 0.10 mol L^{-1} NaOH resulted in the complete desorption of As (V) and 70.3% of As (III). When NaOH was used at a concentration of 0.01 mol L^{-1}, these levels decreased to 79.4% for As (III) and 57.5% for As (V), respectively. Additionally, heavy metal desorption was conducted on the surface of a novel and extremely effective superparamagnetic nanosorbent (Ghasemi et al. 2017). The heavy metals were desorbed using eluents that included varying amounts of HCl (0.001, 0.01, 0.1, 0.5, and 1 mol L^{-1}). The metal ions were liberated when the coordinated cations (heavy metal ions) on EDTA were replaced with H^+ ions in an acidic solution. Regeneration of EDTA functionalized Fe_3O_4 NPs may achieve a maximum of 98% when heavy metal ions are removed simultaneously using 0.5 mol L^{-1} HCl as the eluent. The reuse of Fe_3O_4 NPs functionalized with EDTA was assessed during 30 sequential sorption/desorption cycles. In a similar vein, MNP-MWCNTs were revived by the use of a NaOH solution (0.5 mol L^{-1}). The MNP-MWCNTs were prepared for reuse after desorption by using an external magnetic field to separate them. The sorbents are effective as well as affordable in water treatment because they are simple to reclaim. Soil, spinach, tomatoes, black tea, tobacco, and other water samples were effectively and efficiently used to recover (by adsorption) nickel (II) using immobilized dimethyl glyoxim/sodium dodecyl sulfate immobilized on alumina-coated magnetite NPs (DMG/SDSACMNPs) (Karimi and Kafi 2015). Direct decantation of the magnetite sorbents was achieved after effective magnet recovery. In addition, the eluent of 0.1 mol L^{-1} HNO_3 solution was effective for the maximum extraction of Ni (II). The effectiveness of these may be maintained even after three consecutive uses. Effective disposal of spent sorbents is required by consumers due to their toxicity and the dangers they pose to health and the environment. Regenerating sorbents is the most effective method for controlling their disposal. On the other side, the primary contaminants are the recovered organic pollutants. After filling up certain steel cylinders, they should be buried at great depths.

10.7 Conclusion

Efficient elimination of organic, inorganic, and biological contaminants from wastewater has been achieved by the use of NPs. The efficacy of the sorption approach has been shown to be consistent across all classes of water contaminants. The literature reports that a wide range of NPs, synthesized using simple and environmentally sound methods, are effective for most contaminants. The NPs have the ability to eliminate contaminants even at low levels (e.g., $\mu g\ L^{-1}$). Thermodynamics and pH are two main factors that influence the rate and efficiency of contaminant sorption on NPs. Since most sorption equilibriums were reached within 1–15 min, it is evident from previous research that NP-based sorption is very rapid. Another key element for the broad usage of NPs, both economically and ecologically, is their ability to regenerate. However, due to their high regeneration capacity after sorption and simplicity of synthesis with high yield, metal-based NPs (such as Ag, Fe, Mn, Cu, and Al-NPs) are the most favored and often synthesized NPs. Compared to other iron-based materials, iron NPs are better at eliminating or reducing toxins in both on-site and off-site environmental engineering contexts. Notably, another important issue is the handling of recovered contaminants and used sorbents. The post-regeneration process disposal of spent sorbents and recovered pollutants is mostly unexplored in the existing literature. To solve this problem, experts need to come up with ways to handle waste that are better for the environment. Batch trials, which were the most common method for studying sorption, often produced promising findings. Nevertheless, they were only applicable in a controlled environment in the laboratory. Researchers and water treatment companies must work together to bring these approaches from the lab to the real world, where they may be used on a broad scale. We can only have hope of finding a global solution if that happens, but there were enough citations in the study to show that wastewater treatment using nanosorbents is not only relatively fast and effective, but also environmentally benign, flexible, and easy to scale up to a big scale. Nevertheless, there are plenty of opportunities for improvement in terms of both cost and efficacy via more study and development.

References

Ahmad AA, Hameed BH (2010) Fixed-bed adsorption of reactive azo dye onto granular activated carbon prepared from waste. J Hazard Mater 175:298–303. https://doi.org/10.1016/j.jhazmat.2009.10.003

Ahmaruzzaman M, Gupta VK (2011) Rice husk and its ash as low-cost adsorbents in water and wastewater treatment. Ind Eng Chem Res 50:13589–13613. https://doi.org/10.1021/ie201477c

Al Nafiey A, Addad A, Sieber B, Chastanet G, Barras A, Szunerits S, Boukherroub R (2017) Reduced graphene oxide decorated with Co_3O_4 NPs (rGO-Co_3O_4) nanocomposite: a reusable catalyst for highly efficient reduction of 4-nitrophenol, and Cr(VI) and dye removal from aqueous solutions. Chem Eng J 322:375–384. https://doi.org/10.1016/j.cej.2017.04.039

Anwar J, Shafique U, Waheed-uz-Zaman SM, Dar A, Anwar S (2010) Removal of Pb(II) and Cd(II) from water by adsorption on peels of banana. Bioresour Technol 101:1752–1755. https://doi.org/10.1016/j.biortech.2009.10.021

Asfaram A, Ghaedi M, Agarwal S, Tyagi I, Gupta VK (2015) Removal of basic dye Auramine-O by ZnS: Cu NPs loaded on activated carbon: optimization of parameters using response surface methodology with central composite design. RSC Adv 5:18438–18450. https://doi.org/10.1039/c4ra15637d

Asghar MA, Zahir E, Shahid SM, Khan MN, Asghar MA, Iqbal J, Walker G (2018) Iron, copper and silver NPs: green synthesis using green and black tea leaves extracts and evaluation of antibacterial, antifungal and aflatoxin B1 adsorption activity. LWT 90:98–107. https://doi.org/10.1016/j.lwt.2017.12.009

Babaee Y, Mulligan CN, Rahaman MS (2018) Removal of arsenic (III) and arsenic (V) from aqueous solutions through adsorption by Fe/Cu NPs. J Chem Technol Biotechnol 93:63–71. https://doi.org/10.1002/jctb.5320

Beyersmann D, Hartwig A (2008) Carcinogenic metal compounds: recent insight into molecular and cellular mechanisms. Arch Toxicol 82:493–512

Brammer H, Ravenscroft P (2009) Arsenic in groundwater: a threat to sustainable agriculture in South and South-east Asia. Environ Int 35:647–654

Bushra R (2018) Nanoadsorbents-based polymer nanocomposite for environmental remediation. In: New polymer nanocomposites for environmental remediation. Elsevier Inc., pp 243–260

Cao YR, Liu Z, Cheng GL, Jing XB, Xu H (2010) Exploring single and multi-metal biosorption by immobilized spent Tricholoma lobayense using multi-step response surface methodology. Chem Eng J 164:183–195. https://doi.org/10.1016/j.cej.2010.08.055

Chen LF, Liang HW, Lu Y, Cui CH, Yu SH (2011) Synthesis of an attapulgite clay@carbon nanocomposite adsorbent by a hydrothermal carbonization process and their application in the removal of toxic metal ions from water. Langmuir 27:8998–9004. https://doi.org/10.1021/la2017165

Chen K, He J, Li Y, Cai X, Zhang K, Liu T, Hu Y, Lin D, Kong L, Liu J (2017) Removal of cadmium and lead ions from water by sulfonated magnetic NP adsorbents. J Colloid Interface Sci 494:307–316. https://doi.org/10.1016/j.jcis.2017.01.082

Chowdhury S, Balasubramanian R (2014) Recent advances in the use of graphene-family nanoadsorbents for removal of toxic pollutants from wastewater. Adv Colloid Interface Sci 204:35–56

Darabdhara G, Boruah PK, Hussain N, Borthakur P, Sharma B, Sengupta P, Das MR (2017) Magnetic NPs towards efficient adsorption of gram positive and gram negative bacteria: an investigation of adsorption parameters and interaction mechanism. Colloids Surf A Physicochem Eng Asp 516:161–170. https://doi.org/10.1016/j.colsurfa.2016.12.003

Depault F, Cojocaru M, Fortin F, Chakrabarti S, Lemieux N (2006) Genotoxic effects of chromium(VI) and cadmium(II) in human blood lymphocytes using the electron microscopy in situ end-labeling (EM-ISEL) assay. Toxicol in Vitro 20:513–518. https://doi.org/10.1016/j.tiv.2005.09.003

Diallo MS, Christie S, Swaminathan P, Johnson JH, Goddard WA (2005) Dendrimer enhanced ultrafiltration. 1. recovery of Cu(II) from aqueous solutions using PAMAM dendrimers with ethylene diamine core and terminal NH 2 groups. Environ Sci Technol 39:1366–1377. https://doi.org/10.1021/es048961r

Egger S, Lehmann RP, Height MJ, Loessner MJ, Schuppler M (2009) Antimicrobial properties of a novel silver-silica nanocomposite material. Appl Environ Microbiol 75:2973–2976. https://doi.org/10.1128/AEM.01658-08

El-sayed MEA (2020) Nanoadsorbents for water and wastewater remediation. Sci Total Environ 739:139903. https://doi.org/10.1016/j.scitotenv.2020.139903

Fan HL, Zhou SF, Jiao WZ, Qi GS, Liu YZ (2017) Removal of heavy metal ions by magnetic chitosan NPs prepared continuously via high-gravity reactive precipitation method. Carbohydr Polym 174:1192–1200. https://doi.org/10.1016/j.carbpol.2017.07.050

Fazlzadeh M, Rahmani K, Zarei A, Abdoallahzadeh H, Nasiri F, Khosravi R (2017) A novel green synthesis of zero valent iron NPs (NZVI) using three plant extracts and their efficient application for removal of Cr(VI) from aqueous solutions. Adv Powder Technol 28:122–130. https://doi.org/10.1016/j.apt.2016.09.003

Gehrke I, Geiser A, Somborn-Schulz A (2015) Innovations in nanotechnology for water treatment. Nanotechnol Sci Appl 8

Ghasemi E, Heydari A, Sillanpää M (2017) Superparamagnetic Fe3O4@EDTA NPs as an efficient adsorbent for simultaneous removal of Ag(I), Hg(II), Mn(II), Zn(II), Pb(II) and Cd(II) from water and soil environmental samples. Microchem J 131:51–56. https://doi.org/10.1016/j.microc.2016.11.011

Gupta VK, Saleh TA (2013) Sorption of pollutants by porous carbon, carbon nanotubes and fullerene–an overview. Environ Sci Pollut Res 20:2828–2843

Gupta VK, Agarwal S, Saleh TA (2011) Synthesis and characterization of alumina-coated carbon nanotubes and their application for lead removal. J Hazard Mater 185:17–23. https://doi.org/10.1016/j.jhazmat.2010.08.053

Gupta VK, Ali I, Saleh TA, Siddiqui MN, Agarwal S (2013) Chromium removal from water by activated carbon developed from waste rubber tires. Environ Sci Pollut Res 20:1261–1268. https://doi.org/10.1007/s11356-012-0950-9

Gupta VK, Atar N, Yola ML, Üstündağ Z, Uzun L (2014) A novel magnetic Fe@Au core-shell NPs anchored graphene oxide recyclable nanocatalyst for the reduction of nitrophenol compounds. Water Res 48:210–217. https://doi.org/10.1016/j.watres.2013.09.027

Gupta VK, Chandra R, Tyagi I, Verma M (2016) Removal of hexavalent chromium ions using CuO NPs for water purification applications. J Colloid Interface Sci 478:54–62. https://doi.org/10.1016/j.jcis.2016.05.064

Gupta VK, Jain CK, Ali I, Chandra S, Agarwal S (2002) Removal of lindane and malathion from wastewater using bagasse fly ashFa sugar industry waste

Huang ZH, Zheng X, Lv W, Wang M, Yang QH, Kang F (2011) Adsorption of lead(II) ions from aqueous solution on low-temperature exfoliated graphene nanosheets. Langmuir 27:7558–7562. https://doi.org/10.1021/la200606r

Ji L, Chen W, Duan L, Zhu D (2009) Mechanisms for strong adsorption of tetracycline to carbon nanotubes: a comparative study using activated carbon and graphite as adsorbents. Environ Sci Technol 43:2322–2327. https://doi.org/10.1021/es803268b

Karimi MA, Kafi M (2015) Removal, preconcentration and determination of Ni(II) from different environmental samples using modified magnetite NPs prior to flame atomic absorption spectrometry. Arab J Chem 8:812–820. https://doi.org/10.1016/j.arabjc.2013.05.018

Keshavarz A, Parang Z, Nasseri A (2013) The effect of sulfuric acid, oxalic acid, and their combination on the size and regularity of the porous alumina by anodization. J Nanostructure Chem 3. https://doi.org/10.1186/2193-8865-3-34

Kumari P, Alam M, Siddiqi WA (2019) Usage of NPs as adsorbents for waste water treatment: An emerging trend. Sustain Mater Technol 22:e00128. https://doi.org/10.1016/j.susmat.2019.e00128

Li J, Chen C, Zhu K, Wang X (2016) Nanoscale zero-valent iron particles modified on reduced graphene oxides using a plasma technique for Cd(II) removal. J Taiwan Inst Chem Eng 59:389–394

Luo S, Lu T, Peng L, Shao J, Zeng Q, Gu JD (2014) Synthesis of nanoscale zero-valent iron immobilized in alginate microcapsules for removal of Pb(ii) from aqueous solution. J Mater Chem A Mater 2:15463–15472. https://doi.org/10.1039/c4ta02920h

Martinez-Vargas S, Martínez AI, Hernández-Beteta EE, Mijangos-Ricardez OF, Vázquez-Hipólito V, Patiño-Carachure C, Hernandez-Flores H, López-Luna J (2017) Arsenic adsorption on cobalt and manganese ferrite NPs. J Mater Sci 52:6205–6215. https://doi.org/10.1007/s10853-017-0852-9

Moradi Dehaghi S, Rahmanifar B, Moradi AM, Azar PA (2014) Removal of permethrin pesticide from water by chitosan-zinc oxide NPs composite as an adsorbent. J Saudi Chem Soc 18:348–355. https://doi.org/10.1016/j.jscs.2014.01.004

Nagy A, Harrison A, Sabbani S, Munson RS, Dutta PK, Waldman WJ (2011) Silver NPs embedded in zeolite membranes: release of silver ions and mechanism of antibacterial action. Int J Nanomedicine 6:1833–1852. https://doi.org/10.2147/ijn.s24019

Nasernejad B, Zadeh TE, Pour BB, Bygi ME, Zamani A (2005) Camparison for biosorption modeling of heavy metals (Cr (III), Cu (II), Zn (II)) adsorption from wastewater by carrot residues. Process Biochem 40:1319–1322. https://doi.org/10.1016/j.procbio.2004.06.010

Novoselov KS, Geim AK, Morozov SV, Jiang D, Zhang Y, Dubonos SV, Grigorieva IV, Firsov AA (2000) Electric field effect in atomically thin carbon films. Kluwer

Pacheco S, Medina M, Valencia F, Tapia J (2006) Removal of inorganic mercury from polluted water using structured NPs. J Environ Eng 132:342–349. https://doi.org/10.1061/(asce)0733-9372(2006)132:3(342)

Pan B, Xing B (2008) Adsorption mechanisms of organic chemicals on carbon nanotubes. Environ Sci Technol 42:9005–9013

Pan B, Lin D, Mashayekhi H, Xing B (2008) Adsorption and hysteresis of bisphenol A and 17α-ethinyl estradiol on carbon NMs. Environ Sci Technol 42:5480–5485. https://doi.org/10.1021/es8001184

Patel PK, Pandey LM, Uppaluri RVS (2024a) Cyclic adsorption and desorption characteristics of citric acid-chitosan variant resins for Pb, Fe, and Zn removal from simulated mining and agricultural wastewater system. J Polym Environ. https://doi.org/10.1007/s10924-024-03343-0

Patel PK, Nagireddi S, Uppaluri RVS, Pandey LM (2022) Batch adsorption characteristics of Dowex Marathon MSA commercial resin for Au(III) removal from synthetic electroless plating solutions. In: Materials today: Proceedings. Elsevier Ltd, pp 824–829

Patel PK, Pandey LM, Uppaluri RVS (2023) Cyclic desorption based efficacy of polyvinyl alcohol-chitosan variant resins for multi heavy-metal removal. Int J Biol Macromol 242. https://doi.org/10.1016/j.ijbiomac.2023.124812

Patel PK, Pandey LM, Uppaluri RVS (2023) Synthesized carboxymethyl-chitosan variant composites for cyclic adsorption-desorption based removal of Fe, Pb, and Cu. Chemosphere 340. https://doi.org/10.1016/j.chemosphere.2023.139780

Patel PK, Pandey LM, Uppaluri R (2023) Multi-metal adsorption and cyclic desorption characteristics of Zn^{+2} and Cu^{+2} constituting multi-component synthetic wastewater system using commercial resins. In: Deka D, Majumder SK, Purkait MK (eds) Sustainable environment. NERC 2022. Springer, Singapore. https://doi.org/10.1007/978-981-19-8464-8_1

Patel PK, Pandey LM, Uppaluri RVS (2023) Adsorptive removal of Zn, Fe, and Pb from Zn dominant simulated industrial wastewater solution using polyvinyl alcohol grafted chitosan variant resins. Chemical Engineering Journal 459. https://doi.org/10.1016/j.cej.2023.141563

Patel PK, Pandey LM, Uppaluri RVS (2024) Highly effective removal of multi-heavy metals from simulated industrial effluent through an adsorption process employing carboxymethyl-chitosan composites. Environ Res 240. https://doi.org/10.1016/j.envres.2023.117502

Qu X, Brame J, Li Q, Alvarez PJJ (2013a) Nanotechnology for a safe and sustainable water supply: enabling integrated water treatment and reuse. Acc Chem Res 46:834–843. https://doi.org/10.1021/ar300029v

Qu X, Alvarez PJJ, Li Q (2013b) Applications of nanotechnology in water and wastewater treatment. Water Res 47:3931–3946. https://doi.org/10.1016/j.watres.2012.09.058

Sadegh H, Ali GAM, Gupta VK, Makhlouf ASH, Shahryari-ghoshekandi R, Nadagouda MN, Sillanpää M, Megiel E (2017) The role of NMs as effective adsorbents and their applications in wastewater treatment. J Nanostructure Chem 7:1–14

Sadeghi-Kiakhani M, Arami M, Gharanjig K (2013) Dye removal from colored-textile wastewater using chitosan-PPI dendrimer hybrid as a biopolymer: optimization, kinetic, and isotherm studies. J Appl Polym Sci 127:2607–2619. https://doi.org/10.1002/app.37615

Santhosh C, Velmurugan V, Jacob G, Jeong SK, Grace AN, Bhatnagar A (2016) Role of NMs in water treatment applications: a review. Chem Eng J 306:1116–1137. https://doi.org/10.1016/j.cej.2016.08.053

Saravanan R, Gupta VK, Narayanan V, Stephen A (2013a) Comparative study on photocatalytic activity of ZnO prepared by different methods. J Mol Liq 181:133–141. https://doi.org/10.1016/j.molliq.2013.02.023

Saravanan R, Joicy S, Gupta VK, Narayanan V, Stephen A (2013b) Visible light induced degradation of methylene blue using CeO_2/V_2O_5 and CeO_2/CuO catalysts. Mater Sci Eng, C 33:4725–4731. https://doi.org/10.1016/j.msec.2013.07.034

Savasari M, Emadi M, Bahmanyar MA, Biparva P (2015) Optimization of Cd (II) removal from aqueous solution by ascorbic acid-stabilized zero valent iron NPs using response surface methodology. J Ind Eng Chem 21:1403–1409. https://doi.org/10.1016/j.jiec.2014.06.014

Sen Gupta S, Bhattacharyya KG (2012) Adsorption of heavy metals on kaolinite and montmorillonite: a review. Phys Chem Chem Phys 14:6698–6723

Shimabuku QL, Arakawa FS, Fernandes Silva M, Ferri Coldebella P, Ueda-Nakamura T, Fagundes-Klen MR, Bergamasco R (2017) Water treatment with exceptional virus inactivation using activated carbon modified with silver (Ag) and copper oxide (CuO) NPs. Environ Technol (United Kingdom) 38:2058–2069. https://doi.org/10.1080/09593330.2016.1245361

Sitko R, Zawisza B, Malicka E (2013) Graphene as a new sorbent in analytical chemistry. TrAC–Trends Anal Chem 51:33–43

Tabesh S, Davar F, Loghman-Estarki MR (2018) Preparation of γ-Al_2O_3 NPs using modified sol-gel method and its use for the adsorption of lead and cadmium ions. J Alloys Compd 730:441–449. https://doi.org/10.1016/j.jallcom.2017.09.246

Tiwari D (2008) Sen P (2008) Application of NPs in waste water treatment. World Appl Sci J 3(3):417–433

Verma R, Dwivedi P (2013) Heavy metal water pollution-a case study. Science and Technology 5:98–99

Wang S, Ng CW, Wang W, Li Q, Li L (2012) A comparative study on the adsorption of acid and reactive dyes on multiwall carbon nanotubes in single and binary dye systems. J Chem Eng Data 57:1563–1569. https://doi.org/10.1021/je3001552

Wang X (2012) NMs as sorbents to remove heavy metal ions in wastewater treatment. J Environ Anal Toxicol 02. https://doi.org/10.4172/2161-0525.1000154

Wang S, Shi X (2001) Molecular mechanisms of metal toxicity and carcinogenesis

Xiong C, Wang W, Tan F, Luo F, Chen J, Qiao X (2015) Investigation on the efficiency and mechanism of Cd(II) and Pb(II) removal from aqueous solutions using MgO NPs. J Hazard Mater 299:664–674. https://doi.org/10.1016/j.jhazmat.2015.08.008

Yang K, Xing B (2007) Desorption of polycyclic aromatic hydrocarbons from carbon NMs in water. Environ Pollut 145:529–537. https://doi.org/10.1016/j.envpol.2006.04.020

Yang K, Xing B (2010) Adsorption of organic compounds by carbon NMs in aqueous phase: Polanyi theory and its application. Chem Rev 110:5989–6008. https://doi.org/10.1021/cr100059s

Zare EN, Motahari A, Sillanpää M (2018) Nanoadsorbents based on conducting polymer nanocomposites with main focus on polyaniline and its derivatives for removal of heavy metal ions/dyes: a review. Environ Res 162:173–195

Chapter 11
A Sustainable Approach to Wastewater Remediation Using Nano-adsorbents

Mukesh Bharti, Satish Kumar, and Mihir Kumar Purkait

Abstract Addressing the pressing issue of wastewater pollution, this chapter delves into a sustainable approach for remediation by leveraging the capabilities of nano-adsorbents. Although, various conventional techniques such as ion exchange, coagulation, membrane filtration, chemical precipitation, and electrocoagulation are used for the treatment of wastewater. Nowadays, nano-adsorbents are being used as emerging materials for the effective removal of contaminants due to their high surface area and reactivity, which makes them ideal for adsorbing pollutants and facilitating chemical reactions, leading to pollutant degradation. Additionally, their unique properties and environmentally conscious design make them valuable tools for extracting heavy metals like Chromium (Cr), Mercury (Hg), Cadmium (Cd), Zinc (Zn), Lead (Pb), and Arsenic (As) from wastewater. The increased concentration of these carcinogenic heavy metals in wastewater is owing to harmful environmental impacts. However, this chapter discusses the synthesis and properties of these nano-adsorbents, emphasizing their eco-friendly and sustainable nature. Furthermore, the application of these materials in wastewater treatment processes is explored, highlighting their enhanced adsorption efficiency and selectivity. The environmentally conscious aspect of utilizing nano-adsorbents as a viable solution for wastewater remediation is underscored, emphasizing the potential for mitigating pollution while adhering to sustainability principles. Moreover, through a comprehensive examination of the current research landscape, this chapter aims to contribute valuable insights into the development and implementation of sustainable strategies for wastewater remediation.

Keywords Wastewater · Heavy metals · Adsorption · Nano-adsorbents · Nanomaterials · Sustainable

---

M. Bharti · S. Kumar · M. K. Purkait (✉)
Department of Chemical Engineering, Indian Institute of Technology, Guwahati, Assam, India
e-mail: mihir@iitg.ac.in

N. Agarwal et al. (eds.), *Nanomaterials in Wastewater Research*, Advances in Wastewater Research, https://doi.org/10.1007/978-981-96-3908-3_11

11.1 Introduction

Water, a crucial element for life on Earth, is plentiful and essential. Approximately 70% of freshwater resources are utilized in agriculture, underscoring its pivotal role in the global economy. The world is currently facing a significant environmental challenge in the form of water scarcity, while water contamination has emerged as a global issue with potential repercussions on human health, as well as the well-being of plants, animals, and trees (Yousefi et al. 2018). The utilization of recycled wastewater is increasingly acknowledged as a viable strategy to preserve and broaden the range of accessible water resources amid dwindling freshwater reserves. The reuse of wastewater finds diverse applications, including artificial groundwater recharge, aquaculture, industrial purposes, and various recreational and environmental activities. Recycled wastewater can typically serve the same purposes as fresh water. Therefore, it is important to evaluate certain factors and criteria such as electrical conductivity (EC), levels of dissolved organic matter, post-treatment, and the presence of heavy metals before utilizing wastewater (Rasheed et al. 2019). Nowadays, nanotechnology has opened up possibilities for the creation of advanced materials that can enhance the effectiveness of water and wastewater treatment techniques. These materials include nano-catalysts, modified surfaces, membranes (Deepti et al. 2023), coatings, adsorbents, and chemicals. However, among all the materials, nano-adsorbents are considered the best option for water and wastewater remediation. This is because of their wide availability and ease of use. Nanomaterials (NMs) have proven to be highly effective in purifying water and wastewater owing to their specific features, viz. large surface area and adsorption capability. A study examined by Jiang et al. revealed that NMs have immense potential as the best solution for treating organic and inorganic contaminants. Moreover, the relationship between contaminants and nano-adsorbent materials may involve either chemisorption or physisorption, which is influenced by the functionalization of the NMs (Jiang et al. 2018). Numerous research studies have explored the interaction between nano-sorbent materials and both organic and inorganic contaminants. For example, Uddin et al. investigated the elimination of methyl orange dye using cobalt-oxide nanoparticles (Uddin and Baig 2019), while Dehghani et al. focused on the degradation of diazinon pesticides from water by utilizing carbon nanotubes (Dehghani et al. 2019). Additionally, Khan et al. demonstrated the effectiveness of Fe-Cu binary oxides in eliminating hexavalent chromium from water (Khan et al. 2016), and Mandal et al. highlighted the importance of a novel hybrid material in eliminating arsenic (III) from water (Mandal et al. 2011). Conversely, Kumari et al. showcased the removal of fluoride using activated alumina modified with zirconium and calcium (Kumari et al. 2020). Furthermore, numerous investigations have explored the use of nano-adsorbent materials in wastewater remediation processes. These studies have employed a range of NM sorbents, including metal oxide-based nanoparticles (NPs) such as iron, magnesium, zinc, and titanium oxides, plant nanocomposites (NCs), carbon nanotubes (CNTs) (Singh et al. 2020), and graphite, to eliminate contaminants. Activated carbon presents several limitations, including a restricted ability to regenerate, which can lead to toxicity concerns for different organisms, and its use and reusability often result in significant costs. Therefore, to mitigate expenses and create environmentally friendly adsorbents with improved effectiveness, the adoption of green adsorbents offers significant benefits due to their cost-effectiveness and environmentally sustainable characteristics such as biodegradability, compatibility with biological systems, and renewability. Kumar et al. illustrated that green adsorbents are extensively more efficient than commercial activated carbon (AC) in the degradation of heavy metals from water (Kumar et al. 2023a). Biosorbents, including algae, bacteria, fungi, and yeasts, offer the potential to mitigate water pollutants through a range of biosorption mechanisms, involving the binding of materials derived from diverse biomass sources. The substantial increase in waste generation from various sources has led to their use for environmental objectives (Kumar et al. 2023a). Adsorption has gained significant popularity as a treatment method for water and wastewater, primarily because of its ability to efficiently remove harmful contaminants from industrial and aqueous settings (Changmai and Purkait 2021). To enhance the eco-friendliness and cost-effectiveness of this process, researchers are utilizing waste materials such as agricultural resources to develop adsorbents. Various techniques are being utilized to prepare, modify, and characterize these environmentally friendly adsorbents (El-sayed 2020). However, this chapter discusses the synthesis and properties of various nano-adsorbents, emphasizing their eco-friendly and sustainable nature. Furthermore, the application of these materials in wastewater treatment processes is explored, highlighting their enhanced adsorption efficiency and selectivity.

11.2 Sources of Wastewater and Their Environmental Impacts on Human Health

Wastewater is generated from various sources, including residential, commercial, and industrial activities, as well as surface and groundwater reservoirs. The origin of wastewater plays a crucial role in shaping its composition and the required technique for treatment. For instance, household wastewater typically includes pollutants like organic substances, fats, detergent residues, and particles from both residential and commercial sources. In contrast, industrial wastewater may contain hazardous elements such as noxious chemicals, metals, organic compounds, radioactive materials, substantial sediment loads, and waste products with high temperatures or acidic/basic properties. Contaminants found in water and wastewater are categorized as inorganic, biological, or organic pollutants based on their inherent properties and origins (Dhara et al. 2023).

Inorganic pollutants: According to Abdel-Halim et al., heavy metals present in mine drainage, river sediments, ashes, electronic trash, and industrial wastewater are classified as inorganic pollutants. Heavy metals include cations like chromium, iron, cadmium, copper, lead, mercury, zinc, and arsenic, while anions such as phosphates, sulfates, nitrates, fluorides, chlorides, and oxalates are common inorganic pollutants found in water (Abdel-Halim and Al-Deyab 2011).

Organic pollutants: Organic pollutants are known for their high environmental stability, which is due to the presence of one or more cyclic rings in their chemical structure. Additionally, they lack polar functional groups and have only a limited number of halogen substitutions. An extensive range of substances is included in this category of toxic compounds, which also include fertilizers, plasticizers, formaldehyde, greases, detergents, hydrocarbons, oils, pesticides (including herbicides, fungicides, and insecticides), phenols, fertilizers, formaldehyde, fertilizers, and more. Agricultural and chemical companies are major producers of these organic contaminants.

Biological pollutants: Numerous biological species, including bacteria, fungi, viruses, amoebas, algae, and other parasites, can cause a variety of diseases and lead to water contamination. Significant health problems such as kidney damage, nausea, rheumatoid arthritis, persistent asthma, disorders of the brain and circulatory systems, and, in extreme cases, death are caused by these chemicals in water (Barakat 2011).

11.2.1 Industrial Wastewater

Water plays a vital role in industrial activities, with its accessibility, purity, and dependability profoundly influencing business performance and financial viability. Various sectors such as manufacturing, power generation, agriculture, mining, and oil and gas production rely on water for essential processes. For industries to ensure the sustainability and longevity of their operations, it is therefore essential that they manage water resources properly. Water is essential to the operations of many different industries. For example, the dyeing, printing, and finishing operations used in the textile industry require large amounts of water. According to these operations, a single cotton T-shirt manufacturing might use more than 2500 L of water. According to Wang et al., the paper sector is among the biggest industrial water consumers because the process of producing paper requires bleaching and pulping, all of which require significant amounts of water. Similar to this, significant amounts of water are needed for chemical reactions, solvent extraction, cooling, and petrochemical refining and pharmaceutical production operations. Industrial effluent poses a serious environmental risk because of its high contaminant content, which can harm public health and water purity. As per the United Nations Environment Program (2017), the issue is becoming more serious owing to numerous factors, including the ongoing growth of established companies and the creation of new ones (Man et al. 2018).

The growth in manufacturing operations is associated with a surge in wastewater production, which raises concerns about water pollution. In addition, a lot of companies struggle with inadequate technology and infrastructure for treating wastewater, which results in the discharge of partially or fully treated effluents into water sources. Significant contaminants are present in this circumstance, which could harm the environment and human health. To meet these issues, organizations are implementing a range of wastewater treatment technologies, such as chemical, biological, and physical techniques. Verstraete et al. have identified several common biological treatment techniques, including anaerobic digestion, trickling filters, and activated sludge. Furthermore, coagulants, flocculants, and oxidants are used in chemical treatments to extract contaminants from wastewater (Verstraete et al. 2022). According to Zou et al., these chemical reactions include adsorption, oxidation, and precipitation. Furthermore, physical methods for pollutant removal include membrane separation, filtration, and sedimentation (Zou et al. 2021). Furthermore, for wastewater treatment to target specific contaminants, state-of-the-art technologies such as reverse osmosis, ultrafiltration, and ion exchange are crucial (Singh et al. 2023).

11.2.2 Agricultural Wastewater

Since the middle of the twentieth century, the adoption of intensive agriculture practices has been widely credited with driving up worldwide population and economic growth. For the entire world to meet its food needs, pesticide use is still necessary today. These chemicals have been used extensively in the past few decades all over the world, in a variety of environments such as residences, companies, parks, gardens, and industry. However, the agriculture industry is the main consumer and is responsible for around two-thirds of all pesticide usage worldwide. Despite this dependence, problems about the detrimental health effects resulting from exposure to some pesticides even at low concentrations are becoming more and more serious. These health issues affect many different animals, including humans. They include cancer, neurological impairments, endocrine disturbance, respiratory illnesses, diabetes, and reproductive disorders (Pathak et al. 2022). Furthermore, certain pesticides can decompose into transformation products (TPs), which may be more harmful than the original chemicals. Therefore, the simple fact that farmers are members of environmentally concerned societies does not ensure that pesticides are used safely. To tackle these obstacles, specialized initiatives and strategies that improve farming methods must be put into practice (Beltrán-Flores et al. 2024).

11.2.3 Food Industry Wastewater

In a growing global population and increasing water demands for food production and irrigation, the reutilization of water plays a crucial role in addressing these ongoing challenges and supports sustainable economic growth. There is an urgent call to transition toward wastewater reuse initiatives across all industrial sectors. The food-processing industry stands as one of the largest consumers of potable water globally. The volume of water utilized in food and beverage production varies based on factors such as process type, industrial unit size, cleaning procedures, and equipment employed during production (Thakwani et al. 2023). Consequently, wastewater generated by food industries originates from diverse sources including food-processing units, cleaning operations, and the generation of solid and liquid by-products. Food industry wastewater often contains pathogens such as bacteria, viruses, and parasites, which can contaminate water sources if not properly treated. Consumption of water contaminated with these pathogens can lead to waterborne diseases such as cholera, typhoid fever, and gastroenteritis, causing symptoms like diarrhea, nausea, and vomiting. The majority of food wastewater comprises elevated levels of organic pollutants, including chemical oxygen demand (COD), biological oxygen demand (BOD), total suspended solids (TSS), nitrogen, phosphorus, various chemicals, and nutrients such as carbohydrates, proteins, and lipids. Treating food wastewater is imperative to prevent severe pollution of surrounding land and surface water, as well as to avoid water eutrophication. In line with sustainable development goals, there's a notable shift in focus within the wastewater and organic waste sector, transitioning from a mere "removal and treatment" approach to emphasizing the recovery and reuse of "water, energy, and materials." Due to its nutrient-rich composition, low presence of heavy metals, and reduced toxic pollutants, food wastewater holds significant potential for resource recovery and reuse. By optimizing water utilization and wastewater treatment processes in the food industry, it's possible to not only mitigate environmental pollution but also generate additional economic value through resource recovery, as highlighted by Lee and Stuckey (2022).

11.2.4 Oil and Gas Industry Wastewater

Wastewater generated from oil and gas production remains a challenge for the industry to reduce its detrimental impact on the environment. When oil and gas industry wastewater contaminates water sources used for drinking or agriculture, it can lead to waterborne diseases and health problems. Chemical pollutants can bio accumulate in aquatic organisms, leading to risks for humans who consume contaminated seafood. Wastewater from oil and gas production contains a complex mix of organic compounds, including polycyclic aromatic hydrocarbons, aromatic hydrocarbons, oils, as well as heavy metals and natural radionuclides (NORMs). According to Kassotis et al. reports, each unconventional shale gas well generates approximately 40×10^8 m^3 of wastewater annually. Direct discharge of this wastewater without treatment poses significant environmental risks. Toxic substances enter the land, pollute air and water, endanger wildlife and agriculture, and impact human health (Kassotis et al. 2018). Moreover, substantial water usage during drilling and fracturing processes can lead to water shortages in regions with limited water resources. Achieving compliant discharge and sustainable recycling of oil and gas wastewater has become a challenging issue for the oil industry. Current wastewater treatment methods primarily involve gravity separation, chemical treatments, biodegradation, and other technologies, yet many of these methods are plagued by secondary pollution and high costs (Liu et al. 2021).

11.3 Various Conventional Techniques for Wastewater Remediation

11.3.1 Chemical Precipitation Method

Chemical precipitation and co-precipitation/adsorption are the two primary types into which precipitation technology can be generally divided. These techniques usually consist of three basic steps: nucleation, crystal development, and flocculation. The process of chemical precipitation, which creates conditions of super-saturation, has been thoroughly researched and implemented in industrial facilities. This method has shown notable effectiveness in the treatment of metals like Cd (II), Cu (II), Zn (II), and Mn (II) (Bilal et al. 2013). Moreover, it is regarded as a relatively cost-efficient and traditional method for remedying wastewater polluted with metallic contaminants. This method involves adding chemicals to a solution to cause metal precipitation, which turns a soluble molecule into an insoluble state and creates a supersaturated state above the solubility limit. The metal precipitate can then be separated from the solution through filtration or sedimentation. The main objective of this approach is to reduce the solubility of contaminants. These chemicals modify the pH of the solution, preventing the precipitate from dissolving. However, the application of this method for precipitating organic compounds from solutions is not common. While this method presents numerous benefits, including its simplicity, reliability for osmotic control, economic feasibility for large volumes, and suitability for industrial use, it also has its drawbacks (Kumar et al. 2023b).

11.3.2 Adsorption Method

The adsorption process is recognized as an excellent and very productive approach to removing heavy metal toxicity from different industrial waste streams as compared to earlier technologies. This method is easily accessible, flexible, and effective; it produces high-quality treated wastewater. This process entails mass conversion, where waste material is moved to solid adsorbent surfaces via chemical or physical interactions. The adsorption process shows considerable promise for treating metal ions toxicity in place of conventional techniques, attributed to its cost-effectiveness, scalability, decreased generation of harmful contaminants, straightforward design, ease of use, rapidity, minimal fouling concerns, and lower initial cost and space demands. Furthermore, the adsorption process demonstrates notable economic efficiency, as adsorbents can be regenerated through desorption, thus allowing for recycling and reuse. Different regeneration approaches viz. thermal regeneration, pressure swing, and electrochemical techniques are commonly employed (Rajendran et al. 2022). Thus, the adsorption method is widely recognized as a leading and environmentally sustainable approach in wastewater treatment. The crucial factors for choosing adsorbents include their cost-effectiveness, functional groups, larger surface area, distribution of pore sizes, polarity, and the abundance of binding sites for optimal adsorption. Extensive global research is underway in adsorption-based technology to identify the most suitable adsorbents. These adsorbents, whether organic or inorganic and porous, can effectively capture various heavy metals from aqueous waste streams (Goswami et al. 2012).

11.3.3 Membrane Filtration

Membrane filtration stands out as a highly effective method for removing heavy metal ions from wastewater, playing a crucial role in water treatment processes globally. This technology operates on the principle of selectively allowing certain substances to pass through while retaining contaminants, including heavy metals, based on their charge, solubility, and size. Among the membrane filtration methods commonly employed are Ultrafiltration (UF), Nanofiltration (NF), and Reverse Osmosis (RO). The pore sizes of UF membrane, typically ranging from 5 to 100 nm, effectively capture suspended solids, macromolecules, and larger ions while allowing water molecules and smaller solutes to pass through. NF, with slightly smaller pore sizes in the range of 1–10 nm, can remove divalent ions and some monovalent ions along with suspended particles. RO, with the smallest pore sizes, operates by applying high pressure to force water through a semipermeable membrane, effectively removing nearly all ions, particles, and microorganisms (Deepti et al. 2023). Additionally, emerging membrane technologies like Electrodialysis (ED) are gaining attention for heavy metal removal. In ED, ions are selectively transported via ion exchange membranes while being subjected to an electric field, making it effective for separating and concentrating heavy metals from wastewater streams. The choice among these processes depends on factors such as the operation process and the driving pressure of the membranes. UF and NF stand out as two of the most prominent membranes for treating heavy metal ions (Gurreri et al. 2020).

11.3.3.1 Ultrafiltration

UF is a process that utilizes a permeable membrane, separating substances such as suspended solids, macromolecules, and heavy metals. The size of UF pores typically ranges from 5 to 20 nm, which is greater than the dissolved metal ions in both forms as hydrated ions as well as complexes with common ligands. Consequently, UF requires modification for the effective degradation of heavy metal ions from aqueous solutions. To overcome this limitation using UF, two primary approaches have been developed; 1) polymer-enhanced ultrafiltration (PEUF) and 2) micellar-enhanced UF (MEUF) (Siagian et al. 2021).

11.3.3.2 Nanofiltration

NF stands out as the most efficient pressure-driven technique utilized across diverse chemical and biotech sectors, positioned between UF and RO in terms of functionality. Its benefits include a reduced energy footprint, excellent heavy metal degradation capabilities, simple operation, and lower pressure requirements compared to RO. The effectiveness of NF hinges on several factors like pH, temperature, membrane design, pressure, and feed concentration. Typically composed of synthetic polymers, NF membranes feature surface charges that aid in heavy metal dissociation, thereby enhancing membrane performance through electrostatic interactions with metal ions (Abdel-Fatah 2018).

11.3.4 Electrochemical Treatment

Advanced technologies for treating wastewater involve harnessing electric energy to break down water pollutants. The initial use of electricity for wastewater purification dates back to the late 1900s in the USA, with the first electricity-powered wastewater treatment plant set up in the UK in 1889 (Vik et al. 1984). Initially, widespread adoption was hindered by high capital investment and electricity costs. However, recent years have seen significant research and development in environmental electrochemistry, leading to cleaner treatment of various industrial streams. Electrochemistry deals with the exchange of charge between conductive materials and ionic conductors at their interface, accompanied by reactions occurring within the electrolytes (Naumczyk and Kucharska 2017). Electrochemical processes are extensively employed for removing heavy metal ions toxicity produced by several industrial processes. Additionally, electrochemical-based separation technologies offer numerous advantages over traditional methods, including environmental compatibility through the use of clean reagents like electrons, high-energy efficiency, simple equipment requirements, cost-effectiveness, effective environmental protection, strong reduction capabilities, high yield, adaptability, no need of chemicals, seamless integration with other technologies, and flexibility for automation (Bharti et al. 2023). These technologies are now cost-competitive and potentially more effective than existing alternatives. However, these methods do have certain limitations, such as requiring a high initial investment and facing issues with electrode fouling. Moreover, electrochemical oxidation for pollutant remediation achieves high removal efficiency particularly in wastewater with elevated chloride concentrations due to chlorine generation (Kumar et al. 2023b).

11.4 Synthesis and Properties of Nano-adsorbents

11.4.1 Synthesis of Nano-adsorbents

The synthesis of nano-sorbent materials can be achieved through bottom-up or top-down processes. In the top-down approach, conventional methods are initially used to create nano-adsorbent materials. These materials are then reduced in size through erosion and various techniques, such as reactive milling, high-energy ball milling, and mechanical alloying. However, the bottom-up approach is more modern, involving the construction of a substance from its fundamental components, either atom by atom or molecule by molecule. This method involves sol–gel processes, molecular self-assembly, and physical or chemical vapor deposition. Every synthesis process has its merits and drawbacks. Top-down processes may suffer from damage to surface and crystallographic structures when particle size reduces, yet they remain crucial for nanoparticle synthesis. In contrast, bottom-up processes often require chemical purification and face challenges in large-scale production. However, they are preferred for their ability to tailor materials with specific properties suited for environmental remediation needs (Alharbi et al. 2020).

Furthermore, the chemical and physical processes used to synthesize nano-adsorbent materials often result in materials with specific shapes and sizes. These methods have been found to pose environmental threats owing to the use of toxic chemicals and high temperatures during synthesis processes. Moreover, biological synthesis methods of nano-adsorbent materials, which also produce materials with defined shapes and sizes, require further exploration. This biological synthesis involves using biological entities such as bacteria, fungi, algae, and plant extracts. Microbial synthesis of NMs can occur through intracellular and extracellular mechanisms, with these biological entities known for their high secretion of enzymes and proteins responsible for reducing metal ions and controlling nano-adsorbent materials. Biogenic synthesis methods are considered environmentally friendly and "green" because they do not require harmful chemicals, and the synthesis process typically occurs under normal conditions of pressure and temperature. Consequently, many researchers are now turning to biological synthesis methods for NMs instead of relying solely on physical and chemical processes (El-sayed 2020).

11.4.2 Properties of Nano-adsorbent Material

NMs exhibit remarkable efficiency as adsorbents in wastewater remediation, primarily owing to their chemical and physical characteristics. These characteristics offer a significant surface area, low-temperature modification, adjustable pore size, numerous sorption sites, short intraparticle diffusion distances, and surface chemistry (Rajendran et al. 2022). Besides these attributes, the shape and morphology of

NMs are crucial factors in their effectiveness. The properties of nano-sorbent materials are highly influenced by their intrinsic compositions, apparent sizes, inherent surface properties, and external functionalization, which determine their morphology based on factors such as surface polar charge and elasticity. The functional groups and surface active sites significantly affect the high-quality performance of nano-adsorbents, viz. their surface binding energy, high surface area, adsorptive capability, and chemical activity (Khajeh et al. 2013).

11.5 Adsorption Mechanism by Nano-adsorbents and Their Classifications

Water sources are progressively becoming contaminated due to various anthropogenic activities, adversely impacting the globe. Hence, it is essential to embrace smart, cost-effective, and time-saving technologies to eliminate contaminants from wastewater and ensure its purification. Many traditional methods have been used for degrading pollutants from wastewater viz. chemical precipitation, adsorption, ion exchange, membrane filtration, and electrocoagulation (Das et al. 2023). Among these, nano-adsorbent technology has gained significant recognition for pollutant removal. Nanomaterials have a great potential to effectively adsorb both organic and inorganic contaminants from wastewater due to their high adsorption capability. The distinctive chemical and physical properties of nano-adsorbents make them qualitatively superior in various fields compared to traditional methods.

11.5.1 Type of Adsorption

Adsorption can be classified based on how particles attach to a surface, which includes physical adsorption, chemical adsorption, and ion exchange adsorption.

(a) *Physisorption*, also known as physical adsorption, occurs due to Vander Waals interactions and involves non-covalent bonds that are easily separated. Unlike chemical bonding, electron sharing is not involved in this type of adsorption. This process is a reversible phenomenon.
(b) *Chemisorption*, or chemical adsorption occurs due to chemical bonding between the adsorbate and adsorbent, leading to a strong attachment that is challenging to remove from the solid surface. Therefore, this process is an irreversible phenomenon. This process typically occurs as a monolayer.
(c) Ion exchange adsorption occurs when oppositely charged ions are exchanged between the adsorbate and the surface (adsorbent).

The type of adsorption plays a crucial role in determining its effectiveness, flexibility, and economic viability in application procedures. While traditional adsorbents, viz. clay minerals, carbons, natural zeolites, and chelating materials have been effective in removing heavy metal ions from wastewater but their low absorption ability limits their effectiveness in the purification process. Nano-adsorbents play a vital role in addressing this deficiency and are a powerful tool in water purification (El-sayed 2020).

11.5.2 Mechanism of Metal Ions Adsorption by Nano-adsorbent

Adsorption is a process that involves the interaction of an adsorbent with an adsorbate on a physical and chemical basis, and these interactions are influenced by the surrounding environment. Physical interactions are characterized by weak van der Waals forces between the adsorbent and adsorbate, leading to a reversible process. In physical adsorption, the adsorbate can form monolayer or multilayer adsorption, which is not highly specific. On the other hand, in chemical adsorption, an ionic or covalent bond forms, making the process highly specific. This irreversible chemical interaction results in monolayer adsorption. Various adsorption mechanisms, viz. ion exchange, diffusion, precipitation, electrostatic interaction, and surface adsorption, explain the adherence of metal ions adsorbate, as depicted in Fig. 11.1.

The process of heavy metal adsorption can be thoroughly understood by examining adsorption isotherms, kinetics, and thermodynamics. The mechanisms of adsorption on nano-adsorbents are intricately connected to the chemical nature and

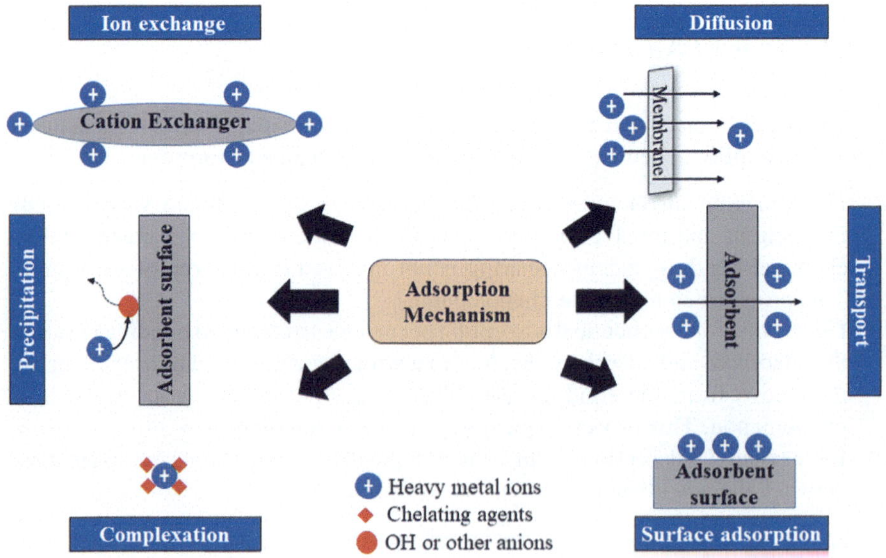

Fig. 11.1 Various adsorption mechanisms of metal ion adsorbate on nano-adsorbent

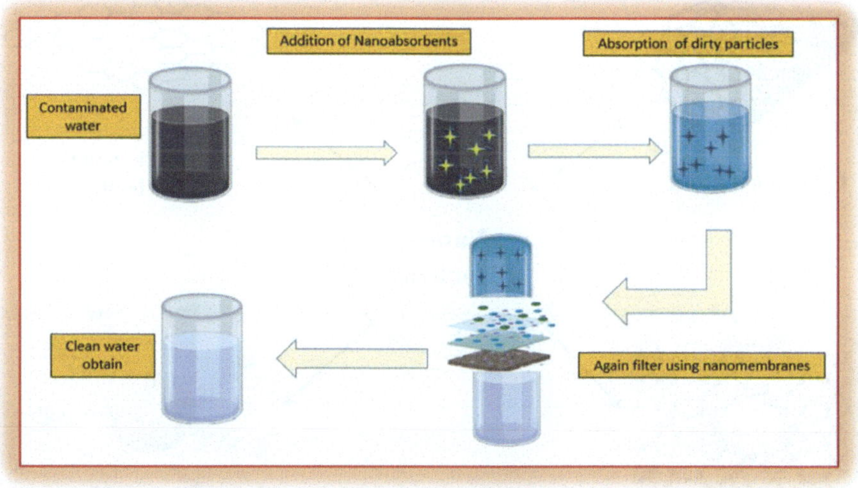

Fig. 11.2 Process involved in the removal of ions Ningthoujam et al. 2022 (Reproduced with permission from Ningthoujam et al. @ *Elsevier* ()

characteristics of the material. Additionally, numerous parameters viz. initial concentration of metal, interaction time, solution pH, adsorbent dosage, and temperature also impact the nature of the adsorption phenomenon (Wang et al. 2020). A process of ion removal using nano-adsorbent is illustrated in Fig. 11.2.

11.5.3 Type of Nano-adsorbents

Various types of nano-adsorbents are depicted in Fig. 11.3 and are discussed in the following sub-sections.

11.5.3.1 Carbon-Based Nano-adsorbents

Carbon nanotubes (CNTs) have been widely investigated owing to their capability to eliminate Potentially Toxic Elements (PTEs) and various organic contaminants from wastewater through adsorption. However, challenges such as their limited disposability, difficulties in separation, and small particle sizes hinder their use as adsorbents. In response to these challenges, researchers have modified conventional CNTs into variants like multi-walled CNTs to overcome these limitations (Daneshvar Tarigh and Shemirani 2013). The enhanced disposability of modified magnetic CNTs allows for easy removal from wastewater or applied media using magnets. Several studies have explored the effectiveness of multi-walled CNTs in removing contaminants like

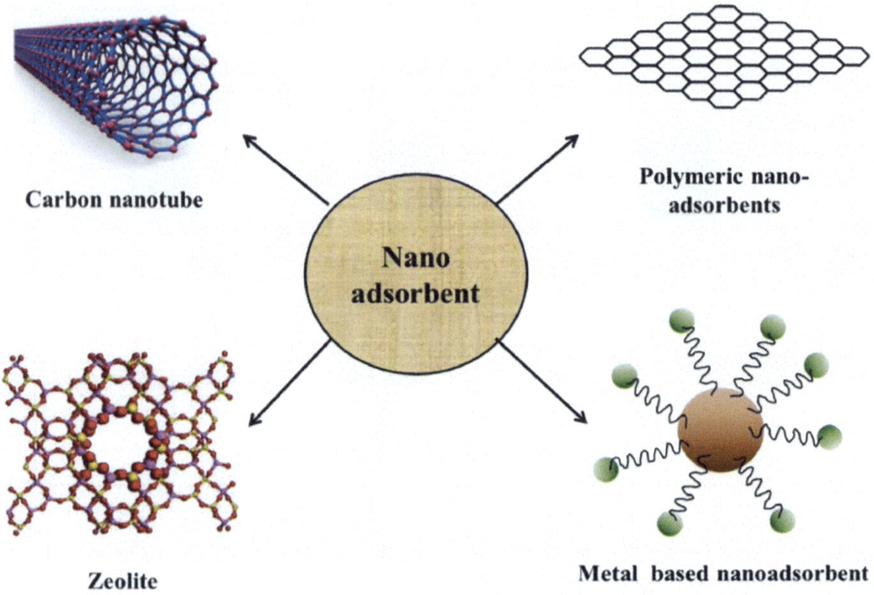

Fig. 11.3 Classification of nano-adsorbents Ningthoujam et al. 2022 (Reproduced with permission from Ningthoujam et al. @ *Elsevier* ()

Pb, Mn, and Cu (Tang et al. 2012). Gupta et al. explored the adsorption capabilities of both coated and uncoated CNTs for aluminum degradation, revealing that coated CNTs exhibited higher degradation capabilities compared to uncoated CNTs. Modifying the surface of CNTs can enhance their overall adsorption capabilities. Several researchers have documented different surface modification methods, such as acid treatment, metal impregnation, and functional molecule or group transplantation. Altering the properties of CNTs is another approach to boost their effectiveness in removing pollutants, which can be achieved through techniques like plasma technology, chemical modification, and microwaves. Among these methods, plasma technology stands out as particularly suitable due to its lower energy utilization and environmentally friendly nature (Gupta et al. 2011). Chen et al. highlighted the effectiveness of modified CNTs infused with various functional groups in treating water contaminated with PTEs (Chen et al. 2012). Moreover, CNTs modified with metals or metal oxides like MnO_2 and Al_2O_3 have demonstrated promising outcomes in their adsorption capabilities (Liang et al. 2015).

11.5.3.2 Metal Oxide-Based Nano-adsorbent

Metal oxide nanoparticles have a greater capability of eliminating heavy metals due to their high surface area and specific affinity for adsorption. These nano-adsorbents are highly favorable for wastewater remediation. These nanoparticles have dimensions between 1 and 100 nm and include various metal oxides such as manganese oxide, zinc oxide, nickel oxide, iron oxide, aluminum oxide, zirconium oxide, titanium oxide, and magnesium oxide. Metal oxide-based nanoparticles have better adsorption capability and extensive surface areas. They are effective in eliminating heavy metal ions from aqueous solutions. These nano-sized metal oxides have been successfully used in various wastewater treatment processes aimed at heavy metal removal. The adsorption processes of metal oxides entail the formation of complexes between dissolved metals and oxygen present within the metal oxides. This process occurs in two steps: initially, metal ions adsorb onto the external surface, followed by rate-limiting intraparticle diffusion within micropores (Wang et al. 2020).

(a) Copper oxide

Copper oxide, as a nano-adsorbent, has demonstrated significant effectiveness in removing heavy metals. Prior studies have indicated its ability to remove Cd and Fe, achieving impressive adsorption capacities of 131.33 mg/g and 94.34 mg/g, respectively. Additionally, copper oxide exhibits a maximum capability to eliminate other heavy metals like Ni and Cd, with adsorption capacities reaching 322.50 mg/g and 64.935 mg/g, respectively (Mohan et al. 2017).

(b) Iron oxide

Iron oxide nanoparticles have been widely investigated and found as an effective nano-adsorbent for eliminating heavy metals. This is due to their high surface area, compact dimensions, excellent adsorption capability, straightforward isolation methods, and magnetic characteristics. Baalousha (2009) investigated the aggregation characteristics of iron oxide under various factors such as particle concentration and pH. In a study on arsenic adsorption, magnetic γ-Fe_2O_3 biochar demonstrated a higher adsorption capability of 3.147 mg/g (Hua et al. 2012).

11.5.3.3 Zeolite

Zeolites are characterized by their microporosity, which includes electrostatic holes formed by alumina and silica minerals, providing a high surface area occupied by cations and water molecules. Their functioning is based on ion exchange and reversible dehydration. Nano-zeolites exhibit a stronger attraction for pollutant adsorption compared to conventional zeolites. Recently, zeolite-based nanocomposite adsorbents have been extensively studied for their ability to adsorb a range of harmful pollutants, including organic dyes (Dutta et al. 2021), emerging contaminants, heavy metals, and metalloids in water. Dyes play essential roles in industries such as printing, food, and textiles. Despite their utility, the wastewater discharged from production containing dyes is extremely toxic, non-biodegradable, intricate, and persistent, posing significant challenges for removal. This pollution can lead to severe environmental degradation and human health concerns. In a separate study, a nanocomposite of polypyrrole/zeolite (PPy/Ze) was employed for the elimination of reactive blue and reactive red dyes. The removal efficiencies reached 88.3% and

82.6% for reactive red and reactive blue, respectively, under optimal conditions with a pH of 9. The influence of pH on the removal process was noted, attributing the high removal rates to the presence of reactive sites within the PPy/Ze nanocomposite and its expansive surface area (Senguttuvan et al. 2022).

11.5.3.4 Polymer-Based Nano-adsorbent

Traditional adsorbents are limited by their lack of specificity, poor adsorption capability, and less recyclability. To overcome these drawbacks, hybrid polymers have been created, which provide superior recyclability, increased thermal stability, and better adsorption capability than traditional adsorbents (Lofrano et al. 2016). Polymer-based nano-adsorbents, characterized by high specific surface areas, porous structures, and functional groups on their surfaces, demonstrate efficient binding capabilities toward organic dyes and heavy metals like Pb, As, Cd, and Zn in wastewater. Various kinds of polymers, categorized by the materials used (e.g., chitosan, dendrimers, cellulose), methods of preparation, and adsorption processes and mechanisms, have been extensively reviewed. These studies have revealed that polymer-based absorbents hold significant potential for eliminating numerous heavy metal ions from wastewater (Baruah et al. 2019). For instance, Dubey et al. examined the effectiveness of chitosan-alginate nanoparticles in Hg elimination, achieving a maximum adsorption capability of 217.39 mg/g under optimal conditions of pH 5 and a 90 min contact time with an initial ion concentration of 4 mg/L (Dubey et al. 2016). Additionally, chitin nanofibrils, a chitosan derivative, were developed by Liu et al. for the removal of a wide range of metal ions, including Cd, Ni, Cu, Zn, Pb, and Cr. In respective to the degradation of pollutants, the merits and limitations of some NMs are shown in Table 11.1.

11.6 Applications of Nano-adsorbents

11.6.1 Wastewater Treatment

Nano-adsorbents have gained significant attention in the field of wastewater treatment due to their unique properties and high efficiency in removing contaminants. These materials, typically in the form of nanoparticles, offer several advantages over traditional adsorbents. Nano-adsorbents are utilized in wastewater remediation due to their efficient capability of heavy metal removal. Their high surface area-to-volume ratio allows for increased adsorption capability. They can selectively adsorb heavy metal ions from wastewater streams, reducing their concentration to safe levels. Additionally, nano-adsorbents are used for the elimination of organic contaminants from wastewater. Their small size and high surface area enhance the adsorption of organic compounds such as dyes, pharmaceuticals (Majumder et al. 2019), and

Table 11.1 Merits and limitations of NMs in pollutant removal (Alharbi et al. 2020)

Types of NMs	Merits	Limitations
Graphene oxides (GO)	The adsorption capability of this material is notably high due to its numerous functional groups and makes surface modification easy. It also provides a large surface area	Its synthesis requires a significant investment and poses challenges in terms of separating from solutions. Additionally, it exhibits drawbacks such as limited selectivity and complexities in large-scale production
Carbon nanotubes (CNTs)	Simply synthesized with a significant external surface area, it also maintains noble stability under both low and high pH conditions	High synthesis investment is required, and it exhibits poor selectivity in adsorption along with limited adsorption capability
Covalent organic frameworks (COF)	Their adsorption capability is exceptionally high, coupled with excellent thermal and chemical stability. Additionally, they possess an ample number of functional groups and are easily modifiable	The synthesis cost is high, and managing the structure and layer stacking poses challenges, making separation difficult
Metal–organic frameworks (MOF)	It's simple to produce in large quantities and can be easily modified due to its numerous functional groups. Moreover, it boasts a high specific surface area and the ability to adjust pore sizes	The production cost is higher, and it exhibits low hydrolytic strength, making it challenging to distinguish from solutions
Carbon nitride (C_3N_4)	The synthesis method is straightforward, and it exhibits excellent photocatalytic properties	They have a low capability for the sorption of metal ions but excel in the photocatalytic reduction of organic contaminants
MXenes	It possesses a plentiful sorption area and exhibits a strong ion interchange capability with metal ions	The synthesis requires a higher investment, and it also shows poor selectivity. Additionally, its stability is low, leading to collapse at high temperatures

organic solvents. This helps in improving water quality by reducing the presence of harmful organic contaminants. Nano-adsorbents also play a role in addressing emerging contaminants like microplastics and nanoparticles themselves. They can effectively adsorb these small particles, preventing their release into the environment and protecting water resources.

11.6.2 Environment Remediation

Nano-adsorbents have emerged as valuable tools in environmental remediation owing to their unique characteristics and high adsorption capabilities. Nano-adsorbents are used for the removal of contaminants from water sources, including heavy metals, organic pollutants, dyes, pesticides, and pharmaceuticals. Their high surface area and functional groups enable efficient adsorption of these pollutants, leading to cleaner water resources. In industrial wastewater treatment, nano-adsorbents play a crucial role in removing toxic substances and pollutants before discharge into water bodies. They can adsorb pollutants from various industries such as textiles, pharmaceuticals, chemicals, and electronics (Gómez-Pastora et al. 2014). Nano-adsorbents are applied in soil remediation to immobilize and remove heavy metals, pesticides, and organic contaminants. They can be used in-situ or ex-situ to clean up contaminated soil, restoring its fertility and reducing environmental risks. Nano-adsorbents are utilized in air purification systems to capture and remove volatile organic compounds (VOCs), particulate matter, and harmful gases. They improve indoor and outdoor air quality, contributing to better respiratory health and reduced environmental pollution. Nano-adsorbents are employed in groundwater treatment to remove contaminants such as arsenic, fluoride, nitrates, and chlorinated compounds. They help in purifying groundwater sources and ensuring access to safe drinking water. Nano-adsorbents can enhance bioremediation processes by immobilizing contaminants, increasing their bioavailability to microorganisms, and facilitating microbial degradation. This synergistic approach accelerates the clean-up of polluted environments (Ningthoujam et al. 2022).

11.6.3 Drug Delivery

Nano-adsorbents have gained significant attention and application in the field of drug delivery owing to their unique characteristics and capabilities. Nano-adsorbents can protect drugs from degradation and improve their stability during storage and transportation. The adsorbent material encapsulates the drug, shielding it from external factors that could affect its efficacy. Nano-adsorbents can be engineered to target specific cells or tissues in the body, allowing for targeted drug delivery (Bassey et al. 2023). Functionalization of the adsorbent surface with ligands or antibodies enables precise targeting, reducing off-target effects and improving therapeutic outcomes. Nano-adsorbents can enhance the bioavailability of poorly soluble drugs by increasing their solubility and dissolution rate. This improvement in bioavailability leads to better absorption and distribution of the drug in the body, improving its overall efficacy. Moreover, targeted delivery and controlled release provided by nano-adsorbents can minimize side effects associated with conventional drug delivery methods. By delivering drugs directly to the site of action, lower doses can often be used, reducing the risk of systemic side effects (Sontakke et al. 2022).

11.6.4 Food Industry

Nano-adsorbents have several promising applications in the food industry, primarily focused on improving food safety, quality, and sustainability. Nano-adsorbents are used in food packaging materials to enhance their barrier properties against oxygen, moisture, and microbial contamination. This helps in extending the shelf life of food products, reducing food waste, and maintaining food freshness. Nano-adsorbents are utilized for the removal of contaminants such as heavy metals, pesticides, toxins, and pathogens from food and water. They can effectively adsorb these contaminants, ensuring food safety and quality (Sahani and Sharma 2021). Nano-adsorbents can be incorporated into food preservatives and coatings to inhibit microbial growth, delay spoilage, and maintain the nutritional content of food products over an extended period. Nano-adsorbents can be used as carriers for flavor compounds, vitamins, antioxidants, and other nutrients. They protect these bioactive compounds from degradation, ensure their controlled release, and enhance their absorption in the body. Nano-adsorbents are used for encapsulating and delivering bioactive compounds, vitamins, minerals, and probiotics in functional foods. This enhances the nutritional value of food products and promotes health benefits (Neme et al. 2021).

11.6.5 Energy Storage

Nano-adsorbents have gained significant attention and application in the field of energy storage owing to their unique characteristics and potential for improving the efficiency and performance of energy storage devices. Nano-adsorbents are used in super-capacitors to enhance their energy storage capacity and charge–discharge kinetics. NMs such as carbon nanotubes, graphene, and metal oxides are employed as electrodes or additives to increase the surface area, conductivity, and ion diffusion rates in super-capacitor devices. Nano-adsorbents play a role in improving the performance of batteries, including lithium-ion batteries (LIBs) and sodium-ion batteries (SIBs). NMs are used in electrode design to enhance ion transport, increase electrode stability, and promote higher energy density and cycling stability (Zheng et al. 2019). Nano-adsorbents are utilized in fuel cell technologies to improve catalytic activity, reduce over potential, and enhance fuel cell efficiency. NMs act as catalyst supports or catalysts themselves, facilitating electrochemical reactions and increasing the overall energy conversion efficiency of fuel cells. Nano-adsorbents are explored for hydrogen storage applications, particularly in nanocomposite materials for reversible hydrogen adsorption and desorption. NMs with high surface area and tailored pore structures offer promising solutions for compact and efficient hydrogen storage systems. Nano-adsorbents have applications in grid-scale energy storage systems, such as redox flow batteries and capacitive deionization devices. NMs enable higher energy density, faster charge–discharge rates, and improved long-term stability, contributing to the efficient storage and management of renewable energy sources in the grid (Pomerantseva et al. 2019).

11.7 Limitations and Future Perspectives

Nano-adsorbents hold great promise for wastewater remediation owing to their large surface area, tunable properties, and efficient adsorption capacities for various contaminants, including heavy metals, organic pollutants, and dyes. However, they also face several limitations that must be addressed for their widespread application in real-world wastewater treatment scenarios. One key limitation is the potential toxicity of certain NMs, raising concerns about their environmental impact and safety during handling and disposal. Additionally, the scalability of nano-adsorbent production and their cost-effectiveness on a large scale remain significant challenges. The stability and recyclability of nano-adsorbents also require improvement to ensure their long-term effectiveness and economic viability in continuous wastewater treatment processes. Furthermore, a cost analysis of nano-adsorbent preparation is essential before their application in actual wastewater treatment. The adsorption capability of each nano-adsorbent should be evaluated alongside cost considerations, as an effective nano-adsorbent should not only be cost-efficient but also demonstrate competitive adsorption capabilities. Another critical factor in developing cost-effective nano-adsorbents is their reusability. Nano-adsorbents with good regeneration properties can maintain high efficiency in removing heavy metals even after multiple recycling cycles. Additional research is required to determine whether nanocomposite adsorbents offer greater advantages compared to using polymers, carbon-based materials, or metal-based nano-adsorbents individually. Despite these limitations, nano-adsorbents offer exciting future perspectives for wastewater remediation. Research efforts are focused on developing eco-friendly and biocompatible NMs that minimize environmental risks and enhance safety. Advancements in nanotechnology aim to improve the synthesis methods of nano-adsorbents, optimizing their performance while reducing production costs. Integration of NMs with advanced technologies such as membrane filtration, electrochemical processes, and photocatalysis holds promise for enhancing the efficiency and versatility of nano-adsorbents in treating complex wastewater streams. Moreover, the development of smart nano-adsorbents with stimuli-responsive properties and enhanced selectivity for specific pollutants could revolutionize the field of wastewater treatment, enabling tailored and efficient pollutant removal strategies.

11.8 Conclusion

Nano-adsorbents represent a promising frontier in wastewater remediation due to their unique properties and remarkable performance in removing pollutants. The use of polymeric-based nano-adsorbents, with their high specific surface areas, porous structures, and functional groups, has shown great potential in efficiently binding organic dyes and heavy metal ions like Pb, As, Cd, Cr and Zn from wastewater. Each nano-adsorbent has its own set of merits and drawbacks regarding its applicability. Combining multiple NM components may lead to alternative approaches for developing highly effective novel materials for pollution purification. Further research such as blending recyclable agro-waste-based NMs (with various nanoparticles to create numerous nanocomposite components) can significantly impact costs. Future advancements will likely focus on creating highly efficient and environmentally friendly nano-based materials and particles with strong biodegradability for remediation, all while maintaining affordability. Developing self-cleaning materials presents a competitive challenge, and evaluating the risks and potential impacts of NMs on human well-being and the environment is imperative.

References

Abdel-Fatah MA (2018) Nanofiltration systems and applications in wastewater treatment: review article. Ain Shams Eng J 9:3077–3092. https://doi.org/10.1016/j.asej.2018.08.001

Abdel-Halim ES, Al-Deyab SS (2011) Removal of heavy metals from their aqueous solutions through adsorption onto natural polymers. Carbohydr Polym 84:454–458. https://doi.org/10.1016/j.carbpol.2010.12.001

Alharbi NS, Hu B, Hayat T, Rabah SO, Alsaedi A, Zhuang L, Wang X (2020) Efficient elimination of environmental pollutants through sorption-reduction and photocatalytic degradation using NMs. Front Chem Sci Eng 14:1124–1135. https://doi.org/10.1007/s11705-020-1923-z

Barakat MA (2011) New trends in removing heavy metals from industrial wastewater. Arab J Chem 4:361–377. https://doi.org/10.1016/j.arabjc.2010.07.019

Baruah A, Chaudhary V, Malik R, Tomer VK (2019) 17–Nanotechnology based solutions for wastewater treatment. In: Ahsan A, A.F.B.T.-N. in W. and Ismail WT (Eds.), Micro nano technol, Elsevier, pp 337–368. https://doi.org/10.1016/B978-0-12-813902-8.00017-4

Bassey VM, Okon EED, Louis H, Benjamin I, Chukwuemeka K, Gber TE, Ezekiel MC, Qader SW, Adeyinka AS (2023) Nano-enhanced drug delivery of dacarbazine using heteroatoms (B, N, P, S) doped Ag-functionalized silicene NMs: Insight from density functional theory. Chem Phys Impact 7:100297. https://doi.org/10.1016/j.chphi.2023.100297

Beltrán-Flores E, Sarrà M, Blánquez P (2024) A review on the management of rinse wastewater in the agricultural sector. Chemosphere 352. https://doi.org/10.1016/j.chemosphere.2024.141283

Bharti M, Das PP, Purkait MK (2023) A review on the treatment of water and wastewater by electrocoagulation process: advances and emerging applications. J Environ Chem Eng 11:111558. https://doi.org/10.1016/j.jece.2023.111558

Bilal M, Shah JA, Ashfaq T, Gardazi SMH, Tahir AA, Pervez A, Haroon H, Mahmood Q (2013) Waste biomass adsorbents for copper removal from industrial wastewater–a review. J Hazard Mater 263:322–333. https://doi.org/10.1016/j.jhazmat.2013.07.071

Changmai M, Purkait MK (2021) Chapter 10–Membrane adsorption. In: M.B.T.-I.S, Ghaedi T (Ed.), Adsorpt Fundam Process Appl, Elsevier, pp 629–653. https://doi.org/10.1016/B978-0-12-818805-7.00007-2

Chen H, Li J, Shao D, Ren X, Wang X (2012) Poly(acrylic acid) grafted multiwall carbon nanotubes by plasma techniques for Co(II) removal from aqueous solution. Chem Eng J 210:475–481. https://doi.org/10.1016/j.cej.2012.08.082

Daneshvar Tarigh G, Shemirani F (2013) Magnetic multi-wall carbon nanotube nanocomposite as an adsorbent for preconcentration and determination of lead (II) and manganese (II) in various matrices. Talanta 115:744–750. https://doi.org/10.1016/j.talanta.2013.06.018

Das PP, Sontakke AD, Purkait MK (2023) Chapter 2–Electrocoagulation process for wastewater treatment: applications, challenges, and prospects. In: M.P.B.T.-D. in W.T.R. and Shah P (Ed.), Elsevier, pp 23–48. https://doi.org/10.1016/B978-0-323-95684-0.00015-4

Deepti, Anweshan, Dhara S, Purkait MK (2023) Membrane and disinfection technologies for industrial wastewater treatment BT–Industrial wastewater reuse: applications, prospects and challenges, in: M.P. Shah (Ed.), Springer Nature Singapore, Singapore, pp 89–112. https://doi.org/10.1007/978-981-99-2489-9_5

Dehghani MH, Kamalian S, Shayeghi M, Yousefi M, Heidarinejad Z, Agarwal S, Gupta VK (2019) High-performance removal of diazinon pesticide from water using multi-walled carbon nanotubes. Microchem J 145:486–491. https://doi.org/10.1016/j.microc.2018.10.053

Dhara S, Das PP, Uppaluri R, Purkait MK (2023) 12–Biological approach for energy self-sufficiency of municipal wastewater treatment plants. In: Sillanpää M, Khadir A, K.B.T.-R.R. in Gurung MWW (eds), Elsevier, pp 235–260. https://doi.org/10.1016/B978-0-323-99348-7.00006-0

Dubey R, Bajpai J, Bajpai AK (2016) Chitosan-alginate nanoparticles (CANPs) as potential nanosorbent for removal of Hg (II) ions. Environ Nanotechnology, Monit Manag 6:32–44. https://doi.org/10.1016/j.enmm.2016.06.008

Dutta S, Gupta B, Srivastava SK, Gupta AK (2021) Recent advances on the removal of dyes from wastewater using various adsorbents: a critical review. Mater. Adv. 2:4497–4531. https://doi.org/10.1039/d1ma00354b

El-sayed MEA (2020) Nanoadsorbents for water and wastewater remediation. Sci Total Environ 739:139903. https://doi.org/10.1016/j.scitotenv.2020.139903

Gómez-Pastora J, Bringas E, Ortiz I (2014) Recent progress and future challenges on the use of high performance magnetic nano-adsorbents in environmental applications. Chem Eng J 256:187–204. https://doi.org/10.1016/j.cej.2014.06.119

Goswami A, Raul PK, Purkait MK (2012) Arsenic adsorption using copper (II) oxide nanoparticles. Chem Eng Res des 90:1387–1396. https://doi.org/10.1016/j.cherd.2011.12.006

Gupta VK, Agarwal S, Saleh TA (2011) Synthesis and characterization of alumina-coated carbon nanotubes and their application for lead removal. J Hazard Mater 185:17–23. https://doi.org/10.1016/j.jhazmat.2010.08.053

Gurreri L, Tamburini A, Cipollina A, Micale G (2020) Electrodialysis applications in wastewater treatment for environmental protection and resources recovery: a systematic review on progress and perspectives. Membranes (Basel). 10:1–93. https://doi.org/10.3390/membranes10070146

Hua M, Zhang S, Pan B, Zhang W, Lv L, Zhang Q (2012) Heavy metal removal from water/wastewater by nanosized metal oxides: a review. J Hazard Mater 211–212:317–331. https://doi.org/10.1016/j.jhazmat.2011.10.016

Jiang Y, Liu Z, Zeng G, Liu Y, Shao B, Li Z, Liu Y, Zhang W, He Q (2018) Polyaniline-based adsorbents for removal of hexavalent chromium from aqueous solution: a mini review. Environ Sci Pollut Res 25:6158–6174. https://doi.org/10.1007/s11356-017-1188-3

Kassotis CD, Nagel SC, Stapleton HM (2018) Unconventional oil and gas chemicals and wastewater-impacted water samples promote adipogenesis via PPARγ-dependent and independent mechanisms in 3T3-L1 cells. Sci Total Environ 640–641:1601–1610. https://doi.org/10.1016/j.scitotenv.2018.05.030

Khajeh M, Laurent S, Dastafkan K (2013) Nanoadsorbents: classification, preparation, and applications (with emphasis on aqueous media). Chem Rev 113:7728–7768. https://doi.org/10.1021/cr400086v

Khan SU, Zaidi R, Hassan SZ, Farooqi IH, Azam A (2016) Application of Fe-Cu binary oxide nanoparticles for the removal of hexavalent chromium from aqueous solution. Water Sci Technol 74:165–175. https://doi.org/10.2166/wst.2016.172

Kumar A, Thakur A, Panesar PS (2023b) A review on the industrial wastewater with the efficient treatment techniques. Versita. https://doi.org/10.1007/s11696-023-02779-3

Kumar S, Mondal P, Purkait MK (2023) Hazardous effects of heavy metals from industrial wastewaters and their remediation through green technology BT–Green technologies for industrial waste remediation. In: Mathuriya AS, Pandit S, Singh NK (Eds.), Springer Nature Switzerland, Cham, pp 29–55. https://doi.org/10.1007/978-3-031-46858-2_3

Kumari U, Siddiqi H, Bal M, Meikap BC (2020) Calcium and zirconium modified acid activated alumina for adsorptive removal of fluoride: performance evaluation, kinetics, isotherm, characterization and industrial wastewater treatment. Adv Powder Technol 31:2045–2060. https://doi.org/10.1016/j.apt.2020.02.035

Lee SY, Stuckey DC (2022) Separation and biosynthesis of value-added compounds from food-processing wastewater: towards sustainable wastewater resource recovery. J Clean Prod 357:131975. https://doi.org/10.1016/j.jclepro.2022.131975

Liang J, Liu J, Yuan X, Dong H, Zeng G, Wu H, Wang H, Liu J, Hua S, Zhang S, Yu Z, He X, He Y (2015) Facile synthesis of alumina-decorated multi-walled carbon nanotubes for simultaneous adsorption of cadmium ion and trichloroethylene. Chem Eng J 273:101–110. https://doi.org/10.1016/j.cej.2015.03.069

Liu X, Ruan W, Wang W, Zhang X, Liu Y, Liu J (2021) The perspective and challenge of NMs in oil and gas wastewater treatment. Molecules 26:1–12. https://doi.org/10.3390/molecules26133945

Lofrano G, Carotenuto M, Libralato G, Domingos RF, Markus A, Dini L, Gautam RK, Baldantoni D, Rossi M, Sharma SK, Chattopadhyaya MC, Giugni M, Meric S (2016) Polymer functionalized nanocomposites for metals removal from water and wastewater: an overview. Water Res 92:22–37. https://doi.org/10.1016/j.watres.2016.01.033

Majumder A, Gupta B, Gupta AK (2019) Pharmaceutically active compounds in aqueous environment: a status, toxicity and insights of remediation. Environ Res 176:108542. https://doi.org/10.1016/j.envres.2019.108542

Man Y, Han Y, Wang Y, Li J, Chen L, Qian Y, Hong M (2018) Woods to goods: water consumption analysis for papermaking industry in China. J Clean Prod 195:1377–1388. https://doi.org/10.1016/j.jclepro.2018.06.026

Mandal S, Padhi T, Patel RK (2011) Studies on the removal of arsenic (III) from water by a novel hybrid material. J Hazard Mater 192:899–908. https://doi.org/10.1016/j.jhazmat.2011.05.099

Mohan S, Kumar V, Singh DK, Hasan SH (2017) Effective removal of lead ions using graphene oxide-MgO nanohybrid from aqueous solution: Isotherm, kinetic and thermodynamic modeling of adsorption. J Environ Chem Eng 5:2259–2273. https://doi.org/10.1016/j.jece.2017.03.031

Naumczyk JH, Kucharska MA (2017) Electrochemical treatment of tannery wastewater—raw, coagulated, and pretreated by AOPs. J Environ Sci Heal Part A Toxic/Hazardous Subst Environ Eng 52:649–664. https://doi.org/10.1080/10934529.2017.1297140

Neme K, Nafady A, Uddin S, Tola YB (2021) Application of nanotechnology in agriculture, postharvest loss reduction and food processing: food security implication and challenges. Heliyon 7:e08539. https://doi.org/10.1016/j.heliyon.2021.e08539

Ningthoujam R, Singh YD, Babu PJ, Tirkey A, Pradhan S, Sarma M (2022) Nanocatalyst in remediating environmental pollutants. Chem Phys Impact 4:100064. https://doi.org/10.1016/j.chphi.2022.100064

Pathak VM, Verma VK, Rawat BS, Kaur B, Babu N, Sharma A, Dewali S, Yadav M, Kumari R, Singh S, Mohapatra A, Pandey V, Rana N, Cunill JM (2022) Current status of pesticide effects on environment, human health and it's eco-friendly management as bioremediation: a comprehensive review. Front Microbiol 13:1–29. https://doi.org/10.3389/fmicb.2022.962619

Pomerantseva E, Bonaccorso F, Feng X, Cui Y, Gogotsi Y (2019) Energy storage: the future enabled by NMs. Science (80) 366:eaan8285. https://doi.org/10.1126/science.aan8285
Rajendran S, Priya AK, Senthil Kumar P, Hoang TKA, Sekar K, Chong KY, Khoo KS, Ng HS, Show PL (2022) A critical and recent developments on adsorption technique for removal of heavy metals from wastewater–a review. Chemosphere 303:135146. https://doi.org/10.1016/j.chemosphere.2022.135146
Rasheed T, Bilal M, Nabeel F, Adeel M, Iqbal HMN (2019) Environmentally-related contaminants of high concern: potential sources and analytical modalities for detection, quantification, and treatment. Environ Int 122:52–66. https://doi.org/10.1016/j.envint.2018.11.038
Sahani S, Sharma YC (2021) Advancements in applications of nanotechnology in global food industry. Food Chem 342:128318. https://doi.org/10.1016/j.foodchem.2020.128318
Senguttuvan S, Janaki V, Senthilkumar P, Kamala-Kannan S (2022) Polypyrrole/zeolite composite—a nanoadsorbent for reactive dyes removal from synthetic solution. Chemosphere 287:132164. https://doi.org/10.1016/j.chemosphere.2021.132164
Siagian UWR, Khoiruddin K, Wardani AK, Aryanti PTP, Widiasa IN, Qiu G, Ting YP, Wenten IG (2021) High-performance ultrafiltration membrane: recent progress and its application for wastewater treatment. Curr Pollut Reports 7:448–462. https://doi.org/10.1007/s40726-021-00204-5
Singh R, Volli V, Lohani L, Purkait MK (2020) Synthesis of carbon nanotubes from industrial wastes following alkali activation and film casting method. Waste Biomass Valorization 11:4957–4966. https://doi.org/10.1007/s12649-019-00827-2
Singh BJ, Chakraborty A, Sehgal R (2023) A systematic review of industrial wastewater management: evaluating challenges and enablers. J Environ Manage 348:119230. https://doi.org/10.1016/j.jenvman.2023.119230
Sontakke AD, Bhattacharjee A, Fopase R, Pandey LM, Purkait MK (2022) One-pot, sustainable and room temperature synthesis of graphene oxide-impregnated iron-based metal-organic framework (GO/MIL-100(Fe)) nanocarriers for anticancer drug delivery systems. J Mater Sci 57:19019–19049. https://doi.org/10.1007/s10853-022-07773-w
Tang WW, Zeng GM, Gong JL, Liu Y, Wang XY, Liu YY, Liu ZF, Chen L, Zhang XR, Tu DZ (2012) Simultaneous adsorption of atrazine and Cu (II) from wastewater by magnetic multi-walled carbon nanotube. Chem Eng J 211–212:470–478. https://doi.org/10.1016/j.cej.2012.09.102
Thakwani Y, Karwa A, BG PK, Purkait MK, Changmai M (2023) A composite starch-date seeds extract based biodegradable film for food packaging application. Food Biosci 54:102818. https://doi.org/10.1016/j.fbio.2023.102818
Uddin MK, Baig U (2019) Synthesis of Co3O4 nanoparticles and their performance towards methyl orange dye removal: characterisation, adsorption and response surface methodology. J Clean Prod 211:1141–1153. https://doi.org/10.1016/j.jclepro.2018.11.232
Verstraete W, Yanuka-Golub K, Driesen N, De Vrieze J (2022) Engineering microbial technologies for environmental sustainability: choices to make. Microb Biotechnol 15:215–227. https://doi.org/10.1111/1751-7915.13986
Vik EA, Carlson DA, Eikum AS, Gjessing ET (1984) Electrocoagulation of potable water. Water Res 18:1355–1360. https://doi.org/10.1016/0043-1354(84)90003-4
Wang L, Shi C, Pan L, Zhang X, Zou JJ (2020) Rational design, synthesis, adsorption principles and applications of metal oxide adsorbents: a review. Nanoscale 12:4790–4815. https://doi.org/10.1039/c9nr09274a
Yousefi M, Ghoochani M, Hossein Mahvi A (2018) Health risk assessment to fluoride in drinking water of rural residents living in the Poldasht city, Northwest of Iran, Ecotoxicol Environ Saf 148:426–430. https://doi.org/10.1016/j.ecoenv.2017.10.057
Zheng G, Chen M, Yin J, Zhang H, Liang X, Zhang J (2019) Metal organic frameworks derived nano materials for energy storage application. Int J Electrochem Sci 14:2345–2362. https://doi.org/10.20964/2019.03.28

Zou Y, Li Q, Tao T, Ye P, Zhang P, Liu Y (2021) Fe-Mn binary oxides activated aluminosilicate mineral and its Tl(I) removal by oxidation, precipitation and adsorption in aqueous. J Solid State Chem 303:122383. https://doi.org/10.1016/j.jssc.2021.122383

Chapter 12
Application of Metal Oxide Nanoparticles for Waste Water Treatment

Azad Yadav and Rajesh Dhankhar

Abstract In this period of water scarcity, wastewater originating due to urbanization and industrialization is the principal cause of water contamination. Pollutants in wastewater present an imminent risk to the health of the environment, accelerating the need for efficient wastewater treatment in developing economies. It is imperative to develop and implement improved wastewater treatment techniques that are both efficient and cost-effective. Among various treatment technologies, nanotechnology has been attracting the attention of researchers because of its potential to improve waste water treatment efficiency. Only a small number of studies emphasized metal oxide NPs and their potential in wastewater decontamination. Nanoparticles (NPs) possess the ability to eliminate heavy metals, organic compounds, and inorganic contaminants from polluted water. Metal oxides are highly distinct category of materials characterized by their exceptional features like improved surface area, relatively low concentration, readily separable after the treatment process, etc. In this chapter, we emphasized iron oxide, copper oxide, titanium oxide, zinc oxide, and silver oxide NPs for waste water treatment processes. This article certainly provides a comprehensive overview of the removal of the contaminants by using above-mentioned metal oxide NPs.

Keywords Nanotechnology · NPs · Metal oxides · Wastewater treatment · Adsorption

---

A. Yadav · R. Dhankhar (✉)
Department of Environmental Sciences, Maharshi Dayanand University, Rohtak, India
e-mail: rajeshdhankhar669@gmail.com

N. Agarwal et al. (eds.), *Nanomaterials in Wastewater Research*, Advances in Wastewater Research, https://doi.org/10.1007/978-981-96-3908-3_12

12.1 Introduction

Pure water is an important component of all living things. Pollution of contemporary water resources has become a worldwide issue owing to the rapid expansion of industry and the substantial increase in population. The need for the utilization of clean water in agriculture has significantly risen. The usage of uncontaminated water in industry, households, and other sectors is equivalent to around 70%, 22%, and 8%, respectively (Helmer and Hespanhol 1997).

The important categories of contaminants are heavy metal ions and dyes. Water having these pollutants must not be consumed. Once the heavy metal ions find their way into water bodies, it becomes very difficult to purify them completely. These aquatic contaminants are harmful to living creatures and negatively impact ecosystems. Therefore, it is necessary to separate these substances from water in order to prevent their negative impacts on life and nature. Globally, around 780 million individuals lack access to safe drinking water (Herschy 2012). Immediate action is required in the afflicted regions, mostly in lower-income countries where wastewater management is typically absent. Nevertheless, current treatment systems and technologies are enhancing their efficiency to provide adequately purified water for human needs and meet other environmental standards. Recent advancements in nanoscience and nanotechnology have promising prospects for enhancing water resources and systems. Nanoscience and nanotechnology are expected to provide highly efficient, integrated, and versatile advancements in wastewater treatment. These advancements are forecasted to deliver superior performance and cost-effective systems in contrast to traditional large-scale infrastructure. The treatment of these contaminants using a sustainable and efficient way is important. A variety of techniques have been explained in the literature to address the treatment and management of polluted water, such as reverse osmosis, ultrafiltration, evaporation, and solvent extraction. Nevertheless, these methods separate contaminants from water while not converting them into innocuous end products. Complete disintegration through oxidation can be easily achieved, either through chemical means or photochemical processes. The purpose of each oxidative reaction is to generate and employ a hydroxyl free radical as a more potent oxidizing agent in order to mitigate the effects of pollutants. When hydrogen peroxide is activated, it can function as an oxidant, comparable to UV radiation, or as a metal ion, or as a Fenton reagent (Slokar and Marechal 1998). NPs are largely utilized to out-pass significant wastewater issues. The idea of a nanomaterial "NM" denotes a size that is one trillionth of a metre, also known as the nanoscale (Auffan et al. 2009). NMs find extensive applications in pollution detection, healthcare sector, telecommunication sector, textile sector, and skin care products. These small NPs cause various alterations in their physical characteristics, such as an increase in the volume to surface–area ratio and the influence of quantum qualities on size of the particles which are different from the normal materials. The characteristics of NMs like their magnetic, visual, and electrical attributes are largely parallel to normal particles. NPs are characterized by properties such as excellent adsorption, catalytic function, and reactivity. NMs have garnered significant interest in recent years and have been successfully utilized in various areas such as biology, sensing applications, catalytic chemistry, and on-going research. NMs are frequently utilized in purification of wastewater. Due to their extensive surface area and compact dimensions, NPs exhibit high reactivity and ability for adsorption. Many contaminants like bacteria, emerging contaminants, and organic and inorganic pollutants have been degraded by multiple types of NPs. NPs, such as zerovalent NPs, metal oxide NPs, carbon nanotubes, and nanocomposites, give exciting prospects for use in a number of wastewater settings. NPs offer innovative approaches to enhance current water infrastructure and non-traditional resources for water. In recent years, various techniques in integration with nanotechnology have emerged for wastewater treatment. Such notable techniques include the use of reverse osmosis, solvent extraction, sedimentation, gravitational separation, flotation, filtration [micro, ultra], procedure of distillation oxidation, adsorption, electrolysis, coagulation process, and the exchange of ions (Yang et al. 2019a).

The existing literature provides a limited overview of research for the use of metal oxide NPs in wastewater treatment setups. Yang et al. examined four categories of NPs—nano ZnO, Ag NPs, nZVI, and nano TiO_2—for the process of purifying wastewater and digestion by anaerobic means (Yang et al. 2013). The conversation revolved around the effect of metal and non-metallic oxide NPs in the treatment of wastewater and the process of anaerobic sludge digesting. The focus of Singh et al. was on utilization of NPs for the purpose of treating wastewater. The authors provided a comprehensive analysis of iron oxide NPs, zinc oxide NPs, and TiO_2 NPs (Singh et al. 2019). Juncheng et al. discussed some metal oxide NPs such as magnesium oxide, titanium dioxide, manganese dioxide, iron oxide, aluminium oxide, and ceric oxide and their use for water remediation (Hu et al. 2010). Furthermore, while certain reviews on NPs mention metal oxides, there is a notable absence of a comprehensive evaluation specifically focussing on the use of metal oxides for the remediation of polluted water (Yang et al. 2019b). In a recent study, industrial utilization of metal oxides (titanium dioxide, iron oxide, and zinc oxide) is brought forth. In order to address the issue and provide a comprehensive analysis, this work is based on several recent advancements and utilization of metal oxide NPs in wastewater remediation and focusses possible usage of such methods to deal with different difficulties experienced by the existing wastewater purification methods. Different kinds of NPs are explained in this study. The rationale to work with these NPs for purification of contaminated water lies in their capacity to undergo oxidation or dissolution in water, hence releasing metal ions and resulting in reduced metal toxicity. Such NPs of metal oxide are chemically inert and do not have any negative effects. They are used in numerous processes like adsorption, photocatalytic processes, antibacterial, and antifungal treatment (Abdelbasir and Shalan 2019).

12.2 Wastewater

Water that is not suitable for use due to the presence of contaminants like metal ions, dyes, bacteria, and organic contaminants is known as wastewater. The natural, biological, and chemical components of contaminated water vary from those of pure water, rendering it unsuitable for prolonged consumption. Contaminants in polluted water are categorized into different types, i.e. biological, organic and inorganic contaminants. Biological contaminants refer to bacteria, viruses, and protozoa that have detrimental effects on the wellness of both aquatic species and humans as well. Therefore, wastewater is a significant contributor to the spread of illnesses that are transmitted through water, such as typhoid fever, diarrhoea, giardia, infections caused by *Escherichia coli*, liver disease, and salmonellosis. Organic pollutants in water consist of approximately 50% oils and fats, 40% carbohydrates, and 10% proteins (Abdelbasir and Shalan 2019). Water contaminated with organic pollutants may also contain basic impurities, surfactants, and other types of toxins. Biological Oxygen Demand (BOD) and Chemical Oxygen Demand (COD) are highly useful parameters for assessing the presence of organic contaminants in water (Chahal et al. 2016). Water may have many inorganic contaminants, like heavy metals, minimal quantities of phosphorus, nitrogen-based compounds, and various toxic inorganic substances (Amil Usmani et al. 2017). Wastewater primarily originates from the agricultural, industrial, residential, and commercial markets. Industrial effluents are the primary cause of contamination of the environment's water supplies. For example, wastewater discharged by tanning industries possesses a large ratio of cadmium, mercury, lead, and silver, which may largely impact human as well as marine ecosystem (Naseem and Durrani 2021). Similarly, phenols, insecticides, lipids, and oils are found in organic pollutants released from drug manufacturing firms, animal husbandries, hospitals, and homes (Amil Usmani et al. 2017). The paper and pulp industries generate a substantial quantity of pollutants associated with BOD, suspended particles, and coloration. Figure 12.1 provides an overview of the primary contributors to wastewater.

When this unfiltered water is released into waterways, it poses a threat to fish and other aquatic life and leads to physical disabilities (Chahal et al. 2016). In the same way, in the clothing sector, the use of aqueous chemical process is employed for the preparation, purification, and finalization of colour products, which therefore leads to the generation of wastewater. The wastewater generated contains microorganisms, hazardous chemicals, trace elements, and radionuclides. Though different traditional methods such as chemical precipitation, flocculation, flotation, biological treatment, ion exchange, evaporation, and filtration, etc., are employed for the treatment of polluted water but these methods have their own limitations, which are mentioned in Table 12.1.

The deleterious consequences of infections and the adverse effects of contaminated water on agriculture and humans necessitated the advancement of treatment technologies of polluted water (Naseem and Durrani 2021).

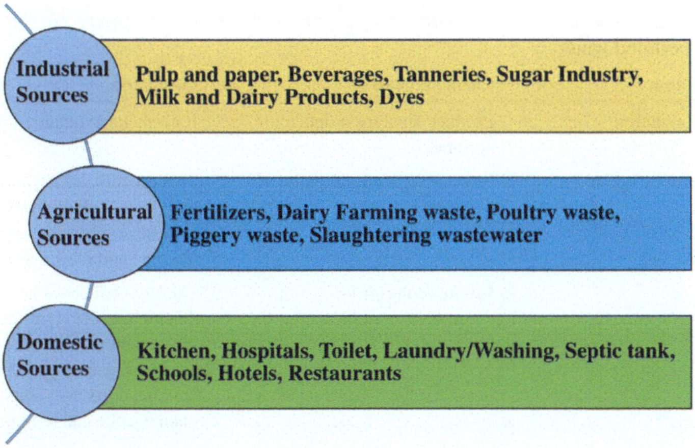

Fig. 12.1 Primary contributors to wastewater

Nanotechnology provides novel water treatment solutions for many processes such as catalysis, adsorption, electrostatics, reactivity, and changeable pore volume. It also enables the development of high aspect ratio sensors and optical electronics, with the ability to control hydrophilic and hydrophobic interactions (Daus et al. 2004). Nanotechnology methods are efficient, adaptable, and diverse, providing cost-effective, high-performance options for water and wastewater treatment. Furthermore, nanotechnology can be efficiently scaled up to purify and rejuvenate unconventional sources of water at low cost.

12.3 Metal Oxide NPs

Metal oxide NPs consist only of metal precursors. NPs have a substantial impact on various fields. Thermal components can generate a diverse array of oxide compounds. These structures can take on a wide range of shapes and have an electrical structure that can display the properties of being an insulator, semiconductor, or metal. The opto-electrical properties of these NPs are distinct due to their widely recognized localized surface plasmon resonance characteristics. The synthesis of modern materials heavily relies on the surface, size, and form of metal NPs, which have crucial functions (Dreaden et al. 2012).

Due to sophisticated optical characteristics, metal NPs find utility in various scientific domains. The application of a gold NP coating is a frequently employed technique in SEM analysis to enhance the electron flow, hence enabling the acquisition of high-resolution SEM images. Nanoscale metal oxides have exceptional characteristics, such as a significant adsorption capability and a strong preference for heavy metals. They possess significant potential as effective adsorbents for heavy metals.

Table 12.1 Characteristics and limitations of different traditional methods employed for the treatment of polluted water

Name of technique	Characteristics	Limitations
Chemical precipitation	Product separation/uptake of pollutants	It needs large quantity of chemicals
Flocculation and coagulation	Product separation/uptake of pollutants	This process, which necessitates an inclusion of alkaline substances to achieve an ideal pH level, is intricate and ineffective
Flotation	Separation process	The key challenges associated with this include the significant upfront investment required, the selectivity being influenced by pH levels, and the high costs associated with maintenance and operation
Biological treatment	Utilization of biological cultures, whether they are mixed or pure	Microorganisms exhibit a high degree of sensitivity to their environment. Intermediate agents can eliminate microbial cells. This approach is costly and requires a significant amount of time
Ion exchange	Non-destructive process	High operating and maintenance expenses as well as high initial capital costs
Filtration (Nano, Ultra, Micro, and Carbon filter)	Utilization of a compact solid substance and a method that does not cause damage	Nano–Pretreatment and high energy for water cleaning are necessary. The retention of salt and univalent ions was limited Ultra–It necessitates a significant amount of energy and is ineffective at eliminating dissolved inorganic substances Micro–It is ineffective in eliminating metals, salt, nitrates, VOCs, colours, and other substances. This approach necessitates frequent cleaning, and there is also a possibility of membrane fouling Carbon filter–It is ineffective at eliminating fluoride, salt, metals, and other substances. It is prone to mould growth and can be blocked by insoluble particles

(continued)

Table 12.1 (continued)

Name of technique	Characteristics	Limitations
Incineration	Combustion-induced destruction	Highly expensive
Evaporation	Methods of separation, heating, and concentration	Costly for large quantities of wastewater. Small and medium-sized industries have significant financial burdens due to the high costs of investing
Liquid–liquid extraction	Separation technology	High cost for equipments
Advanced oxidation processes	Emerging processes, destructive techniques	Medium and small-sized industries find it economically unfeasible due to technical limitations and poor production capacity

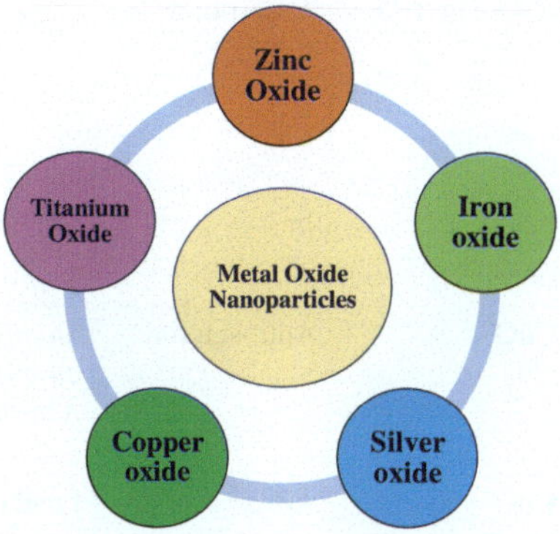

Fig. 12.2 Distinct forms of metal oxide NPs

Metal oxides NPs encompass iron oxides at the nanoscale, titanium oxides, cerium oxides, ZnOs, magnesium oxides, aluminium oxides, and zirconium oxides. This research specifically focuses on the discussion of five distinct categories of metal oxide NPs and their use in process of water decontamination. These five varieties are illustrated in Fig. 12.2, and uses of various metal oxide NPs in wastewater treatment are depicted in Table 12.2.

12.3.1 Zinc Oxide (ZnO) NPs

ZnO NPs are attractive photocatalytic materials for wastewater treatment processes because of their exceptional photocatalytic capacity and great chemical stability. ZnO possesses a significant bandgap of 3.37 electron volts and a notable excitation binding energy of 60 millielectron volts (Yadav et al. 2024). Furthermore, their exceptional resistance to severe circumstances, capacity to catalyse reactions, economical

Table 12.2 Use of various metal oxide NPs for wastewater treatment

NPs	Application	Contaminant	Characteristics	References
ZnO	Antibacterial	*E. coli* *S. aureus*	Complete inhibition of *E. Coli* at 50 mg/L and for *S. aureus* at 100 mg/L	Zare et al. 2018
ZnO	Adsorption	Methyl orange, Amaranth	82.5% and 97.4% removal	Zafar et al. 2019
ZnO	Adsorption	Cadmium (II)	92% removal	Srivastava et al. 2013
Ag_2O	Photocatalysis	Methyl orange	Under visible light irradiation, complete degradation is achieved within 2 min	Jiang et al. 2015
Mn-doped TiO_2 rGO-Mn-TiO_2	Photocatalysis Adsorption	Chromium (VI)	The removal efficiency for Chromium (VI) under sunshine irradiation was 98.68% within 60 min using Mn-doped TiO_2. Similarly, the removal efficiency of 98.72% was achieved using 10 rGO-Mn-TiO_2 NPs	Chen et al. 2016
CuO	Antibacterial	*Staphylococcus aureus* *Aeromonas*	Zone of inhibition for *Staphylococcus aureus* is 3.5 cm and 2.5 cm for *Aeromonas*	Sukumar et al. 2020
CuO	Adsorption	Lead (II)	Showed efficiency of 95%	Sreekala et al. 2019
Magnetite [Fe_3O_4]	Adsorption	Chromium (VI) Lead (II)	Showed efficiency of 34.87 mg/g [sorption] for Chromium (VI) and 53.11 mg/g [sorption] for Lead (II)	Rajput et al. 2016
Hematite [α-Fe_2O_3]	Catalytic degradation	Chromium (VI)	Efficiency of 75.4% after 8 h	Fouda et al. 2021

nature, lack of toxicity, chemical and electrochemical durability, biological compatibility, and ease of synthesis make them very suitable for wastewater treatment. ZnO is the primary and widely explored metal oxide for heterogeneous photocatalysis. Nevertheless, rapid charge carrier recombination in ZnO inhibits the generation of photocurrent by decreasing their photocatalytic efficacy. Many studies are being conducted to enhance ZnO's optical and electrical properties and boost its photocatalytic activity by using metal and non-metal components. Moreover, a range of ZnO nanostructures can be produced, such as complex hybrid structures, NPs, nanowires, nanorods, nanosheets, and nanobelts, which additionally modifies its optoelectronic characteristics. ZnO nanostructures have the ability to absorb ultraviolet (UV) energy and use reactive species such as hydroxyl radicals ($HO^{\bullet}$) or superoxide anion radicals ($O_2^{\bullet-}$) to break down the contaminants into CO_2 and H_2O (Vasantharaj et al. 2021).

ZnO NPs have a wide range of uses in polluted water treatment, including processes such as sonication, adsorption, ozonation, ultrasound, and coagulation, in addition to photocatalysis. Furthermore, ZnO NPs are utilized to protect cells of the skin from detrimental UV radiation. ZnO NPs can be synthesized through various methods including physical, chemical, and biological approaches. Chemical approaches encompass liquid-phase and gas-phase synthesis. Physical synthesis methods involve laser ablation, high-energy ball milling as well as chemical, physical, and solid vapour deposition techniques. Biological approaches entail the utilization of different plants parts, various microbes, as well as bacteria for producing ZnO NPs (Rabani et al. 2021).

Zafar et al. showed that ZnO NPs can effectively remove methyl orange (MO) and amaranth colours from polluted water through the process of adsorption (Zafar et al. 2019). This method involves mixing a dye solution with a predetermined amount of ZnO NPs and then applying agitation using an orbital shaker. After being stirred, the mixture showed a notable decrease in the concentrations of amaranth and methyl orange. The enhanced concentration of ZnO NPs led to breakdown of approximately 82.5% of methyl orange and 97% of amaranth. In addition, they observed the influence of temperature, concentration, pH, and rate of reaction on photocatalytic breakdown (Zafar et al. 2019). Likewise, in another study, researchers achieved the removal of 92% of cadmium from polluted water utilizing ZnO NPs via adsorption, underscoring the effectiveness of ZnO as an adsorbent (Srivastava et al. 2013). Ganapathy and colleagues demonstrated photocatalytic prowess of ZnO NPs on polluted water from domestic sources. In their study, biologically synthesized ZnO NPs were introduced into polluted water of a sewage treatment plant and exposed to sunlight for approximately 30 min. Surprisingly, the Chemical Oxygen Demand of the contaminated water declined significantly from an initial concentration of 750 mg/L to a final concentration of 50 mg/L. It underscores the effectiveness of ZnO NPs as a potent photocatalytic agent for eliminating diverse contaminants found in polluted water (Ganapathy et al. 2019). The process of breaking down industrial dyes using photocatalysis is achieved by utilizing ZnO-polyvinylpyrrolidone (PVP) NPs as the catalyst. These NPs are synthesized utilizing precipitation process. In this study, varying quantities of ZnO-PVP NPs were combined with polluted water containing industrial colour. Subsequently, the resultant mix was introduced into a photoreactor that was linked to a UV light source. It was found that at pH 7, a degradation rate of 90.61% for the industrial dye was achieved (Abu Bakar Sidik et al. 2018). Similarly, studies have reported the use of ZnO NPs for removal of blue cat 41 dyes. The efficacy of decomposing blue cat 41 dye was proven through the use of a mixture of ZnO NPs, ultrasound, and hydrogen peroxide (H_2O_2). Although no deterioration was found in the presence of H_2O_2 alone, the addition of ultrasonication enhanced the degradation efficiency to 55%. The highest achievable efficiency for the breakdown of blue cat 41 dye was 73% when H_2O_2, ultrasound, and ZnO NPs were combined (Golmohammadi et al. 2016). ZnO nanohexagon photocatalysts anchored by La^{3+} ions eliminated the wastewater rhodamine B (RhB) dye with 97% removal efficiency (Karuppasamy et al. 2021). Kadam et al. fabricated Ag–ZnO core–shell nanostructures through a straightforward wet chemical method (Kadam et al. 2018). The core–shell structure exhibited significantly enhanced photocatalytic activity in degrading MO, achieving a degradation rate of 98.40%. This efficiency was six times more in comparison of pure ZnO NPs and four times higher than that of pure TiO_2 NPs. When it comes to *Staphylococcus aureus*, ZnO NPs exhibit superior antibacterial properties compared to five other metal oxide NPs (Kadam et al. 2018). The antibacterial activity of NPs can be altered by their size and the availability of visible light. ZnO NPs can also be triggered by UV light under ambient laboratory conditions. In the visible light spectrum, activated ZnO NPs can be employed as an antibacterial agent (Jones et al. 2008).

12.3.2 Silver Oxide (Ag_2O) NPs

Ag_2O NPs possess the characteristic of being p-type semiconductors with a small band gap ranging from 1.2 to 1.6 eV. This specific band gap allows them to efficiently absorb the entire range of solar radiation wavelengths. Therefore, Ag_2O NPs function very well as photocatalysts for the decomposition of organic contaminants in water. Ag_2O NPs have also found extensive use as antimicrobial applications because of their capacity to release silver ions from surface. The effectiveness of these generated ions is contingent upon precise surface area of NPs (Shume et al. 2020). Ag_2O NPs can be synthesized through various routes:

(1) Gas phase route: This encompasses physical methods such as electrochemical synthesis, evaporation/vapour condensation, arc discharge, etc.
(2) Liquid phase route: This entails employing chemical techniques such as chemical reduction, irradiation-assisted procedures, pyrolysis methods, etc.
(3) Solid phase routes: These procedures involve physical techniques such as high-energy ball milling and physically pulverizing a metal (Madeła 2019).

Sondi et al. first published the initial evidence of antibacterial application of silver NPs towards *Escherichia coli*, which is a notable contaminant of food. According to them, silver NPs accumulate in "pits" within the bacterial cell wall. Subsequently, the removal of free radicals from NPs disrupts the cells, leading to bacterial death. Another mechanism involves the removal of Ag^+ ions from surface of NPs, which also contributes to bacterial cell death (Sondi and Salopek-Sondi 2004). The antibacterial activity of silver ions released is considerably greater than that of direct contact between NPs and bacterial surfaces. Bactericidal activity is attributed to the formation of pits in bacteria's cell wall. Shorter silver NPs have a higher ability to enter these pits in contrast to bigger ones, enhancing their efficiency (Madeła 2019). Gram-positive bacteria, because of their thick cell walls, exhibit minimal susceptibility to Ag_2O NPs. In contrast, gram-negative bacteria, such as *Salmonella enterica*, *Streptococcus pyogenes*, *Enterococcus faecalis*, and other bacteria found in polluted water, are notably impacted by the presence of Ag_2O NPs. Gram-negative bacteria have been found to demonstrate a broader range of inhibition, with a measured value of 44 ± 0.12 at a concentration of 150 μg/ml. Additionally, their growth is more susceptible to Ag_2O NPs in comparison with gram-positive bacteria (Forstner et al. 2020).

Saha et al. demonstrated entire breakdown of methylene blue dye using Ag_2O NPs derived from *G. arborea* extract, through the process of photocatalysis under visible light exposure (Saha et al. 2017). The MO dye is degraded by the modified Ag_2O NPs in about 120 s thanks to their outstanding photocatalytic activity. They discovered that Ag_2O NPs photocatalytic activity was inversely correlated with the light's wavelength (Jiang et al. 2015). Another group observed the remarkable photocatalytic efficacy of Ag_2O NPs in breaking down Azo dyes when exposed to visible light (Bi et al. 2018). By mixing MO dye with Ag_2O NPs, researchers found a substantial decrease in the time needed for the entire photocatalytic degradation of Rhodamine B and Methylene blue dyes. The time was decreased to 18 min from 50 min for Rhodamine B, and to 8 min from 20 min for Methylene blue. These findings indicate that the availability of non-azo dyes could potentially increase the photocatalytic activity of Ag_2O NPs (Bi et al. 2018). Shune and his colleagues conducted an analysis of the photocatalytic activity of Ag_2O NPs. They performed this by using visible light to degrade MO dye. The researchers conducted a comparative analysis of the photocatalytic efficiency of Ag_2O NPs in relation to other NPs, including TiO_2 NPs. The progressive change in colour of the MO dye was detected, suggesting the beginning of deterioration in dye concentration. Their investigation showcased the total degradation of MO dye through the utilization of Ag_2O NPs (Shume et al. 2020). Ag_2O NPs were evaluated against the pathogenic bacteria after being produced from *Zephyranthes Rosea* flower extract. Ag_2O NPs were found to exhibit strong antibacterial action against *Staphylococcus*, mutant *Streptococcus*, and *Escherichia coli* (Maheshwaran et al. 2020). The silver oxide NPs, derived from the prop root of *Ficus benghalensis*, shown exceptional antibacterial properties towards *Lactobacilli sp.* and *Streptococcus* mutans (Manikandan et al. 2017). Li et al. conducted a study where they successfully synthesized Ag_2O NPs using plant powder extracted from *Lippia citriodora* (Li et al. 2019). The NPs demonstrated significant antifungal application towards *S. aureus* and outstanding antibacterial application towards *A. aureus*. In a separate study, Ag_2O NPs synthesized by utilizing *Bacillus thuringiensis* SSV. These NPs exhibited inhibiting properties against both gram-positive and gram-negative bacteria. The sole observed disparity was in the inhibiting properties of Ag_2O NPs on gram-positive and gram-negative bacteria (Vithiya et al. 2014).

12.3.3 Titanium Dioxide (TiO_2) NPs

Recently, TiO_2 NPs have been extensively studied among metal oxide NPs due to their exceptional photocatalytic properties. TiO_2 NPs offer several advantages, including photostability, affordability, and stability in both chemical and biological environments. Usually, TiO_2 shows the charge separation inside its particles because of its significant bandgap energy of 3.2 eV and when it is excited by UV light. TiO_2

NPs possess a wide range of selectivity, which makes them highly effective catalysts for breaking down diverse pollutants, including pesticides, organic chemicals, heavy metals, and polycyclic aromatic hydrocarbons (Kinsinger et al. 2015). When exposed to ultraviolet (UV) radiation, TiO_2 NPs emit reactive oxygen species (ROS), which aid in the breakdown of contaminants that come into contact with the NPs. Moreover, TiO_2 NPs produce hydroxyl radicals when exposed to UV radiation with a wavelength of less than 400 nm. This process enhances the breakdown of cellular structures and functions. Titanium dioxide NPs (TiO_2 NPs) have broad-spectrum antimicrobial applications for gram-positive and gram-negative bacteria, viruses, fungi, protozoa, and algae. There are multiple processes requires in the bactericidal activity of TiO_2 NPs. An essential attribute of TiO_2 NPs is their perpetual lifespan, as they remain unaltered in terms of features and characteristics throughout the decomposition of organic molecules and microbes. NPs are extensively utilized in cleaning of polluted water and anti-fouling due to their less toxicity, biological and chemical stability, and comparably low cost compared to different existing options (Yang et al. 2013).

In their study, Sudarsan et al. employed the wet chemical approach to synthesize TiO_2 NPs. They found that by employing only 30 ppm of these NPs and subjecting them to UV light irradiation, they were able to completely remove chromium and lead (Sudarsan et al. 2015). In another study, polycrystalline TiO_2@Carbon was produced by using *Mentha Aquatica* extract (Bathula et al. 2021). They found that this material achieved over 90% photocatalytic degradation of Rhodamine B (RhB) dye in wastewater. The wastewater underwent prior UV irradiation for duration of 1 h (Bathula et al. 2021). In another study, TiO_2 NPs fabricated using starch as a template, facilitate the production of the anatase phase of TiO_2 NPs. TiO_2 NPs that were created demonstrated photocatalytic properties when exposed to light, leading to the breakdown of Methylene Blue dye (Subhapriya and Gomathipriya 2018). Pena et al. have proven the efficacy of TiO_2 NPs in eliminating arsenate (AsO_4^{3-}) and arsenite (AsO_3^{3-}) (Pena et al. 2005). The researchers noted that TiO_2 NPs showed high efficacy in removing arsenate at a pH of 8 and arsenite at a pH of 7.5. Swathi et al. provided evidence of antibacterial properties of TiO_2 NPs for both gram-positive and gram-negative strains (Swathi et al. 2019). In addition, TiO_2 NPs can be produced using an extract from the *Aspergillus flavus* plant. These NPs have the ability to inhibit the growth of *E. coli* bacteria (Rajakumar et al. 2012).

12.3.4 Iron Oxide NPs

There are three prevalent types of iron oxide: magnetite (Fe_3O_4), maghemite (γ-Fe_2O_3), and hematite (α-Fe_2O_3). Out of these substances, magnetite and maghemite exhibit magnetic characteristics; however, hematite does not possess any magnetic properties. Iron oxide NPs have recently gained considerable attention because of their distinctive characteristics, such as a substantial surface area, superparamagnetism, high surface-to-volume ratio, higher stability, improved adsorption capabilities, low production cost, and straightforward separation processes. Various physical, chemical, and biological methods have been employed to synthesize magnetic NPs with the required surface chemistry. Iron oxide NPs can be readily synthesized with customized changes, which enhances their efficacy in wastewater treatment. The distinctive magnetic characteristics of these NPs distinguish them from other NPs currently on the market. The utilization of magnetic characteristics in conjunction with the adsorption approach significantly improves their efficacy in the elimination of heavy metal ions. Iron oxide NPs are also utilized in the process of photocatalytic degradation of dyes (Aragawal et al. 2021).

Shalini et al. created magnetite NPs by a simple co-precipitation technique to eliminate Chromium (VI) and lead (II) ions. These magnetite NPs had the highest absorption of Pb and Cr during the initial cycle, followed by a decrease in adsorption capacity in subsequent cycles (Rajput et al. 2016). During the initial cycle, a total of 7.91 mg/g of Cr^{+6} was absorbed. This absorption rate reduced to 5.0 mg/g in the second cycle and remained constant in the following cycles. The highest level of adsorption for Pb occurred at pH 5.0, while for Cr, it was recorded at pH 2.0. Sheikholeslami et al. produced magnetite NPs by modifying the co-precipitation process (Sheikholeslami et al. 2019). These NPs were used for the photocatalytic destruction of BTEX (toluene, benzene, ethyl benzene, and xylene) found in polluted water. A breakdown efficacy of 95% was achieved during a period of 5 days when exposed to UV light, while a degradation efficiency of 97% was achieved within duration of 90 min when exposed to visible light. In a study, the elimination of Reactive Black 5, an azo dye, using magnetite NPs was also observed. The study found that when the concentration of Reactive Black 5 increased, the photocatalytic breakdown of the dye reduced. However, the removal efficiency of the dye improved from 30 to 90% by increasing time, temperature, and concentration of the magnetite NPs (Chang and Shih 2018). Magnetic iron oxide NPs are capable of being utilized as adsorbents to eliminate colours found in actual textile and synthetic effluent. Another researcher, investigated the impact of contact duration, acidity, heat, and simultaneous contaminants. The adsorption of dyes in actual textile wastewater occurred rapidly, with equilibrium being reached in less than 125 min. The removal of bromocresol green (BCG) dye was shown to be efficient at a pH of 3, while crystal violet (CV) dye was efficiently eliminated at a pH of 10.5. An increase in temperature was found to boost adsorption efficacy of magnetite NPs for the removal of BCG and CV dye (Nassar et al. 2015). Magnetite NPs were produced by the pyrolysis technique and investigated their ability to degrade methyl orange dye through photocatalysis under visible light exposure (Wang et al. 2017). Farhan et al. produced hematite NPs by the Sol–Gel technique and subsequently subjected these NPs to calcination at a temperature of 400 °C (Farhan et al. 2017). The utilization of hematite NPs was employed to reduce the presence of Cr[VI] ions in wastewater. In their study, E. Veles et al. produced Fe_3O_4 and γ-Fe_2O_3 NPs by the co-precipitation technique (Vélez et al. 2016). The primary component used was iron salt, while sodium hydroxide was employed as the precipitating agent. *Aloe vera* was utilized as the stabilizing agent. The synthesized NPs effectively eliminated almost 87% of mercury ions.

12.3.5 Copper Oxide (CuO) NPs

CuO is classified as a p-type semiconductor due to its electronic properties. It possesses a bandgap, which is the energy difference between the valence band and the conduction band, ranging from 1.2 to 1.9 electron volts [eV]. It is typically encountered in two forms: cupric oxide (CuO) and cuprous oxide (Cu_2O). Copper oxide NPs are highly efficient adsorbents for treating polluted water due to their small dimensions, inexpensive manufacturing cost, large surface area, and natural accessibility for fabrication. CuO NPs have a maximum adsorption capacity of 3152 mg for fluoride, as reported by Ighalo et al. (2021). This suggests that they are highly effective in removing substantial quantities of fluoride from polluted water. The adsorption characteristics of CuO NPs are thermodynamically described as both spontaneous and endothermic (Darwish et al. 2019).

Bacterial disinfection is a notable use of copper and its derivatives. The antibacterial activity of Cu NPs is thought to occur due to the liberation of copper ions from the Cu NPs. The Cu ions that are released contribute to the eradication of bacteria by interfering with their cell membranes. In addition, oxidative stress is another mechanism used to eliminate bacteria, and its efficiency is determined by the pace at which CuO NPs decompose. The antibacterial activity of CuO NPs, synthesized using an extract derived from the *Caesalpinia bonducella* seed, was evaluated against *Aeromonas* and *Staphylococcus* bacteria. The investigation demonstrated that CuO NPs displayed more potent antibacterial properties against *Staphylococcus* in comparison with *Aeromonas*. The antibacterial action was ascribed to the generation of reactive oxygen species (ROSs) from CuO NPs (Sukumar et al. 2020). Zang et al. examined the immediate and prolonged impacts of CuO NPs on partial nitrification. It was noted that when the concentration of CuO NPs is below 1 mg/L, it has a negligible effect on the partial nitrification process (Zhang et al. 2017). Within the range of 3 to 10 mg/L, the presence of CuO NPs enhances the oxidation of ammonia. However, too high concentrations can result in organism mortality. Ruparelia et al. observed that copper oxide NPs (CuO NPs) release reactive oxygen species (ROSs) that cause disruption to bacteria (Ruparelia et al. 2008). The interaction of released ions with bacterial DNA molecules is the primary source of this disruption. Fatima et al. employed a sodium hydroxide alkaline solution as a waste product to remove Amaranth dye from copper oxide NPs by desorption (Fathima et al. 2008).

A group of co-workers produced CuO NPs by employing the electrochemical reduction technique, aided by tetrabutylammonium bromide (Zare et al. 2018). They then studied the antibacterial properties of these CuO NPs for strains of *E. coli* and *S. aureus*. This study demonstrated that CuO NPs had a stronger impact on gram-negative bacteria in comparison with gram-positive bacteria. The presence of CuO NPs greatly hindered the growth of gram-negative bacteria (Jadhav et al. 2011).

12.4 Research Opportunities and Future Perspectives

The studies reveal discrepancies in the degradation time, adsorption efficiency, and removal efficiencies of both organic and inorganic dyes by synthesized NPs for wastewater treatment. These variations underscore the complexities involved in environmental uses for treatment of polluted water and highlight the need for further research to optimize NP-based treatment methods for different pollutants in wastewater. The NPs are ranked according to their benefits and usability in wastewater treatment, as well as their effectiveness in removing various pollutants, colours, effluents, and hazardous waste water (HWW). This information is depicted in Fig. 12.3. The outcome will be determined by the synthesis techniques employed for the desired NPs, the biomolecules that are adsorbed onto surface of the NPs during synthesis, the size and shape of the generated NPs, their ability to disperse, and various parameters of the process such as pH, precursor concentration, and surfactants used, calcination temperatures, etc. Additionally, factors such as crystallinity, purity, phase formations, and their transformations will affect the stability of the NPs under different temperature and pressure conditions.

It is suggested that more optimization studies be conducted as well as the necessity of quickly extending the production scale with consistent repeatability through an easy-to-use, affordable method that eliminates toxicity and is environmentally benign. Future opportunities for young researchers will be realized through the optimization of reaction parameters within a specified procedure to produce synthesized NPs in dry powder form with consistency. While non-linear isotherms were infrequently seen, the majority of study focussed on corresponding linear isotherms.

When choosing an adsorption or photocatalytic degradation isotherm and modelling the kinetics of the data, the non-linear (NLR) analysis must be used instead of the linear (LR) approach. Because of this, adsorption parameter estimate using isotherm equations in their original form is more accurate (Osagie et al. 2021). Additionally, it was designed to minimize the gap between the experimental data

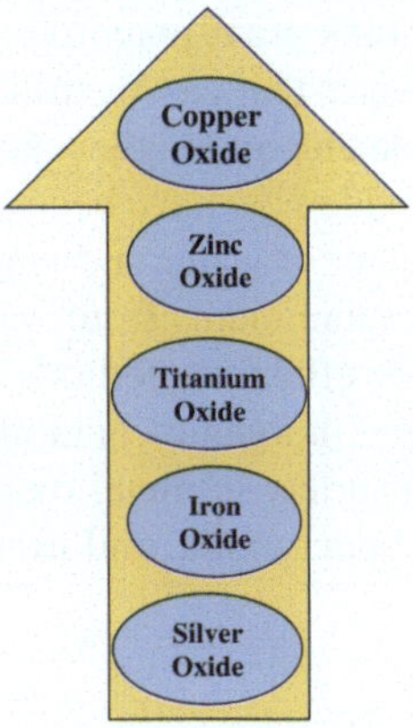

Fig. 12.3 Ranking of different NPs in the context of their use in wastewater treatment

that was actually collected and the isotherms that were predicted. The anticipated isotherms are always generated by NLR fitting to converge. It forecasted the potential degradation process and the generated NPs maximal capacity to remove dyes, wastewater, heavy metals, and other contaminants (Osagie et al. 2021; Muneer et al. 2019; Vasantharaj et al. 2021). To forecast the immediate degradation intermediates, the breakdown process should be examined using appropriate chromatograms, such as HP-LC, LC–MS, GC–MS, etc. It might provide control over the division of groups and chromophoric group disintegration of different dyes with molecular weight. It causes intermediate fragments of degradation to develop, which causes low molecular weight fragments of mineralization and decolorization (Muneer et al. 2019).

12.5 Conclusion

Indeed, wastewater from various sources can contain a range of harmful contaminants that pose risks to both human health and the environment. Heavy metals are particularly concerning due to their toxicity and persistence. Additionally, suspended particles, total dissolved solids, and organic matter further contribute to the pollution load. While conventional methods such as sedimentation, filtration, and flocculation/coagulation have been used for wastewater treatment, they have limitations including high energy requirements, large spatial footprint for infrastructure, and significant capital costs. These constraints have led to the exploration of alternative technologies, including NM-based approaches, which offer promising solutions for more efficient and cost-effective wastewater treatment. Because of their distinctive qualities, which include high surface area-to-volume ratio, high adsorption capacity, and superior catalysis, NMs efficiently eliminate the contaminant in an environmentally responsible way. Because of their tiny size, high catalytic activity, vast surface area, good adsorption qualities, and great selectivity towards heavy metals, metal oxide NPs are extensively accepted for application in contaminated water treatment. Wastewater holds great potential as a viable source for acquiring potable and uncontaminated water that can be utilized for various purposes. Reusing wastewater would result in cleaner and higher-quality water availability, than what is currently available. Metal oxide NPs have been shown to have promising prospective applications in the treatment of water and wastewater. The most widely utilized techniques for eliminating contaminants from wastewater are photocatalysis and adsorption procedures. These are effective methods for getting rid of hazardous materials. The removal of organic and inorganic contaminants from wastewater has been shown in the literature using a variety of metal oxide NPs including silver dioxide, titanium dioxide, zinc oxide, copper oxide, and iron oxide NPs.

References

Abdelbasir SM, Shalan AE (2019) An overview of NMs for industrial wastewater treatment. Korean J Chem Eng 36:1209–1225. https://doi.org/10.1007/s11814-019-0306-y

Abu Bakar Sidik D, Hairom NH, Zainuri NZ, Desa AL, Misdan N, Yusof N, Mohd Aripen NS (2018) Photocatalytic degradation of industrial dye wastewater using zinc oxide polyvinylpyrrolidone NPs. Malays J Anal Sci 22(4):1–9. https://doi.org/10.17576/mjas-2018-2204-16

Amil Usmani M, Khan I, H Bhat A, Pillai R, Ahmad N, K Mohamad Haafiz M, & Oves M (2017). Current trend in the application of NPs for waste water treatment and purification: a review. Current Organic Synthesis 14(2): 206–226. https://doi.org/10.2174/1570179413666160928125328

Aragawal TA, Bogale FM, Aragawal BA (2021) Iron-based NPs in wastewater treatment: a review on synthesis methods, applications, and removal mechanisms. J Saudi Chem Soc 25(8):101280. https://doi.org/10.1016/j.jscs.2021.101280

Auffan M, Rose J, Bottero JY, Lowry GV, Jolivet JP, Wiesner MR (2009) Towards a definition of inorganic NPs from an environmental, health and safety perspective. Nat Nanotechnol 4(10):634–641. [Online]. Available. https://doi.org/10.1038/nnano.2009.242

Bathula C, Rabani I, Sekar S, Youi HK, Choy JY, Kadam A, Kim HS (2021) Enhanced removal of organic dye by activated carbon decorated TiO_2 NPs from Mentha Aquatica leaves via ultrasonic approach. Ceram Int 47(6):8732–8739. https://doi.org/10.1016/j.ceramint.2020.12.282

Bi N, Zheng H, Zhu Y, Jiang W, Liang B (2018) Visible-light-driven photocatalytic degradation of non-azo dyes over Ag_2O and its acceleration by the addition of an azo dye. J Environ Chem Eng 6(2):3150–3160. https://doi.org/10.1016/j.jece.2018.04.047

Chahal C, Van Den Akker B, Young F, Franco C, Blackbeard J, Monis P (2016) Pathogen and particle associations in wastewater: significance and implications for treatment and disinfection processes. Adv Appl Microbiol 97:63–119. https://doi.org/10.1016/bs.aambs.2016.08.001

Chang M, Shih YH (2018) Synthesis and application of magnetic iron oxide NPs on the removal of Reactive Black 5: reaction mechanism, temperature and pH effects. J Environ Manage 224:235–242. https://doi.org/10.1016/j.jenvman.2018.07.021

Chen Z, Li Y, Guo M, Xu F, Wang P, Du Y, Na P (2016) One-pot synthesis of Mn-doped TiO_2 grown on graphene and the mechanism for removal of Cr (VI) and Cr (III). J Hazard Mater 310:188–198. https://doi.org/10.1016/j.jhazmat.2016.02.034

Darwish AAA, Rashad M, AL-Aoh HA (2019) Methyl orange adsorption comparison on NPs: isotherm, kinetics, and thermodynamic studies. Dye Pigment 160:563–571. https://doi.org/10.1016/j.dyepig.2018.08.045

Daus B, Wennrich R, Weiss H (2004) Sorption materials for arsenic removal from water: a comparative study. Water Res 38(12):2948–2954. [Online]. http://www.sciencedirect.com/science/article/pii/S0043135404001678

Dreaden EC, Alkilany AM, Huang X, Murphy CJ, El-Sayed MA (2012) The golden age: gold NPs for biomedicine. Chem Soc Rev 41(7):2740–2779. https://doi.org/10.1039/C1CS15237H

Farhan AM, Abdullah HI, Ali AJ (2017) Sol-Gel synthesis of hematite NPs and photo degradation of cibacron red FN-R Dye. Baghdad Sci J 14(3):0539–0539. https://doi.org/10.21123/bsj.2017.14.3.0539

Fathima NN, Rajaram A, Sreedhar B, Mandal AB (2008) The formation of copper oxide nanorods in the presence of various surfactant micelles. Indian J Sci Technol, pp 1–6. https://doi.org/10.17485/ijst/2008/v1i7/29592

Forstner C, Orton T G, Wang P, Kopittke PM, Dennis PG (2020) Wastewater treatment processing of silver NPs strongly influences their effects on soil microbial diversity. Environ Sci Technol 54(21):13538–13547. https://doi.org/10.1021/acs.est.0c01312

Fouda A, Hassan SED, Abdel-Rahman MA, Farag MM, Shehal-deen A, Mohamed AA, Azab MS (2021) Catalytic degradation of wastewater from the textile and tannery industries by green synthesized hematite (α-Fe_2O_3) and magnesium oxide (MgO) NPs. Curr Res Biotechnol 3:29–41. https://doi.org/10.1016/j.crbiot.2021.01.004

Ganapathy NRV, Krishman V, Veeraraghavan AJ, Devaraj G (2019) Biosynthesis of zinc oxide NPs from Plectranthus amboinicus and its photocatalytic effect on wastewater treatment. Int J Recent Technol Eng 8(4S2):660–663. https://doi.org/10.35940/ijrte.d1140.1284s219

Golmohammadi S, Ahmadpour M, Mohammadi A, Alinejad A, Mirzaei N, Ghaderpoori M, Ghaderpoori A (2016) Removal of blue cat 41 dye from aqueous solutions with ZnO NPs in combination with US and US-H2O2 advanced oxidation processes. https://doi.org/10.15171/ehem.2016.08

Helmer R, Hespanhol I (1997) Water pollution control: a guide to the use of water quality management principles. CRC Press. [Online]. Available: https://extranet.who.int/iris/restricted/handle/10665/41967. https://doi.org/10.1016/j.jece.2021.105772

R.W. Herschy, Water quality for drinking: WHO guidelines, Encycl Earth Sci Ser, pp 876–883. https://doi.org/10.1007/978-1-4020-4410-6_184

Hu J, Song Z, Chen L, Yang H, Li J, Richards R (2010) Adsorption properties of MgO (111) nanoplates for the dye pollutants from wastewater. J Chem Eng Data 55(9):3742–3748

Ighalo JO, Sagboye PA, Umenweke G, Ajala OJ, Omoarukhe FO, Adeyanju CA, Adeniyi AG (2021) CuO NPs (CuO NPs) for water treatment: a review of recent advances. Environ Nanotechnol, Monit Manag 15:100443. https://doi.org/10.1016/j.enmm.2021.100443

Jadhav S, Gaikwad S, Nimse M, Rajbhoj A (2011) Copper oxide NPs: synthesis, characterization and their antibacterial activity. J Cluster Sci 22:121–129. https://doi.org/10.1007/s10876-011-0349-7

Jiang W, Wang X, Wu Z, Yue X, Yuan S, Lu H, Liang B (2015) Silver oxide as superb and stable photocatalyst under visible and near-infrared light irradiation and its photocatalytic mechanism. Ind Eng Chem Res 54(3):832–841

Jones N, Ray B, Ranjit KT, Manna AC (2008) Antibacterial activity of ZnO NP suspensions on a broad spectrum of microorganisms. FEMS Microbiol Lett 279(1):71–76. https://doi.org/10.1111/j.1574

Kadam AN, Bhopate DP, Kondalkar VV, Majhi SM, Bathula CD, Tran AV, Lee SW (2018) Facile synthesis of Ag-ZnO core–shell nanostructures with enhanced photocatalytic activity. J Ind Eng Chem 61:78–86. https://doi.org/10.1016/j.jiec.2017.12.003

Karuppasamy K, Rabani I, Vikraman D, Bathula C, Theerthagiri J, Bose R, Kim HS (2021) ZIF-8 templated assembly of La^{3+}-anchored ZnO distorted nano-hexagons as an efficient active photocatalyst for the detoxification of rhodamine B in water. Environ Pollut 272:116018. https://doi.org/10.1016/j.envpol.2020.116018

Kinsinger N, Honda R, Keene V, Walker SL (2015) Titanium dioxide NP removal in primary prefiltration stages of water treatment: role of coating, natural organic matter, source water, and solution chemistry. Environ Eng Sci 32(4):292–300. https://doi.org/10.1089/ees.2014.0288

Li R, Chen Z, Ren N, Wang Y, Wang Y, Yu F (2019) Biosynthesis of silver oxide NPs and their photocatalytic and antimicrobial activity evaluation for wound healing applications in nursing care. J Photochem Photobiol, B 199:111593

Madeła M (2019). Impact of silver NPs on wastewater treatment in the SBR. In: E3S web of conferences, vol 86, p 00027. EDP Sciences

Maheshwaran G, Bharathi AN, Selvi MM, Kumar K, Kumar RM, Sudhahar S (2020) Green synthesis of Silver oxide NPs using Zephyranthes Rosea flower extract and evaluation of biological activities. J Environ Chem Eng 8(5):104137. https://doi.org/10.1016/j.jece.2020.104137

Manikandan V, Velmurugan P, Park JH, Chang WS, Park YJ, Jayanthi P, Oh BT (2017) Green synthesis of silver oxide NPs and its antibacterial activity against dental pathogens. 3 Biotech 7:1–9. https://doi.org/10.1007/s13205-017-0670-4

Muneer M, Saeed M, Bhatti IA, Haq AU, Khosa MK, Jamal MA, Ali S (2019) Radiation induced degradation of Congo red dye: a mechanistic study. Nukleonika 64(2):49–53

Naseem T, Durrani T (2021) The role of some important metal oxide NPs for wastewater and antibacterial applications: a review. Environ Chem Ecotoxicol 3:59–75. https://doi.org/10.1016/j.enceco.2020.12.001

Nassar NN, Marei NN, Vitale G, Arar LA (2015) Adsorptive removal of dyes from synthetic and real textile wastewater using magnetic iron oxide NPs: thermodynamic and mechanistic insights. Can J Chem Eng 93(11):1965–1974. https://doi.org/10.1002/cjce.22315

Osagie C, Othmani A, Ghosh S, Malloum A, Esfahani ZK, Ahmadi S (2021) Dyes adsorption from aqueous media through the nanotechnology: a review. J Market Res 14:2195–2218. https://doi.org/10.1016/j.jmrt.2021.07.085

Pena ME, Korfiatis GP, Patel M, Lippincott L, Meng X (2005) Adsorption of As (V) and As (III) by nanocrystalline titanium dioxide. Water Res 39(11):2327–2337. https://doi.org/10.1016/j.watres.2005.04.006

Rabani I, Lee SH, Kim HS, Yoo J, Park YR, Maqbool T, Seo YS (2021) Suppressed photocatalytic activity of ZnO based Core@ Shell and RCore@ Shell nanostructure incorporated in the cellulose nanofiber. Chemosphere 269:129311. https://doi.org/10.1016/j.chemosphere.2020.129311

Rajakumar G, Rahuman AA, Roopan SM, Khanna VG, Elango G, Kamaraj C, Velayutham K (2012) Fungus-mediated biosynthesis and characterization of TiO^2 NPs and their activity against pathogenic bacteria. Spectrochim Acta Part A Mol Biomol Spectrosc 91:23–29. https://doi.org/10.1016/j.saa.2012.01.011

Rajput S, Pittman CU, Mohan D (2016) Magnetic magnetite (Fe_3O_4) NP synthesis and applications for lead (Pb_{2+}) and chromium (Cr_{6+}) removal from water. J Colloid Interface Sci 468:334–346. https://doi.org/10.1016/j.jcis.2015.12.008

Ruparelia JP, Chatterjee AK, Duttagupta SP, Mukherji S (2008) Strain specificity in antimicrobial activity of silver and copper NPs. Acta Biomater 4(3):707–716. https://doi.org/10.1016/j.actbio.2007.11.006

Saha J, Begum A, Mukherjee A, Kumar S (2017) A novel green synthesis of silver NPs and their catalytic action in reduction of Methylene Blue dye. Sustain Environ Res 27(5):245–250. https://doi.org/10.1016/j.serj.2017.04.003

Sheikholeslami Z, Yousefi Kebria D, Qaderi F (2019) Investigation of photocatalytic degradation of BTEX in produced water using γ-Fe_2O_3NP. J Therm Anal Calorim 135:1617–1627. https://doi.org/10.1007/s10973-018-7381-x

Shume WM, Murthy HA, Zereffa EA (2020) A review on synthesis and characterization of Ag2O NPs for photocatalytic applications. J Chem 2020:1–15. https://doi.org/10.1155/2020/5039479

Singh S, Kumar V, Romero R, Sharma K, Singh J (2019) Applications of NPs in wastewater treatment. Nanobiotechnology in Bioformulations 395:418

Slokar YM, Le Marechal AM (1998) Methods of decoloration of textile wastewaters. Dyes Pigm 37(4):335–356. https://doi.org/10.1016/S0143-7208(97)00075-2

Sondi I, Salopek-Sondi B (2004) Silver NPs as antimicrobial agent: a case study on E. coli as a model for Gram-negative bacteria. J Colloid Interface Sci 275(1):177–182. https://doi.org/10.1016/J.JCIS.2004.02.012

Sreekala G, Beevi F A, Beena B (2019) Adsorption of lead (II) ions by ecofriendly copper oxide NPs. https://doi.org/10.13005/ojc/350615

Srivastava V, Gusain D, Sharma YC (2013) Synthesis, characterization and application of zinc oxide NPs (n-ZnO). Ceram Int 39(8):9803–9808. https://doi.org/10.1016/j.ceramint.2013.04.110

Subhapriya S, Gomathipriya PJMP (2018) Green synthesis of titanium dioxide (TiO_2) NPs by Trigonella foenum-graecum extract and its antimicrobial properties. Microb Pathog 116:215–220. https://doi.org/10.1016/j.micpath.2018.01.027

Sudarsan JS, Prasanna K, Baskar G, George RB (2015) Role of titanium oxide NP on heavy metal reduction in electroplating waste water treatment. Int J ChemTech Res 7:547–553

Sukumar S, Rudrasenan A, Padmanabhan Nambiar D (2020) Green-synthesized rice-shaped copper oxide NPs using Caesalpinia bonducella seed extract and their applications. ACS Omega 5(2):1040–1051. https://doi.org/10.1021/acsomega.9b02857

Swathi N, Sandhiya D, Rajeshkumar S, Lakshmi T (2019) Green synthesis of titanium dioxide NPs using Cassia fistula and its antibacterial activity. Int J Res Pharm Sci 10(2):856–860. https://doi.org/10.26452/ijrps.v10i2.261

Vasantharaj S, Sathiyavimal S, Senthilkumar P, Kalpana VN, Rajalakshmi G, Alsehli M, Pugazhendhi A (2021) Enhanced photocatalytic degradation of water pollutants using bio-green synthesis of zinc oxide NPs (ZnO NPs). J Environ Chem Eng 9(4):105772

Vélez E, Campillo G E, Morales G, Hincapié C, Osorio J, Arnache O, Jaramillo F (2016) Mercury removal in wastewater by iron oxide NPs. J Phys: Conf Ser 687(1):012050. IOP Publishing. https://doi.org/10.1088/1742-6596/687/1/012050

Vithiya K, Kumar R, Sen S (2014) Antimicrobial activity of biosynthesized silver oxide NPs. J. Pure Appl. Microbiol 4:3263–3268

Wang J, Shao X, Zhang Q, Tian G, Ji X, Bao W (2017) Preparation of mesoporous magnetic Fe_2O_3 NP and its application for organic dyes removal. J Mol Liq 248:13–18. https://doi.org/10.1016/j.molliq.2017.10.026

Yadav A, Narwal S, Deswal T, Sindhu A, Langyan R, Kumar N, Dhankhar R (2024) Synthesis and nanotechnological applications of multi-efficient zinc oxide NPs-a review. Asian J Chem 36(2):263–274. https://doi.org/10.14233/ajchem.2024.28417

Yang Y, Zhang C, Hu Z (2013) Impact of metallic and metal oxide NPs on wastewater treatment and anaerobic digestion. Environ Sci Process Impacts 15(1):39–48. https://doi.org/10.1039/C2EM30655G

Yang J, Hou B, Wang J, Tian B, Bi J, Wang N, Huang X (2019a) NMs for the removal of heavy metals from wastewater. Nms 9(3):424. https://doi.org/10.3390/nano9030424

Yang L, Meng F, Qu X, Xia L, Huang F, Qin S, Liu H (2019b) Multiple-twinned silver NPs supported on mesoporous graphene with enhanced antibacterial activity. Carbon 155:397–402. https://doi.org/10.1016/j.carbon.2019.09.002

Zafar MN, Dar Q, Nawaz F, Zafar MN, Iqbal M, Nazar MF (2019) Effective adsorptive removal of azo dyes over spherical ZnO NPs. J Market Res 8(1):713–725. https://doi.org/10.1016/j.jmrt.2018.06.002

Zare M, Namratha K, Byrappa K, Surendra DM, Yallappa S, Hungund B (2018) Surfactant assisted solvothermal synthesis of ZnO NPs and study of their antimicrobial and antioxidant properties. J Mater Sci Technol 34(6):1035–1043. https://doi.org/10.1016/j.jmst.2017.09.014

Zhang X, Zhou Y, Yu B, Zhang N, Wang L, Fu H, Zhang J (2017) Effect of copper oxide NPs on the ammonia removal and microbial community of partial nitrification process. Chem Eng J 328:152–158. https://doi.org/10.1016/j.cej.2017.07.028

Chapter 13
Harnessing Sustainable Nanomaterials for Next-Generation Wastewater Treatment: Challenges and Opportunities

Deeptimayee Pal and Abhishek Kumar

Abstract The present study examines the potential of sustainable nanomaterials (NMs) to address the critical challenges of wastewater remediation, with a specific emphasis on developing environmentally benign and highly efficient solutions. Researchers are studying many environmentally friendly NMs, such as nanoparticles (NPs), nanometal oxides, nanofibers, and nanocomposites, to determine their effectiveness in adsorbing pollutants from wastewater. This study evaluates the efficacy of these environmentally friendly NMs in capturing heavy metals, organic pollutants, and other harmful substances, emphasizing their potential for real-world use in water purification processes. The research also examines the biodegradability and recyclability of these NMs, guaranteeing that their use enhances the overall sustainability of the wastewater treatment process. Furthermore, this article examines the difficulties and potential benefits of incorporating environmentally friendly NP into current methods of treating wastewater. Factors to be considered include the potential to expand, cost efficiency, and compliance with regulations, focusing on implementing approaches that support both environmental sustainability and technical efficiency. Using sustainable NMs in wastewater treatment demonstrates a commitment to achieving a harmonious equilibrium between safeguarding the environment, conserving resources, and fostering technological advancement. The results highlight the capacity of sustainable NMs to serve as crucial elements in developing environmentally friendly and highly efficient methods for treating wastewater.

Keywords Wastewater remediation · Nanotechnology · NMs · Sustainability

D. Pal (✉) · A. Kumar
Department of Chemical Engineering, National Institute of Technology Rourkela, Rourkela, Odisha, India
e-mail: deeptimayeenitrkl@outlook.com

N. Agarwal et al. (eds.), *Nanomaterials in Wastewater Research*, Advances in Wastewater Research, https://doi.org/10.1007/978-981-96-3908-3_13

13.1 Introduction

The environment is continuously stressed due to growing industry and urbanization. Water shortage has become the primary environmental concern facing mankind and is among the most urgent global challenges. Over the next several decades, the population in developing areas is projected to rise rapidly, increasing the need for clean water in many sectors such as households, agriculture, industry, and energy (Pendergast and Hoek 2011). According to projections from the World Health Organization in 2014, almost 50% of the global population is expected to live in places with water scarcity by 2025. According to the United Nations (2016), in 2015, only around 20% of worldwide wastewater underwent adequate treatment, indicating a substantial deficiency in wastewater management. The issue is considerably more severe in less economically developed nations since almost 70% of industrial effluent is released into the environment without sufficient treatment. This unprocessed sewage presents significant hazards to the quality of the environment and the well-being of the general populace (Boretti and Rosa 2019).

The existing infrastructure for wastewater treatment and the generation of safe, easily accessible water faces challenges in meeting the ever-increasing standards and the rising need for top-notch water in both developed and emerging countries (Pal and Sen 2024a). The difficulty is worsened by deteriorating infrastructure, restricted financial resources, and technical limitations. Therefore, there is a pressing need for sophisticated water treatment methods that are extremely effective and economical. The development of such technology is crucial for tackling water shortages and maintaining the sustainability of water management. Nanotechnology innovations provide prospective answers by strengthening the elimination of pollutants from water, raising the effectiveness of treatment procedures, and decreasing operating expenses. These technological improvements may assist in reducing the disparity between existing water supply capabilities and future water needs, therefore contributing to the overarching objective of attaining sustainable water security for everyone (Zhang et al. 2016).

13.2 NMs

Nanotechnology is the scientific study, examination, and use of materials at the nanoscale, where molecules and atoms display distinct properties compared to their bigger forms. The term "nanoscale" precisely delineates NMs as materials with dimensions ranging from 1 to 100 nm (Tesh and Scott 2014). Nevertheless, the broad term may include materials with diameters ranging from a few nanometers to several hundred nanometers. The emergence of this subject occurred in the late nineteenth century with the fabrication and use of NP in several scientific fields. The introduction of innovative tools, such as the scanning tunneling microscope (STM), had a crucial role in identifying and investigating NMs, signifying the beginning of fundamental research and methodologies in nanotechnology.

NMs have several crucial characteristics that make them very promising for diverse applications. These characteristics include a large surface area, increased reactivity, modifiable pore volume, and notable electrostatic, hydrophilic, and hydrophobic interactions. Moreover, their innate antibacterial qualities contribute to their adaptability. The distinctive characteristics of NMs make them very useful as adsorption catalysts (Ali 2012) and in photocatalysis (Ayati et al. 2014), especially for eliminating contaminants from the environment. NMs may be classified into many categories, including NP, nanometal oxides, nanofibers, and nanocomposites. Each category includes diverse materials for particular purposes, such as water purification, wastewater, and groundwater. Effective management of polluted water is essential due to hazardous substances such as heavy metal ions, radioactive nuclides, organic and inorganic chemicals, pathogens, and microbes. These pollutants pose significant risks to the quality of the environment, the health of humans, and the overall visual appeal. Hence, using NMs in water treatment procedures tackles these difficulties by capitalizing on their distinct characteristics to eliminate a diverse range of contaminants effectively, thus ensuring the protection of the environment and public well-being (Zhang et al. 2016).

13.2.1 NPs

NP have received considerable interest in recent decades and have been extensively studied and produced for use in various sectors, including biology, sensing, healthcare, and catalytic chemistry. One application that has gained significant attention is the use of NP in wastewater treatment, primarily because of their distinct characteristics. NP possess a tiny size and a big surface area, which results in a significant capacity and reactivity for adsorption (Mauter et al. 2018). This attribute makes them very efficient in adsorption and eliminating impurities from wastewater.

13.2.1.1 Zinc NPs

The presence of Zinc NP (Zn NPs) in ecosystems is growing due to their extensive use in various products. NPs are used in sunscreen, packaging materials, paints, plastics, dietary supplements, and cosmetic goods, which leads to their presence in the environment. The production of Zn NPs in the European Union has increased to 1.6 metric tons per year, indicating their expanding use in industry. Zn NPs are often considered a preferable substitute for iron (Fe) NP due to many factors. Zinc is a stronger reducing agent than iron due to its high negative reduction potential. This attribute makes it very efficient in reducing impurities. As a result, Zn NPs have been widely used in wastewater treatment procedures. Zn NPs have a much greater rate of degradation of pollutants in wastewater compared to zerovalent iron NP, indicating the superior effectiveness of Zn NPs in this context. Multiple studies show that Zn NPs primarily function by catalyzing the dehalogenation process in wastewater treatment. Dehalogenation eliminates halogen atoms from organic molecules, an essential step in detoxifying several contaminants. In addition to dehalogenation, studies have investigated the function of Zn NPs in traditional activated sludge procedures used in wastewater treatment facilities. These investigations have shown that Zn NPs can impact several facets of these processes (Puay et al. 2015).

Nevertheless, the influence of Zn NPs is not uniformly advantageous. Zn NPs at a low concentration of 5.0 mg/L have been seen to hinder the development of nitrifying bacteria in sequencing batch reactors (SBRs), a specific form of wastewater treatment procedure. This inhibition decreases the elimination of ammonium nitrogen (NH_4^+-N), an essential process in the nitrogen cycle and wastewater treatment. Additional investigation has shown that the 50% inhibitory concentration (IC50) of Zn NPs for ammonium-oxidizing bacteria is 13.1 mg/L, demonstrating a significant harmful impact at this concentration. Zn NPs have also shown the capability to break down intricate contaminants under normal environmental settings. For example, studies have shown that they can efficiently degrade octachlorodibenzo-p-dioxin, a hazardous and long-lasting organic pollutant, into less damaging chlorinated congeners (Bokare et al. 2013). This feature highlights the capacity of Zn NPs to effectively tackle some of the most difficult pollutants in the environment.

13.2.1.2 Zerovalent Iron (ZVI) NPs

ZVI NPs have shown considerable potential in advancing remediation technology for on-site wastewater treatment. Extensive empirical research has shown their capacity to neutralize a wide range of dangerous substances, such as hexachlorocyclohexane (lindane), carbon tetrachloride, vinyl chloride, chlorinated organic compounds (COCs), and trichloroethane. These NP are highly effective and cost-efficient as electron donors, making them essential for wastewater treatment. The primary emphasis of most research on using iron NP in wastewater treatment has been their capacity to induce reductive dehalogenation processes (Anjum et al. 2019). Reductive dehalogenation is a chemical reaction in which halogen atoms in organic compounds are substituted with hydrogen atoms, detoxifying the molecules. Under circumstances without oxygen, ZVI NP undergo oxidation by water molecules or protons, producing Fe (II) and hydrogen gas (H_2). Fe (II) and H_2 are effective reducing agents that may degrade different pollutants found in wastewater.

When Fe (II) interacts with impurities, it undergoes oxidation and transforms into Fe (III), resulting in the formation of ferric hydroxide ($Fe(OH)_3$). This chemical functions as a very efficient flocculants, facilitating the aggregation and elimination of organic and inorganic impurities from wastewater by precipitation. Flocculation is crucial in physically eliminating contaminants from water, improving the treatment process's overall effectiveness. ZVI NP have exceptional efficacy in the oxidation and degradation of diverse organic pollutants. They transfer electrons to an oxygen molecule, generating hydrogen peroxide (H_2O_2). H_2O_2 undergoes further reduction to become water (H_2O). In addition, the combination of H_2O_2 with Fe(II) produces hydroxyl radicals (•OH), which are very reactive and have powerful oxidizing properties against many organic molecules (Hao et al. 2015). The hydroxyl radicals can efficiently decompose complex organic contaminants into less hazardous and more straightforward compounds. ZVI NP are extensively used to eliminate various pollutants due to their versatile characteristics, such as oxidation, precipitation, reduction, and adsorption. The substances mentioned include nitroaromatic compounds, inorganic anions, phosphates, radioactive elements, nitrates, phenols, organic dyes, and chlorinated and halogenated organic compounds. ZVI NPs are versatile and effective, making them a vital tool for global environmental clean-up efforts.

13.2.1.3 Silver NPs (Ag NPs)

Ag NPs have received significant interest due to their powerful antibacterial characteristics, which make them particularly efficient in water disinfection. These NP have shown effectiveness against several diseases, such as bacteria, viruses, and fungi. The primary mechanism by which Ag NPs exert their antibacterial effect is via their contact with the bacterial cell membrane. Ag NPs connect to the membrane and enhance permeability by producing reactive oxygen species or free radicals. This, in turn, undermines the integrity of the cell membrane. The impairment of the cell membrane causes a disturbance in the organization of the cell, eventually resulting in cell death, also known as apoptosis. Ag NPs also engage with the phosphorus and sulfur components in the DNA of bacterial cells, resulting in substantial harm to these genetic resources and thus impeding the bacteria's ability to reproduce and survive. Besides directly interacting with the cell membrane and DNA, the breakdown of Ag NPs results in the release of silver ions (Ag^+), which have potent antibacterial capabilities. The ions engage with the thiol groups of vital enzymes in bacterial cells, causing them to lose their function and interrupt important biological processes critical for the cells' survival (Rudakiya and Pawar 2017).

Nevertheless, a drawback linked to Ag NPs is their inclination to form aggregates in liquid media, decreasing their long-term efficacy as antibacterial agents. To tackle this problem, many approaches have been devised to immobilize Ag NPs, hence improving their stability and efficacy. A technique that may be used includes the immobilization of Ag NPs on cellulose fibers. Research has shown that these cellulose sheets coated with Ag NPs had potent antibacterial characteristics against *Enterococcus faecalis* and *Escherichia coli* suspensions, successfully rendering these bacteria inactive in filtration systems. In addition, Ag NPs have been added to polyethersulfone (PES) microfiltration membranes using several chemical reduction techniques. Integrating Ag NPs into PES membranes effectively inhibits the growth of microorganisms on the membrane surface. The PES-Ag NP membranes possess strong antibacterial characteristics, making them favorable for wastewater. The antimicrobial effect of Ag NPs in these membranes is achieved through multiple mechanisms, such as causing physical disturbance through oxidative stress, interrupting specific microbial processes through oxidation, and interfering with essential cellular components or structures in the microbial cell membrane (Choi et al. 2014).

In the last two decades, there has been a notable rise in research dedicated to immobilizing Ag NPs on different membranes and ceramic materials. The increased interest in Ag NPs is motivated by their ability to decrease biofouling and provide efficient disinfection in sewage and wastewater treatment systems. Biofouling, the process of bacteria accumulating on surfaces, is a significant obstacle in wastewater treatment, resulting in decreased effectiveness and higher maintenance expenses. The bactericidal characteristics of Ag NPs may effectively combat this problem by suppressing the development of microorganisms on treated surfaces. Researchers have developed high-porosity filters made of materials such as sawdust and clay to improve the effectiveness of Ag NP-based therapies. The efficacy of these filters in eliminating pathogens, such as E. coli, from wastewater is exceptional. The filters' high porosity facilitates increased surface area interaction between the wastewater and the filter material, resulting in enhanced removal efficiency of pollutants. The use of colloidal chemistry has significantly enhanced the efficiency of these filters. When colloidal NPs are added to the filter matrix, they greatly improve the filter's capacity to eliminate harmful microorganisms. Research has shown that filters enhanced with colloidal NP may effectively eliminate E. coli bacteria, with removal rates ranging from 97.8 to 100% (Abebe et al. 2014). Colloidal NPs enhance the antimicrobial surface area and provide more active sites for bacterial interaction, resulting in increased disinfection rates.

13.2.2 Nano Metal Oxides

Nano metal oxides (NMOs) have shown significant promise as an eco-friendly, cost-efficient, and sustainable solution for treating wastewater. These NPs possess several favorable characteristics that make them appropriate for this specific purpose. Firstly, NMOs possess a limited distance for intraparticle diffusion and a large specific surface area. Consequently, the reduced distance between pollutants and particles accelerates adsorption rates. In addition, NMOs may undergo compression without seeing a substantial decrease in their surface area, thereby maintaining a high adsorption capacity even when subjected to mechanical pressure. In addition, they have many adsorption sites, which enhances their efficiency in absorbing contaminants from wastewater. Similarly, some metal oxide NP display superparamagnetism. This characteristic enables easy extraction from the treated water by applying magnetic fields, hence facilitating the retrieval and recycling of the NP. NMOs have an advantage over activated carbon as adsorbents due to their ability to be separated from water more easily (Corsi et al. 2018). The main mechanism by which NMOs adsorb involves forming complexes between dissolved metals and oxygen molecules in metal oxide groups. The surface of NMOs facilitates the efficient adsorption of heavy metals by intraparticle diffusion along the walls, which is often the step that limits the adsorption rate.

The significant adsorption capacity and fast kinetics of NMOs are due to their abundant surface reaction sites, such as edges, vacancies, and corners. These sites provide many active sites for interacting with contaminants. The nanoscale phenomenon, observed when the size of nanomagnetite particles is decreased, greatly amplifies the ability to adsorb substances, reaching levels up to 100 times greater than those of bigger particles. The reason for this is the augmentation of the particular surface area and the formation of fresh adsorption sites on the altered surface of the NP. Titanium, zinc, aluminum, and iron are metal oxides that are efficient and cost-effective adsorbents for removing heavy metals and radionuclides from water. Extensive research has comprehensively demonstrated the efficacy of NMOs in removing several heavy metals, such as chromium, nickel, cadmium, copper, mercury, arsenic, and lead. The NMOs may be regenerated and reused simply by adjusting the pH of the solution, which restores their ability to adsorb substances (Abdulhamid et al. 2024). The regeneration's simplicity, affordability, high adsorption capacity, and simple separation procedure give magnetic NP (MNPs) a significant economic and technical advantage in wastewater treatment applications.

13.2.2.1 Titanium Dioxide (TiO_2)

TiO_2 is widely used as a photocatalytic, especially in wastewater treatment. Their significant chemical stability, cost-effectiveness, and capacity to produce reactive oxygen species (ROS) when exposed to light and their little harm to people significantly contribute to this desire. TiO_2 has been widely researched and used as a catalyst in ozonation processes, renowned for its efficacy in attaining comprehensive and total mineralization of organic pollutants in wastewater. Researchers have shown great interest in TiO_2 in recent years due to their powerful antiviral effects. TiO_2 has shown efficacy in deactivating several viruses, such as Herpes simplex virus, MS2 bacteriophages, hepatitis B virus, and poliovirus. The optimal concentration of TiO_2 needed to eradicate microorganisms in wastewater efficiently generally falls within the range of 0.1 to 1 g/L. The effectiveness of this process is determined by variables such as the strength of the light frequency and the dimensions of the TiO_2 (Ibhadon and Fitzpatrick 2013).

The primary reason for the antibacterial properties of TiO_2 is its ability to produce ROS, including hydrogen peroxide (H_2O_2) and hydroxyl radicals (OH^-). The production of these ROS occurs when exposed to ultraviolet (UV)-A radiation via reductive and oxidative processes. The photocatalytic activity demonstrates a high level of effectiveness in deactivating a wide range of pathogens. Research has shown that plastic containers treated with TiO_2 may effectively deactivate fecal coliforms at a concentration of 3000 colony-forming units per 100 mL in a brief timeframe. It is a common practice to include silver molecules as dopants. to improve the ability of TiO_2 to deactivate viruses and bacteria by photocatalysis (Gelover et al. 2006).

Doping greatly enhances their ability to remove pathogens, increasing their effectiveness in treating wastewater. Moreover, the progress made in the solar disinfection units, which exploit the photocatalytic characteristics of TiO_2, has been a noteworthy breakthrough. These devices are very advantageous in delivering uncontaminated and secure drinking water in developing countries, where the availability of clean water is a crucial concern.

13.2.2.2 Zinc Oxide (ZnO)

ZnO is receiving considerable interest as a potential technique for the photocatalytic breakdown of pollutants. This is mostly because of their significant transparency to visible light and outstanding effectiveness in absorbing UV radiation. The unique attributes of these NP, such as their strong oxidation ability, good photocatalytic capabilities, and broad bandgap in the UV spectral region, make them particularly successful in wastewater treatment applications. ZnO is introduced into the environment via several uses in formulations, paints, packaging, and cosmetics (Pal and Sen 2023a). Studies have shown that ZnO notably impacts the effectiveness of biological nitrogen treatment in wastewater systems. Additionally, they have exceptional antibacterial activity against a wide range of disease-causing microorganisms. Nevertheless, the exact processes responsible for these antimicrobial actions remain incompletely understood. The prevailing belief is that the main process includes the photocatalytic production of ROS, which can strongly harm bacterial cells. In addition, ZnO NP may disrupt the structure of the cell membrane and enter the cell envelope, impeding bacterial development (Bhuyan et al. 2015).

Metal doping is often used to increase the efficiency of ZnO NP. To enhance the photocatalytic capabilities of ZnO, several metal dopants, including dopants, rare-earth dopants, cationic dopants, and anionic dopants, are included. Several studies have shown that combining ZnO NPs with other semiconductors, such as reduced graphene oxide (RGO), graphene oxide (GO), TiO_2, SnO_2, CeO_2, and CdO, is an effective method to improve their ability to degrade light-induced reactions (Azarang et al. 2015). Although ZnO NPs have potential uses, there is a significant absence of information addressing their changes and destiny throughout the wastewater treatment process. Furthermore, there is a lack of knowledge on the impact of product formulation chemistry on the behavior and stability of ZnO NPs during waste treatment. The lack of this knowledge is crucial since ZnO NPs are often added to specific formulations to improve their stability and enable their successful incorporation into goods such as sunscreens and cosmetics.

13.2.2.3 Iron Oxide NPs (IONPs)

The study of IONPs has been extensively researched because of their distinctive characteristics, such as a high surface-to-volume ratio, superparamagnetism, and nanoscale size. These NP can be synthesized using various methods such as chemical co-precipitation (Kumar and Gangawane 2022a), sol–gel, hydrothermal, and sonochemical (Kumar and Gangawane 2022b). These NP are desirable due to their ease of synthesis, modification, and coating, enabling fine manipulation and atomic adjustment. Their versatility makes them very flexible for many purposes. In addition, IONPs have chemical inertness, biocompatibility, and low toxicity, which greatly boost their potential when used in conjunction with biotechnology. As shown by laboratory and field on-site studies, IONPs have shown significant effectiveness in eliminating different pollutants. Their present uses in wastewater treatment mainly utilize two primary technologies: adsorptive and photocatalytic (Kumar et al. 2022). Iron oxide NPs are used in adsorptive technologies as immobilization carriers or nanosorbents. They help improve the effectiveness of pollutant removal. Photocatalytic technologies use NP as semiconductor photocatalysts to transform pollutants into less harmful chemicals.

The processes via which IONPs capture pollutants from locations or wastewater include electrostatic interactions, magnetic selective adsorption, ligand combination, and surface binding. The adsorption process for both inorganic and organic pollutants mainly includes surface binding processes. Initially, contaminants adhere to the surface of the nanoparticle. Subsequently, they either spread into the adsorbent or are attracted to it by electrostatic interactions, enabling further interactions with the functional groups on the nanoparticle surface. To improve the usefulness and effectiveness of IONPs in removing contaminants, it is frequently essential to subject them to chemical treatment or modification. The NP may experience substantial enhancements in their adsorption capacities via these alterations, increasing their effectiveness in various environmental applications (Das et al. 2017).

13.2.3 Nanocomposites

Several research investigations have been carried out to improve the versatility and combined effects of inorganic or polymeric membranes by integrating NP. NMs, such as photocatalytic NMs (e.g., TiO_2 and bimetallic NP), antimicrobial NP (e.g., carbon nanotubes (CNTs) and silver NP (nano-Ag)), and metal oxide NP (e.g., zeolite, aluminum oxide (Al_2O_3), and TiO_2), have diverse applications in various fields including drug delivery systems, soil quality enhancement, wastewater treatment, and biological sciences. Nanocomposites, formed by incorporating NMs into membranes, have recently gained considerable interest for their capacity to effectively adsorb diverse pollutants such as dyes, heavy metal ions, and pesticides from wastewater. Nanocomposites possess expansive surface areas and distinctive features, making them exceptionally efficient for eliminating pollutants. When photoactive or antimicrobial NP are added to these nanocomposite membranes, they become more responsive, which reduces fouling, a major problem in membrane technology (Rodrigues et al. 2017). Nanocomposites have various applications in wastewater treatment, as illustrated in Fig. 13.1.

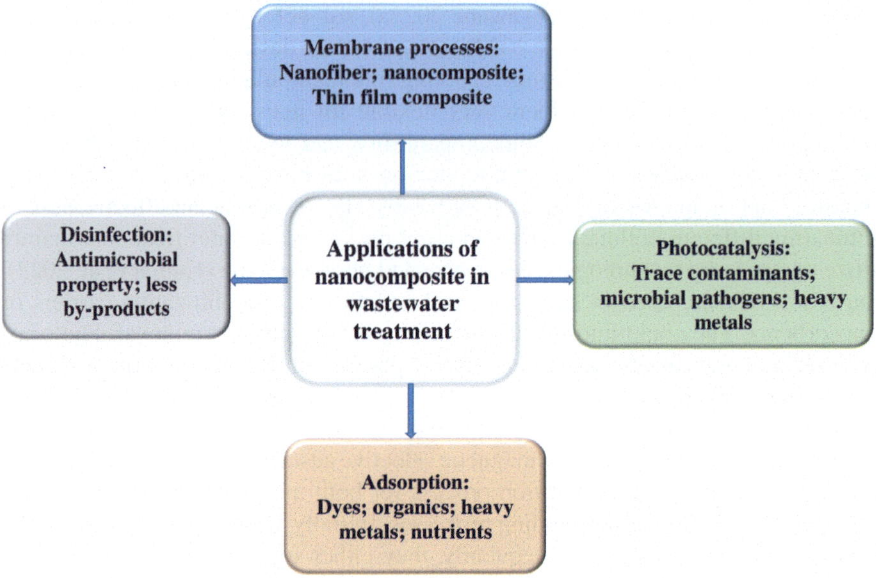

Fig. 13.1 Applications of nanocomposites in wastewater treatment (Naseem and Waseem 2022)

Membranes made of ceramic and polymer materials that include TiO_2 have shown effective removal of pathogens and pollutants upon exposure to UV-A light. TiO_2's photocatalyst characteristics allow for separating organic pollutants and neutralizing harmful microorganisms, making nanocomposite biomembranes less prone to biological and organic fouling. In addition, applying a chitosan layer over nanocomposite membranes composed of polyethylene glycol diacrylate or a polyacrylic acid layer has been discovered to prevent the growth of Gram-negative and Gram-positive bacteria successfully. Furthermore, there is potential for future advancement in incorporating diverse functional NP, such as antibacterial, photocatalytic, and catalytic NP, into membranes used for wastewater treatment. The treatment system can use UV light by incorporating photocatalytic NP into nanocomposite materials. An external submerged ultrafiltration (UF) membrane or microfiltration (MF) reactor setup allows for convenient exposure to photocatalytic light via UV sources. Using different optical fibers in submerged UF membranes or MF reactors is a promising approach to enhance photocatalytic efficiency (Oller et al. 2018). This integration aims to enhance wastewater treatment operations' overall efficiency and efficacy by using advanced characteristics of NMs.

The technique used in the manufacture of clay–polymer nanocomposites may have a substantial impact on the effectiveness of water treatment. For example, using in situ polymerization processes might result in the formation of nanocomposites with bead-like structures, which have a smaller surface area when compared to clay alone. Little research is available on polymer nanocomposites produced using melt-blending techniques for water treatment. The melt-blending process is advantageous since it does not need a solvent, decreasing environmental risks (Madaleno et al. 2010). The process of immobilizing clay particles on polymers by coating allows for preserving the characteristics of the clay particles, resulting in the development of high-performance adsorbents. The main function of the polymer in clay-based polymer nanocomposites is to provide support for the clay particles, which include adsorption sites. However, depending on the kind of polymer and the hydration state of the heavy metals, the polymer may also adsorb pollutants. The pH of the solution may considerably influence the adsorption characteristics of polymer nanocomposites. Research on the adsorption of lead ions on clay-based polymer nanocomposites shows that raising the pH from 3 to 4.8 enhances the absorption of Pb^{2+}. Equilibrating the levels of H_3O^+ and OH^- ions leads to an elevation in the $Pb(OH)^+$ concentration, establishing an adsorption equilibrium. The nanocomposite's edges and planar sites play a role in adsorption, although the unsaturated edges exhibit greater capacity and reactivity than the planar sites (Bahramian et al. 2008). The variability in the characteristics of the edges and planar sites, together with parameters such as the kind of metal ion, its concentration, pH level, and ionic strength, may impact the process of adsorption and desorption.

During batch adsorption experiments, the mixture is vigorously shaken. The rate at which adsorption occurs is mainly influenced by two key processes: (i) the movement of ions through the thin liquid layer on the surface of the polymer nanocomposite and (ii) the diffusion of ions between particles to reach the active sites and bind to the surface of the polymer nanocomposite. The wettability of polymer nanocomposites may influence their stability. Hydrophobic polymer nanocomposites in aqueous environments enhance chemical and mechanical stability. The shape and structural characteristics of polymer nanocomposites may impact their ability to adsorb efficiently. Increased presence of free-volume holes may facilitate the interaction between clay particles and lead ions, hence mitigating issues related to hydrophobic polymer nanocomposites. Increased hydrophobicity results in less water sorption and, as a result, decreased metal adsorption. Research has shown that including polyurethane, which has a strong attraction to water, enhances the effectiveness of polymer nanocomposites produced using the melt-blending method (Dlamini et al. 2012). A study found that bentonite/polyurethane/ethylene vinyl acetate nanocomposites successfully eliminated 90% of lead ions from aqueous solutions. In contrast, when granular bed filters composed of polymer nanocomposites were employed in potable and wastewater purification processes, only 10% of Pb^{2+} was removed.

Clay–polymer nanocomposites have a significant adsorption capacity. However, the tiny dimensions and limited porosity of clay particles might lead to hydraulic resistance and reduced efficiency, resulting in poor flux. Clay–polymer nanocomposites created using in situ polymerization procedures, resulting in the formation of bead-like composites, show great promise for water filtration systems. After reconfiguration, polymer nanocomposites produced using the melt-blending technique may be used as adsorbents in granular bead filters. The thermal treatment of these composites has little impact on the adsorption capabilities. Disrupting the polymer's crystalline regions might enhance the nanocomposite's fragility, resulting in higher water permeability (Alexandre and Dubois 2000). The limited moisture resistance of clay-based polymer nanocomposites results in increased thermal resistance, enabling water treatment at elevated temperatures. Although clay-based polymer nanocomposites have several benefits, the polymer matrix may cover the active adsorption sites of the clay particles, resulting in reduced adsorption efficiency (Mukhopadhyay et al. 2020). This issue is particularly prone to arise in polymer nanocomposites generated using the melt-blending technique. The clay particles' significant swelling ability increases weight, which causes the polymer nanocomposites to settle in water. To avoid the process of settling, it is necessary to use a greater amount of energy to agitate the solution aggressively.

Natural polymer-based nanocomposites have shown favorable outcomes in eliminating residual organic pollutants from surface, groundwater, and wastewater. Multiple investigations have shown the existence of pharmaceutical residues (ranging from nanograms per liter to milligrams per liter) in water after treatment. Prolonged exposure to these contaminants may harm human and aquatic life, possibly resulting in endocrine disruption. Kyzas et al. developed a nanocomposite material by combining graphite oxide with poly(acrylic acid)-grafted chitosan. This material was designed to remove dorzolamide from biomedical wastewater effectively. The findings indicated that when utilized individually, the removal effectiveness of graphene oxide and chitosan-grafted-poly(acrylic acid) at pH 3 was 52% and 68%, respectively. However, when combined as a nanocomposite, the removal efficiency was significantly enhanced to 92%. The researchers observed that reducing the pH level from 9 to 3 significantly increased the percentage of dorzolamide removal, increasing from 23 to 92%. This effect can be attributed to the interaction between the protonated dorzolamide groups in an acidic environment and the eCOOH, eOH, and eNH_2 groups found in the polymer nanocomposite. Graphene oxide and chitosan have functional groups like eCOOH, eOH, and eNH_2, which make them very suitable for creating various polymer nanocomposites. These nanocomposites have great potential as effective adsorbents for organic pollutants, especially dyes (Kyzas et al. 2014). Travlou et al. synthesized a nanocomposite of graphene oxide and magnetic chitosan to eliminate Reactive Black 5 (RB5) from wastewater. The maximum adsorption capacity at ambient temperature was seen at pH 3, resulting in the removal of 81% of RB5, with an adsorption capacity of 391 mg/g. In an acidic environment, the eNH_2 groups in chitosan get protonated and transform into eNH^{3+} cations. These cations have a strong affinity for attracting the dissociated anions of RB5. The inclusion of magnetic Fe_3O_4 NP in chitosan boosted the effectiveness of polymer nanocomposites by increasing the surface area and the attraction between Fe_3O_4 particles and oxygen atoms in dye molecules (Travlou et al. 2013).

Rajabi et al. prepared nanocomposites of poly(methyl methacrylate)/graphene oxide and poly(methyl methacrylate)/graphene oxide-Fe_3O_4 to eliminate malachite green dye. The results indicated that the polymer nanocomposite with Fe_3O_4 particles exhibited a much greater removal percentage (~90%) than the nanocomposite without Fe_3O_4 particles (~45%). Adding magnetic NP such as Fe_3O_4 may enhance the adsorption efficiency, reduce toxicity, and improve the reusability of polymer nanocomposites. Additional NP, including TiO_2, ZrO_2, and ZnO, has been investigated as nanofillers or additives in various polymer nanocomposites. These additions enhance thermal stability, mechanical strength, chemical stability, sorption capacity, and anti-fouling capabilities (Rajabi et al. 2019). Hosseini and Toosi developed a polysulfone membrane that incorporates a graphene oxide-TiO_2 nanocomposite. This membrane effectively eliminates herbicides such as glyphosate, trifluralin, and butachlor from water. The researchers said that including a hydrophilic graphene oxide–TiO_2 nanocomposite led to a significant increase in adsorption efficiency due to the strong attraction formed by hydrogen bonds with the glyphosate herbicide. The hydrophobic character of butachlor and trifluralin herbicides resulted in lower adsorption since they had less contact with the hydrophilic graphene oxide–TiO_2 nanocomposite. Polymer nanocomposites can manipulate and alter their hydrophobic/hydrophilic characteristics, making them very sophisticated materials for treating oily wastewater (Hosseini and Toosi 2019).

13.2.4 Nanofibers

Nanofibers have become an ideal product for environmental purposes, namely in water purification. Nanofibers are polymer fibers with sizes ranging from the nanometer to submicron scale. They may be produced by combining different polymeric precursors with appropriate solvents. Polymeric solutions, including natural and synthetic polymers, produce nanofibers via processing (Reneker and Yarin 2008). Nanofibers have several exceptional characteristics, such as an exceptionally large specific surface area, low density, high porosity, and the capacity to create a highly porous structure with great interconnection among its pores. Nanofibers have features that allow precise adjustment and customization to fulfill particular process needs via manipulating synthesis parameters. The nanofibers' extensive specific surface area enables the integration of specialized functions via immobilizing NP, zeolites, metal–organic frameworks, and other materials. As a result, this leads to improved specificity, reaction rates, and ability to remove contaminants in various processes. Nanofiber-based applications provide ease in contact mode, as they may be packed in fixed-bed columns or used as membrane reactors, depending on the separation process needs. Nanofibers possess features that make them ideal for various applications across several domains of human existence. These applications include medication delivery, the biomedical sector, energy storage, protective apparel, and environmental cleanup. Nanofibers have shown great potential in water treatment for effectively eliminating targeted pollutants, including heavy metals, persistent organic pollutants (POPs), contaminants of emergent concern (CECs), and greasy compounds, as shown in Fig. 13.2. CECs contain endocrine-disrupting compounds, surfactants, flame retardants, and per- and poly-fluoroalkyl substances (PFAS) (Pal and Sen 2023b). These compounds are highly toxic and adversely affect human health and the environment (Pal and Sen 2023b). Moreover, they exhibit

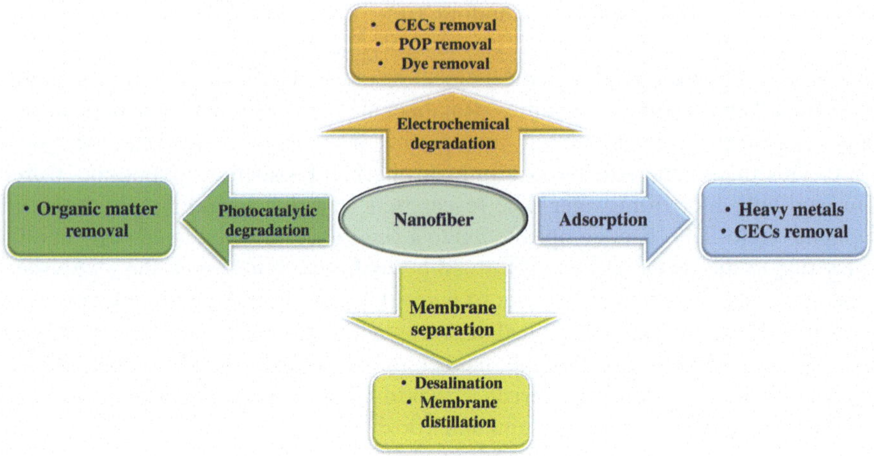

Fig. 13.2 Applications of NFs for wastewater treatment (Agrawal et al. 2021)

efficacy in desalination processes and in eliminating particulate matter and organic contaminants from air filtering systems (Kenry 2017).

Nanofibers predominantly remove heavy metals from wastewater by adsorption, further enhanced by ion exchange, photocatalytic processes, electrostatic interactions between metal ions, diffusion dialysis, and metal–organic framework crystals or polymers (Zhu et al. 2021). Adsorption is a mass transfer process in which heavy metals from wastewater stick to nanofibers. This may occur by physical adsorption, driven by van der Waals forces, or chemical adsorption, which includes chemical reactions on the surface of the adsorbent. At first, heavy metals are transported from the wastewater to the surface of the nanofiber. Subsequently, the metals permeate into the nanofiber matrix, constituting the step determining the overall pace. Ultimately, the heavy metals and the nanofibers engage in chemical or physical interactions. The presence of important functional groups on the surface of the nanofiber is essential for the selective adsorption of heavy metals. These functional groups allow for a stable concentration of adsorbed metals in the wastewater, reaching equilibrium. The Langmuir and Freundlich isotherms often investigate these equilibrium conditions (Ligneris et al. 2020).

Extensive research has been dedicated to fabricating nanofibers with surface changes to optimize their adsorption properties. Many adsorbents, including metal oxides, polymer solutions, zeolites (Pal and Sen 2024b), activated carbon, biomaterials, chelating compounds, and functionalized chitosan, have eliminated heavy metals (Almasian et al. 2018). Nanofibers are produced using electrospinning, melt blowing, polymerization, melt mix extrusion, centrifugal force spinning, and phase separation. Out of all the methods mentioned, electrospinning is the most adaptable and direct technique for developing nanofibers with widths ranging from hundreds of nanometers to microns. Surface alterations on electrospun fibers, such as doping, dispersion, cross-linking, and grafting with different chemicals, have been used to improve the ability of nanofibers to adsorb substances (Sankararamakrishnan et al. 2019).

Several varieties of nanofibers have been investigated to eliminate heavy metals. Cellulose nanofibers, characterized by their environmentally friendly nature and ability to decompose naturally, may be produced from sugarcane bagasse. Incorporating nano MgS into these nanofibers has developed bio-nanocomposite fibers that exhibit enhanced adsorption ability for Cd(II) (Sankararamakrishnan et al. 2019). Chitosan (CS) nanofibers have environmental benefits due to their capacity to degrade, lack of toxicity, compatibility with living organisms, affordability, and widespread availability. These nanofibers effectively bind heavy metals because of amine and hydroxyl groups in their polymer structure (Jiang et al. 2018). Modifications have enhanced the electrospinnability of CS nanofibers, resulting in increased binding capacities for heavy metals. An instance of crosslinking between chitosan (CS) and polyacrylic acid sodium (PAAS) has resulted in the formation of a uniformly crosslinked chitosan structure that is efficient in eliminating Cr(VI) (Jiang et al. 2018). The adsorption capacity of CS membranes for heavy-metal ions has been enhanced by grafting polyethyleneimine (PEI) onto them. Poly(vinyl alcohol) (PVA) is a remarkable substance known for its ability to dissolve in water, lack of toxicity, capacity to break down naturally, strong structure, and compatibility with living organisms. Polyvinyl alcohol's exceptional capacity to produce fibers at a nanoscale level makes it a highly suitable choice for manufacturing nanofibers. An instance of a composite nanofiber, composed of PVA and sodium alginate (SA), was created using electrospinning. Thermal crosslinking improved its capacity for extracting Cd(II) from aqueous solutions (Ebrahimi et al. 2019). In addition, Cu(II) has been effectively removed using PVA-co-PE nanofibers made using melt extrusion and functionalized with iminodiacetic acid. EVOH nanofibers, modified with PPy by in situ oxidative polymerization, have effectively removed Cr(VI) from aqueous solutions.

Polyacrylonitrile (PAN) is a man-made polymer often used to produce nanofibrous materials using electrospinning. Polyacrylonitrile (PAN) nanofibers can include several functional groups, such as amidoxime, hydrazine, amine, and phosphoryl. These functional groups enable the nanofibers to bind heavy metals via chelation. The researchers have produced various PAN nanofibers using chemical grafting with diethylenetriamine (DTA), ethylenediamine (EDA), and ethylene glycol (EG) to eliminate Zn(II), Cu(II), and Pb(II). Yang et al. created a CS–PGMA–PEI electrospun membrane that effectively adsorbed Cr(VI), Cu(II), and Co(II) with maximum capacities of 138.96 mg/g, 69.27 mg/g, and 68.31 mg/g, respectively. The ideal pH values for adsorption were 2, 4, and 6. The membrane exhibited a greater affinity for adsorbing Cr(VI) in comparison to Co(II) and Cu(II). However, there was a reduction in adsorption capabilities by 17.80%, 11.30%, and 13.50% for Cr(VI), Cu(II), and Co(II) after five adsorption–desorption cycles (Yang et al. 2019). On the other hand, CS/PAAS nanofibers exhibited an initial adsorption capacity for Cr(VI) of 26.02 mg/g, which was subsequently boosted to 78.92 mg/g by post-modification with chelating agents, hence improving their chelating capabilities (Jiang et al. 2018).

Cellulose nanofibers doped with MgS have shown exceptional adsorption and repeatability for Cd(II), with a maximum adsorption capacity of 333.33 mg/g. The adsorption of cadmium was strongly influenced by the solution pH. Increasing the pH from 5 to 5.5 resulted in a 17% improvement in cadmium removal. However, additional increases in pH caused a 40% reduction in adsorption. Ebrahimi et al. synthesized nanofibers composed of polyvinyl alcohol (PVA) and sodium alginate (SA) that showed an enhanced ability to adsorb Cd(II). The maximum adsorption capacity was 163.93 mg/g when the initial cadmium concentration was 120 mg/L (Ebrahimi et al. 2019). On the other hand, PVA/TiO_2 nanofibers treated with sulfhydryl showed the ability to adsorb Th(VI) at a rate of 238.1 mg/g. Additionally, including ZnO in the system further improved the adsorption capacity of PVA/TiO_2/ZnO to 333.3 mg/g (Alipour et al. 2016). The electrospinning and deposition of polypropylene glycol (PPG) on the surface of mesoporous MgO nanofibers resulted in significant adsorption capacities for Pb(II), Cd(II), and Cu(II), with values of 2500.48 mg/g, 2407.74 mg/g, and 2415.74 mg/g, respectively. The adsorption capacities of Pb(II), Cd(II), and Cu(II) on the mesoporous MgO nanofiber alone were 378.58 mg/g, 311.47 mg/g, and 270.11 mg/g, respectively. However, these capacities were much greater than the mesoporous MgO nanofiber alone. In addition, scientists have synthesized zonal silica nanofibers by dissolving PAN templates with DMF to remove Hg^{2+}, resulting in an adsorption capacity of 77.49 mg/g. In contrast, solely silica nanofibers exhibited a lower adsorption capacity of 1.36 mg/g (Li et al. 2011).

Nanofibers have been successfully used as adsorbents for CECs. Their polymeric matrix can trap target pollutants via adsorption, making them appropriate for subsequent targeted oxidation utilizing AOPs or electrochemical techniques. The adjustable characteristics of nanofibers enable their modification with different groups, catalysts, and additives that exhibit a high degree of selectivity toward certain CECs, hence improving their efficacy in removing such contaminants, even in samples with a high concentration of other substances. One additional benefit of nanofibers is their capacity to include NP, which solves the problem of clumping and excessive cloudiness often seen with dispersed nanoparticle adsorbents/catalysts in water treatment. Additionally, this enhances the ability to retrieve and recycle NP, increasing the process's sustainability. In addition, nanofibers can potentially immobilize macrocyclic hosts such as cyclodextrins (CD). This immobilization results in a large surface area and increased permeability, which improves the removal of CECs by adsorption. The enhanced localized accumulation of CECs on nanofibers also enhances the effectiveness of targeted therapies such as oxidation processes (Agrawal et al. 2021).

Multiple studies have shown that nanofibers are useful in treating CECs by either adsorbing them alone or in conjunction with catalytic, electrochemical, or plasma oxidation methods. Camiré et al. examined the elimination of fluoxetine by using nanofibers produced from alkali lignin (AL) and polyvinyl alcohol (PVA) solutions. The researchers discovered that a 50:50 AL ratio resulted in the highest fluoxetine removal (70%) from a 50 mg/L solution in a batch system, with hydrogen bonding being the primary removal method. Nevertheless, the research also emphasized problems such as decreased solution viscosity when lignin was added, necessitating a slower flow rate during electrospinning. This resulted in longer processing times and increased expenses. Furthermore, the durability of the nanofibers in water-based solutions was a matter of concern, requiring heat stabilization to prevent the lignin from dissolving (Camiré et al. 2020). Peter et al. conducted electrospinning research to create a composite material consisting of carbon nanofiber (CNF) and carbon nanotube (CNT). This composite material was then used to remove atrazine and sulfamethoxazole from solutions containing water effectively. The presence of CNTs greatly enhanced the ability of the nanofibers to adsorb substances by increasing their specific surface area and mechanical strength, hence improving both the adsorption capacity and kinetics. The composite nanofibers successfully eliminated over 90% of all evaluated CECs in a flow-through system, with atrazine and sulfamethoxazole exhibiting absorption capabilities of 30 mg/g and 20 mg/g, respectively (Peter et al. 2016). Similarly, Chabalala et al. improved the elimination of atrazine by altering polyacrylonitrile (PAN) nanofibers with β-cyclodextrin (β-CD), resulting in a removal effectiveness of 91.46% as opposed to 51% with unaltered PAN nanofibers. The enhancement was ascribed to the augmented specific surface area and the formation of a complex between atrazine and β-CD (Chabalala et al. 2021).

In addition, Lv et al. (2021) enhanced cellulose nanofibers by including β-CD to efficiently eliminate bisphenol A (BPA), bisphenol S (BPS), and bisphenol F (BPF) from water-based solutions. The research determined that the adsorption process is influenced by parameters such as the amount of adsorbent used, temperature, and pH. The maximum adsorption capacities were 50.37 mg/g, 48.52 mg/g, and 47.25 mg/g for BPA, BPS, and BPF, respectively. The removal method included hydrophobic interactions, hydrogen bonding, and π–π interactions. The nanofibers could be reused for up to five cycles after being removed using water and ethanol (Lv et al. 2021). In their study, Khalil and Schäfer examined the use of polyethersulfone (PES) nanofibers with β-CD deposited onto PES ultrafiltration membranes. The objective was to assess their effectiveness in removing steroid hormones such as estrone, β-estradiol, and testosterone from water samples. The primary method for elimination relied on size exclusion, with the diameter of the β-CD cavity playing a critical role (Khalil and Schäfer 2021). Similarly, Khalil et al. investigated the influence of several β-CD crosslinking agents on eliminating estradiol. They discovered that nanofibers crosslinked with trimethylolpropane triglycidyl ether (TPTE) demonstrated the best efficacy in removing estradiol (Khalil et al. 2021).

Nanofibers have been used in combination with photocatalysts to facilitate the degradation of CECs. For instance, Nor et al. showed the deterioration of BPA by using TiO_2 nanofibers that were compressed over a polyvinylidene difluoride (PVDF) membrane and exposed to UV radiation. The TiO_2 photocatalytic activity led to a breakdown of BPA ranging from 63 to 85%, much greater than the 27.62% elimination obtained by adsorption alone (Nor et al. 2016). In their study, Gadisa et al. examined using ZnO–carbon composite nanofibers to remove caffeine. They discovered that the selection of the polymeric precursor impacted the characteristics of the charge carriers and the efficacy of the photocatalytic process. Combining ZnO–polystyrene (PS) nanofibers exhibited the most effective removal rates for caffeine and diclofenac (Gadisa et al. 2020). Ramasundaram et al. investigated the elimination of cimetidine using a photocatalytic stainless-steel filter coated with TiO_2 nanofibers. Their findings demonstrated that increasing the nanofiber layer's thickness enhanced removal effectiveness (Ramasundaram et al. 2013). In their study, Chen et al. used Fe-doped PAN nanofibers to catalyze peroxymonosulfate (PMS) and effectively eliminate BPA. The combination of adsorption and oxidation processes eliminated BPA within 6 min. Researchers have studied PAN nanofibers combined with TiO_2 NP to remove different CECs. These systems have shown excellent performance in removing CECs when used in flow-through systems. Electrochemical devices that use nanofibers have shown potential for remediation of contaminants in contaminated environmental medium (Chen et al. 2021). Kim et al. investigated the use of antimony tin oxide-doped ruthenium oxide (ATO-RO) NP in PVP nanofibers for the electrochemical oxidation of BPA. They successfully achieved total degradation of BPA within 20 min (Kim et al. 2019). Xie et al. examined the use of Fe/Co alloy on PVP nanofibers for the degradation of tetracycline. They found that this method achieved high clearance rates for different concentrations of tetracycline (Xie et al. 2021). Yang and Hen investigated using tubular carbon nanofibers combined with activated alumina in an electrocoagulation system. They achieved significant removal rates for caffeine, sulfamethoxazole, and acetaminophen (Yang and Yen 2013).

13.3 Conclusions and Future Perspectives

NMs, such as NP, nanometal oxides, nanofibers, and nanocomposites, have shown great promise in improving wastewater treatment technology. Advanced materials include distinctive characteristics, including a large surface area, enhanced reactivity, and the capacity to be tailored for particular purposes. These attributes make them very efficient in eliminating various pollutants from wastewater. NMs in wastewater treatment are characterized by their capacity to adapt and flexibility. These contaminants may be effectively targeted and treated using various methods, including adsorption, filtration, and photocatalysis. In addition, NP may be modified with specific chemical groups to target certain impurities, providing high customization and effectiveness selectively. The ability to recycle and reuse NMs increases their attractiveness, as it helps to provide sustainable and economical solutions for treating wastewater. Although NP provide various benefits, some obstacles must be overcome to fully exploit their potential in wastewater treatment. These concerns include issues about NP' environmental ramifications and toxicity, the capacity to scale up manufacturing processes, and the cost efficiency of treatment systems based on NMs. Current research and development efforts are focused on addressing these obstacles by developing environmentally benign and economically feasible NMs, enhancing comprehension of their interactions with contaminants, and maximizing their incorporation into pre-existing treatment infrastructures.

NMs have shown considerable potential in wastewater treatment, while their complete capabilities are still being revealed. The advancement of NP in wastewater treatment relies on developing versatile NMs that can effectively target many pollutants simultaneously. These materials can incorporate characteristics such as adsorption, photocatalysis, and antibacterial action into a single structure. For instance, nanocomposites may be engineered to include various metal oxides, carbon-based materials, and polymers. This allows them to simultaneously target heavy metals, organic contaminants, and pathogens, enhancing treatment efficiency and reducing process complexity. Progress in producing and modifying NMs will improve the ability to target and remove particular pollutants. NMs may be modified regarding surface chemistry and porosity to enhance their selectivity toward certain contaminants, enabling more accurate and effective removal. This is especially crucial for the management of complex wastewater streams that consist of a combination of pollutants. Environmental concerns and the need for sustainable practices will motivate the development of more environmentally friendly synthesis techniques for NMs. Using natural precursors, plant extracts, and environmentally friendly solvents throughout the synthesis process may decrease the environmental impact of NM manufacturing. Furthermore, there will be a shift toward recycling and reusing NP used in wastewater treatment, which will enhance the overall sustainability of these technologies.

By combining NMs with advanced technologies like sensors and IoT (Internet of Things) devices, real-time monitoring and managing wastewater treatment operations will be possible. Intelligent NMs may be engineered to react to environmental changes, allowing for flexible and adaptable treatment abilities. For instance, NMs with integrated sensors might identify the presence of certain pollutants and activate suitable reactions, such as launching photocatalytic processes or modifying their ability to adsorb substances. Machine learning and artificial intelligence (AI) will substantially impact enhancing the use of NMs in wastewater treatment. Artificial intelligence algorithms can evaluate large datasets to forecast the behavior and effectiveness of different NMs under varied settings. This analysis helps in directing the development of more efficient treatment systems. Furthermore, these technologies may assist in formulating proactive maintenance timetables for treatment systems that use NMs, guaranteeing their durability and dependability. With the increasing use of NMs in wastewater treatment, there will be a corresponding need for strong regulatory frameworks to guarantee their secure and efficient application. Future studies will prioritize comprehending the enduring ecological and physiological consequences of NMs, ultimately resulting in the development of protocols and benchmarks for their use. Conducting thorough and transparent evaluations of risks will be crucial to establishing public confidence and getting regulatory authorization. For NP to substantially influence wastewater treatment, it is crucial to develop manufacturing techniques that can be scaled up and cost-effective commercialization strategies.

Future research will focus on developing large-scale synthesis methods that preserve the distinctive characteristics of NMs while still being economically feasible. Collaborations among research institutions, industry, and government agencies will be essential in converting laboratory achievements into feasible and commercially viable solutions. In future, there will be a rise in the use of hybrid and composite NMs, which include combining several kinds of nanostructures to exploit their benefits. For example, integrating carbon nanotubes with metal oxides or combining magnetic NP with polymer matrices might improve the overall effectiveness, durability, and capacity to reuse the NMs used in wastewater treatment. Additional field applications and pilot projects will be carried out to showcase the practical feasibility of using NMs for wastewater treatment. These studies will provide useful data about the efficacy, difficulties, and advantages of using NMs in practical situations. The knowledge acquired from these applications will direct future research and development, enhancing and streamlining NM technology. Disseminating information to the general public and those involved on the advantages and safety of NMs in treating wastewater is essential for universal approval and implementation. Upcoming efforts will prioritize enhancing awareness via outreach activities, openly communicating research results, and collaborating with community groups to showcase the beneficial effects of NMs on water quality and environmental health.

References

Abdulhamid MA, Alqadhi N, Osornio C, Szekely G (2024) Green nanotechnology for water purification, Elsevier Inc., 2024. https://doi.org/10.1002/9783527838059.ch4

Abebe LS, Smith JA, Narkiewicz S, Oyanedel-Craver V, Conaway M, Singo A, Amidou S, Mojapelo P, Brant J, Dillingham R (2014) Ceramic water filters impregnated with silver NP as a point-of-use water-treatment intervention for HIV-positive individuals in Limpopo Province, South Africa: a pilot study of technological performance and human health benefits. J Water Health 12:288–300. https://doi.org/10.2166/wh.2013.185

Agrawal S, Ranjan R, Lal B, Rahman A, Singh SP, Selvaratnam T, Nawaz T (2021) Synthesis and water treatment applications of nanofibers by electrospinning. Processes 9. https://doi.org/10.3390/pr9101779

Alexandre M, Dubois P (2000) Polymer-layered silicate nanocomposites: preparation, properties and uses of a new class of materials. Mater. Sci. Eng. R Reports. 28:1–63. https://doi.org/10.1016/S0927-796X(00)00012-7

Ali I (2012) New generation adsorbents for water treatment. Chem Rev 112:5073–5091. https://doi.org/10.1021/cr300133d

Alipour D, Keshtkar AR, Moosavian MA (2016) Adsorption of thorium(IV) from simulated radioactive solutions using a novel electrospun PVA/TiO_2/ZnO nanofiber adsorbent functionalized with mercapto groups: study in single and multi-component systems. Appl Surf Sci 366:19–29. https://doi.org/10.1016/j.apsusc.2016.01.049

Almasian A, Najafi F, Maleknia L, Giahi M (2018) Mesoporous MgO/PPG hybrid nanofibers: synthesis, optimization, characterization and heavy metal removal property. New J Chem 42:2013–2029. https://doi.org/10.1039/c7nj03200e

Anjum M, Miandad R, Waqas M, Gehany F, Barakat MA (2019) Remediation of wastewater using various nano-materials. Arab J Chem 12:4897–4919. https://doi.org/10.1016/j.arabjc.2016.10.004

Ayati A, Ahmadpour A, Bamoharram FF, Tanhaei B, Mänttäri M, Sillanpää M (2014) A review on catalytic applications of Au/TiO_2 NP in the removal of water pollutant. Chemosphere 107:163–174. https://doi.org/10.1016/j.chemosphere.2014.01.040

Azarang M, Shuhaimi A, Yousefi R, Jahromi SP (2015) One-pot sol–gel synthesis of reduced graphene oxide uniformly decorated zinc oxide NP in starch environment for highly efficient photodegradation of Methylene Blue. RSC Adv 5:21888–21896. https://doi.org/10.1039/C4RA16767H

Bahramian AR, Kokabi M, Famili MHN, Beheshty MH (2008) High temperature ablation of kaolinite layered silicate/phenolic resin/asbestos cloth nanocomposite. J Hazard Mater 150:136–145. https://doi.org/10.1016/j.jhazmat.2007.04.104

Bhuyan T, Mishra K, Khanuja M, Prasad R, Varma A (2015) Biosynthesis of zinc oxide NP from Azadirachta indica for antibacterial and photocatalytic applications. Mater Sci Semicond Process 32:55–61. https://doi.org/10.1016/j.mssp.2014.12.053

Bokare V, lim Jung J, Chang YY, Chang YS (2013) Reductive dechlorination of octachlorodibenzo-p-dioxin by nanosized zero-valent zinc: modeling of rate kinetics and congener profile. J Hazard Mater, pp 250–251, 397–402. https://doi.org/10.1016/j.jhazmat.2013.02.020

Boretti A, Rosa L (2019) Reassessing the projections of the world water development report. Npj Clean Water. 2:15. https://doi.org/10.1038/s41545-019-0039-9

Camiré A, Espinasse J, Chabot B, Lajeunesse A (2020) Development of electrospun lignin nanofibers for the adsorption of pharmaceutical contaminants in wastewater. Environ Sci Pollut Res 27:3560–3573. https://doi.org/10.1007/s11356-018-3333-z

Chabalala MB, Al-Abri MZ, Mamba BB, Nxumalo EN (2021) Mechanistic aspects for the enhanced adsorption of bromophenol blue and atrazine over cyclodextrin modified polyacrylonitrile nanofiber membranes. Chem Eng Res des 169:19–32. https://doi.org/10.1016/j.cherd.2021.02.010

Chen S, Li M, Zhang M, Wang C, Luo R, Yan X, Zhang H, Qi J, Sun X, Li J (2021) Metal organic framework derived one-dimensional porous Fe/N-doped carbon nanofibers with enhanced catalytic performance. J Hazard Mater 416:126101. https://doi.org/10.1016/j.jhazmat.2021.126101

Choi J, Seo Y, Hwang J, Kim J, Jeong Y, Hwang M (2014) Antibacterial activity and cytotoxicity of multi-walled carbon nanotubes decorated with silver NP. Int J Nanomedicine 9:4621. https://doi.org/10.2147/IJN.S69561

Corsi I, Winther-Nielsen M, Sethi R, Punta C, Della Torre C, Libralato G, Lofrano G, Sabatini L, Aiello M, Fiordi L, Cinuzzi F, Caneschi A, Pellegrini D, Buttino I (2018) Ecofriendly nanotechnologies and NMs for environmental applications: key issue and consensus recommendations for sustainable and ecosafe nanoremediation. Ecotoxicol Environ Saf 154:237–244. https://doi.org/10.1016/j.ecoenv.2018.02.037

Das R, Vecitis CD, Schulze A, Cao B, Ismail AF, Lu X, Chen J, Ramakrishna S (2017) Recent advances in NMs for water protection and monitoring. Chem Soc Rev 46:6946–7020. https://doi.org/10.1039/C6CS00921B

Dlamini DS, Mishra AK, Mamba BB (2012) Adsorption Behaviour of Ethylene Vinyl Acetate and Polycaprolactone-Bentonite Composites for Pb^{2+} Uptake. J Inorg Organomet Polym Mater 22:342–351. https://doi.org/10.1007/s10904-011-9640-1

des Ligneris E, Dumée LF, Kong L (2020) Nanofibers for heavy metal ion adsorption: correlating surface properties to adsorption performance, and strategies for ion selectivity and recovery. Environ Nanotechnology, Monit Manag 13:100297. https://doi.org/10.1016/j.enmm.2020.100297

Ebrahimi F, Sadeghizadeh A, Neysan F, Heydari M (2019) Fabrication of nanofibers using sodium alginate and Poly(Vinyl alcohol) for the removal of Cd_{2+} ions from aqueous solutions: adsorption mechanism, kinetics and thermodynamics. Heliyon. 5:e02941. https://doi.org/10.1016/j.heliyon.2019.e02941

Gadisa BT, Kassahun SK, Appiah-Ntiamoah R, Kim H (2020) Tuning the charge carrier density and exciton pair separation in electrospun 1D ZnO-C composite nanofibers and its effect on photodegradation of emerging contaminants. J Colloid Interface Sci 570:251–263. https://doi.org/10.1016/j.jcis.2020.03.002

Gelover S, Gómez LA, Reyes K, Teresa Leal M (2006) A practical demonstration of water disinfection using TiO_2 films and sunlight. Water Res 40:3274–3280. https://doi.org/10.1016/j.watres.2006.07.006

Hao C, Feng F, Wang X, Zhou M, Zhao Y, Ge C, Wang K (2015) The preparation of Fe_2O_3 NP by liquid phase-based ultrasonic-assisted method and its application as enzyme-free sensor for the detection of H_2O_2. RSC Adv 5:21161–21169. https://doi.org/10.1039/C4RA17226D

Hosseini N, Toosi MR (2019) Removal of 2,4-D, glyphosate, trifluralin, and butachlor herbicides from water by polysulfone membranes mixed by graphene oxide/TiO_2 nanocomposite: Study of filtration and batch adsorption. J Environ Heal Sci Eng 17:247–258. https://doi.org/10.1007/s40201-019-00344-3

Ibhadon A, Fitzpatrick P (2013) Heterogeneous photocatalysis: recent advances and applications. Catalysts 3:189–218. https://doi.org/10.3390/catal3010189

Jiang M, Han T, Wang J, Shao L, Qi C, Zhang XM, Liu C, Liu X (2018) Removal of heavy metal chromium using cross-linked chitosan composite nanofiber mats. Int J Biol Macromol 120:213–221. https://doi.org/10.1016/j.ijbiomac.2018.08.071

Kenry CT (2017) Lim, Nanofiber technology: current status and emerging developments. Prog Polym Sci 70:1–17. https://doi.org/10.1016/j.progpolymsci.2017.03.002

Khalil AM, Schäfer AI (2021) Cross-linked β-cyclodextrin nanofiber composite membrane for steroid hormone micropollutant removal from water. J. Memb. Sci. 618:118228. https://doi.org/10.1016/j.memsci.2020.118228

Khalil AM, Hashem T, Gopalakrishnan A, Schäfer AI (2021) Cyclodextrin composite nanofiber membrane: impact of the crosslinker type on steroid hormone micropollutant removal from water. ACS Appl. Polym. Mater. 3:2646–2656. https://doi.org/10.1021/acsapm.1c00231

Kim JC, Oh SI, Kang W, Yoo HY, Lee J, Kim DW (2019) Superior anodic oxidation in tailored Sb-doped SnO_2/RuO_2 composite nanofibers for electrochemical water treatment. J Catal 374:118–126. https://doi.org/10.1016/j.jcat.2019.04.025

Kumar A, Gangawane KM, Ramanjaneyulu B (2022) Ferrofluids for waste-water treatment, In: International conference on chemical, bio and environmental engineering, Springer, pp 723–744. https://doi.org/10.1007/978-3-030-96554-9_48

Kumar A, Gangawane KM (2022a) Effect of precipitating agents on the magnetic and structural properties of the synthesized ferrimagnetic NP by co-precipitation method. Powder Technol 401:117298. https://doi.org/10.1016/j.powtec.2022.117298

Kumar A, Gangawane KM (2022b) Synthesis and effect on the surface morphology and magnetic properties of ferrimagnetic NP by different wet chemical synthesis methods. Powder Technol 410:117867. https://doi.org/10.1016/j.powtec.2022.117867

Kyzas GZ, Bikiaris DN, Seredych M, Bandosz TJ, Deliyanni EA (2014) Removal of dorzolamide from biomedical wastewaters with adsorption onto graphite oxide/poly(acrylic acid) grafted chitosan nanocomposite. Bioresour Technol 152:399–406. https://doi.org/10.1016/j.biortech.2013.11.046

Li S, Yue X, Jing Y, Bai S, Dai Z (2011) Fabrication of zonal thiol-functionalized silica nanofibers for removal of heavy metal ions from wastewater. Colloids SurfS Physicochem Eng Asp 380:229–233. https://doi.org/10.1016/j.colsurfa.2011.02.027

Lv Y, Ma J, Liu K, Jiang Y, Yang G, Liu Y, Lin C, Ye X, Shi Y, Liu M, Chen L (2021) Rapid elimination of trace bisphenol pollutants with porous β-cyclodextrin modified cellulose nanofibrous membrane in water: adsorption behavior and mechanism. J Hazard Mater 403:123666. https://doi.org/10.1016/j.jhazmat.2020.123666

Madaleno L, Schjødt-Thomsen J, Pinto JC (2010) Morphology, thermal and mechanical properties of PVC/MMT nanocomposites prepared by solution blending and solution blending+melt compounding. Compos Sci Technol 70:804–814. https://doi.org/10.1016/j.compscitech.2010.01.016

Mauter MS, Zucker I, Perreault F, Werber JR, Kim JH, Elimelech M (2018) The role of nanotechnology in tackling global water challenges. Nat Sustain 1:166–175. https://doi.org/10.1038/s41893-018-0046-8

Mukhopadhyay R, Bhaduri D, Sarkar B, Rusmin R, Hou D, Khanam R, Sarkar S, Kumar Biswas J, Vithanage M, Bhatnagar A, Ok YS (2020) Clay–polymer nanocomposites: progress and challenges for use in sustainable water treatment. J Hazard Mater 383:121125. https://doi.org/10.1016/j.jhazmat.2019.121125

Naseem T, Waseem M (2022) A comprehensive review on the role of some important nanocomposites for antimicrobial and wastewater applications. Int J Environ Sci Technol 19:2221–2246. https://doi.org/10.1007/s13762-021-03256-8

Nor NAM, Jaafar J, Ismail AF, Mohamed MA, Rahman MA, Othman MHD, Lau WJ, Yusof N (2016) Preparation and performance of PVDF-based nanocomposite membrane consisting of TiO_2 nanofibers for organic pollutant decomposition in wastewater under UV irradiation. Desalination 391:89–97. https://doi.org/10.1016/j.desal.2016.01.015

Oller I, Miralles-Cuevas S, Aguera A, Malato S (2018) Monitoring and removal of organic micro-contaminants by combining membrane technologies with advanced oxidation processes. Curr Org Chem 22:1103–1119. https://doi.org/10.2174/1385272822666180404152113

Pal D, Sen S (2023) Emerging petroleum pollutants and their adverse effects on the environment. In: Impact of petroleum waste on environmental pollution and its sustainable management through circular economy, pp 103–137. https://doi.org/10.1007/978-3-031-48220-5_5

Pal D, Sen S (2023) In-depth coverage of petroleum waste sources, characteristics, environmental impact, and sustainable remediation process. In: Impact of petroleum waste on environmental pollution and its sustainable management through circular economy, pp 1–38. https://doi.org/10.1007/978-3-031-48220-5_1

Pal D, Sen S (2024a) Valorization of sponge iron industrial waste into iron-modified zeolite X for ciprofloxacin removal: a multi-parameter optimization study. Environ Sci Pollut Res 31:53381–53398. https://doi.org/10.1007/s11356-024-34730-8

Pal D, Sen S (2024b) Optimal synthesis of dolochar derived faujasite zeolite X for highly effective Cd(II) removal. Environ Res 240:117494. https://doi.org/10.1016/j.envres.2023.117494

Pendergast MM, Hoek EMV (2011) A review of water treatment membrane nanotechnologies. Energy Environ Sci 4:1946. https://doi.org/10.1039/c0ee00541j

Peter KT, Vargo JD, Rupasinghe TP, De Jesus A, Tivanski AV, Sander EA, Myung NV, Cwiertny DM (2016) Synthesis, optimization, and performance demonstration of electrospun carbon nanofiber-carbon nanotube composite sorbents for point-of-use water treatment. ACS Appl Mater Interfaces 8:11431–11440. https://doi.org/10.1021/acsami.6b01253

Puay NQ, Qiu G, Ting YP (2015) Effect of Zinc oxide NP on biological wastewater treatment in a sequencing batch reactor. J Clean Prod 88:139–145. https://doi.org/10.1016/j.jclepro.2014.03.081

Rajabi M, Mahanpoor K, Moradi O (2019) Preparation of PMMA/GO and PMMA/GO-Fe_3O_4 nanocomposites for malachite green dye adsorption: kinetic and thermodynamic studies. Compos Part B Eng 167:544–555. https://doi.org/10.1016/j.compositesb.2019.03.030

Ramasundaram S, Yoo HN, Song KG, Lee J, Choi KJ, Hong SW (2013) Titanium dioxide nanofibers integrated stainless steel filter for photocatalytic degradation of pharmaceutical compounds. J Hazard Mater 258–259:124–132. https://doi.org/10.1016/j.jhazmat.2013.04.047

Reneker DH, Yarin AL (2008) Electrospinning jets and polymer nanofibers. Polymer (Guildf) 49:2387–2425. https://doi.org/10.1016/j.polymer.2008.02.002

Rodrigues SM, Demokritou P, Dokoozlian N, Hendren CO, Karn B, Mauter MS, Sadik OA, Safarpour M, Unrine JM, Viers J, Welle P, White JC, Wiesner MR, Lowry GV (2017) Nanotechnology for sustainable food production: promising opportunities and scientific challenges, Environ. Sci. NANO 4:767–781. https://doi.org/10.1039/c6en00573j

Rudakiya DM, Pawar K (2017) Bactericidal potential of silver NP synthesized using cell-free extract of Comamonas acidovorans: in vitro and in silico approaches, 3 Biotech. 7(92). https://doi.org/10.1007/s13205-017-0728-3

Sankararamakrishnan N, Singh R, Srivastava I (2019) Performance of novel MgS doped cellulose nanofibres for Cd(II) removal from industrial effluent—mechanism and optimization. Sci Rep 9:1–8. https://doi.org/10.1038/s41598-019-49076-2

Tesh SJ, Scott TB (2014) Nano-composites for water remediation: a review. Adv Mater 26:6056–6068. https://doi.org/10.1002/adma.201401376

Travlou NA, Kyzas GZ, Lazaridis NK, Deliyanni EA (2013) Functionalization of graphite oxide with magnetic chitosan for the preparation of a nanocomposite dye adsorbent. Langmuir 29:1657–1668. https://doi.org/10.1021/la304696y

Xie W, Shi Y, Wang Y, Zheng Y, Liu H, Hu Q, Wei S, Gu H, Guo Z (2021) Electrospun iron/cobalt alloy NP on carbon nanofibers towards exhaustive electrocatalytic degradation of tetracycline in wastewater. Chem Eng J 405:126585. https://doi.org/10.1016/j.cej.2020.126585

Yang J, Hou B, Wang J, Tian B, Bi J, Wang N, Li X, Huang X (2019) NMs for the removal of heavy metals from wastewater. NMs 9. https://doi.org/10.3390/nano9030424

Yang GCC, Yen CH (2013) The use of different materials to form the intermediate layers of tubular carbon nanofibers/carbon/alumina composite membranes for removing pharmaceuticals from aqueous solutions. J Memb Sci 425–426:121–130. https://doi.org/10.1016/j.memsci.2012.09.011

Zhang Y, Wu B, Xu H, Liu H, Wang M, He Y, Pan B (2016) NMs-enabled water and wastewater treatment. NanoImpact. 3–4:22–39. https://doi.org/10.1016/j.impact.2016.09.004

Zhu F, Zheng YM, Zhang BG, Dai YR (2021) A critical review on the electrospun nanofibrous membranes for the adsorption of heavy metals in water treatment. J Hazard Mater 401:123608. https://doi.org/10.1016/j.jhazmat.2020.123608

Chapter 14
Nanotechnology Revolution: Advancing Wastewater Remediation with Nano Adsorbents

Deeptimayee Pal and Abhishek Kumar

Abstract Wastewater contamination is a significant global challenge that calls for innovative and sustainable technologies for remediation. The effective removal of a wide variety of pollutants from wastewater is a critical environmental constraint. The application of nano adsorbents is of utmost importance in wastewater remediation because of their distinctive and favorable characteristics, such as excellent reactivity, high surface area-to-volume ratio, and tunable properties. Nano adsorbents, such as graphene oxide, carbon nanotubes, metal–organic frameworks, nano zero-valent iron, nanoclays, and nano zeolites, have been used extensively for wastewater remediation. This study explores the unique characteristics of nano adsorbents, underlying mechanisms, and the complex processes that influence the adsorption of contaminants. The practical value of nano adsorbents in collecting diverse pollutants, including heavy metals, dyes, surfactants, pharmaceuticals, and personal care items, is shown by applications that take place in real-world applications. These applications also demonstrate the effectiveness of nano adsorbents in complicated wastewater matrices. The scope of the study includes the present difficulties and the potential for the future. It specifically addresses the possibility of scaling up, the economic viability, and the interaction with the existing wastewater treatment systems. This scientific investigation improves our knowledge of nano adsorbents as powerful instruments for environmentally responsible and technologically enhanced wastewater treatment. The on-going progress of these materials offers the potential for a more environmentally friendly and efficient method for addressing the complex challenges posed by wastewater contamination.

Keywords Wastewater contamination · Nanotechnology · Adsorption · Nano adsorbents · Environment · Remediation

D. Pal (✉) · A. Kumar
Department of Chemical Engineering, National Institute of Technology Rourkela, Rourkela, Odisha, India
e-mail: deeptimayeenitrkl@outlook.com

N. Agarwal et al. (eds.), *Nanomaterials in Wastewater Research*, Advances in Wastewater Research, https://doi.org/10.1007/978-981-96-3908-3_14

14.1 Introduction

Water's critical significance was underscored in the inaugural United Nations World Water Development Report titled "Water for People, Water for Life." Helver et al. have emphasized the importance of water in maintaining ecosystems and the perpetuation of all life on Earth (Helmer and Hespanhol 1997). However, providing pure water remains a substantial global challenge. The increasing number and variety of pollutants such as heavy metals, pharmaceuticals, personal care products, dyes, petroleum wastes, and polluting water sources have been influenced by various factors, including rapid population growth, substantial industrialization, increased agricultural activities to meet food demand, and other geological, environmental, and global changes (Pal and Sen 2023a). The presence of these pollutants has adverse effects on the environment and health, as these pollutants are highly toxic and carcinogenic (Pal and Sen 2023b). As these pollutants can enter the water bodies through point and non-point sources, the purity of water sources for industrial, agricultural, and human consumption is declining globally. This decline emphasizes the necessity of water quality preservation and enhancement, which has become a critical issue for governmental agencies, water regulatory authorities, and scientists. Inorganic, organic, and biological are the three common categories of water pollutants (Karimi-Maleh et al. 2020).

Over the past few decades, various water purification and recycling technologies have been developed and documented in the literature. These technologies employ various physical, chemical, biological, thermal, and electrical principles. Screening, coagulation-flocculation, filtration, biological treatment, oxidation, reverse osmosis, ion exchange processes, distillation, electrochemical dialysis, and adsorption are among the most frequently employed water purification technologies (Blanco et al. 2019). Adsorption processes are distinguished from these purification techniques by their universality, simplicity of operation, and low construction and maintenance costs. The adsorption process can obtain up to 99.9% removal efficiencies for various inorganic, organic, and biological pollutants in wastewater samples. The adsorbents' porosity, pore size, and surface area affect the adsorption efficiency. Clays, activated alumina, silica gel, limestone, activated carbon, zeolites, and chitosan are currently used as adsorbents for effluent purification. One of the most frequently employed adsorbents is activated carbon. However, activated carbon has numerous deficiencies despite its adaptability. Its adsorption capacity is rapidly diminished, demonstrating restricted porosity, surface area, and pore volume. Furthermore, the recovery of adsorption capacity following regeneration is inadequate. Surface functionalization of conventional adsorbents enhances performance; however, these functionalized materials frequently necessitate advanced, multistep synthesis processes, which renders large-scale production prohibitively expensive (Bhatnagar et al. 2013). In addition, conventional adsorbents encounter substantial obstacles in conveniently separating the adsorbent from purified water and removing pollutants at parts per billion (ppb) levels.

In recent nanotechnology advancements, mechanical, sol–gel, combustion, sonochemical, chemical, microwave, and other methods have been employed to synthesize nanomaterials (NMs) from various sources, such as biomass residues, agro wastes, and residues. Researchers have developed nano adsorbents, the next-generation adsorbents, to address the constraints of conventional adsorbents in water treatment systems. Nano adsorbents exhibit extraordinary physical and chemical characteristics, including superior selectivity, high surface area, high chemical reactivity, conductivity, catalytic activity, and magnetic and optical properties. Compared to their bulk counterparts, nano adsorbents are superior due to their high surface area, which provides more active sites for interactions with various pollutants (Kalfa et al. 2009). The market value of nano adsorbents in environmental industries is anticipated to exceed €1.6 trillion in the next decade, which could provide over one million jobs.

14.2 Adsorption

Adsorption, a surface phenomenon, is widely used to remove various pollutants. This approach is based on the interaction between an adsorbable solute and a solid adsorbent with a highly porous surface structure. The adsorbate concentration on the adsorbent is increased due to the solute molecules that concentrate on the adsorbent surface due to the attractive forces between the solute and the adsorbent. Adsorption is widely regarded as a highly effective, cost-effective, versatile, and easy-to-manage method for the purification of water and effluent (Pal and Sen 2024a). In some situations, the adsorbent is cost-effective and precludes the necessity of a pretreatment step before application. Consequently, adsorption is frequently the preferable method for water remediation and is commonly employed by environmental scientists worldwide to eliminate hazardous inorganic and organic contaminants. Adsorption methods can be classified according to the type of surface attachment, which may be physical, chemical, or exchange adsorption:

- Physical Adsorption (Physisorption): This process is facilitated by the van der Waals forces, which make the separation process more accessible. Adsorption that is physical in nature is reversible and does not involve sharing or transferring electrons.
- Chemistry Adsorption (Chemisorption): Chemisorption is the process by which the adsorbate and adsorbent form a chemical bond. This process yields a monolayer of adsorbate on the surface and is typically irreversible, rendering it challenging to remove the adsorbed materials.

The charge attraction between the adsorbate and the adsorbent surface is the cause of exchange adsorption. Physical and chemical adsorption can occur simultaneously or alternately under favorable conditions, increasing overall adsorption efficiency (Pal and Sen 2024b).

The comprehension of the interaction between an adsorbate and an adsorbent is essential for optimizing the performance of any adsorption process, and adsorption isotherms are essential for this purpose. The adsorption isotherm's form offers valuable insights into the adsorbate molecules' adsorption affinity and the stability of interactions between the adsorbate and adsorbent. Various mathematical models can be used to represent these isotherms. Some of these models are empirical models designed to accurately fit experimental data, while others are based on simplified physical descriptions of the adsorption process. The Langmuir, Freundlich, and Brunauer–Emmett–Teller (BET) isotherms are among the most well-known adsorption isotherm models. The Langmuir isotherm is predicated on monolayer adsorption on a homogeneous surface with a finite number of identical sites, each occupied by a single adsorbate molecule. This model is particularly beneficial for comprehending adsorption processes that reach a saturation point. On the other hand, the Freundlich isotherm is an empirical model that is particularly effective for modeling adsorption processes over a broad range of concentrations and describes adsorption on heterogeneous surfaces. The Langmuir model is extended to multilayer adsorption by the BET isotherm, frequently employed to characterize adsorbents' surface area and porosity (Obaid 1664).

Adsorption kinetics is critical in determining the efficacy and effectiveness of the adsorption process and isotherms. The rate at which the analyte is adsorbed onto the adsorbent and the residence time of the adsorbate on the solid–liquid interface are both described by adsorption kinetics. The velocity of adsorption is contingent upon a variety of factors, such as the frequency of collisions between adsorbate molecules and the adsorbent surface within a specific area per unit of time and the number of adsorbate particles that adsorb onto the adsorbent surface per unit of time. Understanding the dynamics of the adsorption process is a critical component of the study of adsorption kinetics. Kinetic models, including the pseudo-first-order and pseudo-second-order models, are frequently employed to describe this process. In the pseudo-first-order model, the occupancy rate of adsorption sites is assumed to be proportional to the number of unoccupied sites. Conversely, the pseudo-second-order model presupposes that the adsorption rate is proportional to the square of the number of unoccupied sites, which is a more suitable fit for adsorption processes in which chemisorption is the rate-limiting phase (Musah et al. 2022).

14.3 Nano Adsorbents

Nano adsorbents are advanced substances designed at the nanoscale, with a typical dimension of less than 100 nm in at least one direction. They are employed for their remarkable adsorption properties. These materials are intended to eliminate diverse contaminants from water, air, and other environments through adsorption. Specifically, contaminants are attracted to the adsorbent's surface. The high surface area-to-volume ratio of nano adsorbents results from their compact dimension, providing many active adsorption sites. This significantly improves their ability to capture and retain contaminants. Nano adsorbents are highly effective in binding with various pollutants, such as heavy metals, organic compounds, pharmaceuticals, personal care products, dyes, and microorganisms, due to the increased reactivity resulting from the high surface area. Nano adsorbents possess distinctive physical and chemical characteristics not present in their bulk counterparts. These properties, which can be customized for specific applications through surface functionalization, include enhanced mechanical strength, electrical conductivity, and catalytic activity. The efficacy and effectiveness of nano adsorbents in removing pollutants from complex mixtures can be improved by designing them to target specific contaminants selectively through chemical modifications and functionalization. Various nano adsorbents such as graphene and its derivatives, zeolites, nano zero-valent iron (nZVI), nanoclays, and carbon nanotubes (CNTs) have been discussed in this chapter.

14.3.1 Graphene

Graphene, a commonly used nanosorbent, has one or more atomic layers of carbon atoms. This material's mechanical, thermal, and electrical properties are exceptional and have a distinctive two-dimensional structure. The adsorption of dyes onto few-layered graphene nanosheets is facilitated by π-π stacking interactions and van der Waals forces, as shown in Fig. 14.1. The initial incorporation of graphene oxide nanosheets (GONSs) into composite materials can improve manufacturability and enhance the chemical and physical properties of reduced graphene oxide nanosheets (rGONSs). Both rGONSs and solitary GONSs layers demonstrate extensive π-electronic surfaces and high aspect ratios, which facilitate the establishment of robust intermolecular forces among adsorbates. Compared to CNTs, rGONSs exhibit significantly quicker adsorption kinetics due to their open-layer structure (Yu et al. 2014). The adsorption capacities of rGONSs for organic compounds such as phenanthrene and biphenyl have been superior to those of a variety of carbon-based materials, including coal-based activated carbon (AC), single-walled CNTs (SWCNTs), and multi-walled CNTs (MWCNTs). Notably, rGONSs are significantly more cost-effective than SWCNTs. These materials have been utilized as adsorbents to eliminate cationic red X-GRL, methylene blue (MB), methyl orange, Congo red (CO), and other organic substances from aqueous solutions. At 303 K, the maximum adsorption capacities of p-toluenesulfonic acid, 1-naphthalenesulfonic acid, and MB on graphene nanosheets are reported to be 1430, 1460, and 1520 mg/g, respectively (Wu et al. 2011).

Furthermore, GONSs and rGONSs have demonstrated unparalleled adsorption capacities for pesticides, including chlorpyrifos, endosulfan, and malathion, with values up to 1200, 1100, and 800 mg/g, respectively. These rGONSs have been employed as sorbents for removing Cd(II) and Co(II) ions from aqueous solutions, and the synthesis of few-layered GONSs has been reported using a modified Hummers' method. It has been noted that the sorption of heavy metal ions on

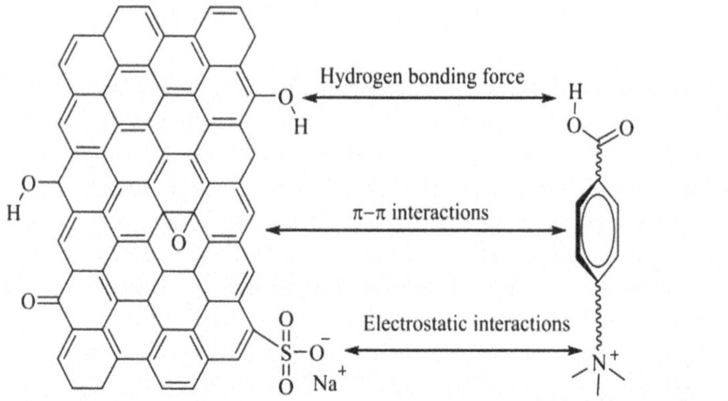

Fig. 14.1 Possible adsorption mechanisms between any contaminant and GONSs (Yu et al. 2015)

nanosheets is influenced by ionic strength and pH, with numerous oxygen-containing functional groups on the surfaces of graphene oxide nanosheets being particularly effective (Zhao et al. 2011a). High binding capacities for arsenic species (As(III) and As(V)) have been demonstrated by magnetite-graphene composite adsorbents with particle sizes of approximately 10 nm, which are attributed to the increased adsorption sites in the graphene composite. The presence of functional solid groups on the surface of graphene oxide (GO) makes it a possible adsorbent for the complexation of metal ions. Both the electrostatic and coordinate approaches can accomplish this. While rGONSs can adsorb cationic and anionic metals, GO is often better at adsorbing cationic metals. Organic materials or metal oxides can also modify GO composites, effectively removing anionic metals through functionalizations. Cu(II) and GO interaction in aqueous solutions leads to GO sheets' folding and large aggregates' formation. This is primarily due to the coordination between Cu(II) and oxygen atoms on GO, which has an adsorption capacity of 46.6 mg/g, surpassing that of CNTs (28.5 mg/g) and AC (4–5 mg/g) (Yang et al. 2010).

A study has been conducted to investigate the removal of Cd(II), Co(II), Pb(II), and U(VI) ions from aqueous solutions using few-layered GONSs. The results indicate that the oxygen-containing functional groups on GONS surfaces are essential for metal sorption. The presence of humic acid was observed to reduce sorption at pH 8, while Cd(II) and Co(II) sorption on GONSs is strongly dependent on pH and faintly dependent on ionic strength. At 303 K and pH 6.0, Cd(II) and Co(II) maximal sorption capacities on GONSs were approximately 106.3 and 68.2 mg/g, respectively. At 293, 313, and 333 K, the maximum adsorption capacities for Pb(II) were approximately 842, 1150, and 1850 mg/g, respectively (Zhao et al. 2011b). Conversely, the sorption capacity for U(VI) was 97.5 mg/g at pH 5.0, and 293 K. Organic materials can modify GO to improve the adsorption of various metal ion species by altering the surface functional groups. Compared to unmodified GO and AC, GO modified with thiol groups through diazonium chemistry has been reported to adsorb six times higher concentrations of Hg(II) ions. Furthermore, the adsorption of Pb(II) was rapid and complete in GO functionalized with N-(trimethoxysilylpropyl) ethylenediamine-triacetic acid (EDTA-silane) within 20 min, with a capacity of 479 mg/g at pH 6.8 (Madadrang et al. 2012). GO and GNS are significantly more effective for environmental remediation than AC and CNTs due to their superior adsorption capabilities for various water pollutants.

14.3.2 Zeolites

Zeolites are three-dimensional aluminosilicates that occur naturally or can be synthesized using various sources such as fly ash, red mud, slag, and dolochar (Pal and Sen 2024a). The hydrothermal crystallization method was employed to synthesize nano zeolite X which was then used as an adsorbent for the simultaneous sono-adsorption of methyl orange, Reactive Blue 15 (RB-15), and Reactive Red 239 (RR-239) dyes. The response surface methodology (RSM) was employed to investigate the effect of various independent variables on the adsorption efficacy of the dye mixture. All models achieved fit convergence ($R^2 > 98\%$) by employing the central composite design within RSM to optimize process parameters. The responses to changes in the input variable were accurately described by a quadratic polynomial mathematical model, as indicated by the analysis of variance. The second-order quadratic model predicted the adsorption response for each dye. RSM confirmed the independent variables' significance and interactions ($R^2 > 0.95$, $P < 0.05$). The optimal conditions were determined to be 0.4 g NX, pH 6, 5 mg/L dye concentration, and ultrasound exposure for 1200 s, which resulted in a dye removal efficiency of over 93% (Shojaei et al. 2021).

Nano zeolite X, synthesized from sodium aluminate salt and silica sol, was employed in a study to eliminate three nitrophenol isomers: Meta, ortho, and para. The adsorption experiments' pH, contact time, adsorbent dose, and initial nitrophenol concentration varied. Zeolite demonstrated exceptional efficacy, achieving adsorption equilibrium within 150 min. Meta, ortho, and para-nitrophenols exhibited maximal adsorption capacities of 125.7, 143.8, and 156.7 mg/g, respectively, more consistent with the Freundlich isotherm than the Langmuir isotherm. The regeneration studies demonstrated that zeolite maintained over 70% of its initial removal efficiency after five adsorption–desorption cycles, with adsorption costs 46.6% lower than those of activated carbon (Pham et al. 2016). Zeolite has been implemented to eliminate methylene blue from effluent by utilizing a packed bed column. The diameter was determined to be 95 nm through XRD and AFM characterization. The adsorption of methylene blue was investigated in a packed bed column of a pilot plant. The breakthrough periods were analyzed at varying flow rates (0.5–1.5 ml/min), initial concentrations (15–35 mg/L), and bed heights (4–8 cm). The results showed that the breakthrough time increased as the bed height increased and decreased as the initial concentrations and flow rates increased (Gadhban et al. 2020).

Zeolite X, synthesized to remove Pb^{2+} from aqueous solutions, was influenced by pH, adsorbent dose, contact time, initial Pb^{2+} concentration, and temperature. The adsorption efficiency of Pb^{2+} was enhanced by an increase in the dosage of zeolite, with the optimum removal occurring at pH 3–5.5 for an initial concentration of 15 mg/L. The pseudo-first-order and pseudo-second-order kinetic models accurately predicted the adsorption process and the equilibrium data was consistent with the Langmuir and Freundlich isotherm models. The Langmuir model demonstrated that the maximal monolayer adsorption capacity increased from 863.79 to 909.09 mg/g as the temperature increased from 298 to 318 K. The thermodynamic evaluations demonstrated endothermic and spontaneous adsorption (Rasouli et al. 2012). Heavy metal removal was achieved by employing a nano-sized clinoptilolite zeolite generated through planetary ball milling in sodium hexametaphosphate. At pH 7.5 and 60 °C, the adsorption efficiency of Ni^{2+}, Cd^{2+}, and Cu^{2+} increased with pH, initial metal concentration, and temperature, respectively, reaching maximum efficiencies of 74.20%, 97.60%, and 99.50%. The Langmuir model was more suitable for Ni^{2+} and Cu^{2+}, while the Freundlich model was more appropriate for Cd^{2+}, as indicated by the thermodynamic evaluations. Spontaneous adsorption behavior was also observed. The pseudo-second-order kinetic model most accurately interpreted the experimental data (Kamran Haghighi et al. 2021).

To ascertain the primary variables that influence the adsorption capacity and removal of organic dyes with nano zeolites, a machine learning study was implemented. Four machine learning algorithms, such as Light Gradient Boosting (LGB), Xtreme Gradient Boosting (XGB), Random Forest (RF), and Artificial Neural Network (ANN), were evaluated. XGB demonstrated the most exceptional performance, accurately anticipating the adsorption capacities of nano zeolites for cationic dyes. Exploratory analysis and hypothesis testing verified the substantial influence of pH, contact time, and concentrations of nano adsorbents and dyes on the adsorption process. XGB could predict adsorption capacity from limited datasets, making it a valuable tool for preliminary assessments before experimental procedures and scale-up in wastewater treatment (Oviedo et al. 2023). Batch experiments were conducted to evaluate the efficacy of nano zeolite, produced from coal fly ash through alkaline hydrothermal treatment, in removing ammonium and phosphate. The procedure achieved equilibrium in less than 60 min, suggesting a significant potential for the rapid immobilization of pollutants. To achieve a 71% removal efficacy, the ammonium concentration decreased from 118 to 35 mg/L, while the phosphate concentration decreased from 52 to 45 mg/L. The efficiencies of ammonium removal were 36.6%, 51.8%, and 70.9% at pH 3, 7, and 10, respectively. By contrast, the phosphate removal efficiency was 92.1%, 47.3%, and 12.3% at pH 3, 7, and 10, respectively. This indicates that nano zeolite could effectively immobilize ammonium but not phosphate. Modifying the surface of nano zeolite is advised to improve pollutant removal efficacy (Gao et al. 2018).

Cesium adsorption was investigated using four distinct varieties of chabazite. Nano zeolite exhibited exceptional efficacy, obtaining adsorption equilibrium within one minute. As demonstrated by a robust positive correlation between the initial cesium concentrations ($R^2 = 0.99$) and the rate constant ratio (k_2, bulk/k_2, nano), the cesium adsorption rate by nano zeolite was orders of magnitude greater than that of more extensive zeolite phases. Nano zeolite had the potential to be precipitated from high-salt solutions with ease using a ferric flocculant despite the difficulty in particle settling and high dispersivity. The flocculation index reached a stable state within 10 min. These findings emphasize the applicability of nano zeolite for rapid processing in emergencies through effective flocculation and adsorption (Lee et al. 2016). The feasibility of employing synthesized magnetic nano zeolite to remove cesium (Cs^+) and strontium (Sr^{2+}) from liquid radioactive waste streams was investigated. The Cs^+/Na^+ and Sr^{2+}/Na^+ binary ion exchange experiments were performed in batch mode at 298, 303, and 313 K temperatures, with an optimized pH, equilibrium time, and constant total normality. The isotherm models of Freundlich, Langmuir, and Redlich-Peterson were employed to analyze the experimental data. The Redlich-Peterson and Langmuir models demonstrated a superior fit to the data compared to the Freundlich model, as evidenced by the correlation coefficients. Confirming the effectiveness of magnetic zeolite as a sorbent, the separation parameters for both metals were less than unity. The intricate calculation of Wilson's parameters necessitated using Excel's trial-and-error procedure to predict activity coefficients in the solid phase. The convex shape of the equilibrium isotherms indicated that the entry of Cs^+ and Sr^{2+} ions was more selective than that of the host Na^+ ions in the magnetic zeolite. The exchange process was endothermic with a positive entropy change, as indicated by the thermodynamic parameters ($\Delta H°$, $\Delta G°$, $\Delta S°$). This suggests that magnetic nano zeolite is feasible and effective for removing metal ions (Moamen et al. 2016). The application of nano zeolite for the removal of various contaminants has been listed in Table 14.1.

Table 14.1 Nano zeolites for wastewater treatment

Nano zeolite	Pollutants	References
Zeolite X	CV	Sivalingam and Sen (2019a)
EU-12	Rhodamine B	Maharana and Sen (2022)
ZSM-5	CV	Sivalingam and Sen (2020)
Zeolite X	Methylene blue	Sivalingam et al. (2019)
Zeolite X	Zn^{2+}, Cu^{2+}, Cd^{2+}, Pb^{2+}, Ni^{2+}, Ca^{2+}, Mg^{2+}, methylene blue, CV, indigo carmine, and Congo red	Sivalingam and Sen (2018)
Zeolite A	Ca^{2+} and Mg^{2+}	Manna and Sen (2023)
Zeolite Y	Arsenic	Sivalingam and Sen (2019b)

14.3.3 NZVI

Iron oxide nanoparticles can be synthesized using a dry and wet chemical synthesis approach (Kumar and Gangawane 2022a, 2022b). Iron oxide nanoparticles can act as adsorbents and photocatalysts in wastewater treatment (Kumar et al. 2022). nZVI refers to the iron nanoparticles where the iron is in a zero-valent state (Fe^0). nZVI can effectively adsorb Ni^{2+} ions; a complete removal can be achieved within 3 h with a 5 g/L nZVI and Ni^{2+} concentration of 100 mg/L. nZVI could remove 88–95% of phosphate within the first 10 min, achieving 96–100% removal at 12 pH after 30 min. The process involved the simultaneous adsorption and chemical precipitation of crystalline vivianite. Nevertheless, the efficacy of phosphate elimination decreased as the pH increased, and the performance of nZVI synthesized in a CMC–water solution was superior to that of nZVI synthesized in an ethanol–water solution (Almeelbi and Bezbaruah 2012). nZVI swiftly adsorbed Co^{2+} ions, with absorption increasing as pH increased. Kaolinite primarily facilitated Cu^{2+} adsorption and decreased nZVI aggregation through redox mechanisms. nZVI effectively adsorbs cadmium, adhering to pseudo-second-order kinetics and suiting the Langmuir isotherm. The efficiency of nZVI in removing 99.9% of As(III) within 10 min at pH 7 decreased as the arsenic concentration and pH increased (Uzum et al. 2009). pH and ionic strength were the primary factors influencing nitrate removal by nZVI. The optimal performance was observed at neutral pH. nZVI supported by polystyrene polymers exhibited improved nitrate reduction. nZVI could reduce soluble Se(VI) to insoluble Se(II) faster than Fe^0 powder. nZVI can effectively remove Cr(VI) from water and sediment, with the efficacy of this process increasing as the pH level lowers. Improved Cr(VI) removal efficacy was observed in stabilized nZVI composites, including those supported by bentonite or CMC. Various nZVI-supported materials, such as PAA-stabilized iron, kaolin-supported nZVI, and zeolite-nZVI composites, effectively removed Pb(II) (Esfahani et al. 2014).

The biological degradation of contemporary azo dyes is challenging due to their high resistance in natural environments. Nevertheless, it is well-established that zero-valent iron (ZVI) assists in reducing azo bonds, facilitating the subsequent biological treatment of the degradation products. Fan et al. demonstrated that nZVI particles can achieve complete decolorization of the azo dye methyl orange within 10 min, demonstrating their ability to decolorize it swiftly. The degradation products, which GC–MS identified, consisted of *N*,*N*-dimethyl-*p*-phenylenediamine, sulfanilic acid, and *N*-methyl-*p*-phenylenediamine. nZVI was investigated for the decolorization of Reactive Black 5 (RB5) and Reactive Red 198 (RR198) (Fan et al. 2009). They discovered that the reduction efficacy was improved by a bimetallic system (with added Pd and TiO_2/light), pitting corrosion, and a lower pH. When nZVI was supported on sand, silica, and biological sediment, the removal efficiency of both pigments was significantly enhanced. Cation exchange resin-supported nZVI was employed to decolorize an azo dye solution containing Acid Blue 113 (AB 113). Using an initial dye concentration of 100 mg/L, a 50.8 mg/g nZVI load, and a 20 g/L resin load, they achieved nearly 100% pigment removal and 12.6% total organic carbon (TOC) removal within 30 min. The decolorization was conducted following a modified pseudo-first-order reaction. A study was conducted on effectively eliminating Acid Blue 24 dye using nZVI and the factors influencing the removal rate. They discovered that the primary mechanism of dye removal was reduction at pH levels below 6, while adsorption was the dominant process at pH levels above 6. The ability of nZVI to decolorize Orange (II) dye when supported on bentonite was investigated. The Fe^0-clay material demonstrated superior reactivity toward the dye than purified nZVI, decreasing the initial dye concentration from 96.5 mg/L to 3 mg/L. The rapid decolorization of water-soluble azo dyes using nZVI immobilized on cation exchange resin underscores the efficacy of nZVI in the treatment of resistant dye pollutants (Xi et al. 2011).

Through batch experiments, the interactions between tetracycline (TC) and nZVI modified by polyvinylpyrrolidone (PVP-K30) were investigated, with variables including reactant concentration, pH, temperature, and the presence of competitive anions. Both pH and temperature were discovered to influence the substantial degradation of TC. Higher temperatures and lower pH levels facilitated more efficient degradation. Competitive anions, including phosphate and silicate, significantly impeded the removal of TC, while acetate and sulfate had only a minor inhibitory effect. The treated solutions' liquid chromatography-mass spectrometry (LC–MS) analysis demonstrated that the degradation products of TC resulted from the elimination of functional groups from the TC ring structure. These degradation products were detected on the surface of PVP-nZVI and in the solution after a 4-h interaction period, suggesting that PVP-nZVI can adsorb both TC and its degradation products (Chen et al. 2011). Upon contact with both ZVI and nZVI, the complete removal of amoxicillin (AMX) and ampicillin (AMP) was attained in another study that involved zero-valent iron (ZVI). The removal mechanisms employed were as follows: (a) Rapid reduction through the rupture of the β-lactam ring, (b) adsorption of AMX and AMP onto iron corrosion products, and (c) sequestration of AMX and AMP within the matrix of precipitating iron hydroxides, a process referred to as coprecipitation with iron corrosion products (Ghauch et al. 2009). At an initial solution, pH of 5.60, and an nZVI dose of 0.1 g/L, metronidazole (MNZ) was swiftly eliminated by nZVI within 5 min at a concentration of 80 mg/L. Following a pseudo-first-order kinetics model, the removal process was implemented. The removal of MNZ was marginally more efficient in the presence of air than in nitrogen, underscoring the importance of oxygen in the removal process. Under the same dosage conditions, the MNZ removal efficiency of nZVI was approximately 49 times greater than that of commercial iron powder. In addition, metronidazole was degraded using nano zero-valent metal (nZVM) synthesized from pickling waste fluid discharged by the steel industry. The surface area-normalized rate coefficient (kSA) for nZVM was 0.254 L/min/m^2, 375.2 times greater than that of commercial iron powder (6.67×10^{-4} L/min/m^2). This result suggests that nZVM degrades antibiotics more efficiently (Fang et al. 2010).

The extensive use of chlorinated nitroaromatic compounds in producing pharmaceuticals, explosives, pigments, and pesticides has resulted in their widespread presence in the environment. Environmental scientists are concerned about these compounds due to their potential carcinogenic, mutagenic, or deleterious effects. Research has demonstrated that trichloroethylene (TCE) can be effectively degraded by nZVI and palladium-doped nZVI immobilized on an alginate substrate. When 50 g/L of Fe/Pd-alginate (3.7 g Fe/L) was introduced into an aqueous solution, the removal efficiency of TCE exceeded 99.8% in one study. The primary final products were ethane and butane, and the degradation of TCE by Fe/Pd-alginate followed pseudo-first-order kinetics. The final reaction products comprised 80% ethane and approximately 20% C3–C6 molecules at a low TCE concentration (4.4 mg/L) and excess nZVI (approximately 1.9 g/L). In contrast, the addition of a small quantity of nZVI (approximately 0.36 g/L) to a high TCE concentration (290 mg/L) resulted in the formation of even-numbered saturated alkanes (C2, C4, C6) and other saturated and unsaturated hydrocarbons (Hojeong, 2010). Although intermediates such as vinyl chloride and cis-dichloroethylene (cis-DCE) were detected during the reaction, they were present at deficient levels and dissipated rapidly (Liu et al. 2005). The degradation pathways for TCE dechlorination by nZVI and tetrachloroethylene (PCE) dechlorination by surface-modified nZVI involve the reductive beta elimination of PCE to acetylene, followed by the transformation of acetylene to ethene and ethane.

The remediation of dichlorodiphenyltrichloroethane (DDT) from contaminated water using nZVI was investigated. The addition of 30 mg/L Fe0 to an aqueous solution of 3 mg/L DDT resulted in a 99.2% reduction of DDT within 4 h. Diphenylmethane (DPE), an entirely dechlorinated hydrocarbon skeleton of DDT, was confirmed by GC/MS analysis. This confirms the removal of all five chlorine atoms (three alkyls and two aryls). This investigation determined that nanoscale Fe0 effectively dechlorinated DDT, dichlorodiphenyldichloroethane (DDD), and dichlorodiphenylmonochloroethane (DDMS). Another application involved the use of nZVI in conjunction with hydrogen peroxide to remediate soil that had been contaminated with pentachlorophenol. The decay rate of pentachlorophenol in Chengchun (Cf) series soil was increased from 37 to 81% and in Pianchen (Pc) series soil from 41 to 75%, respectively, by the addition of 5% calcium carbonate for 40 h. This novel treatment method is a benchmark for on-site applications (Liao et al. 2007). The sequential dehalogenation trend of hexachlorobenzene (HCB) and polychlorinated biphenyls (PCB) was demonstrated by nZVI in the dehalogenation of these compounds. The end product of HCB reductive dehalogenation was 1,2-dichlorobenzene (DCB). Another investigation investigated the reductive dechlorination of TCE that was sorbed in two model soils (Smith Farm soil and potting soil) using carboxymethyl cellulose (CMC)-stabilized Fe–Pd bimetallic nanoparticles. Although the degradation kinetics were inhibited by the soil organic matter (approximately 8.2%) in the potting soil, the nanoparticles effectively degraded TCE. Four prototype surfactants were evaluated for their impact on TCE's desorption and degradation rates. These surfactants included two anionic surfactants (sodium dodecyl sulfate [SDS] and sodium dodecyl benzene sulfonate [SDBS]), a cationic surfactant (hexadecyltrimethylammonium [HDTMA] bromide), and a nonionic surfactant (Tween-80). The anionic surfactant SDS was the most effective, although all four surfactants enhanced TCE desorption at concentrations either below or above the critical micelle concentration (Zhang et al. 2011).

14.3.4 Nanoclays (NCs)

Phyllosilicates comprise NCs primarily composed of oxygen, silicon, and additional elements such as aluminum, calcium, magnesium, and sodium. Various industries pretreat NCs to optimize their performance for particular applications. The layered structure of NCs is a notable characteristic, as it enables them to accumulate water, resulting in a unique swell-shrink behavior. This characteristic renders them especially advantageous in applications requiring materials that can expand and contract in response to moisture content. Another significant attribute of NCs is their ability to absorb water. The volume of NCs can increase by up to six times when water is added, resulting in the formation of stable gels (Gahlot et al. 2020). This behavior is essential for applications in pharmaceuticals, cosmetics, and environmental engineering industries, where the controlled formation of gels and expansion is advantageous. Figure 14.2 illustrates that NCs comprise mineral silicates arranged in strata, as their chemical composition indicates. These minerals' unique properties result from their fine-grained structure with a sheet-like geometry. Talc, Mica, Kaolin, Montmorillonite, Serpentine, and Sepiolite are among the numerous minerals classified as NCs. The dimensions, chemical formulae, and structural aspects of these minerals, including the arrangement of the silicon-oxygen tetrahedron and embedded ions, differ (Mahdavinia et al. 2013; Sethy et al. 2023). NCs are classified according to the structure of the silicon-oxygen tetrahedron, the type and distribution of embedded ions, and the properties of their constituents, primarily their size. The structures of various forms of NCs are highly complex and diverse. Therefore, it is essential to comprehend these properties to optimize NCs for various industrial applications.

Gahlot et al. explained the efficacy of NCs in removing heavy metals and dyes, obtaining maximal removal rates of 85–90% within 10–60 min for various cationic and anionic dyes (Gahlot et al. 2020). This improved removal efficacy is primarily due

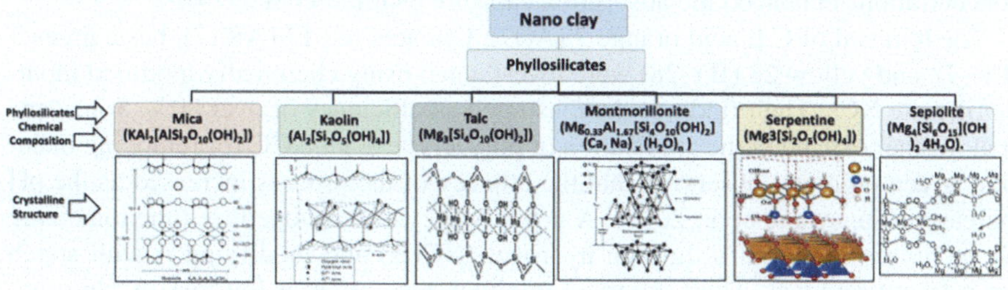

Fig. 14. 2 Different forms of Nano clay and their chemical structure used for the various purposes. Generally, the Nano clay found to be long chain chemical structure (Sethy et al. 2023)

to the substantial increase in NC's specific surface area (SSA) following exfoliation. The increased cation exchange capacity of NCs in an ultrasound-assisted environment can further enhance the adsorption rate. The efficacy of exfoliation techniques, which elevate the SSA, is responsible for this enhanced capacity. Although other adsorbents have demonstrated higher rates of dye removal, modified boron-based NCs have been significantly more effective. NCs were developed by Chen et al. from natural soil clay to remove Cu^{2+} (Chen et al. 2015). They reported that NCs outperformed natural soil clay by a factor of three. Kaolinite clay was modified with polyvinyl alcohol (PVA) to produce kaolinite clay-based NCs. This modification substantially enhanced the adsorption capacity, resulting in removal efficiencies of 92% for Pb^{2+} and 78% for Cd^{2+}. The removal of Cu^{2+} from acid mine drainage was achieved using montmorillonite NCs treated with hexadecyltrimethylammonium (HDTMA), resulting in a 96.3% removal rate. The montmorillonite NCs were treated with hydroxyl iron to remove As^{3+}. They stated that an optimal pH would lead to an elimination rate exceeding 90%. Sharififard et al. demonstrated that TiO_2-modified NCs demonstrated 90% efficacy in removing Cd^{2+} under optimal pH and dosage conditions (Sharififard et al. 2018).

The environmental impact of dyes disposed of by industries without adequate treatment has been addressed using NC composites (NCCs) to remove them from effluent. Dyes are capable of being highly toxic and carcinogenic, and they are resistant to degradation by physicochemical or biological processes. Dyes, which include auxochromes and chromophores, are classified according to their application: Cationic, direct, reactive, sulfur, dispersive, solvent, anionic, and vat dyes. Researchers have evaluated the effectiveness of Na-montmorillonite NCs in the removal of crystal violet (CV) and reactive blue 29 (RB-29) pigments, and they have determined that NCs are only effective in an acidic environment (Mahdavinia et al. 2013). Sonawane et al. investigated the efficacy of organophilic bentonite NCs modified by HDTMA, tetrabutylammonium chloride, and *N*-acetyl-*N*, *N*,*N* trimethyl ammonium bromide in the removal of methyl blue. They discovered substantial sonication effects (Sonawane et al. 2009). In contrast, poly(acrylic acid)-bentonite-FeCo hydrogel NCC was employed to eliminate CV pigments, observing that the maximal removal occurred at higher pH levels. Shruthi et al. synthesized NCC for CV dye removal using silane-modified NCs, observing that increased initial dye concentrations enhanced the adsorption capacity (Shruthi et al. 2016).

The removal of C.I. acid orange 7 (AO7), C.I. acid red 17 (AR17), basic green 4 (BG-4), and yellow 28 (BY-28) were investigated using chemically modified montmorillonite NCs. Their findings demonstrated an efficiency of over 80%. Sonawane et al. further expanded the application of polyacrylic acid–NCC for removing malachite green pigment, observing that the efficacy of the process increased as the pH increased (Sonawane et al. 2009). A novel NCC was developed by functionalizing Na-montmorillonite NCs through hydrogel (NCHs) and modifying it with starch for safranin removal. They observed minimal adsorption at low pH. A chitosan-organically modified NCs (Cloisite 30B) nano-bio composite has substantial removal rates for reactive blue-21 and reactive red-141 dyes at low pH. In contrast to other studies that reported that acidic conditions were more effective, El Haouti et al.

reported that the removal efficiency of toluidine blue and crystal violet dyes from effluent could be improved by increasing the pH and using an appropriate NC dosage (Haouti et al. 2019).

Although nano adsorbents are highly effective in effluent treatment, they have significant drawbacks. The synthesis and processing are initially costly, rendering them less economically viable for large-scale applications. The complex and expensive materials and methods frequently employed in the synthesis processes often elevate the overall treatment costs. Furthermore, nano adsorbents exhibit substantial environmental and health hazards. Drinking nanoparticles into the environment can be deleterious to aquatic life and ecosystems, and their potential path into potable water sources raises significant public health concerns. Further complicating their use, handling these materials necessitates specialized knowledge and apparatus due to their reactivity and compact size. The regeneration and reuse of nano adsorbents present an additional obstacle. In contrast to conventional adsorbents, nanoparticles are challenging to recover and regenerate without compromising their adsorption efficiency, which reduces their cost-effectiveness. Another concern is scalability; although nano adsorbents demonstrate satisfactory performance in controlled laboratory environments, the technology is frequently impractical and difficult to scale-up for the treatment of large volumes of effluent. Moreover, the efficacy of nanoparticles in adsorbing contaminants can be diminished due to their agglomeration or degradation over time. Despite their exceptional efficacy in pollutant removal, these factors hinder the widespread adoption and practical implementation of nano adsorbents in wastewater treatment (El-sayed 2020).

14.3.5 Carbon Nanotubes (CNTs)

Surface-modified CNTs have been extensively investigated for their ability to remove heavy metals from wastewater and water. Contaminant adsorption primarily occurs at four prospective sites on the CNTs: Outer and inner fissures, interstitial channels, and the interstitial region. However, the interior region of the nanotubes is less adsorbed (Ali et al. 2019). Functionalized MWCNTs are effective for heavy metal adsorption because of the numerous adsorption sites available on the tubes' surfaces, as demonstrated (Bahgat et al. 2011). More oxygen-containing functional groups in plasma-treated CNTs resulted in higher adsorption of Zn^{2+} ions in another study. This suggests that the surface complexation mechanism significantly influences cationic ions' adsorption, as deprotonated functional groups increase the number of available adsorption sites. Metal oxide-modified CNTs and heavy metal ions have been observed to exhibit both physical and chemical adsorption mechanisms. The specific hydration energy, ionic radius, and hydrolysis potential of the contaminants also affect the adsorption capacities of CNTs. For example, Hu et al. discovered that modified MWCNTs with iron oxide demonstrated a higher removal efficacy for Pb^{2+} than Cu^{2+} under the same experimental procedures. The presence of other organic substances can influence the adsorption efficacy of CNTs for heavy metals (Hu et al.

2011). For instance, the removal of Cd^{2+} ions was facilitated by the presence of 1-naphthol when oxidized MWCNTs were employed. However, the removal of 1-naphthol itself was not significantly influenced by the presence of Cd^{2+} ions. This phenomenon is the result of unique interaction mechanisms that are responsible for the adsorption of various pollutants.

The interactions between the functional groups on the CNT surface and specific contaminants, rather than the nanotube size, determine the efficacy of heavy metal adsorption by functional CNTs. Addo Ntim and Mitra demonstrated zirconia nanocomposites' superior arsenic adsorption capacity by reporting that zirconia-modified CNTs exhibited higher As(III) and As(V) adsorption than iron oxide-coated CNTs. The adsorption capacity and mechanisms of carbon nanotubes (CNTs) are significantly influenced by pH. The interaction of CNTs with various contaminants is influenced by the surface charge of CNTs, which is influenced by the pH of the solution. The charge density of functionalized CNTs becomes more negative at higher pH levels, which generates alkaline conditions. The adsorption of cationic species is facilitated by the increased negative charge, which maximizes the electrostatic attraction between the negatively charged CNT surfaces and positively charged metal ions. Conversely, the positive charge density increases when the surface functional groups of CNTs become protonated, which occurs at lower pH levels or under acidic conditions. This positive charge repels cationic metal ions, which decreases the adsorption efficacy of CNTs for these species (Yang et al. 2012).

Typically, organic contaminants in water and effluent are dissolved and/or suspended particulate matter, often due to human activities and animal or plant degradation. Functionalization can substantially improve the adsorption efficiencies of CNTs for organic dyes. For example, oxidized MWCNTs have been demonstrated to effectively remove methylene blue and orange dyes from water matrices, surpassing other modified adsorbents. Duman et al. effectively eliminated crystal violet dye by employing a novel nanocomposite composed of MWCNTs, carrageenan, and Fe_3O_4 (Duman et al. 2016). New nanocomposites, including MWCNT-COOH-cysteamine, have been created to enhance the adsorption capacity of cationic dyes. Sadegh and colleagues reported a maximum adsorption capacity of 131 mg/g for MWCNT-COOH-cysteamine, significantly higher than the 90 mg/g of MWCNT-COOH. Nevertheless, the adsorption capacities of CNT composites can be substantially diminished due to the competition between a variety of organic contaminants present in both water and effluent. Ali et al., however, reported that the initial concentration of inorganic contaminants did not substantially influence the removal capacities of CNTs (Ali et al. 2019). This underscores the significance of customizing and modifying the surface of CNTs to facilitate the selective adsorption of organic contaminants. Wang et al. demonstrated that CNT composites are approximately 95% effective in the removal of pharmaceutical and personal care products (PPCPs) by increasing the specific surface area and aromatic ring content of CNTs (Wang et al. 2015). Jahangiri-Rad et al. demonstrated that oxidized single-walled carbon nanotubes (SWCNTs) with a large specific surface area had an adsorption capacity of 496 mg/g to remove blue 29 dye (Jahangiri-Rad et al. 2013).

Furthermore, the pH of the solution influences the interaction mechanisms between functionalized CNTs and organic contaminants. For example, the adsorption of natural organic matter (NOM) can be reduced by oxygen-containing functional groups on the surface of CNTs (Engel and Chefetz 2016). This is because these functional groups enhance electrostatic repulsion, diminishing the π–π interactions between the large specific surface area of functionalized CNTs and NOM. The effective NOM adsorption on CNTs depends on these π–π interactions. The efficiency of NOM adsorption is further diminished by the more pronounced electrostatic repulsion at higher pH levels. Yang et al. reported comparable results, suggesting that organic contaminants, including aromatic compounds, are predominantly adsorbed onto the CNT surface through π–π interactions (Yang et al. 2012). These interactions occur between the CNTs' surface and organic molecules' aromatic rings, such as 1–naphthol. The electrostatic environment surrounding the CNTs and the contaminants is influenced by the pH of the solution, which in turn impacts the efficacy of these interactions.

Additionally, CNTs have been successfully employed to eliminate pesticides. The removal efficiency of diuron was exceptional for both oxidized and as-prepared MWCNTs at pH levels of 7.0 or higher. The adsorption of diuron is enhanced by the oxidation of CNTs, which increases their surface area and pore size. Adding amino-functionalized CNTs decreased the uptake of chlordane and *p,p′*-dichlorodiphenyldichloroethylene in lettuce roots from 78 to 23% (Hamdi et al. 2015). Semiconductor CNTs were discovered to absorb pesticides such as 1-pyrenebutyric acid, 2,4-dichlorophenoxyacetic acid, and diquat dibromide more effectively (up to 70.6%) than metallic CNTs. This is likely due to the reduced electron density on the surface of semiconductor CNTs, which enables greater adsorption. Nevertheless, pesticide removal was restricted in batch mode systems, even though SWCNTs and MWCNTs achieved complete pesticide adsorption in fixed bed systems by extending the contact duration. Bacteriological contaminants, which have various effects on the adjacent environment and degrade the assimilative capacity of water bodies, pose a substantial challenge. Surface waters, wastewater, and treatment facilities are frequently contaminated with these substances. CNTs demonstrate high-affinity interactions with biological contaminants due to their diverse surface and functional characteristics. The large surface area of these sorbents has resulted in higher adsorption capacities, inactivation efficiencies for viral and bacterial particles, and a greater antimicrobial potential than conventional sorbents.

The capacity of SWCNTs to penetrate the bacterial cell wall is the reason for their inactivation of E. coli. Surface-modified CNTs can more effectively destroy the cell wall of microorganisms with various chemical groups than unprocessed and polymer-grafted CNT membranes. Bacterial cell walls' metabolism and morphological structure can be adversely affected by the direct interaction of microorganisms, such as E. coli, with functionalized CNTs. The primary cause for the increase in inactivation affinity of CNTs is their penetration. CNTs functionalized with silver nanoparticles have demonstrated an exceptional capacity to inactivate microorganisms. Silver-doped CNTs were successfully synthesized and employed to inactivate E. coli bacteria, achieving a 100% bacterial inactivation rate (Ihsanullah et al. 2015).

Nanotubes' diameter can also be a critical factor in the inactivation of pathogenic microorganisms in effluent and water. The single kinetics of CNTs demonstrate their capacity to eliminate pathogens through deep filtration mechanisms, with microorganisms remaining on the CNT surface. A polyvinylidene fluoride-based SWCNT filter successfully eradicates E. coli bacteria at low pressure (Brady-Estévez et al. 2008). The filter completely captured the cells. Furthermore, the MS2 virus was effectively eliminated by filtering samples through a controlled nanoporous CNT-based filter at an 8 to 11 bar pressure. In addition to filtration, CNTs are extensively employed as antimicrobial agents in water and wastewater treatment, offering a viable alternative to chemical disinfectants to prevent pathogens. Due to the low oxidation state and solubility of CNTs in water, this method prevents the formation of hazardous disinfectant by-products such as trihalomethanes, aldehydes, and haloacetic acids. Surfactants or polymers, including sodium dodecylbenzenesulfonate and polyvinylpyrrolidone, are employed to enhance the dispersibility of these compounds. CNTs' exceptional mechanical properties enable them to function as scaffolds for antimicrobial agents, including silver nanoparticles and antibacterial lysozyme, thereby increasing their efficacy in disinfection treatments.

14.4 Conclusions and Future Perspectives

Human health, ecosystems, and economies are all at risk due to water contamination, which continues to be a significant global issue. The accelerated pace of industrialization, urbanization, and agricultural activities has facilitated the proliferation of contaminants in water bodies, such as heavy metals, organic compounds, pigments, and pathogens. The efficient elimination of these diverse contaminants can often pose a challenge for conventional water treatment methods, so exploring advanced technologies is necessary for effective wastewater purification. A promising frontier in water treatment has emerged with the introduction of innovative materials and techniques to address water contamination challenges, namely nanotechnology. Nano adsorbents have attracted considerable attention due to their exceptional efficacy in eliminating a wide range of pollutants and their distinctive characteristics. Adsorption, a surface phenomenon, is particularly well-suited for the purification of water and effluent due to its cost-effectiveness, versatility, and simplicity. The process entails the adhesion of contaminants to the surface of solid adsorbents, which is caused by various forces, including van der Waals forces, electrostatic interactions, and chemical bonding. Nano adsorbents can capture contaminants due to their high surface area-to-volume ratio and nanoscale dimensions, providing many active adsorption sites. This chapter has examined the applications and advantages of various nano adsorbents, such as graphene, zeolites, NCs, and nZVI, and has emphasized their respective mechanisms of action and advantages.

The large surface area, high aspect ratio, and tunable surface chemistry enable CNTs to demonstrate exceptional adsorption capabilities. They can efficiently adsorb organic contaminants, heavy metals, and dyes. By introducing specific chemical groups that interact powerfully with target contaminants, the functionalization of CNTs improves their selectivity and adsorption efficiency. Nevertheless, the ubiquitous application of CNTs is restricted by obstacles such as high production costs, potential environmental and health dangers, and the difficulty of uniformly disseminating CNTs in aqueous solutions. To address these concerns, future research should concentrate on developing safer handling practices and more cost-effective synthesis methods. Graphene and its derivatives, including GO, exhibit exceptional adsorption capabilities due to their two-dimensional structure, high surface area, and simplicity of functionalization. These materials effectively remove pathogens, organic contaminants, and heavy metals. Despite these advantages, the potential environmental impact, high production costs, and aggregation issues must be resolved to facilitate large-scale implementation. The broader adoption of graphene in wastewater treatment applications is contingent upon innovations in graphene production, including reducing synthesis costs and enhancing scalability. Zeolites are known for their thermal stability and high ion exchange capacities, which are attributed to their crystalline aluminosilicate framework. They can effectively adsorb organic pollutants, ammonium, and heavy metals. Their selectivity and efficacy are improved by the capacity to adjust pore diameters and surface chemistry. Nevertheless, the adsorption capacity of natural zeolites is frequently enhanced through modification, and their regeneration can be energy-intensive. The practical utility of zeolites in wastewater treatment could be improved through research into more efficient modification techniques and sustainable regeneration methods.

NCs, including montmorillonite and kaolinite, exhibit exceptional adsorption properties and high specific surface areas due to their layered structures. They can effectively eliminate organic pollutants, heavy metals, and dyes. Adding organic compounds or polymers improves the adsorption capacity and selectivity of nanoclays. Nevertheless, the adsorption capacity of nanoclays is relatively lower than that of other nano adsorbents, and they have the propensity to enlarge in water. Consequently, they present a challenge. Future research should investigate innovative modification strategies to address these constraints and enhance the performance of nanoclays in various environmental conditions. Multiple contaminants, such as chlorinated compounds, heavy metals, and nitrates, are effectively degraded by nZVI through reductive dechlorination and adsorption. The rapid removal of pollutants is facilitated by the high reactivity and extensive surface area of nZVI. Nevertheless, the practical implementation of nZVI particles is restricted by factors such as particle agglomeration, potential toxicity to aquatic life, and the rapid oxidation of the particles. The use of stabilizers and surface modifications can alleviate these issues; however, they also increase the complexity and cost. It is imperative to conduct ongoing research on more environmentally favorable and stable formulations of nZVI to expand its use in effluent treatment.

Nano adsorbents are characterized by their high adsorption capacity, rapid kinetics, and capacity to target a wide range of pollutants. The adsorption mechanisms are contingent upon the nature of the pollutant and the type of nano adsorbent, and they include ion exchange, chemisorption, and physical adsorption (physisorption).

Physisorption is typically reversible and facilitates the regeneration of the adsorbent, as van der Waals forces drive it. Chemisorption is characterized by the formation of stronger chemical bonds, which results in increased adsorption capacities. However, regeneration is frequently more difficult. Electrostatic interactions are the foundation of ion exchange mechanisms, particularly effective in removing charged pollutants such as heavy metals. Nano adsorbents substantially improve wastewater treatment, improving selectivity and efficacy in removing contaminants. CNTs, graphene, zeolites, nanoclay, and nZVI can effectively address various pollutants due to their distinctive properties. Nevertheless, addressing the challenges associated with their synthesis, application, and potential environmental impacts is imperative to ensure their successful integration into large-scale wastewater treatment systems. The protection of public health and the environment will be facilitated by developing more sustainable and effective water treatment solutions through on-going research and innovation in this field.

A dynamic and swiftly evolving discipline is represented by the on-going development and application of nano adsorbents for wastewater treatment. The high surface area, tunable surface chemistry, and enhanced reactivity of NMs are among the many unique properties that provide substantial advantages over conventional materials. Nevertheless, many obstacles must be overcome to realize the potential of nano adsorbents. One of the primary obstacles to the widespread adoption of nano adsorbents is the high cost and complexity of their synthesis. Developing more cost-effective and scalable methods for producing NMs should be the primary focus of future research. This may entail the investigation of alternative raw materials, the optimization of production processes, and the implementation of green chemistry strategies to reduce the environmental impact. For instance, the potential for cost reduction and sustainability enhancements could be achieved by utilizing biogenic and waste-derived materials to synthesize nano adsorbents. The potential environmental and health hazards associated with NMs are a significant concern. To comprehend the behavior, fate, and toxicity of nano adsorbents in aquatic environments, it is imperative to implement thorough risk assessments. Implementing strategies to prevent nanoparticle discharge into the atmosphere, exploring the long-term effects of NMs on human health and ecosystems, and developing safer NMs with reduced ecotoxicity should be the primary focus of future studies. The potential remedy to these concerns lies in developing biodegradable and biocompatible NMs.

The economic viability and environmental sustainability of nano adsorbents are contingent upon their capacity to regenerate and reuse. Future research should focus on innovative regeneration techniques that preserve nano adsorbents' structural integrity and adsorption capacity. Potential strategies include the development of thermal, pharmacological, or biological regeneration methods. Investigating using external stimuli (e.g., magnetic fields, light) to facilitate regeneration and developing nano adsorbents with self-cleaning properties. It is essential to improve the efficiency of nano adsorbents by increasing their selectivity toward specific contaminants. Functionalization techniques can introduce specific chemical groups onto the surface of NMs to target specific pollutants. Future research should explore molecular imprinting and other advanced techniques to improve selectivity, designing multifunctional nano adsorbents that can remove multiple contaminants simultaneously and optimizing functionalization processes to achieve high selectivity without compromising adsorption capacity. The efficacy can be improved by combining the benefits of various adsorbents by developing hybrid and composite NMs. For instance, the adsorption capacity, stability, and regeneration potential of conventional adsorbents (e.g., activated carbon, zeolites) can be enhanced by the integration of NMs. Future research should investigate the interactions between various components in composite materials, design and synthesize novel hybrid materials with synergistic properties, and evaluate the performance of hybrid nano adsorbents in real-world wastewater treatment scenarios.

Optimizing the design and application of nano adsorbents necessitates a more profound comprehension of the adsorption mechanisms at the nanoscale. Valuable insights into the interactions between NMs and contaminants can be obtained through advanced characterization techniques, including molecular simulations, spectroscopy, and high-resolution microscopy. Future research should examine the influence of surface chemistry, morphology, and structural properties on adsorption processes. Comprehending the thermodynamics and kinetics of adsorption at the nanoscale, along with the influence of environmental factors (e.g., pH, temperature, and ionic strength), is essential for optimizing adsorption behavior. The successful implementation of nano adsorbents necessitates translating laboratory-scale research into real-world applications. Pilot-scale studies and field trials are essential to assess nano adsorbents' performance, reliability, and scalability in real-world scenarios. Future research should concentrate on implementing exhaustive experimental studies to evaluate the feasibility and efficacy of nano adsorbents in various wastewater treatment scenarios. To optimize their effectiveness, nano adsorbents should be implemented in current wastewater and water treatment systems. This integration can potentially improve the overall efficacy and cost-effectiveness of treatment processes. Future research should investigate the development of modular nano adsorbent systems that can be seamlessly integrated into existing infrastructure. To guarantee the safe and sustainable use of nano adsorbents, they must be supported by appropriate policies and regulations during the development and implementation process. Future research should concentrate on establishing regulatory frameworks that address the production, application, and disposal of NMs. Promoting public awareness and stakeholder engagement to resolve concerns and facilitate the acceptance of nanotechnology in water treatment and developing guidelines for safely handling and using nano adsorbents in wastewater treatment. The field of nano adsorbents in wastewater treatment is poised for advancement due to several emergent trends and innovations. These include the development of smart nano adsorbents that can adapt to changing environmental conditions and self-regenerate, integration of biological systems with NMs to create bio-inspired and bio-functionalized adsorbents, utilization of three-dimensional printing technology to design and fabricate custom nano adsorbent structures with optimized properties, and utilizing nano adsorbents in decentralized and point-of-use treatment systems to present clean water in remote and underserved areas.

References

Ali S, Rehman SAU, Luan HY, Farid MU, Huang H (2019) Challenges and opportunities in functional carbon nanotubes for membrane-based water treatment and desalination. Sci Total Environ 646:1126–1139. https://doi.org/10.1016/j.scitotenv.2018.07.348

Almeelbi T, Bezbaruah A (2012) Aqueous phosphate removal using nanoscale zero-valent iron. In: Nanotechnology for sustainable development. Springer International Publishing, Cham, pp 197–210. https://doi.org/10.1007/978-3-319-05041-6_16

Bahgat M, Farghali AA, El Rouby WMA, Khedr MH (2011) Synthesis and modification of multi-walled carbon nano-tubes (MWCNTs) for water treatment applications. J Anal Appl Pyrolysis 92:307–313. https://doi.org/10.1016/j.jaap.2011.07.002

Bhatnagar A, Hogland W, Marques M, Sillanpää M (2013) An overview of the modification methods of activated carbon for its water treatment applications. Chem Eng J 219:499–511. https://doi.org/10.1016/j.cej.2012.12.038

Blanco M, Monteserín C, Angulo A, Pérez-Márquez A, Maudes J, Murillo N, Aranzabe E, Ruiz-Rubio L, Vilas JL (2019) TiO_2-doped electrospun nanofibrous membrane for photocatalytic water treatment. Polymers (Basel) 11:1–11. https://doi.org/10.3390/polym11050747

Brady-Estévez AS, Kang S, Elimelech M (2008) A single-walled-carbon-nanotube filter for removal of viral and bacterial pathogens. Small 4:481–484. https://doi.org/10.1002/smll.200700863

Chen H, Luo H, Lan Y, Dong T, Hu B, Wang Y (2011) Removal of tetracycline from aqueous solutions using polyvinylpyrrolidone (PVP-K30) modified nanoscale zero valent iron. J Hazard Mater 192:44–53. https://doi.org/10.1016/j.jhazmat.2011.04.089

Chen Y, Tsao T, Wang M, Yu S, Liu C, Li H, Chiu C, Wang L (2015) Kinetic and thermodynamic studies on removal of Cu(II) from aqueous solutions using soil nanoclays. Water Environ Res 87:88–95. https://doi.org/10.2175/106143014X14062131179159

Duman O, Tunç S, Polat TG, Bozoğlan BKI (2016) Synthesis of magnetic oxidized multiwalled carbon nanotube-κ-carrageenan-Fe_3O_4 nanocomposite adsorbent and its application in cationic methylene blue dye adsorption. Carbohydr Polym 147:79–88. https://doi.org/10.1016/j.carbpol.2016.03.099

El Haouti R, Ouachtak H, El Guerdaoui A, Amedlous A, Amaterz E, Haounati R, Addi AA, Akbal F, El Alem N, Taha ML (2019) Cationic dyes adsorption by Na-montmorillonite nano clay: experimental study combined with a theoretical investigation using DFT-based descriptors and molecular dynamics simulations. J Mol Liq 290:111139. https://doi.org/10.1016/j.molliq.2019.111139

El-sayed MEA (2020) Nanoadsorbents for water and wastewater remediation. Sci Total Environ 739:139903. https://doi.org/10.1016/j.scitotenv.2020.139903

Engel M, Chefetz B (2016) Adsorption and desorption of dissolved organic matter by carbon nanotubes: effects of solution chemistry. Environ Pollut 213:90–98. https://doi.org/10.1016/j.envpol.2016.02.009

Esfahani AR, Firouzi AF, Sayyad G, Kiasat A, Alidokht L, Khataee AR (2014) Pb(II) removal from aqueous solution by polyacrylic acid stabilized zero-valent iron nanoparticles: process optimization using response surface methodology. Res Chem Intermed 40:431–445. https://doi.org/10.1007/s11164-012-0975-1

Fan J, Guo Y, Wang J, Fan M (2009) Rapid decolorization of azo dye methyl orange in aqueous solution by nanoscale zerovalent iron particles. J Hazard Mater 166:904–910. https://doi.org/10.1016/j.jhazmat.2008.11.091

Fang Z, Qiu X, Chen J, Qiu X (2010) Degradation of metronidazole by nanoscale zero-valent metal prepared from steel pickling waste liquor. Appl Catal B Environ 100:221–228. https://doi.org/10.1016/j.apcatb.2010.07.035

Gadhban MY, Abdulmajed YR, Ali FD, Al-Sharify ZT (2020) Preparation of nano zeolite and its application in water treatment. IOP Conf Ser Mater Sci Eng 870:012054. https://doi.org/10.1088/1757-899X/870/1/012054

Gahlot R, Taki K, Kumar M (2020) Efficacy of nanoclays as the potential adsorbent for dyes and metal removal from the wastewater: a review, Environ Nanotechnol Monit Manage 14:100339. https://doi.org/10.1016/j.enmm.2020.100339

Gao F, Xiao L, Zhang H (2018) Potential of using synthesized nano-zeolite for ammonium and phosphate immobilization in dairy wastewater. J Dairy Res 85:375–378. https://doi.org/10.1017/S0022029918000560

Ghauch A, Tuqan A, Assi HA (2009) Antibiotic removal from water: elimination of amoxicillin and ampicillin by microscale and nanoscale iron particles. Environ Pollut 157:1626–1635. https://doi.org/10.1016/j.envpol.2008.12.024

Hamdi H, De La Torre-Roche R, Hawthorne J, White JC (2015) Impact of non-functionalized and amino-functionalized multiwall carbon nanotubes on pesticide uptake by lettuce (*Lactuca sativa* L.), Nanotoxicology 9:172–180. https://doi.org/10.3109/17435390.2014.907456

Helmer R, Hespanhol I (1997) Water pollution control—a guide to the use of water quality management principles. CRC Press, Boca Raton

Hu J, Zhao D, Wang X (2011) Removal of Pb(II) and Cu(II) from aqueous solution using multiwalled carbon nanotubes/iron oxide magnetic composites. Water Sci Technol 63:917–923. https://doi.org/10.2166/wst.2011.270

Ihsanullah HA, Asmaly TA, Saleh T, Laoui VK, Gupta MA (2015) Atieh, Enhanced adsorption of phenols from liquids by aluminum oxide/carbon nanotubes: comprehensive study from synthesis to surface properties. J Mol Liq 206:176–182. https://doi.org/10.1016/j.molliq.2015.02.028

Jahangiri-Rad M, Nadafi K, Mesdaghinia A, Nabizadeh R, Younesian M, Rafiee M (2013) Sequential study on reactive blue 29 dye removal from aqueous solution by peroxy acid and single wall carbon nanotubes: experiment and theory. Iran J Environ Heal Sci Eng 10:1–8. https://doi.org/10.1186/1735-2746-10-5

Kalfa OM, Yalçinkaya Ö, Türker AR (2009) Synthesis of nano B_2O_3/TiO_2 composite material as a new solid phase extractor and its application to preconcentration and separation of cadmium. J Hazard Mater 166:455–461. https://doi.org/10.1016/j.jhazmat.2008.11.112

Kamran Haghighi H, Irannajad M, MohammadJafari A (2021) Thermodynamic and kinetic studies of heavy metal adsorption by modified nano-zeolite. Geosystem Eng 24:101–113. https://doi.org/10.1080/12269328.2021.1873197.

Karimi-Maleh H, Shafieizadeh M, Taher MA, Opoku F, Kiarii EM, Govender PP, Ranjbari S, Rezapour M, Orooji Y (2020) The role of magnetite/graphene oxide nano-composite as a high-efficiency adsorbent for removal of phenazopyridine residues from water samples, an experimental/theoretical investigation. J Mol Liq 298:112040. https://doi.org/10.1016/j.molliq.2019.112040

Kumar A, Gangawane KM, Ramanjaneyulu B (2022) Ferrofluids for waste-water treatment. In: Advances in Chemical, bio and environmental engineering. Springer, Berlin, pp 723–744. https://doi.org/10.1007/978-3-030-96554-9_48

Kumar A, Gangawane KM (2022a) Effect of precipitating agents on the magnetic and structural properties of the synthesized ferrimagnetic nanoparticles by co-precipitation method. Powder Technol 401:117298. https://doi.org/10.1016/j.powtec.2022.117298

Kumar A, Gangawane KM (2022b) Synthesis and effect on the surface morphology & magnetic properties of ferrimagnetic nanoparticles by different wet chemical synthesis methods. Powder Technol 410:117867. https://doi.org/10.1016/j.powtec.2022.117867

Lee K-Y, Kim K-W, Park M, Kim J, Oh M, Lee E-H, Chung D-Y, Moon J-K (2016) Novel application of nanozeolite for radioactive cesium removal from high-salt wastewater. Water Res 95:134–141. https://doi.org/10.1016/j.watres.2016.02.052

Liao C-J, Chung T-L, Chen W-L, Kuo S-L (2007) Treatment of pentachlorophenol-contaminated soil using nano-scale zero-valent iron with hydrogen peroxide. J Mol Catal A Chem 265:189–194. https://doi.org/10.1016/j.molcata.2006.09.050

Liu Y, Majetich SA, Tilton RD, Sholl DS, Lowry GV (2005) TCE dechlorination rates, pathways, and efficiency of nanoscale iron particles with different properties. Environ Sci Technol 39:1338–1345. https://doi.org/10.1021/es049195r

Madadrang CJ, Kim HY, Gao G, Wang N, Zhu J, Feng H, Gorring M, Kasner ML, Hou S (2012) Adsorption behavior of EDTA-graphene oxide for Pb (II) removal. ACS Appl Mater Interfaces 4:1186–1193. https://doi.org/10.1021/am201645g

Maharana M, Sen S (2022) Industrial solid waste based EU-12 nanozeolite: synthesis and characterisation. Waste and Biomass Valorization. 13:1695–1703. https://doi.org/10.1007/s12649-021-01553-4

Mahdavinia GR, Aghaie H, Sheykhloie H, Vardini MT, Etemadi H (2013) Synthesis of CarAlg/MMt nanocomposite hydrogels and adsorption of cationic crystal violet. Carbohydr Polym 98:358–365. https://doi.org/10.1016/j.carbpol.2013.05.096

Manna M, Sen S (2023) Tuning crystallization for controlled morphology of zeolite A by an eco-friendly sonochemical precursor-less method. Mater Chem Phys 309:128378. https://doi.org/10.1016/j.matchemphys.2023.128378

Moamen OAA, Ibrahim HA, Abdelmonem N, Ismail IM (2016) Thermodynamic analysis for the sorptive removal of cesium and strontium ions onto synthesized magnetic nano zeolite. Microporous Mesoporous Mater 223:187–195. https://doi.org/10.1016/j.micromeso.2015.11.009

Musah M, Azeh Y, Mathew J, Umar M, Abdulhamid Z, Muhammad A (2022) Adsorption kinetics and isotherm models: a review, caliphate. J Sci Technol 4:20–26. https://doi.org/10.4314/cajost.v4i1.3

Obaid SA (2020) Langmuir, Freundlich and Tamkin adsorption isotherms and kinetics for the removal Aartichoke Tournefortii straw from agricultural waste. J Phys Conf Ser 1664. https://doi.org/10.1088/1742-6596/1664/1/012011

Oviedo LR, Oviedo VR, Dalla Nora LD, da Silva WL (2023) Adsorption of organic dyes onto nanozeolites: a machine learning study. Sep Purif Technol 315:123712. https://doi.org/10.1016/j.seppur.2023.123712

Pal D, Sen S (2023) In-Depth coverage of petroleum waste sources, characteristics, environmental impact, and sustainable remediation process. In: Impact of petroleum waste on environmental pollution and its sustainable management through circular economy, pp 1–38. https://doi.org/10.1007/978-3-031-48220-5_1

Pal D, Sen S (2023) Emerging Petroleum pollutants and their adverse effects on the environment. In: Impact of petroleum waste on environmental pollution and its sustainable management through circular economy, pp 103–137. https://doi.org/10.1007/978-3-031-48220-5_5

Pal D, Sen S (2024a) Optimal synthesis of dolochar derived faujasite zeolite X for highly effective Cd(II) removal. Environ Res 240:117494. https://doi.org/10.1016/j.envres.2023.117494

Pal D, Sen S (2024b) Valorization of sponge iron industrial waste into iron-modified zeolite X for ciprofloxacin removal: a multi-parameter optimization study. Environ Sci Pollut Res 31:53381–53398. https://doi.org/10.1007/s11356-024-34730-8

Pham T-H, Lee B-K, Kim J (2016) Improved adsorption properties of a nano zeolite adsorbent toward toxic nitrophenols. Process Saf Environ Prot 104:314–322. https://doi.org/10.1016/j.psep.2016.08.018

Rasouli M, Yaghobi N, Hafezi M, Rasouli M (2012) Adsorption of divalent lead ions from aqueous solution using low silica nano-zeolite X. J Ind Eng Chem 18:1970–1976. https://doi.org/10.1016/j.jiec.2012.05.014

Sethy SK, Kishore MV, Bhagat C, Kumar M (2023) Periodic monitoring of nano clay as the potential adsorbent to remove metal and dyes from wastewater: a review. Total Environ. Res. Themes. 7:100067. https://doi.org/10.1016/j.totert.2023.100067

Sharififard H, Shahraki ZH, Rezvanpanah E, Rad SH (2018) A novel natural chitosan/activated carbon/iron bio-nanocomposite: sonochemical synthesis, characterization, and application for cadmium removal in batch and continuous adsorption process. Bioresour Technol 270:562–569. https://doi.org/10.1016/j.biortech.2018.09.094

Shojaei S, Nouri A, Baharinikoo L, Davoodabadi Farahani M, Shojaei S (2021) Removal of the hazardous dyes through adsorption over nanozeolite-X: simultaneous model, design and analysis of experiments. Polyhedron 196:114995. https://doi.org/10.1016/j.poly.2020.114995

Shruthi SB, Bhat C, Bhaskar SP, Preethi G, Sailaja RRN (2016) Microwave assisted synthesis of Guar Gum grafted acrylic acid/nanoclay superabsorbent composites and its use in crystal violet dye absorption. Green Sustain Chem 06:11–25. https://doi.org/10.4236/gsc.2016.61002

Sivalingam S, Sen S (2018) Rapid ultrasound assisted hydrothermal synthesis of highly pure nanozeolite X from fly ash for efficient treatment of industrial effluent. Chemosphere 210:816–823. https://doi.org/10.1016/j.chemosphere.2018.07.091

Sivalingam S, Sen S (2019a) Efficient removal of textile dye using nanosized fly ash derived zeolite-x: kinetics and process optimization study. J Taiwan Inst Chem Eng 96:305–314. https://doi.org/10.1016/j.jtice.2018.10.032

Sivalingam S, Sen S (2019b) Sono-assisted adsorption of As(V) from water by rice-husk-ash-derived iron-modified mesoporous zeolite Y: a cradle-to-cradle solution to a problematic solid waste material. Ind Eng Chem Res 58:14073–14087. https://doi.org/10.1021/acs.iecr.9b01785

Sivalingam S, Sen S (2020) Rice husk ash derived nanocrystalline ZSM-5 for highly efficient removal of a toxic textile dye. J. Mater. Res. Technol. 9:14853–14864. https://doi.org/10.1016/j.jmrt.2020.10.074

Sivalingam S, Kella T, Maharana M, Sen S (2019) Efficient sono-sorptive elimination of methylene blue by fly ash-derived nano-zeolite X: process optimization, isotherm and kinetic studies. J Clean Prod 208:1241–1254. https://doi.org/10.1016/j.jclepro.2018.10.200

Sonawane SH, Chaudhari PL, Ghodke SA, Parande MG, Bhandari VM, Mishra S, Kulkarni RD (2009) Ultrasound assisted synthesis of polyacrylic acid-nanoclay nanocomposite and its application in sonosorption studies of malachite green dye. Ultrason Sonochem 16:351–355. https://doi.org/10.1016/j.ultsonch.2008.10.008

Uzum C, Shahwan T, Eroglu A, Hallam K, Scott T, Lieberwirth I (2009) Synthesis and characterization of kaolinite-supported zero-valent iron nanoparticles and their application for the removal of aqueous Cu2+ and Co2+ ions. Appl Clay Sci 43:172–181. https://doi.org/10.1016/j.clay.2008.07.030

Wang S, Liang S, Liang P, Zhang X, Sun J, Wu S, Huang X (2015) In-situ combined dual-layer CNT/PVDF membrane for electrically-enhanced fouling resistance. J Memb Sci 491:37–44. https://doi.org/10.1016/j.memsci.2015.05.014

Wu T, Cai X, Tan S, Li H, Liu J, Yang W (2011) Adsorption characteristics of acrylonitrile, p-toluenesulfonic acid, 1-naphthalenesulfonic acid and methyl blue on graphene in aqueous solutions. Chem Eng J 173:144–149. https://doi.org/10.1016/j.cej.2011.07.050

Xi Y, Megharaj M, Naidu R (2011) Dispersion of zerovalent iron nanoparticles onto bentonites and use of these catalysts for orange II decolourisation. Appl Clay Sci 53:716–722. https://doi.org/10.1016/j.clay.2011.06.010

Yang S-T, Chang Y, Wang H, Liu G, Chen S, Wang Y, Liu Y, Cao A (2010) Folding/aggregation of graphene oxide and its application in Cu2+ removal. J Colloid Interface Sci 351:122–127. https://doi.org/10.1016/j.jcis.2010.07.042

Yang S, Guo Z, Sheng G, Wang X (2012) Investigation of the sequestration mechanisms of Cd(II) and 1-naphthol on discharged multi-walled carbon nanotubes in aqueous environment. Sci Total Environ 420:214–221. https://doi.org/10.1016/j.scitotenv.2012.01.018

Yu J-G, Zhao X-H, Yang H, Chen X-H, Yang Q, Yu L-Y, Jiang J-H, Chen X-Q (2014) Aqueous adsorption and removal of organic contaminants by carbon nanotubes. Sci Total Environ 482–483:241–251. https://doi.org/10.1016/j.scitotenv.2014.02.129

Yu J-G, Yu L-Y, Yang H, Liu Q, Chen X-H, Jiang X-Y, Chen X-Q, Jiao F-P (2015) Graphene nanosheets as novel adsorbents in adsorption, preconcentration and removal of gases, organic compounds and metal ions. Sci Total Environ 502:70–79. https://doi.org/10.1016/j.scitotenv.2014.08.077

Zhang M, He F, Zhao D, Hao X (2011) Degradation of soil-sorbed trichloroethylene by stabilized zero valent iron nanoparticles: effects of sorption, surfactants, and natural organic matter. Water Res 45:2401–2414. https://doi.org/10.1016/j.watres.2011.01.028

Zhao G, Li J, Ren X, Chen C, Wang X (2011a) Few-layered graphene oxide nanosheets as superior sorbents for heavy metal ion pollution management. Environ Sci Technol 45:10454–10462. https://doi.org/10.1021/es203439v

Zhao G, Ren X, Gao X, Tan X, Li J, Chen C, Huang Y, Wang X (2011b) Removal of Pb(ii) ions from aqueous solutions on few-layered graphene oxide nanosheets. Dalt. Trans. 40:10945–10952. https://doi.org/10.1039/c1dt11005e

Chapter 15
Design and Development of Novel Membrane Technologies for Futuristic Wastewater Treatment

Harsh Prajapati, Mansi Desai, Kamlesh Prajapati, and Balanagulu Busupalli

Abstract A composite is a blend of two or more materials with distinct chemical and physical characteristics and an identifiable interface. Composite materials provide several benefits over other metal compounds. These benefits include superior toughness, high specific stiffness, high specific strength, gas barrier properties, corrosion resistance, low density, and thermal insulation. The continuous phase; known as the matrix, and the scattered phase; known as reinforced materials, comprise composite materials. Numerous fields, including the biological sciences, medication delivery systems, and wastewater treatment, have found use for nanomaterials (NMs), particularly nanocomposites. Membrane separation technologies have been widely used in wastewater treatment. Owing to their mechanical strength, structural and chemical durability, polysulfone (PSF) membranes are the most often employed membranes in wastewater ultrafiltration. Membranes are constructed from hydrophobic polymers such as polyvinylidene fluoride (PVDF), polytetrafluoroethylene (PTFE), or polypropylene (PP), utilizing a variety of processes such as phase inversion and electrospinning. Several studies have been carried out to improve the hydrophilic qualities of the membrane's polymeric surface; more recently, metal nanoparticles (NPs) have been introduced to the polymer matrix to minimize fouling potential and improve membrane performance. Recent advances in nanotechnology assist in outstripping techniques for the removal of pollutants by overcoming the constraints of conventional treatment technologies and providing low-cost treatment methods with high capacity. This chapter introduces the classification of membrane, NPs, and nanocomposites and their efficacy in removal of contaminants. The benefits, limitations, and new developments in water, wastewater, and groundwater treatments are also evaluated and scrutinized.

H. Prajapati · M. Desai · B. Busupalli (✉)
Department of Chemistry, Pandit Deendayal Energy University (PDEU), Gandhinagar, Gujarat, India
e-mail: baalanaagulu@gmail.com

K. Prajapati
Department of Chemistry, V S Patel College of Arts and Science, Bilimora, Gujarat, India

N. Agarwal et al. (eds.), *Nanomaterials in Wastewater Research*, Advances in Wastewater Research, https://doi.org/10.1007/978-981-96-3908-3_15

Keywords Membrane · Nanocomposite · Nanotechnology · Nanoparticles · Low cost · Pollutant

15.1 Introduction

Water is the most essential element for life on Earth, and having clean, potable water is a global challenge for the twenty-first century. Pure, unpolluted water is a necessity for all living things. Water covers more than 71% of the earth's surface, but only about 1% is potable according to international standards due to various contaminations. Water contamination is primarily caused by wastewater discharge from industries, agricultural activities, municipal wastewater, and environmental contaminants (Fig. 15.1). Even in trace amounts, the presence of heavy metals, dyes, and microorganisms is hazardous to human health, aquatic systems, and the environment. Water pollution is a worldwide issue that endangers the entire environment and affects the lives of millions of people.

The discharge of large amounts of metal-contaminated wastewater into the water stream, as well as industries containing heavy metals such as Cd, Cr, Cu, Ni, As, Pb, and Zn, are hazardous to human beings and the environment. Due to high solubility in water, heavy metals can be absorbed by living organisms. Many heavy metals may accumulate in the human body once they enter the food chain. As a result, before being discharged into the environment, contaminated wastewater must be treated. Heavy metal ions can be removed from effluent using conventional treatment methods such as chemical precipitation, ion exchange, and electrochemical reduction. All of these processes have drawbacks, such as incomplete removal, high energy requirements, and the production of toxic sludge.

Membrane-based water treatment technology provides the finest choices for "drought-proofing" humanity on an increasingly thirsty globe by purifying seawater or being employed in wastewater treatment (Fig. 15.2). Although reverse osmosis

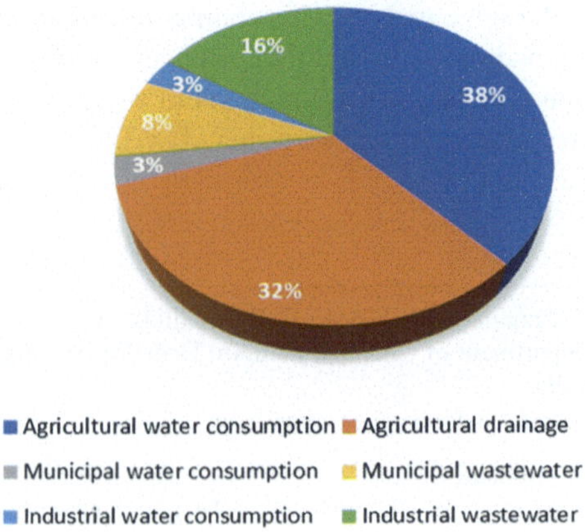

Fig. 15.1 Water consumption and wastewater production by major sectors (WWAP U 2017)

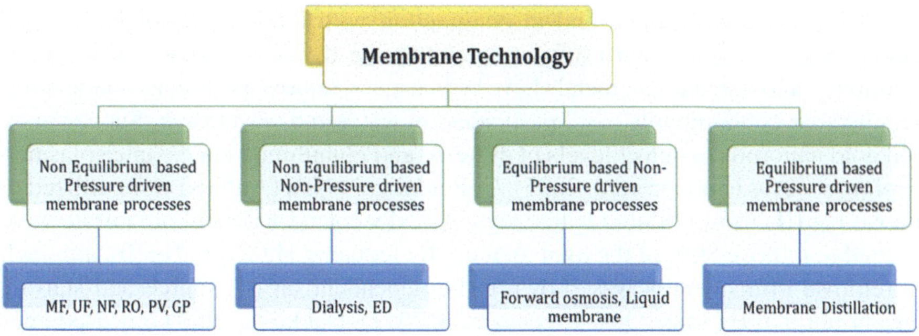

Fig. 15.2 Schematic representation of membrane processes (Jhaveri and Murthy 2016)

(RO), desalination, and membrane bioreactor wastewater treatment are widely established, numerous membrane technologies still require significantly better separation characteristics, energy demand, and prices. However, improved device design and membrane engineering must keep pace with membrane expansion. Fluid mechanics and mass transport limitations will most likely define the highest limits of membrane performance. Nonetheless, major advancements and growth can be expected over the next 20 years, with RO retaining the key to desalination and reclamation.

The earliest commercially available membranes were built of cellulose, ethyl cellulose, and cellulose acetate (CA). Since these membrane materials were derived from natural sources, they were inexpensive. Their limited chemical resistance, however, and the rising demand for membrane processes in demanding applications under harsh circumstances made them less appealing. Consequently, synthetic membranes were explored and commercialized for industrial and home applications for the last two decades. Membranes can be made of organic (polymers) or inorganic (ceramics) materials; however, polymeric materials are more commonly investigated due to their mechanical strength, flexibility, and chemical stability. Due to its exceptional chemical and thermal stability, polysulfone (PSF) is the most extensively utilized membrane material. Polymers being researched and utilized include polyvinylidene fluoride (PVDF), polyether sulfone (PES), polyacrylonitrile (PAN), polyvinyl alcohol (PVA), polyvinyl chloride (PVC), polyethylene (PE), polypropylene (PP), polyamide, and chitosan.

Nano-sized materials are being researched and developed to provide innovative functional materials for the wastewater treatment and water production industries. Due to their intriguing features, NMs have shown a remarkable ability to overcome various existing challenges in several separation and purification technologies. Nano-adsorbents outperform traditional adsorbents due to their large surface area, strong reactivity, electronic characteristics, high sorption capacity, and structural qualities (Wang et al. 2012; Lu et al. 2016). Carbon nanotubes (CNTs), fullerene-based NMs, and metal-based NPs have all been used as adsorbents in water purification procedures.

Metallic NPs, both alone and in conjunction with other forms of NMs, have demonstrated promising performance in combating the issues presented by water pollution. Nano-sized noble metals have been used to remove pollutants and identify biomolecular contaminants (e.g., pesticides) in water and wastewater. Sudeep et al. used gold nanorods to detect levels of cysteine and glutathione via the inter-plasmon coupling process (Sudeep et al. 2005). Colloidal gold NPs (AuNPs) were utilized to remove Hg (II) from polluted waterways. Au NPs coated with sodium citrate were reported to remove 80% of the total mercury by reducing Hg (II) to Hg (0), although the removal effectiveness was shown to be dependent on the source and starting concentration of Hg ions (Ojea-Jiménez et al. 2012). Dankovich and Gray used filter paper and cellulosic materials (e.g., cotton fabric, bacterial cellulose) to inactivate two harmful bacteria in drinking water, *Escherichia coli* and *Enterococcus faecalis* (Dankovich and Gray 2011). The paper with the largest number of silver (Ag) NPs was found to have the highest bacteria inactivation (5.9 mg Ag per 1 g of paper). The least inhibitory concentration of Ag NPs was discovered to be dependent on NP size, bacteria concentration, sorption period, and nutritional environment. As the average diameter of the NPs reduced from 62 nm to 9.2 nm, the least inhibitory dose for Ag NPs decreased from 10 to 5.4 ppm.

This study summarizes current advances in membrane science and nanotechnologies for water filtration. Water filter design, development, and application are among the topics covered, with an emphasis on emerging novel membrane materials. We go over manufacturing methods for managing the filter size regime, analytical tools for performance testing, and molecular modeling for transport and separation to give readers a good foundation. Membrane chemical stability, fouling, and environmental impact are also discussed as outstanding questions.

15.2 Current Water Challenges

The current global water challenges are multifaceted and severe, impacting billions of people and exacerbating social, economic, and environmental issues worldwide. As of 2024, approximately 2.2 billion people lack access to safe drinking water, and 3.6 billion people live without safely managed sanitation, highlighting a stark disparity in water security and basic health standards (WHO 2011). Climate change intensifies these problems by altering precipitation patterns and increasing the frequency of extreme weather events, such as droughts and floods, which further deplete water resources and affect their distribution (WHO 2011).

Water stress is most pronounced in regions like the Middle East and North Africa, where limited rainfall and high demand create a precarious balance. Even in water-rich regions, inadequate infrastructure and poor management can lead to significant shortages, as seen in parts of Sub-Saharan Africa and urban centers in developed countries like the United States.

Water scarcity not only drives conflicts over resources but also exacerbates health crises by impairing hygiene and sanitation. This situation leads to the spread of diseases such as cholera and typhoid, which disproportionately affect vulnerable populations, including women and children, who often bear the burden of water collection and face increased risks of gender-based violence during water shortages. Moreover, inadequate water supply can lead to severe agricultural impacts, jeopardizing food security and exacerbating malnutrition, particularly in rural and impoverished communities.

Efforts to mitigate these challenges include international cooperation on transboundary water management, investments in water infrastructure, and initiatives to promote sustainable water use and sanitation practices. For instance, organizations like World Vision are working to provide clean water access to healthcare facilities and communities, aiming to reach millions by 2030 through the development of sustainable water systems and hygiene education (World Vision). However, achieving global water security requires significant financial investments, estimated at $114 billion annually for safe drinking water and sanitation in low- to middle-income countries.

Overall, addressing the global water crisis necessitates a comprehensive approach that combines technological innovation, robust infrastructure development, effective governance, and international collaboration to ensure equitable and sustainable access to this vital resource (Gupta and Ali 2013).

15.2.1 Water Scarcity and Pollution

Water scarcity and pollution are pressing global issues that significantly impact human health, ecosystems, and economic development. Water scarcity arises from both natural and human-induced factors, including climate change, over-extraction of water resources, and inefficient water management practices. Currently, more than 40% of the world's population is affected by water scarcity, a figure projected to rise with increasing population growth and environmental changes (Shemer et al. 2023). Compounding this issue, water pollution severely reduces the availability of clean, safe water for drinking, agriculture, and industrial uses.

Industrial discharge, agricultural runoff, untreated sewage, and improper waste disposal are primary sources of water pollution, introducing contaminants such as heavy metals, microplastics, pharmaceuticals, and pathogens into water bodies (Babuji et al. 2023). These pollutants pose significant risks to human health and ecosystems. For instance, exposure to heavy metals can lead to severe health problems, including neurological disorders, reproductive issues, and cancer. Agricultural runoff containing pesticides and fertilizers often leads to eutrophication, depleting oxygen in water bodies and killing aquatic life, thus disrupting ecosystems (Pandya et al. 2021).

Addressing these intertwined issues requires a comprehensive approach. Integrated Water Resources Management (IWRM) is essential for balancing water demand and supply, promoting sustainable use, and protecting water quality. Techniques such as rainwater harvesting, desalination, and wastewater recycling are crucial for augmenting water supplies. Desalination, particularly, has shown promise in converting seawater to potable water, especially in arid regions, though it remains energy-intensive and poses environmental challenges such as brine disposal (Shemer et al. 2023).

Pollution control measures are equally vital. Strengthening regulations to limit pollutant discharge, improving sewage treatment infrastructure, and encouraging environmentally friendly agricultural practices can significantly reduce water pollution. Public awareness and community engagement are crucial for fostering a culture of water stewardship, where educational campaigns can highlight the importance of water conservation and the health risks associated with polluted water (Babuji et al. 2023).

In addition to technological solutions, natural and engineered interventions such as riparian buffer zones and constructed wetlands can effectively mitigate pollution and enhance water quality. These interventions help filter pollutants, stabilize soil, and provide habitat for wildlife, contributing to overall ecosystem health (Mekonnen and Hoekstra 2016). Water scarcity and pollution have profound socioeconomic implications. Inadequate water supply and poor water quality can lead to significant healthcare costs, loss of productivity, and reduced economic development. In developing countries, the health impacts are particularly severe, with waterborne diseases remaining a leading cause of morbidity and mortality. For instance, in regions with poor sanitation and water treatment infrastructure, diseases such as diarrhea, cholera, and hepatitis are rampant, disproportionately affecting children and vulnerable populations.

Pollution worsens the water scarcity issue by contaminating available water resources and making them unsafe for consumption or use. Improper waste disposal introduces a complex array of pollutants including heavy metals, chemicals, and organic compounds. Traditional water treatment methods often fall short of removing these contaminants which gives rise to the development of advanced approaches for efficient removal.

Effective wastewater treatment is crucial to address these challenges. Advanced treatment technologies such as membrane filtration, activated carbon, and biological treatments are often employed to address the wide array of contaminants. These technologies not only improve the quality of treated water but also make it safe for reuse in various sectors.

15.2.2 Limitation of Traditional Treatment Methods

Traditional water treatment methods have been the stepping stone of water purification approaches for decades. Techniques like coagulation, sedimentation, filtration, and chlorination have been very efficient at removing a wide range of contaminants from water. However, as the complexity of water pollution increases with rapid growth in industrialization, and urbanization, traditional methods fall short and the limitations of techniques are becoming more evident.

One of the primary limitations of conventional water treatment approaches is their inefficiency in eliminating emerging contaminants. Many contaminants are present in water at a trace level but have considerably harmed various beings' ecological and health. Traditional approaches are not tailored to meet the need to eliminate emerging contaminants. For example, multiple pharmaceuticals can pass through filtration systems, posing a significant risk to human health and the ecosystem. To identify the presence of these contaminants, advancements in treatment technologies become a necessity.

15.3 Current Water Treatment Methods Used in Industries

Water filtration is necessary when we live off the grid or need clean water for drinking. In this, we need to ensure that the water filtration system we use; effectively cleanses water. Water filtration is a process of cleaning water contamination through a home-based filtration system, using chemical treatments or biological treatments. We need to know that filtered water looks cleaner but many chemicals and viruses are still left. Purification works as a chemical process that usually uses either iodine or chlorine. It is necessary to know that simple water purification cannot remove all impurities found in the water. Different commercially used membrane filtration technologies are ultrafiltration, nanofiltration, reverse osmosis, and electrodialysis, and these could be better techniques for water purification. A comparative account of these commercially available membrane filtration technologies is given in Table 15.1.

15.3.1 Ultrafiltration

According to the molecular size of dissolved molecules, heavy metal ions, and other pollutants, wastewater is filtered using a membrane. Larger molecules and heavy metals cannot flow through certain types of membranes and must instead be sorted out. Different membrane types only permit the passage of small molecular solutes. Ultrafiltration has been divided into subcategories like complexation–ultrafiltration (Molinari et al. 2008), chelating-enhanced ultrafiltration, and micellar-enhanced ultrafiltration (Kryvoruchko et al. 2002; Landaburu-Aguirre et al. 2012).

15.3.2 Nanofiltration

The technology of membrane separation known as nanofiltration is utilized to remove heavy metals from aqueous solutions (Murthy and Chaudhari 2008; Muthukrishnan and Guha 2008). It is used to remove any heavy metals, including nickel (Cséfalvay et al. 2009), chromium (Ahmad and Ooi 2010), copper (Nguyen et al. 2009; Figoli

Table 15.1 Comparative features of different commercially used membrane filtration technologies (Ahn et al. 2008)

Microfiltration	Particles and dissolved macromolecules larger than 100 nm are rejected
	Contains macropores (> 50–500 nm)
	Made up of polymers such as polypropylene, polysulfone, polyurethane, and so on
	Works with an operating pressure of > 1.0 bar
	Poor self-cleaning ability with high membrane fouling
Ultrafiltration	Solutes of > 1–100 nm in diameter are removed from the solvent
	Contains mesopores (> 2–50 nm)
	Made up of polymers such as polysulfone, acrylic, cellulose, and others
	Works with an operating pressure of 1–10 bar
	Poor self-cleaning ability with high membrane fouling
Nanofiltration	Particles and dissolved macromolecules smaller than 2 nm are rejected
	Contains mesopores (> 2 nm)
	Made up of organic polymers such as polyamide, polyester, and other porous polymers
	Works with an operating pressure of 20–40 bar
	Moderate self-cleaning ability with considerable membrane fouling
RO	A transmembrane pressure allows the selective movement of solvent against the osmotic pressure difference
	Contains micropores (0.3–0.6 nm)
	Made up of organic polymers such as polyamide, polysulfone and polyether sulfone
	Works with an operating pressure of 30–60 bar
	Moderate self-cleaning ability with considerable membrane fouling

et al. 2010), and arsenic (Dialynas and Diamadopoulos 2009; Mohammadi et al. 2005). It is comparatively more effective, fairly simple to use, and uses less energy than other methods.

15.3.3 Reverse Osmosis (RO)

The separation and fractionation of organic and inorganic compounds and heavy metals in aqueous and nonaqueous solutions is primarily accomplished using RO technology. Chemical, textile, petrochemical, electrochemical, culinary, paper, and tannery industries are among the industries that can employ RO technology to clean their industrial effluents. This method is effective for high-level metal removal when combined with the pilot membrane reactor (Sadrzadeh et al. 2008). It also has several drawbacks, such as high power consumption during membrane repair and pumping pressure.

15.3.4 Electrodialysis

In the separation process known as electrodialysis (ED), dissolved ions are transferred from one solution to another through a charged membrane while being induced by an electric field (Sadrzadeh et al. 2009; Nataraj et al. 2007). It is utilized for wastewater treatment, producing drinking water from saltwater, separating and recovering heavy metal ions, and producing salt (Cifuentes et al. 2009). Numerous researchers use it to remove heavy metals, including chromium, copper, and iron (Das et al. 2014; Yu et al. 2009).

15.4 Nanocomposite Membranes and Its Applications

Nanocomposite membranes are made by embedding nanofilters into a polymeric or inorganic matrix and are used extensively in wastewater treatment. NPs have recently been used to improve the hydrophilicity, strength, selectivity, and permeability of polymeric membranes. For example, when silica NPs are incorporated into polyvinylidene fluoride membranes, the membrane has improved temperature resistance as well as increased selectivity and diffusivity (Maximous et al. 2009). Zinc oxide NPs in a chitosan membrane demonstrated strong mechanical properties as well as a high level of antibacterial activity (Ahn et al. 2008); and silica NPs in polysulfone membranes resulted in increased gas permeability (Hu et al. 1997). When polyether sulfone membranes were incorporated with aluminum oxide NPs, it resulted in greater porosity, a low decline in flux, and pseudo-steady-state permeability (Singh et al. 2017). When silica NPs were incorporated into polybenzimidazole membranes, it resulted in greater selectivity during gas separation as well as enhanced permeability (Ebrahiminezhad et al. 2018). Adding particles to a polymer matrix may also stabilize the polymeric membrane in terms of variations in perm-selectivity when temperatures fluctuate.

15.4.1 Metal and Metal Oxide-Based NPs in Nanocomposite Membrane

15.4.1.1 Iron-Based NPs for Nanocomposite Membrane

Modern civilization is supported mainly by iron, the most frequent transition metal and the fourth most frequent element among those that make up the earth's crust. Iron oxides and other materials like cobalt, gold, platinum, and nickel have received more attention than NPs. Iron's reactivity is crucial in macroscopic applications (mainly rusting), but it is quite problematic in nanoscale applications. Iron NPs have not been the focus of more extensive research since it has long been known that finely split iron is pyrophoric. The breakdown of iron NPs is complex due to iron's strong reactivity, and they are not considered suitable for practical use.

Nevertheless, iron offers several benefits at the nanoscale, notably concerning its potent catalytic and magnetic capabilities. This study will examine iron NPs' use in membrane separation procedures in the light of recent research that has begun to identify their potential. Iron is inappropriate for use as pure metal NPs due to its strong reactivity, as was already indicated. So, iron complexes have often been used in polymeric membranes rather than pure iron NPs. In essence, adding iron compound NPs improves the functionality of membranes for a particular application.

Iron oxide NPs have been successfully employed to eliminate dyes from effluents (Chernousova and Epple 2013). A group of co-workers created the super magnetic Fe_3O_4 NPs and coated them with green tea polyphenols to remove the color from an aqueous solution. Magnetic iron oxide NPs were also formed by and utilized to absorb azo dye (Zhu et al. 2010). It has been demonstrated that the produced Fe_3O_4 NPs operate well in removing dye over a wide pH range. Another group prepared NPs from a mesoporous Fe_3O_4 carbon shell was employed to break down the color methylene blue with a remarkable efficiency (Basri et al. 2011). According to a recent study, processed iron NPs had an excellent capacity for dye removal (Shi et al. 2012). After six hours, 95% of the methyl orange was removed with decolorization (Liu et al. 2016). Another study looked into applying brilliant green and malachite green pigments onto Mn-doped Fe_3O_4 carbon NPs (Kim et al. 2008).

15.4.1.2 Silver-Based NPs for Nanocomposite Membrane

Numerous inorganic antibacterial materials, including metal, semiconductor oxide, metallic oxide, and their nanocomposites, have been extensively studied for their potential in the preparation of antibacterial membranes. Silver (Ag), one of the precious metals, is among the most promising substances due to its corrosion resistance, low toxicity, and antibacterial qualities. It has been used for thousands of years to make tableware and coins (Kang et al. 2006). The ability of AgNPs to successfully suppress aquatic microbes, including bacteria, fungi, and algae, makes them useful in water filtration systems. Given that is thermally and chemically stable, fights a wide range of bacteria, has good catalytic activity, and has no over-toxic effect on humans, AgNPs or ions have recently received a lot of attention in the possibility of being used when fabricating antibacterial membranes for water treatment purposes (Damm et al. 2007; Deng et al. 2008; Chou et al. 2005). As a result of their well-known antibacterial qualities, AgNPs are often used in the polymer matrix of membranes. The interactions between sulfur and phosphorus, especially with the thiol groups (S–H) found in cysteine and other chemicals, have been suggested to be the source of silver's antibacterial effect. Ionic silver, which derives from $_{n}$Ag, interacts with thiol groups. When S–Ag or disulfide connections are formed, they have the potential to damage bacterial proteins, dimerize DNA, and disrupt the electron transport chain. The antiviral effects of silver ions are comparable and interact with the thiol groups in viral DNA and proteins (Gil et al. 2017). The ability of silver NPs to function well as selective barriers and to aid in the transit of certain elements is also believed to be possible. Due to their distinct characteristics, AgNP-infused membranes may be used in various applications requiring the transportation of expensive goods.

Researchers have focused on Ag ions and NPs for various water treatment applications, including membranes for water filtration. To create membranes, nano-silver has been introduced into poly(2-ethyl-2-oxazoline) (Lee et al. 2008), polyamide (Ben-Sasson et al. 2014), polyimide (Rajesh et al. 2012), and cellulose acetate (Borkow and Gabbay 2009). Nevertheless, there haven't been a lot of studies done to determine how well $_{n}$Ag-infused membranes are used for continuous filtering work to avoid biofouling. Ultrafiltration, microfiltration, and nanofiltration membranes made of PSF that have been combined with $_{n}$Ag are often employed to filter water; however, there is little study on their properties.

15.4.1.3 Titanium-Based NPs for Nanocomposite Membrane

Titanium dioxide (TiO_2) has received much attention due to its high photocatalytic capabilities that help decompose organic compounds and kill microorganisms. Most of the research has been on employing TiO_2 particles suspended in water as a catalyst (Ingle et al. 2014). TiO_2 has also been researched to remedy various environmental issues and purify water and air. Numerous studies have been conducted on reducing membrane biofouling issues using inorganic NPs like TiO_2. These fillers may adjust the membranes by altering the polymerization process or how the polymer is arranged in its network. TiO_2 has attracted a lot of attention from researchers because it possesses photocatalytic characteristics that may break down organic compounds and bacterial cells, which gives the possibility of self-cleaning and, therefore, a decrease in fouling during membrane separation (Flemming and Trevors 1989). TiO_2 is often used as a coating for or an ingredient in membranes since adding it directly to feed water during the desalination process may need a significant amount of material and not be economically feasible (Georgopoulos et al. 2001). Hydrophilic poly (poly amide imide) hydrated the membrane surface when used with TiO_2 NPs, reducing the number of fouls that bonded with it. As a result, fouling-induced flux reduction was less pronounced in the membrane, and it was simpler to use sheer force to remove the adsorbed foulant blood serum albumin (BSA) than it had been with the developing membrane (Kanninen et al. 2008).

15.4.1.4 Copper-Based (Cu) NPs for Nanocomposite Membrane

For many years, copper has been used as an antimicrobial agent. In ancient Egypt, copper was used to sanitize wounds and water (2000 BCE). For sanitary reasons, cooking utensils were made of copper throughout the Roman era. Phoenician shipbuilders placed copper strips on the hulls of their vessels to reduce fouling and increase speed and maneuverability. Japanese soldiers added copper to their canteens to prevent dysentery during World War II (Sarkar and Chakraborty 2021). Copper sulfate/lime/water or copper sulfate/sodium carbonate mixtures were used as fungicides in the USA and France in the late nineteenth century (Li et al. 2007). Remember that copper compounds and copper ions may be hazardous to humans, the environment, and microbes [59].

CuNPs may act as a reservoir for the carefully regulated release of copper ions, reducing the toxicity. The process of oxidizing CuNPs to produce copper oxides is straightforward. The use of organic coatings and an inert environment made of argon or nitrogen is required to prevent oxidization (Naseem and Durrani 2021). Using the Cambridge Engineering Selector (CES) tool, the impact of copper loadings on the mechanical and physical characteristics of PVDF has been produced. Young's modulus, shear modulus, and thermal conductivity may all significantly improve in the volume percent (vf) 1- 10% range, as shown. Flexural strength is one of the qualities that diminish when the amount of Cu particles increases. In the addition of Cu, other mechanical characteristics showed a minor improvement.

15.5 Different Domains of Application of Nanocomposite Membranes

Nanocomposite membranes are a significant representation in the field of wastewater treatment offering a superior performance compared to conventional membranes (Miteva and Stoyanova 2020). These membranes are engineered to incorporate NMs into the polymeric matrices, thereby enhancing mechanical, thermal, and chemical properties. The inclusion of NMs like carbon nanotubes (Collivignarelli et al. 2019), graphene oxide (Koc-Jurczyk 2014), metal and metal oxide (Alila and Boufi 2009), and zeolites (Peters 2010) increase efficiency, and selectivity, and impart unique functionalities. Their versatility and enhanced properties make them effective in addressing the complex and modern challenges of wastewater treatment.

15.5.1 Food Industry

The food industry encompasses a wide array of sectors, including those involved in the production of fish, dairy, livestock, vegetables, and beverages. Each of these sectors generates wastewater with distinct characteristics and substantial organic loads. Notably, these wastewaters also contain valuable compounds, such as phenols, carotenoids, pectin, lactose, and proteins which can be recovered and utilized (Johnson and Hilal 2022).

Membrane technology has proven to be an effective solution for treating wastewater from various food production processes. It has been successfully applied in industries such as potato starch production, fruit juice processing, and seafood processing, among others. The diversity in the types of waste and their composition presents both a challenge and an opportunity, as the extraction of high-value substances from wastewater can contribute to more sustainable practices within the food industry.

15.5.2 Textile Industry

The textile processing industry (TPI) relies heavily on water for dyeing, applying finishing agents, and removing impurities, resulting in significant wastewater generation. A modern approach to managing this wastewater focuses on energy recovery and reuse. Combining anaerobic membrane bioreactor (AnMBR) and aerobic membrane bioreactor (MBR) technologies offers an effective treatment solution.

Initially, the AnMBR system facilitates energy recovery by anaerobically digesting organic pollutants, producing biogas, and reducing the organic load. The subsequent aerobic MBR treatment further degrades residual contaminants and reduces color through aerobic microbial activity. This dual-stage process ensures a high-quality effluent suitable for reuse (Hameed et al. 2023).

The integration of AnMBR and aerobic MBR technologies not only enhances wastewater treatment efficiency but also supports sustainability by recovering energy and recycling water within the textile processing industry, aligning with eco-friendly practices and resource conservation goals.

15.5.3 Industrial Wastewater Treatment

Leachate, a wastewater rich in organic matter and ammonia nitrogen, forms when rainwater percolates through landfill waste, absorbing moisture and contaminants. Its chemical composition varies with the landfill's age; younger leachate contains higher organic content compared to older, more mature leachate. Effective leachate treatment involves multiple steps: stripping to remove volatile compounds, flocculation to aggregate and remove suspended particles, followed by MBR treatment for biological degradation, and reverse osmosis (RO) for final purification. This comprehensive approach successfully reduces contaminants, ensuring safer environmental discharge or potential reuse (Camargo et al. 2009).

Oil and petrochemical wastewater pose significant environmental challenges due to their toxic and resistant nature, stemming from various sources including crude oil extraction, refining, petrochemical manufacturing, metalwork, lubricants, coolants, and car washing. To address this, a customized large-scale facility transitioned from chemical demassification to a UF (Ultrafiltration) process coupled with MBR technology (Zodrow et al. 2009). This method effectively treated oil-contaminated wastewater, achieving a remarkable 90% removal of COD (Chemical Oxygen Demand) and complete elimination of tar, grease, and phenolic compounds.

15.5.4 Pharmaceutical Wastewater Treatment

The pharmaceutical industry generates a wide variety of compounds with differing structures, functions, and impacts. Notably, wastewater containing cephalosporin, a common pharmaceutical ingredient, undergoes improved degradation when treated with Membrane Bioreactor (MBR) technology, largely due to bioaugmentation techniques. By introducing specialized microorganisms, bioaugmentation enhances the breakdown of contaminants, thereby enhancing the efficiency of MBR systems. Consequently, MBRs equipped with these tailored microbial populations are emerging as strong contenders to traditional methods for treating pharmaceutical wastewater (Yin and Deng 2015).

This innovative approach holds promise for addressing the complexities associated with pharmaceutical wastewater treatment, offering potential advancements in both efficacy and sustainability. By leveraging the capabilities of MBR technology and bioaugmentation, the pharmaceutical industry can make significant strides toward mitigating its environmental impact and promoting more responsible waste management practices.

15.6 Nanocomposite Membranes: Future Prospects

Nanocomposite membranes have demonstrated numerous impressive properties but their long-term stability is still a big challenge and this stability has become a part of research (Do-Quang et al. 1998). A study showed a 10% loss of AgNPs from a nanocomposite membrane after a relatively short filtration period, leading to a notable decline in antifouling and antibacterial performance due to the release of Ag^+ ions from the membrane surface (Pearce 2007). The primary mechanisms for NP leakage from nanocomposite membranes include direct leaching and the release of trapped NPs during polymer matrix degradation. The stability of NPs within these membranes strongly depends upon the polymer matrix, which helps to strengthen the

NPs and serves as a protective barrier against environmental factors. For example, well-dispersed carbon nanotubes are tightly bound to polyamide membranes and provide strong stability (Al-Asheh et al. 2021). Reports of carbon nanotube leaching from such membranes are rare, suggesting that significant leaching is unlikely unless the polymer matrix undergoes structural or chemical changes. Compared to unmodified membranes, polyamide nanocomposite membranes with carbon nanotubes show enhanced stability under UV radiation and high temperatures. Carbon nanotubes also behave as nucleating agents that affect the polyamide membrane's lamellar orientation, dispersion, and viscosity, consequently improving its mechanical and thermal properties. However, chlorine disinfection-induced chemical erosion is considered to be the most likely cause of NP release from degraded membranes.

Looking toward the future of wastewater treatment, nanocomposites have become increasingly crucial in meeting the needs and challenges. Expanding upon the groundwork laid by the early research and current trends, the journey of the utilization of nanocomposite in wastewater treatment is envisioned to make advancements. The future work is a profound transformation driven by technological trends, automation, and early groundwork.

15.6.1 Computational Approach

Exploration of novel materials including nanofibers, quantum dots, and hybrid structures to enhance catalytic activities and adsorption capacities is underway. One of the newest approaches is the utilization of computational modeling to gain precise control over the size and composition of the nanocomposite. It provides valuable insights about structural properties and density function calculations and helps to optimize material with tailored properties. Computational fluid dynamics (CFD) is employed to optimize nanocomposite membranes.

15.6.2 Real-Time Sensors and Electrochemical Sensors

Another future approach would be smart nanocomposites, which offer real-time monitoring and a controlled treatment. This approach is usually incorporated with a sensing mechanism, responsive materials, and nanocomposites are employed to detect contaminant concentration, pH levels, and concentration of pollutants. Metals like Pb, Cd, and Zn are employed in nanocomposite-based smart sensors. A class of electrochemical sensors has also been in the recent trends. It provides rapid detection, high sensitivity, and real-time monitoring of pollutants.

15.6.3 Modular and Scalable Approach

Membranes, adsorption column, and catalytic reactors are a class of modular and scalable approaches. These are employed to provide need-based treatment. Membranes Ace recently came to the play. Membrane Filtration Process (MF), Membrane Bioreactors (MBRs), and Membrane Distillation (MD) are often integrated with nanocomposites and 2D materials like graphene and MXenes.

15.6.4 Machine Learning (ML)

ML when integrated with nanocomposite for wastewater treatment shows enhanced efficiency and adaptability. ML algorithms recently have been used to optimize the synthesis of NMs. ML models identify precursors, predict properties, and identify synthesis conditions which enable researchers to develop nanocomposites with desired properties.

15.6.5 Cellulose Nanocomposite Material

Composite materials are broadly classified into fiber-reinforced composites and particle-reinforced composites, with nanocomposites falling under the latter category. Within cellulose nanocomposites, reinforcement can be achieved using cellulose nanofibers or cellulose nanocrystals. Cellulose, as the most abundant natural biopolymer, is found in various plant fibers and exhibits a hierarchical structure comprising microfibrils surrounded by other components like hemicellulose and lignin. Acid treatment of cellulose fibers breaks down amorphous regions, releasing individual crystallites.

Cellulosic materials are highly prized in nanotechnology for their nano-fibrillar structure and multifunctional capabilities. Nano-whiskers derived from sources like cotton and wood serve as effective reinforcement agents in nanocomposites, with properties influenced by constituent characteristics and morphology. Micro-fibrillated celluloses are utilized in composites, coatings, and films due to their mechanical properties, produced via homogenization techniques from pulp containing various components. The properties of nanocrystalline cellulose are influenced by factors such as acid type, cellulose source, and reaction conditions.

Additionally, various cellulosic waste materials such as corncob and sawdust are repurposed to create nanocomposites and nanostructured ceramics, investigating their impact on product properties. Nanocomposite films using bacterial cellulose and polyurethane resin were fabricated and characterized, while cellulose nanofibril-reinforced polycarbonate composites were produced via compression molding, showcasing innovative approaches to utilize cellulosic waste and improving material properties.

15.7 Conclusion

In conclusion, the design and development of novel membrane technologies for futuristic water treatment is a step toward addressing and solving the issues of wastewater treatment. Throughout the chapter, we have explored cutting-edge advancements like nanocomposite membranes, zeolite-based materials, metal and metal-oxide-based membranes, and polymeric membranes while considering the economic viability of these methods. The incorporation of nanocomposite materials into membranes has broadened the scope of wastewater treatment.

Furthermore, the utilization of zeolite within the membrane has proven to treat water containing heavy metals and organic compounds as well as increasing the efficacy of the treatment process. This incorporation highlights the importance of innovation in sustainable solutions. Additionally, the evolution of polymeric membranes has unlocked new potentials and also contributed in providing economic solutions. As we look to the future, we can continue to push the boundaries and strive to provide breakthrough, sustainable, economic solutions. By collaborating across disciplines and sectors, we can provide real-world solutions and strive to provide impactful solutions.

The future of nanocomposites for wastewater treatment holds a remarkable opportunity to address the challenges. Nanocomposites synthesized with sensing abilities, adaptable treatment, modular approach, machine learning, and computational modeling approach are expected to revolutionize wastewater treatment. Employing nanotechnology will allow researchers all over the globe to have the capacity to transform and expand the range of possibilities. By embracing innovation, we can revolutionize approaches as an alternative to traditional methods.

References

Ahmad AL, Ooi BS (2010) A study on acid reclamation and copper recovery using low pressure nanofiltration membrane. Chem Eng J 156:257–263. https://doi.org/10.1016/j.cej.2009.10.014

Ahn J, Chung W-J, Pinnau I, Guiver MD (2008) Polysulfone/silica nanoparticle mixed-matrix membranes for gas separation. J Membr Sci 314:123–133. https://doi.org/10.1016/j.memsci.2008.01.031

Al-Asheh S, Bagheri M, Aidan A (2021) Membrane bioreactor for wastewater treatment: a review. Case Stud Chem Environ Eng 4:100109. https://doi.org/10.1016/j.cscee.2021.100109

Alila S, Boufi S (2009) Removal of organic pollutants from water by modified cellulose fibres. Ind Crops Prod 30:93–104. https://doi.org/10.1016/j.indcrop.2009.02.005

Babuji P, Thirumalaisamy S, Duraisamy K, Periyasamy G (2023) Human health risks due to exposure to water pollution: a review. Water 15:2532. https://doi.org/10.3390/w15142532

Basri H, Ismail AF, Aziz M (2011) Polyethersulfone (PES)–silver composite UF membrane: effect of silver loading and PVP molecular weight on membrane morphology and antibacterial activity. Desalination 273:72–80. https://doi.org/10.1016/j.desal.2010.11.010
Ben-Sasson M, Zodrow KR, Genggeng Q, Kang Y, Giannelis EP, Elimelech M (2014) Surface functionalization of thin-film composite membranes with copper nanoparticles for antimicrobial surface properties. Environ Sci Technol 48:384–393. https://doi.org/10.1021/es404232s
Borkow G, Gabbay J (2009) Copper, an ancient remedy returning to fight microbial, fungal and viral infections. Curr Chem Biol 3:272–278. https://doi.org/10.2174/187231309789054887
Camargo PHC, Satyanarayana KG, Wypych F (2009) Nanocomposites: synthesis, structure, properties and new application opportunities. Mat Res 12:1–39. https://doi.org/10.1590/S1516-14392009000100002
Chernousova S, Epple M (2013) Silver as antibacterial agent: ion, nanoparticle, and metal. Angew Chem Int Ed 52:1636–1653. https://doi.org/10.1002/anie.201205923
Chou W-L, Yu D-G, Yang M-C (2005) The preparation and characterization of silver-loading cellulose acetate hollow fiber membrane for water treatment. Polym Adv Technol 16:600–607. https://doi.org/10.1002/pat.630
Cifuentes L, García I, Arriagada P, Casas JM (2009) The use of electrodialysis for metal separation and water recovery from $CuSO_4$–H_2SO_4–Fe solutions. Sep Purif Technol 68:105–108. https://doi.org/10.1016/j.seppur.2009.04.017
Collivignarelli MC, Abbà A, Carnevale Miino M, Damiani S (2019) Treatments for color removal from wastewater: state of the art. J Environ Manage 236:727–745. https://doi.org/10.1016/j.jenvman.2018.11.094
Cséfalvay E, Pauer V, Mizsey P (2009) Recovery of copper from process waters by nanofiltration and reverse osmosis. Desalination 240:132–142. https://doi.org/10.1016/j.desal.2007.11.070
Damm C, Münstedt H, Rösch A (2007) Long-term antimicrobial polyamide 6/silver-nanocomposites. J Mater Sci 42:6067–6073. https://doi.org/10.1007/s10853-006-1158-5
Dankovich TA, Gray DG (2011) Bactericidal paper impregnated with silver nanoparticles for point-of-use water treatment. Environ Sci Technol 45:1992–1998. https://doi.org/10.1021/es103302t
Das R, Ali MdE, Hamid SBA, Ramakrishna S, Chowdhury ZZ (2014) Carbon nanotube membranes for water purification: a bright future in water desalination. Desalination 336:97–109. https://doi.org/10.1016/j.desal.2013.12.026
Deng Y, Dang G, Zhou H, Rao X, Chen C (2008) Preparation and characterization of polyimide membranes containing Ag nanoparticles in pores distributing on one side. Mater Lett 62:1143–1146. https://doi.org/10.1016/j.matlet.2007.08.002
Dialynas E, Diamadopoulos E (2009) Integration of a membrane bioreactor coupled with reverse osmosis for advanced treatment of municipal wastewater. Desalination 238:302–311. https://doi.org/10.1016/j.desal.2008.01.046
Do-Quang Z, Cockx A, Liné A, Roustan M (1998) Computational fluid dynamics applied to water and wastewater treatment facility modeling. Environ Eng Policy 1:137–147. https://doi.org/10.1007/s100220050015
Ebrahiminezhad A, Taghizadeh S, Ghasemi Y, Berenjian A (2018) Green synthesized nanoclusters of ultra-small zero valent iron nanoparticles as a novel dye removing material. Sci Total Environ 621:1527–1532. https://doi.org/10.1016/j.scitotenv.2017.10.076
Figoli A, Cassano A, Criscuoli A, Mozumder MSI, Uddin MT, Islam MA, Drioli E (2010) Influence of operating parameters on the arsenic removal by nanofiltration. Water Res 44:97–104. https://doi.org/10.1016/j.watres.2009.09.007
Flemming CA, Trevors JT (1989) Copper toxicity and chemistry in the environment: a review. Water Air Soil Pollut 44:143–158. https://doi.org/10.1007/BF00228784
Georgopoulos G, Roy A, Yonone-Lioy MJ, Opiekun RE, Lioy P (2001) Environmental copper: its dynamics and human exposure issues. J Toxicol Environ Health Part B 4:341–394. https://doi.org/10.1080/109374001753146207
Gil A, García AM, Fernández M, Vicente MA, González-Rodríguez B, Rives V, Korili SA (2017) Effect of dopants on the structure of titanium oxide used as a photocatalyst for the removal

of emergent contaminants. J Ind Eng Chem 53:183–191. https://doi.org/10.1016/j.jiec.2017.04.024

Gupta VK, Ali I (2013) Water Treatment by Membrane Filtration Techniques. In: Environmental Water. Elsevier, pp 135–154

Hameed S, Hayat A, Alghamdi EA, Bano N, Jubeen F, Khurram AA (2023) A review on recent advances in covalent organic frameworks-based membranes: synthesis, modification, and applications in liquid phase separation. J Sci Adv Mater Dev 8:100632. https://doi.org/10.1016/j.jsamd.2023.100632

Hu Q, Marand E, Dhingra S, Fritsch D, Wen J, Wilkes G (1997) Poly(amide-imide)/TiO_2 nanocomposite gas separation membranes: fabrication and characterization. J Membr Sci 135:65–79. https://doi.org/10.1016/S0376-7388(97)00120-8

Ingle AP, Duran N, Rai M (2014) Bioactivity, mechanism of action, and cytotoxicity of copper-based nanoparticles: a review. Appl Microbiol Biotechnol 98:1001–1009. https://doi.org/10.1007/s00253-013-5422-8

Jhaveri JH, Murthy ZVP (2016) A comprehensive review on anti-fouling nanocomposite membranes for pressure driven membrane separation processes. Desalination 379:137–154. https://doi.org/10.1016/j.desal.2015.11.009

Johnson DJ, Hilal N (2022) Nanocomposite nanofiltration membranes: state of play and recent advances. Desalination 524:115480. https://doi.org/10.1016/j.desal.2021.115480

Kang SW, Kim JH, Char K, Won J, Kang YS (2006) Nanocomposite silver polymer electrolytes as facilitated olefin transport membranes. J Membr Sci 285:102–107. https://doi.org/10.1016/j.memsci.2006.08.005

Kanninen P, Johans C, Merta J, Kontturi K (2008) Influence of ligand structure on the stability and oxidation of copper nanoparticles. J Colloid Interface Sci 318:88–95. https://doi.org/10.1016/j.jcis.2007.09.069

Kim JY, Lee C, Cho M, Yoon J (2008) Enhanced inactivation of *E. coli* and MS-2 phage by silver ions combined with UV-A and visible light irradiation. Water Res 42:356–362. https://doi.org/10.1016/j.watres.2007.07.024

Koc-Jurczyk J (2014) Removal of refractory pollutants from landfill leachate using two-phase system. Water Environ Res 86:74–80. https://doi.org/10.2175/106143013X13807328848810

Kryvoruchko A, Yurlova L, Kornilovich B (2002) Purification of water containing heavy metals by chelating-enhanced ultrafiltration. Desalination 144:243–248. https://doi.org/10.1016/S0011-9164(02)00319-3

Kyzas GZ, Deliyanni EA, Matis KA (2014) Graphene oxide and its application as an adsorbent for wastewater treatment. J Chem Tech Biotech 89:196–205. https://doi.org/10.1002/jctb.4220

Landaburu-Aguirre J, Pongrácz E, Sarpola A, Keiski RL (2012) Simultaneous removal of heavy metals from phosphorous rich real wastewaters by micellar-enhanced ultrafiltration. Sep Purif Technol 88:130–137. https://doi.org/10.1016/j.seppur.2011.12.025

Lee HS, Im SJ, Kim JH, Kim HJ, Kim JP, Min BR (2008) Polyamide thin-film nanofiltration membranes containing TiO2 nanoparticles. Desalination 219:48–56. https://doi.org/10.1016/j.desal.2007.06.003

Li YH, Zhao YM, Hu WB, Ahmad I, Zhu YQ, Peng XJ, Luan ZK (2007) Carbon nanotubes—the promising adsorbent in wastewater treatment. J Phys: Conf Ser 61:698–702. https://doi.org/10.1088/1742-6596/61/1/140

Liu X, Foo L-X, Li Y, Lee J-Y, Cao B, Tang CY (2016) Fabrication and characterization of nanocomposite pressure retarded osmosis (PRO) membranes with excellent anti-biofouling property and enhanced water permeability. Desalination 389:137–148. https://doi.org/10.1016/j.desal.2016.01.037

Lu H, Zhou X, Li R (2016) An optimization algorithm for trajectory planning of a 7-DOF redundant manipulator. In: 2016 IEEE Chinese guidance, navigation and control conference (CGNCC). IEEE, Nanjing, pp 2016–2021

Maximous N, Nakhla G, Wan W, Wong K (2009) Preparation, characterization and performance of Al2O3/PES membrane for wastewater filtration. J Membr Sci 341:67–75. https://doi.org/10.1016/j.memsci.2009.05.040

Mekonnen MM, Hoekstra AY (2016) Four billion people facing severe water scarcity. Sci Adv 2:e1500323. https://doi.org/10.1126/sciadv.1500323

Miteva A, Stoyanova V (2020) Zeolites application in terrestrial and space industry—a review. ARB 32:209–223. https://doi.org/10.3897/arb.v32.e17

Mohammadi T, Moheb A, Sadrzadeh M, Razmi A (2005) Modeling of metal ion removal from wastewater by electrodialysis. Sep Purif Technol 41:73–82. https://doi.org/10.1016/j.seppur.2004.04.007

Molinari R, Poerio T, Argurio P (2008) Selective separation of copper(II) and nickel(II) from aqueous media using the complexation–ultrafiltration process. Chemosphere 70:341–348. https://doi.org/10.1016/j.chemosphere.2007.07.041

Murthy ZVP, Chaudhari LB (2008) Application of nanofiltration for the rejection of nickel ions from aqueous solutions and estimation of membrane transport parameters. J Hazard Mater 160:70–77. https://doi.org/10.1016/j.jhazmat.2008.02.085

Muthukrishnan M, Guha BK (2008) Effect of pH on rejection of hexavalent chromium by nanofiltration. Desalination 219:171–178. https://doi.org/10.1016/j.desal.2007.04.054

Naseem T, Durrani T (2021) The role of some important metal oxide nanoparticles for wastewater and antibacterial applications: a review. Environ Chem Ecotoxicol 3:59–75. https://doi.org/10.1016/j.enceco.2020.12.001

Nataraj SK, Hosamani KM, Aminabhavi TM (2007) Potential application of an electrodialysis pilot plant containing ion-exchange membranes in chromium removal. Desalination 217:181–190. https://doi.org/10.1016/j.desal.2007.02.012

Nguyen CM, Bang S, Cho J, Kim K-W (2009) Performance and mechanism of arsenic removal from water by a nanofiltration membrane. Desalination 245:82–94. https://doi.org/10.1016/j.desal.2008.04.047

Ojea-Jiménez I, López X, Arbiol J, Puntes V (2012) Citrate-Coated gold nanoparticles as smart scavengers for mercury(II) removal from polluted waters. ACS Nano 6:2253–2260. https://doi.org/10.1021/nn204313a

Pandya A, Shah K, Prajapati H, Vishwakarma GS (2021) GQD embedded bacterial cellulose nanopaper based multi-layered filtration membranes assembly for industrial dye and heavy metal removal in wastewater. Cellulose 28:10385–10398. https://doi.org/10.1007/s10570-021-04174-1

Pearce G (2007) Introduction to membranes: Filtration for water and wastewater treatment. Filtr Sep 44:24–27. https://doi.org/10.1016/S0015-1882(07)70052-6

Peters Th (2010) Membrane technology for water treatment. Chem Eng Technol 33:1233–1240. https://doi.org/10.1002/ceat.201000139

Rajesh S, Fauzi Ismail A, Mohan DR (2012) Structure-property interplay of poly(amide-imide) and TiO_2 nanoparticles impregnated poly(ether-sulfone) asymmetric nanofiltration membranes. RSC Adv 2:6854. https://doi.org/10.1039/c2ra20265d

Sadeghi M, Semsarzadeh MA, Moadel H (2009) Enhancement of the gas separation properties of polybenzimidazole (PBI) membrane by incorporation of silica nano particles. J Membr Sci 331:21–30. https://doi.org/10.1016/j.memsci.2008.12.073

Sadrzadeh M, Mohammadi T, Ivakpour J, Kasiri N (2008) Separation of lead ions from wastewater using electrodialysis: comparing mathematical and neural network modeling. Chem Eng J 144:431–441. https://doi.org/10.1016/j.cej.2008.02.023

Sadrzadeh M, Mohammadi T, Ivakpour J, Kasiri N (2009) Neural network modeling of Pb2+ removal from wastewater using electrodialysis. Chem Eng Process 48:1371–1381. https://doi.org/10.1016/j.cep.2009.07.001

Sarkar S, Chakraborty S (2021) Nanocomposite polymeric membrane a new trend of water and wastewater treatment: a short review. Groundw Sustain Dev 12:100533. https://doi.org/10.1016/j.gsd.2020.100533

Shemer H, Wald S, Semiat R (2023) Challenges and solutions for global water scarcity. Membranes 13:612. https://doi.org/10.3390/membranes13060612

Shi Z, Zhou H, Qing X, Dai T, Lu Y (2012) Facile fabrication and characterization of poly(tetrafluoroethylene)@polypyrrole/nano-silver composite membranes with conducting and antibacterial property. Appl Surf Sci 258:6359–6365. https://doi.org/10.1016/j.apsusc.2012.03.040

Singh KK, Senapati KK, Sarma KC (2017) Synthesis of superparamagnetic Fe_3O_4 nanoparticles coated with green tea polyphenols and their use for removal of dye pollutant from aqueous solution. J Environ Chem Eng 5:2214–2221. https://doi.org/10.1016/j.jece.2017.04.022

Sudeep PK, Joseph STS, Thomas KG (2005) Selective detection of cysteine and glutathione using gold nanorods. J Am Chem Soc 127:6516–6517. https://doi.org/10.1021/ja051145e

Wang X, Chen Z, Yang S (2012) Forecasting modeling and simulation analysis of a power system in China, based on a class of semi-parametric regression approach. Sajie 23:154. https://doi.org/10.7166/23-3-518

WHO G (2011) Guidelines for drinking-water quality. World Health Organization 216:303–304

WWAP U (2017) The United Nations World Water Development Report 2017: Wastewater. The untapped resource. United Nations World Water Assessment Programme Paris

Yin J, Deng B (2015) Polymer-matrix nanocomposite membranes for water treatment. J Membr Sci 479:256–275. https://doi.org/10.1016/j.memsci.2014.11.019

Yu S, Zuo X, Bao R, Xu X, Wang J, Xu J (2009) Effect of SiO_2 nanoparticle addition on the characteristics of a new organic–inorganic hybrid membrane. Polymer 50:553–559. https://doi.org/10.1016/j.polymer.2008.11.012

Zhu X, Bai R, Wee K-H, Liu C, Tang S-L (2010) Membrane surfaces immobilized with ionic or reduced silver and their anti-biofouling performances. J Membr Sci 363:278–286. https://doi.org/10.1016/j.memsci.2010.07.041

Zodrow K, Brunet L, Mahendra S, Li D, Zhang A, Li Q, Alvarez PJJ (2009) Polysulfone ultrafiltration membranes impregnated with silver nanoparticles show improved biofouling resistance and virus removal. Water Res 43:715–723. https://doi.org/10.1016/j.watres.2008.11.014

Chapter 16
Nanotechnology-Driven Solutions for Wastewater Treatment: Overcoming Challenges and Envisioning the Future

Kanchan Yadav and Jais Kumar

Abstract The escalating global population and intensified exploitation of natural resources have given rise to numerous environmental issues, with water pollution being a prominent concern. Industrialization, in particular, significantly contributes to the degradation of natural water resources, generating a substantial proportion of wastewater. The increasing volume of wastewater on earth necessitates urgent technological interventions to address this formidable challenge. Over the past three decades, nanotechnology has emerged as a versatile tool for tackling various issues related to health, energy, and the environment. While nanotechnology holds promise for wastewater treatment, there remains ample room for in-depth exploration and the translation of laboratory findings to larger-scale applications. This chapter focuses primarily on the utilization of nanomaterials (NMs) in wastewater treatment, spanning from zero-dimensional to three-dimensional structures. These NMs encompass carbon quantum dots, single-walled carbon nanotubes (SWCNTs), multi-walled carbon nanotubes (MWCNTs), organic composites, as well as metal and metal oxide-based frameworks. Additionally, graphene, graphene oxide, and graphene-based nanoparticles (NPs) have been extensively employed in treatment processes. The functional properties and surface characteristics of NMs play a pivotal role in sensing and treating wastewater. Various metal oxide-based materials, including ferric oxide and pyrrolidone-conjugated oxides, have been utilized as absorbents for metal ions in filtration processes. This chapter offers a comprehensive analysis of recent advancements in the application of various NMs and technologies for wastewater treatment.

Keywords Nanotechnology · Nanomaterials · Nanoparticles · Wastewater · Photocatalysts

K. Yadav (✉)
Department of Chemical Engineering, University of Seoul, Seoul, South Korea
e-mail: kanchany10038@gmail.com

J. Kumar
Department of Physics, K.N. Government P.G. College, Gyanpur, Bhadohi 221304, India

N. Agarwal et al. (eds.), *Nanomaterials in Wastewater Research*, Advances in Wastewater Research, https://doi.org/10.1007/978-981-96-3908-3_16

16.1 Introduction

Water, being essential for the sustenance of all living organisms, faces a critical challenge as nearly 800 million people globally lack access to potable water for domestic purposes (Makarigakis and Jimenez-Cisneros 2019). The regulation of water scarcity and quality has emerged as a paramount concern for sustainable development due to the pervasive anthropogenic contamination of water resources. Organic pollutants, including pesticides, viruses, bacteria and anhydrous phase liquids, alongside inorganic contaminants, such as radioactive materials, metals, salts, metals, are introduced into the environment through human activities, notably industrial expansion and population growth (Hassan Al-Taai 2021; Mishra et al. 2023; Shah et al. 2023). This influx of pollutants into the environment (water–soil–plant) continuum, primarily via water pollution, poses significant threats to human health. Contaminated water adversely affects community health, disrupts aquatic ecosystems, undermines environmental sustainability, and impacts societal well-being and economic prosperity (Lin et al. 2022; Babuji et al. 2023). Grimly, inadequate water supply and substandard hygiene practices contribute to approximately 30,000 daily deaths worldwide, with rural areas, particularly infants, bearing the brunt of these consequences. The global demand for clean and safe water is escalating, yet conventional treatment methods fall short in meeting the requisite quantity of high-quality water (Salehi 2022; Tarrass and Benjelloun 2012; Toivettula et al. 2023). Thus, the imperative lies in the exploration of novel, energy-efficient technologies for water purification (Grzegorzek et al. 2023; Murugan and Ganesh Ram 2018; Nawaz and Sengupta 2021). The utilization of nanomaterials (NMs) in wastewater treatment has emerged as a significant advancement, facilitating the rapid and significant removal of contaminants through adsorption or remediation mechanisms. In wastewater treatment, NMs play a crucial role in eliminating substances, such as toxic metal ions, dyes, and phenols. Magnetic NMs offer additional benefits, including facile recovery via external magnetic fields and the potential for multiple reuses (Jangid and Prabhu Inbaraj 2021; Yang et al. 2019; Chhetri et al. 2022). However, it is imperative to address concerns regarding the toxicity of certain NPs, which pose new environmental and health risks (Malhotra et al. 2020). Among magnetic NPs (MNPs), transition metal oxides with spinel structures, known as ferrites, hold particular significance. Ferrites exhibit diverse crystalline structures and magnetic properties, including spinel, garnet, hexaferrite, and orthoferrite phases (Shukla et al. 2021). The wastewater produced by urban communities and industries serves as a significant carrier of pollutants and necessitates proper management and disposal (Singh et al. 2023). Addressing the presence of harmful pollutants, both organic and inorganic, along with complex compounds, has become a focal point of research efforts. Implementing effective technologies for wastewater treatment stands as a key strategy to improve water quality (Limousy et al. 2022; Lin et al. 2019; Ahmed et al. 2022).

This chapter endeavors to furnish a comprehensive overview of the multifaceted application of NMs/NPs in wastewater treatment, spanning a diverse array of methodologies including adsorption, membrane modification, analyte extraction, photocatalysis, pathogen disinfection, and metal ion elimination. Additionally, it encompasses an exhaustive analysis of the various techniques employed in NP preparation. By conducting an in-depth exploration of the intrinsic properties, merits, and demerits inherent in different types of NMs, this study aims to furnish a comprehensive understanding to a wide-ranging readership regarding their pivotal role in augmenting and optimizing water treatment processes.

16.2 Spectrum of Pollutants in Wastewater

Water pollution, stemming from a diverse range of organic and inorganic pollutants, has emerged as a paramount global environmental issue, posing significant threats to human life and ecosystems alike. These pollutants are primarily categorized into three classes—inorganic, organic, and biological pollutants—each presenting its own unique set of hazards to the environment and living organisms (Shah et al. 2023; Wasewar et al. 2020; Farhan Hanafi and Sapawe 2020). Inorganic pollutants, such as heavy metals and chemicals, can persist in water bodies for extended periods, leading to long-term ecological damage. Organic pollutants, including pesticides, solvents, and petroleum products, contribute to water contamination and can have adverse effects on aquatic life and human health. Biological pollutants, such as bacteria, viruses, and pathogens, pose risks of waterborne diseases and environmental degradation.

16.2.1 Organic Pollutants

A diverse array of organic pollutants pervades wastewater, encompassing a range of chemical compounds originating from various sources. Pesticides, utilized extensively in agriculture to control pests and enhance crop yields, constitute a significant portion of organic pollutants in wastewater. These chemicals, designed to be biologically active, can persist in the environment and have adverse effects on aquatic ecosystems and human health. Pharmaceutical residues, emerging as a growing concern, enter wastewater through human excretion and improper disposal of unused medications (Wilkinson et al. 2022). These compounds, designed to exert specific physiological effects within the body, may disrupt endocrine function and exert toxic effects on aquatic organisms (Ojha and Tiwary 2021). Additionally, industrial activities contribute a plethora of organic pollutants to wastewater, including polychlorinated biphenyls (PCBs) (Cui et al. 2020), polycyclic aromatic hydrocarbons (PAHs) (Vijayanand et al. 2023), surfactants, solvents, and dyes. PCBs, once widely used in electrical equipment and industrial applications, are highly toxic and persist in the environment due to their chemical stability. PAHs, generated from incomplete combustion of organic matter, pose carcinogenic and mutagenic risks to aquatic life and humans. Surfactants, employed in detergents and cleaners, can disrupt aquatic ecosystems by altering surface tension and interfering with the functioning of aquatic organisms. Solvents and dyes, originating from industrial processes, such as textile manufacturing and chemical production, introduce complex organic compounds into wastewater, posing challenges for treatment and removal.

16.2.2 Inorganic Pollutants

Inorganic contaminants in wastewater comprise a diverse array of substances, including industrial waste, sediments, heavy metals, and various ions, such as fluoride, sulfate, nitrate, and chloride, all of which can detrimentally affect water quality when present in elevated concentrations. Among these contaminants, arsenic emerges as one of the most hazardous pollutants, alongside heavy metals like lead, chromium, and cadmium, which pose significant threats to both environmental integrity and human health. While certain inorganic pollutants occur naturally in the environment, human activities, such as mining, industrial processes, and agricultural practices, have substantially augmented their presence and concentration levels (Kurwadkar 2019). For instance, mining activities can introduce elevated levels of heavy metals like mercury and copper into water bodies, while industrial discharges may contain toxic substances, such as mercury, nickel, and zinc. Agricultural runoff, on the other hand, can contribute to the accumulation of nitrate and phosphate ions in water sources, leading to eutrophication and harmful algal blooms. Unlike organic pollutants, which may degrade over time, inorganic contaminants often persist in the environment indefinitely, accumulating in soil, sediment, and aquatic ecosystems. This persistence poses a significant risk to human health, as these contaminants can enter the food chain and bio-accumulate in organisms, leading to long-term health consequences (Kumar et al. 2023; Verma and Ratan 2020).

16.2.3 Biological Contaminants

Biological contaminants in wastewater encompass a wide array of microorganisms, including bacteria, viruses, protozoa, fungi, and algae, which can pose significant health risks to humans and the environment. These contaminants originate from various sources, including human and animal waste, agricultural runoff, and industrial effluents. Pathogenic bacteria, such as *Escherichia coli*, *Salmonella* spp., and *Campylobacter* spp., are of particular concern due to their ability to cause gastrointestinal illnesses in humans upon ingestion of contaminated water (Pandey et al. 2014). Similarly, viruses, such as norovirus, hepatitis A virus, and rotavirus, can lead to severe gastrointestinal and systemic infections. Protozoa, such as *Giardia lamblia* and *Cryptosporidium parvum*, are known for causing diarrheal diseases, especially in immunocompromised individuals. Additionally, fungal contaminants like Aspergillus spp. and Candida spp. may proliferate in wastewater, posing risks of respiratory and skin infections. Algal blooms, fueled by nutrient-rich wastewater discharges, can produce toxins harmful to aquatic organisms and humans, leading to water quality degradation and ecosystem disruptions.

16.3 Role of NPs in Wastewater Treatment

The distinctive properties of NMs have transformed the landscape of wastewater treatment, offering innovative solutions for wastewater treatment. NMs, including NPs, nanocomposites, and carbonaceous NMs, exhibit a diverse array of characteristics that make them well suited for addressing the complex challenges posed by contaminants in water systems. NPs, composed of metal and metal oxide molecules, possess remarkable reactivity, particularly on their surfaces, allowing them to serve as efficient catalysts for the degradation of various pollutants, such as pesticides, dyes, heavy metals, and hydrocarbons. Metal oxide NPs like Fe_2O_3, CuO, and TiO_2 have garnered attention for their stability, porosity, and light absorption properties, enabling effective decomposition of harmful substances in wastewater (Li et al. 2022; Ighalo et al. 2021; Dutta 2020). MNPs offer the advantage of facile physical removal of contaminants due to their magnetic properties, while nanocomposites, which incorporate NPs into supportive materials, enhance catalytic reactivity and provide increased surface area for pollutant adsorption and degradation. Semiconductor NP-based nanocomposites, when coated with metals, exhibit enhanced photocatalytic activity, enabling the efficient breakdown of pollutants under light irradiation (Imtiaz et al. 2019). Carbonaceous NMs, such as fullerene and carbon nanotubes (CNTs), possess unique attributes including high surface area, hydrophobicity, and strong adsorption capabilities, making them valuable tools for wastewater treatment. CNTs, renowned for their stability, mechanical strength, and interaction with pollutants, play a crucial role in adsorption and removal processes, contributing significantly to the purification of water resources (Nasrollahzadeh et al. 2021). By harnessing the multifaceted capabilities of NMs, wastewater treatment technologies can achieve higher efficiency, reliability, and sustainability in safeguarding water quality and protecting the environment.

16.4 Methods to Synthesize NPs

The synthesis of NPs, structures with at least one dimension less than 100 nm, encompasses physical, chemical, and biological methods (Baig et al. 2021). Physical methods involve vaporization and condensation processes, including inert gas condensation, sputtering, spark discharge, and laser ablation (Kumari et al. 2023).

Chemical methods rely on chemical reactions to generate NPs from precursor materials, such as co-precipitation, sol-gel, hydrothermal synthesis, microwave-assisted synthesis, and microemulsion. Biological methods utilize biological entities like microorganisms, enzymes, and plant extracts for NP synthesis, offering environmentally friendly approaches. These methods enable the production of NPs with tailored properties suited for diverse applications, ranging from electronics to medicine, by leveraging different mechanisms to control NP size, shape, composition, and surface characteristics. In the subsequent sections, each method will be thoroughly examined, elucidating their principles, advantages, limitations, and applications in NP synthesis (Nam and Luong 2019).

16.4.1 Physical Methods to Prepare NPs

Laser ablation: Laser ablation refers to the process of removing material from a surface through laser irradiation. When exposed to low laser power, the material undergoes sublimation and evaporation as a result of absorbing laser energy. Conversely, high laser power typically converts the material into plasma. This method has been utilized for the production of metal NPs (Kim et al. 2017).

Inert gas condensation: The technique, commonly used for the rapid production of metal-based NPs, is inert gas condensation. In this process, metals evaporate under high pressure within a vacuum chamber filled with argon or helium gas. The highly energetic metal atoms, upon colliding with the gas molecules in the medium, lose their kinetic energy, leading to the formation of fine particles. These small particles then aggregate through coagulation, gradually growing into nanocrystals (Suryanarayana and Prabhu 2007).

Sputtering: The phenomenon of atoms being ejected from the surface of a material by energetic particles is known as sputtering. During sputtering, ions bombard the cathode or target, causing rapid atom transfer. The ejected atoms then travel until they reach the substrate, where they form the desired layer. This process has been employed in the production of TiO2 NPs and the synthesis of nanocatalysts, such as Pd and Pt (Käufer et al. 2024).

Template synthesis: Template synthesis is recognized as a green and environmentally friendly approach for preparing NMs. This method is particularly promising for the synthesis of monodispersed inorganic NPs, utilizing singular cavities of cellular materials as hosts to facilitate NP synthesis. The template serves a dual purpose: enabling the structure to achieve maximum size and serving as a scaffold for organizing various functionalities, active components, and interfaces of the device (Li et al. 2022).

Ultrasound: Ultrasound has become a common method for preparing NPs nowadays. In this technique, ultrasonic waves cause the formation of ultrasonic cavitation in liquid solutions, resulting in various changes, such as high pressure, cooling rate, and temperature, which provide favorable conditions for chemical reactions. Ultrasound-assisted synthesis is an effective approach for producing NPs with controlled morphology. This method has been successfully utilized for the synthesis of silver NPs (AgNMs) and gold NPs (AuNPs)(Bao et al. 2023).

16.4.2 Chemical Methods to Prepare NPs

Precipitation: In precipitation techniques, nucleation, agglomeration, and growth processes occur simultaneously. Reactions associated with precipitation exhibit distinct characteristics, with products typically being insoluble and formed under saturation conditions. Nucleation, the primary step in precipitation, leads to the formation of numerous fine particles. Additional processes like Ostwald ripening and aggregation significantly influences the properties and size of NPs. Conditions of supersaturation are essential to induce precipitation. This method has been utilized for the synthesis of various NPs, including ZnO, Co_3O_4, and MNPs (Aragaw et al. 2021).

Microemulsion: The microemulsion method is often considered ideal for preparing inorganic NPs, although the exact formation process within microemulsions remains unclear. However, researchers have proposed a mechanism for NP production using this method. During the mixing process, reagents undergo rapid exchange facilitated by the collision of water particles, leading to a precipitation reaction and the formation of NPs. Subsequently, particle coagulation and nucleation growth occur, resulting in the formation of NPs surrounded by water or surfactant materials. This method has been employed for size-controlled synthesis of silver and starch NPs (Lakshmanan et al. 2014).

Hydrothermal synthesis: Hydrothermal synthesis relies solely on the absolute solubility of minerals in hot water and elevated pressure to produce individual crystals of atoms. The process involves crystal growth within an iron pressure-maintaining vessel called an autoclave, where raw materials are combined with water. Temperature gradients are maintained at the corners of the chamber to control crystal growth, with hot areas facilitating the dissolution of nutrients and cold areas promoting the multiplication of seeds (Lone et al. 2023).

Microwave-assisted technique: Microwave technology has found extensive applications in both nano-technological and biochemical processes. Compared to traditional convection heating methods, chemical reactions conducted using microwaves are often faster and exhibit higher conductivity, resulting in fewer byproducts. Microwave reactors are capable of withstanding high temperatures and pressures, providing precise control over reaction mixtures and ensuring repeatability from one reaction to another. One notable advantage of microwave techniques is the ability to engineer a high-degree separation between NP growth and nucleation processes, even starting the reaction at room temperature. This selective heating capability enables the activation of specific precursor molecules or solvents, crucial for scaling up production. Microwave synthesis allows for the selective heating of solvent or precursor molecules, facilitating the preparation of NMs (Adeola et al. 2023).

Sol-gel: The sol-gel method enables the production of fine solid particles using small molecules. In this process, two different solutions are mixed to form a gel-like material, ranging from continuous polymer networks to discrete particles in morphology. The production involves several steps: firstly, preparing solutions of the required materials, followed by poly-condensation and hydrolysis reactions to form a transparent and stable sol system. Over time, particles in colloidal suspension gradually aggregate as the sols are left undisturbed, forming gels with three-dimensional structures. Subsequent processes, such as sintering and drying, lead to the formation of nano- and microstructures (Johnson et al. 2023).

16.4.3 Biological Synthesis

This method adheres to the principles of green chemistry, combining biotechnology and nanotechnology. However, the synthesis rate of NPs using this process is typically slow, and the resulting NPs are often not monodispersed. Nucleation depends on the concentration of various components involved in macromolecule synthesis, leading to polydispersity in the NPs and affecting the synthesis rate. Optimization of multiple processes is required to address persistent issues in NP formation, including improvements in microbial culture and extraction methods. Comprehensive methodologies, such as photo-biological techniques, can be employed to enhance NP synthesis. Additionally, studying mechanisms at the molecular, cellular, and biochemical levels is necessary to improve NP properties and synthesis potential. Further exploration of the potential of biological materials for NP synthesis is warranted, given the diversity of microbes and plants available for research (Rathod et al. 2024).

16.5 Strategies for Purifying Wastewater

Industries, such as non-ferrous metal smelting, discharge vast volumes of highly toxic heavy metal wastewater (HMW) worldwide each year, posing significant challenges to conventional wastewater treatment plants and environmental well-being. Traditionally, HMW has been treated through chemical precipitation methods involving lime, caustic, or sulfide, yet these methods often fail to meet increasingly stringent discharge standards (Qasem et al. 2021). NMs demonstrate exceptional attributes stemming from their nanoscale dimensions, such as amplified surface area-to-volume ratios, augmented reactivity, and versatile surface chemistry. Recent progress in nanotechnology has led to the creation of novel NMs tailored to combat water pollution. Among these innovations are nano-sorbents, nanomotors, nano-membranes, nano-photocatalysts, and others, each engineered to address specific challenges associated with contaminated water sources (Epelle et al. 2022). These advancements signify groundbreaking solutions poised to revolutionize the treatment of polluted water bodies. All the methods have been thoroughly explored in the preceding sections.

16.5.1 NM-Based Adsorbents and Membranes

Adsorption is a widely recognized and extensively studied technique for purifying aqueous media. Its high efficiency and simplicity make it a preferred method for decontaminating water from a diverse range of pollutants. Compared to other available methods, adsorption stands out for its effectiveness, ease of implementation, cost-effectiveness, and flexibility. NPs (both organic and inorganic) with high adsorption capacity and the ability to isolate a wide range of contaminants owe their efficacy to their high porosity, large surface area, and active surfaces, earning them the title of nanoadsorbents. These nanoadsorbents work swiftly and effectively, boasting the additional advantage of regeneration. This means that NPs can be reused after being separated from the pollutants. Desorption experiments have been conducted to regenerate carbon NPs by removing adsorbed hydrocarbons. Various types of nanoadsorbents, including metal oxide-based, carbon-based, metal NPs, and NP composites, are employed for environmental decontamination purposes. Presently, water pollution encompasses various pollutants categorized into inorganic, organic, and biological categories. According to literature findings, nanoadsorbents exhibit equal efficiency in adsorbing inorganic, organic, and biological pollutants alike. The adsorption technique has a strong track record in removing inorganic pollutants, particularly heavy metals like arsenic (As). Manganese ferrite (Katubi et al. 2021) and cobalt ferrite NPs (Mmelesi et al. 2021) were utilized for As adsorption, with cobalt ferrite NPs proving more effective. Hematite NPs showed high adsorption capacity for both As species (+ 5 and + 3), with micro-fibrillated cellulose-activated magnetite NPs also demonstrating exceptional adsorption properties. Additionally, iron oxide/alumina nanocomposites exhibited superior adsorption capacity for As compared to iron oxide NPs alone. Chromium (Cr) and cadmium (Cd) are toxic metals found in water due to industrial activities. Cr(VI) is particularly hazardous, known to cause cancer, while Cd is non-biodegradable and accumulates in the body, causing bone demineralization and kidney damage. Various methods have been developed to remove these contaminants. Graphene oxide (GO)/Co_3O_4 nanocomposites (Pourzare et al. 2017) have shown high adsorption capacities for Cr. Magnetic carbon nanocomposites and sandwiched nanocomposites GO/MnO2/Fe_3O_4/PPy (Liu et al. 2018) effectively remove Cr from water. Heavy metal contamination, particularly lead (Pb) and copper (Cu), poses significant health risks. Zeolite/PEG/GO nanocomposites are also effective for removing Cd and Pb from water. These methods offer efficient and economical solutions for heavy metal removal from contaminated water sources. Fe_3O_4/CSNPs and hydroxyapatite NPs (Vahdat et al. 2019) synthesized from eggshells are effective for Pb removal, while agricultural waste-derived nanoadsorbents, such as OSH, offer cost-effective and eco-friendly solutions. For Cu removal, nanocomposites like PNCA and hybrid nanoadsorbents combining polyethylenimine with magnetic silica exhibit high adsorption capacities. Carbon nanofibers and carbon quantum dots supported on activated carbon thin films also show promise in removing Cu from wastewater. In one study, ultrafine mesoporous Fe_3O_4 NPs (UFMNPs) (Fato et al. 2019) effectively removed Ni^{+2}, Cd^{+2}, Pb^{+2}, and Cu^{+2} from river water, achieving removal efficiencies ranging from 54 to 86%. Another investigation utilized activated carbon derived from Cucumis melo peel to successfully adsorb Pb^{+2}, Cd^{+2}, Cr^{+6}, and Ni^{+2} from industrial wastewater, demonstrating the effectiveness of this adsorbent in metal ion removal. Additionally, Fe_3O_4-based nanoadsorbents, including Fe_3O_4, Fe_3O_4-GO, and Fe_3O_4-NH_2, exhibited significant adsorption capacities for various metal cations, indicating their potential for wastewater treatment. These findings underscore the efficacy of single adsorbents in simultaneously removing multiple metal contaminants from water sources. These NMs present efficient and versatile solutions for heavy metal removal from contaminated water sources. Several studies have explored the removal of multiple metal ions using single adsorbents, indicating promising results.

Researchers have also developed efficient nanoadsorbents for the removal of various organic contaminants commonly found in wastewater, such as dyes, pesticides, and pharmaceuticals. For instance, nanoadsorbents loaded with ZnS:CuNPs (Isac and Enesca 2022) achieved high removal rates of auramine-O (AO) and other dyes. CS-coated Fe_3O_4 NPs immobilized on dextran gel columns exhibited excellent adsorption capacities for Evans blue dye. Hybrid nanoadsorbents combining magnetic-carrageenan/SiO_2 showed promising results in removing methylene blue (MB) and other dyes from wastewater. In addressing pesticide contamination, nanoadsorbents like MWCNTs and CS-ZnONPs composites demonstrated high efficiency in removing various pesticides, such as malathion and permethrin. For pharmaceutical contaminants, nanoadsorbents based on MWCNTs-GO nanosheets and silica NPs from rice husks showed remarkable removal capacities for antibiotics and other pharmaceuticals. Overall, these nanoadsorbents offer effective solutions for the removal of organic contaminants from wastewater, contributing to environmental protection and public health. Conventional methods often struggle to effectively filter cyanobacterial contaminants, necessitating alternative approaches. While adsorbents like activated carbon (AC) can remove contaminants, various factors can affect their efficiency. In addressing microbial contaminants, Fe_3O_4 NPs synthesized via top-down and bottom-up approaches have shown promise in removing both Gram-positive and Gram-negative bacteria. Similarly, amine-modified Fe_3O_4 NPs and Fe_3O_4-SiO_2-NH_2 nanocomposites have demonstrated efficient removal of disease-causing bacteria (Zhan et al. 2014). For biological toxin removal, RNA-GO nanosheets have been utilized, albeit with reduced efficiency under extreme conditions. Additionally, the combination of ultrafiltration (UF) and powdered activated carbon (PAC) has proven effective in removing cyanobacterial toxins from natural blooms, providing valuable insights into water treatment strategies.

16.5.2 Photocatalysts for Advanced Oxidation Processes (AOPs)

Combining the Greek words "photo" and "catalysis" yields "photocatalysis," a process wherein compounds degrade in the presence of light. Although there is no universally agreed definition for photocatalysis, it generally involves stimulating or activating materials with light, be it visible, sunlight, or UV. Nano-photocatalysts, favored for treating contaminated water, leverage their shape-dependent features and high surface ratio to enhance catalytic activity. The unique surface properties and quantum effects of NPs set them apart from bulk materials, augmenting their optical, magnetic, electric, mechanical, and chemical reactivity. Notably, nano-photocatalysts amplify the production of oxidizing species on their surfaces, thereby expanding oxidation capabilities and facilitating pollutant degradation in water. In addressing environmental pollutants like chlorpyrifos, azo dyes, and organochlorine pesticides, semiconductors, zero-valence metal-based materials, and bimetallic NPs prove effective. TiO_2 nanotubes are commonly employed to eliminate organic pollutants from water due to TiO_2's low cost, nontoxic nature, easy availability, and chemical stability. Among TiO_2's crystalline forms, anatase is particularly renowned for its photocatalytic prowess (Rojviroon et al. 2021). Additionally, ZnO NPs serve as effective photocatalysts for wastewater treatment and are reusable (Roy and Chakraborty 2021). The synergistic effect of composite NMs like CdS/TiO_2 enhances photocatalytic degradation under visible light. Furthermore, incorporating materials like iron into photocatalysts enables easy recycling due to their ferromagnetic properties. While metal oxides are primarily targeted for enhancing photocatalytic abilities, further modifications involving metal ions, dyes, and carbonaceous materials are imperative for continued improvement.

16.5.3 Membrane Filtration

Specialized nano-membranes composed of various nanofibers serve to effectively remove small particles from aqueous media, functioning at high exclusion speeds and often utilized as a pretreatment method for reverse osmosis. They are categorized into three types: organic, inorganic, and organic-inorganic hybrids. Organic nano-membranes consist of polymeric materials, such as cellulose acetates, polyethersulfone, and polyvinyl alcohol, while inorganic nano-membranes are composed of metals, metal oxides, carbon nanotubes, and graphene. The combination of inorganic materials with polymers yields hybrid nano-membranes, enhancing their performance and versatility. These membranes play crucial roles in various filtration processes, including ultrafiltration, reverse osmosis, and nanofiltration, contributing to the purification of water and removal of undesirable contaminants. Innovations in nano-membrane technology have revolutionized water treatment processes. Researchers have developed an aquaporin Z-based nano-membrane, enhancing water flux by 65% and reducing $MgCl_2$ and NaCl concentrations by 88.1% and 66.2%, respectively (Ryu et al. 2018). Similarly, another study implemented a hydrophobic nano-membrane using polyvinylidene fluoride-co-hexafluoropropylene and carbon nanotubes, resulting in a 33–59% improvement in water flux and > 99.99% salt removal efficiency. Additionally, a nano-membrane based on a zirconium (VI) metal-organic framework exhibited remarkable removal rates for multivalent ions, such as Al+3 (99.3%), Ca^{+2} (86.3%), and Mg^{+2} (98%) without performance decline over 170 hours of testing (Liu et al. 2015). In arsenic removal, amyloid fibril-based membranes demonstrated a remarkable 99.6% removal efficiency. One study utilized cellulose nanofibrils to create an ultra-thin nano-membrane, achieving significant NP removal with a water flux of 1.14 L/h m^2 bar (Hassan et al. 2017). Similarly, a carbon nanotube sponge-based nano-membrane effectively removed various NPs and organic molecules from water. Organic contaminant removal was addressed by synthesizing composite nano-membranes, effectively removing dyes and other pollutants. Nano-membranes prepared with vertically aligned carbon nanotubes and immobilized silver nanoclusters demonstrated efficient oil removal from water across diverse environmental conditions. To address waterborne diseases caused by biological contaminants like viruses, bacteria, plankton, and algae, compact nano-membranes offer promising solutions. For instance, a TiO_2 nanowire-based UF membrane, boasting a pore size of around 20 nm, effectively filters contaminants while exhibiting antibacterial properties, antifouling capabilities, and photocatalytic oxidation. Similarly, a highly selective nanoporous membrane composed of copolymers demonstrates efficient virus filtration, featuring a pore size of approximately 17 nm and maintaining stability even under pressures of up to 2 bars. This membrane efficiently filters viruses like human rhinovirus type 14, typically with a diameter of about 30 nm. Additionally, innovative nano-fibrous scaffolds efficiently remove viruses and bacteria from water sources. Another compact nano-membrane, coated with ɣ-alumina and titanium nanocrystals, exhibits remarkable performance, achieving 100% removal of microorganisms and around 25% removal of ions from water samples (Deepa and Arthanareeswaran 2022). Nano-membranes, specialized membranes composed of various nanofibers, serve to effectively remove small particles from aqueous media, functioning at high exclusion speeds and often utilized as a pretreatment method for reverse osmosis.

16.5.4 Nanoscale Zero-Valent Iron (nZVI) Remediation

There has been a surge in research and development of innovative treatment technologies, with NP-based approaches garnering significant attention. Among these, the utilization of nanoscale zero-valent iron (nZVI) stands out as a well-developed solution (Li et al. 2023). While nZVI is renowned for its effectiveness in site remediation, this perspective highlights its application in treating HMW while simultaneously enabling metal recovery. Several advantages of nZVI in this context are elucidated, including its ability to sequester a wide range of metal ions (over 30 species), its capacity to capture and concentrate metal ions at low concentrations (with a removal capacity of up to 500 $mg{\cdot}g^{-1}$ nZVI), and its operational simplicity due to its unique hydrodynamics. These advantages stem from nZVI's small NP size and unique iron chemistry. Additionally, the engineering implementation of this application is presented, having successfully treated millions of cubic meters of HMW and recovered substantial amounts of valuable metals, such as copper and gold. In conclusion, nZVI emerges as a potent reagent for treating HMW, offering an eco-friendly solution to address this hazardous waste. Similarly, NZVI supported on reduced GO (NZVI/rGO) (Alam et al. 2022) and MoS_2/SH-MWCNT nanocomposites exhibit high adsorption capacities for Cd and Pb (Gusain et al. 2019).

16.5.5 Nano- and Micromotors

In recent years, nanotechnology has witnessed significant advancements, introducing innovative approaches to water treatment, including the development of nano/micromotors (Verma et al. 2022). These motors have the remarkable ability to harness energy from diverse sources and convert it into mechanical force. These motors exhibit versatility in achieving diverse objectives through various mechanisms and can operate with or without external energy sources, such as electric fields, magnetic fields, or acoustics. Characterized by specific control over movement, self-mixing ability, high speed, and power, these advanced motors hold promise for addressing environmental sustainability and stability concerns by removing pollutants from contaminated water sources. Nano/micromotors, however, possess inherent self-propulsion capabilities, enabling them to overcome diffusion limitations through energetic mixing. By integrating with nano/microstructured materials, nano/micromotors enhance water treatment efficiency by offering greater workability and surface area (Shivalkar et al. 2021). Engineered micro- and nanomotors utilize a variety of NMs to achieve their functionalities. For instance, metal NPs, such as platinum, gold, or nickel, are commonly used in catalytic motors fueled by chemical reactions like hydrogen peroxide decomposition. Carbon-based NMs like carbon nanotubes and graphene offer exceptional mechanical strength and electrical conductivity, making them ideal components for electrically powered motors. Magnetic NPs, such as iron oxide, enable magnetically propelled motors, responding to external magnetic fields for controlled movement. These examples demonstrate the versatility of NMs in designing micro- and nanomotors for wastewater treatment applications (Wang and Villa 2021).

16.6 Advantages of NPs Over Conventional Treatments

A recent report by WHO highlighted the staggering impact of poor-quality drinking water, revealing that approximately 783 million people worldwide are affected. Alarmingly, waterborne diseases contribute to around 1.87 million deaths of children. Particularly in underdeveloped countries, conventional treatment methods are often impractical due to their high costs and maintenance issues. Compounding this challenge, many people in developing nations resort to collecting and storing water from various sources for drinking, increasing the risk of contamination during transportation and storage.

16.6.1 Tailored Physical Properties

Tailored physical properties of NMs offer several advantages in wastewater treatment processes. One key advantage is their tunable properties, allowing for precise control over characteristics, such as size, shape, surface area, and porosity. This tunability enables the design of NMs with enhanced adsorption capacity, catalytic activity, and selectivity for specific contaminants present in wastewater. Additionally, NMs exhibit high surface-to-volume ratios, providing a large active surface area for interactions with contaminants, thereby improving treatment efficiency. Their small size facilitates penetration into water matrices and interactions with pollutants at the molecular level, leading to more effective removal. Furthermore, NMs can be engineered to possess tailored surface functionalities, enabling targeted adsorption or degradation of specific pollutants. Overall, the tailored physical properties of NMs offer promising opportunities for the development of advanced and efficient wastewater treatment technologies (Ahmed et al. 2022).

16.6.2 Selective Functionality and Targeting

Selective functionality and targeting of NMs play a crucial role in wastewater treatment, offering several advantages. One key benefit is the ability to design NMs with specific functional groups or surface modifications that can selectively adsorb or degrade target contaminants while leaving other components unaffected. This selectivity allows for the efficient removal of pollutants without impacting beneficial compounds or natural water constituents. Additionally, NMs can be engineered to target particular contaminants present in wastewater through various mechanisms, such as electrostatic interactions, chemical affinity, or specific binding sites. This targeted approach enhances treatment efficiency and minimizes the use of resources by focusing on the removal of priority pollutants. Furthermore, NMs can be functionalized with ligands or receptors that enable recognition and binding to specific contaminants, enabling highly selective removal even at low concentrations. Overall, the selective functionality and targeting of NMs offer a promising strategy for the development of tailored and efficient wastewater treatment solutions (Ahmad et al. 2021).

16.6.3 Reduced Energy and Resource Consumption

NPs contribute to reducing energy consumption in wastewater treatment primarily through their enhanced adsorption and catalytic properties. Unlike conventional treatment methods that often rely on energy-intensive processes, such as aeration, agitation, and high-pressure pumping, NPs can achieve efficient contaminant removal using minimal energy input (Masłoń et al. 2024). Their high surface area-to-volume ratio allows for greater adsorption capacity, enabling effective removal of pollutants from wastewater without the need for extensive mechanical agitation (Ghimire et al. 2021). Additionally, NPs can serve as highly efficient catalysts for various treatment reactions, such as oxidation or reduction processes, which can be driven by natural or low-energy light sources like sunlight. By harnessing these catalytic properties, NPs enable the degradation of contaminants without the need for energy-intensive chemical treatments or high temperatures (Isac and Enesca 2022). Furthermore, the development of NM-based membranes and filtration systems enables energy-efficient separation processes, reducing the energy requirements for wastewater treatment compared to traditional filtration methods. Overall, NPs offer a promising avenue for achieving energy savings in wastewater treatment processes through their unique adsorption, catalytic, and filtration capabilities (Singh et al. 2023).

16.7 Limitations Associated with Nanotechnology

The commercial viability of NPs for wastewater treatment hinges on their impact on the aqueous environment. Various studies, including life cycle analyses, pathway assessments, NP dispersal evaluations, technology assessments, and toxicity tests, are conducted to gauge the health risks associated with NMs. These findings offer insights into the behavior of different NP types in aqueous environments, aiding in informed decision-making regarding their application in wastewater treatment.

16.7.1 Environmental and Health Concerns

A thorough investigation conducted by the US Environmental Protection Agency (EPA) in 2010 examined the effects of titanium dioxide (TiO_2) NPs on various organisms and the environment. The study revealed that nano-TiO_2 had adverse effects on organisms like daphnia, leading to reduced reproduction and behavioral changes, as well as pathological variations in their intestines, gills, and respiratory systems. Algae were also negatively impacted by nano-TiO_2, with the severity of effects dependent on concentration and type (Galletti et al. 2016; Shah et al. 2017). The duration of NP exposure emerged as a crucial factor, with prolonged exposure exacerbating toxicity. Studies on other NPs like silver (AgNPs) and zinc oxide (ZnONPs) highlighted their potential to induce oxidative stress and toxicity in aquatic organisms, including zebrafish embryos and protozoa (Lee et al. 2022; Dianová et al. 2022). Quantum dots of cadmium (CdSe/ZnS and CdTe) also exhibited toxicity to various organisms, while carbon nanotubes (CNTs) affected the activity of Na^+ K^+-ATPase in rainbow trout and caused respiratory dysfunction (Amiri, et al. 2017; Saleemi et al. 2021). Fullerene NPs and C_{60} NPs were implicated in oxidative damage and cell death in daphnia. These findings underscore the need to assess NP environmental impacts and implement measures to mitigate their adverse effects.

16.7.2 NP Stability and Integration

NPs can infiltrate the environment through diverse avenues, spanning from nonpoint origins like clothing, washing machines, and NP-based materials, to point sources, such as wastewater treatment plants and production facilities. For example, substantial amounts of AgNPs might be discharged into the environment during laundry cycles under specific conditions (Bisi-Johnson 2023). The properties of NPs, encompassing their solubility in water, chemical reactivity, and affinity for other components, profoundly dictate their mobility. Upon contact with water, NPs swiftly adsorb and accumulate due to their elevated specific surface area. Within aquatic systems, polymer surface traits can foster NP agglomeration and protein interaction, leading to persistence in sediments within water bodies like rivers and lakes, where they may lay dormant until environmental shifts, such as pH reduction and temperature elevation, instigate dissolution. Despite their limited stability, research endeavors are actively underway to bolster the stability of metal oxide NPs (Tso et al. 2010).

NPs can degrade within aquatic systems via diverse pathways, including photoreaction spurred by sunlight, contingent upon water turbidity and NP chemical attributes. Degradation pathways may also entail abiotic and biotic routes, with specific metallic and organic NMs undergoing transformations under anaerobic conditions (Klekotka et al. 2021). In wastewater treatment, conventional processes like sedimentation, flocculation, and filtration are prevalent in municipal treatment plants. Biological treatment approaches are often harnessed for NP removal, harnessing interactions with microbial communities to amplify sedimentation, integration, and degradation of NPs. Early agglomeration holds pivotal importance in expediting NP settling during the sedimentation process, as larger particles exhibit swifter settling rates. For instance, C_{60} aggregates can be efficiently eliminated through such methods, as delineated by prior studies (Tong et al. 2016).

16.7.3 Regulatory Challenges

The regulatory approval process poses significant barriers to the widespread adoption of NM-based treatment methods due to the lack of comprehensive guidelines tailored to nanotechnology applications in wastewater treatment. Navigating this complex and time-consuming landscape requires thorough risk assessments and stringent safety evaluations, especially concerning potential environmental and health impacts. Addressing these challenges necessitates collaborative efforts among stakeholders, including researchers, policymakers, regulatory agencies, and industry players. Establishing clear regulatory frameworks and standards, along with investing in research initiatives to understand long-term implications, is crucial. By overcoming regulatory hurdles, we can unlock the full potential of NM-based treatment methods to address global water quality challenges (Satyam and Patra 2024).

16.8 Future Prospects

Nanotechnology plays a crucial role in water reutilization and wastewater treatment, offering solutions to traditional challenges in water purification. Unique characteristics of NMs, such as catalytic properties, high surface area-to-volume ratio, and antimicrobial effects, enable their effective use in processes like desalination and detection. By harnessing solar energy, engineered NMs can serve as efficient photocatalysts for water purification. Moreover, nanotechnology addresses issues of water quality degradation in distribution networks, particularly beneficial in disaster-prone areas. With its capacity to utilize alternative water sources while minimizing energy consumption, nanotechnology presents a promising solution to water contamination problems, especially in developing regions. As demand grows for innovative water treatment technologies, nanotechnology-based purification systems are poised to become cost-effective, efficient, and capable of removing a wide array of water pollutants, offering sustainable solutions for the future.

16.9 Conclusions

This chapter highlights the significant implications of untreated wastewater across various sectors, underscoring the global challenge posed by its production and the diverse array of contaminants it carries, including organic, inorganic, and biological substances. The adverse impacts of contaminated wastewater extend to plants, animals, and humans, both directly and indirectly, while also posing threats to soil microbial communities and contributing to soil degradation. Various treatment methods are discussed, with a focus on the utilization of nanotechnology to enhance efficiency. Nanoadsorbents, characterized by their high surface area and reactivity, have revolutionized pollutant removal through adsorption techniques. Nanomembrane filtration, although effective, is noted for its energy-intensive and costly nature. Photocatalysis employing nanocatalysts is highlighted as a promising method for degrading organic contaminants, albeit with limitations that hinder widespread application. The use of antimicrobial NMs, such as silver NPs, is also explored for water decontamination, although concerns regarding their environmental toxicity persist. Despite these challenges, NMs are anticipated to offer practical solutions for treating contaminated wastewater in the near future.

References

Adeola AO, Duarte MP, Naccache R (2023) Microwave-assisted synthesis of carbon-based NMs from biobased resources for water treatment applications: emerging trends and prospects. Front Carbon 2:1220021. https://doi.org/10.3389/frcrb.2023.1220021

Ahmad NA, Goh PS, Zulhairun AK, Wong TW, Ismail AF (2021) The role of functional NMs for wastewater remediation. In: Ismail AF, Goh PS (ed) Functional hybrid NMs for environmental remediation. The Royal Society of Chemistry, pp 1–28. https://doi.org/10.1039/9781839165283-00001

Ahmed SF et al (2022) NMs as a sustainable choice for treating wastewater. Environ Res 214:113807. https://doi.org/10.1016/j.envres.2022.113807

Alam MM, Masud A, Scharf B, Bradley I, Aich N (2022) Long-term exposure and effects of rGO–nZVI nanohybrids and their parent nms on wastewater-nitrifying microbial communities. Environ Sci Technol 56(1):512–524. https://doi.org/10.1021/acs.est.1c02586

Amiri G et al (2017) Comparison of toxicity of CdSe: ZnS quantum dots on male reproductive system in different stages of development in mice. Int J Fertil Steril 9(4). https://doi.org/10.22074/ijfs.2015.4610

Aragaw TA, Bogale FM, Aragaw BA (2021) Iron-based NPs in wastewater treatment: a review on synthesis methods, applications, and removal mechanisms. J Saudi Chem Soc 25(8):101280. https://doi.org/10.1016/j.jscs.2021.101280

Babuji P, Thirumalaisamy S, Duraisamy K, Periyasamy G (2023) Human health risks due to exposure to water pollution: a review. Water 15(14):2532. https://doi.org/10.3390/w15142532

Baig N, Kammakakam I, Falath W (2021) NMs: a review of synthesis methods, properties, recent progress, and challenges. Mater Adv 2(6):1821–1871. https://doi.org/10.1039/D0MA00807A

Bao J, Guo S, Fan D, Cheng J, Zhang Y, Pang X (2023) Sonoactivated NMs: a potent armament for wastewater treatment. Ultrason Sonochem 99:106569. https://doi.org/10.1016/j.ultsonch.2023.106569

Bisi-Johnson M (2023) Monitoring the behaviour of silver NPs as emerging contaminants in urban wastewaters. Cape Peninsula University of Technology, p. 2622080 Bytes. https://doi.org/10.25381/CPUT.22325020.V1

Chhetri T et al (2022) Wastewater treatment using novel magnetic nanosponges. Water 14(3):505. https://doi.org/10.3390/w14030505

Cui X et al (2020) Polychlorinated biphenyls in the drinking water source of the Yangtze River: characteristics and risk assessment. Environ Sci Eur 32(1):29. https://doi.org/10.1186/s12302-020-00309-6

Deepa K, Arthanareeswaran G (2022) Influence of various shapes of alumina NP in integrated polysulfone membrane for separation of lignin from woody biomass and salt rejection. Environ Res 209:112820. https://doi.org/10.1016/j.envres.2022.112820

Dianová L et al (2022) Effects of selected metal NPs (Ag, ZnO, TiO_2) on the structure and function of reproductive organs. Toxics 10(8):459. https://doi.org/10.3390/toxics10080459

Dutta S (2020) Wastewater treatment using TiO_2-based photocatalysts. In: Handbook of smart photocatalytic materials. Elsevier, Amsterdam, pp 303–323. https://doi.org/10.1016/B978-0-12-819051-7.00010-5

Epelle EI, Okoye PU, Roddy S, Gunes B, Okolie JA (2022) Advances in the applications of nms for wastewater treatment. Environments 9(11):141. https://doi.org/10.3390/environments9110141

Farhan Hanafi M, Sapawe N (2020) A review on the water problem associate with organic pollutants derived from phenol, methyl orange, and remazol brilliant blue dyes. Mater Today Proc 31:A141–A150. https://doi.org/10.1016/j.matpr.2021.01.258

Fato TP, Li D-W, Zhao L-J, Qiu K, Long Y-T (2019) Simultaneous removal of multiple heavy metal ions from river water using ultrafine mesoporous magnetite NPs. ACS Omega 4(4):7543–7549. https://doi.org/10.1021/acsomega.9b00731

Galletti A, Seo S, Joo SH, Su C, Blackwelder P (2016) Effects of titanium dioxide NPs derived from consumer products on the marine diatom *Thalassiosira pseudonana*. Environ Sci Pollut Res 23(20):21113–21122. https://doi.org/10.1007/s11356-016-7556-6

Ghimire U, Sarpong G, Gude VG (2021) Transitioning wastewater treatment plants toward circular economy and energy sustainability. ACS Omega 6(18):11794–11803. https://doi.org/10.1021/acsomega.0c05827

Grzegorzek M, Wartalska K, Kaźmierczak B (2023) Review of water treatment methods with a focus on energy consumption. Int Commun Heat Mass Transf 143:106674. https://doi.org/10.1016/j.icheatmasstransfer.2023.106674

Gusain R, Kumar N, Fosso-Kankeu E, Ray SS (2019) Efficient removal of Pb(II) and Cd(II) from industrial mine water by a hierarchical MoS_2 /SH-MWCNT nanocomposite. ACS Omega 4(9):13922–13935. https://doi.org/10.1021/acsomega.9b01603

Hassan M, Abou-Zeid R, Hassan E, Berglund L, Aitomäki Y, Oksman K (2017) Membranes based on cellulose nanofibers and activated carbon for removal of *Escherichia coli* bacteria from water. Polymers 9(12):335. https://doi.org/10.3390/polym9080335

Hassan Al-Taai SH (2021) Water pollution its causes and effects. IOP Conf Ser Earth Environ Sci 790(1):012026. https://doi.org/10.1088/1755-1315/790/1/012026

Ighalo JO et al (2021) CuO NPs (CuO NPs) for water treatment: a review of recent advances. Environ Nanotechnol Monit Manag 15:100443. https://doi.org/10.1016/j.enmm.2021.100443

Imtiaz F, Rashid J, Xu M (2019) Semiconductor nanocomposites for visible light photocatalysis of water pollutants. In: Rahman M, Khan A, Asiri A, Inamuddin I (eds) Concepts of semiconductor photocatalysis. IntechOpen. https://doi.org/10.5772/intechopen.86542

Isac L, Enesca A (2022) Recent Developments in ZnS-based nanostructures photocatalysts for wastewater treatment. Int J Mol Sci 23(24):15668. https://doi.org/10.3390/ijms232415668

Jangid P, Prabhu Inbaraj M (2021) Applications of NMs in wastewater treatment. Mater Today Proc 43:2877–2881. https://doi.org/10.1016/j.matpr.2021.01.126

Johnson E, Krishnan RR, Chandran SR, Prema KH (2023) Green mediated sol-gel synthesis of copper oxide NP: an efficient candidate for waste water treatment and antibacterial agent. J Sol-Gel Sci Technol 107(3):697–710. https://doi.org/10.1007/s10971-023-06172-0

Katubi KMM, Alsaiari NS, Alzahrani FM, Siddeeg SM, Tahoon MA (2021) Synthesis of manganese ferrite/graphene oxide magnetic nanocomposite for pollutants removal from water. Processes 9(4):589. https://doi.org/10.3390/pr9040589

Käufer F, Quade A, Kruth A, Kahlert H (2024) Magnetron sputtering as a versatile tool for precise synthesis of hybrid iron oxide-graphite nm for electrochemical applications. Nms 14(3):252. https://doi.org/10.3390/nano14030252

Kim M, Osone S, Kim T, Higashi H, Seto T (2017) Synthesis of NPs by laser ablation: a review. KONA Powder Part J 34:80–90. https://doi.org/10.14356/kona.2017009

Klekotka U, Zambrzycka-Szelewa E, Satuła D, Kalska-Szostko B (2021) Stability studies of magnetite NPs in environmental solutions. Materials 14(17):5069. https://doi.org/10.3390/ma14175069

Kumar V et al (2023) Removal of inorganic pollutants from wastewater: innovative technologies and toxicity assessment. Sustainability 15(23):16376. https://doi.org/10.3390/su152316376

Kumari S et al (2023) A comprehensive review on various techniques used for synthesizing NPs. J Mater Res Technol 27:1739–1763. https://doi.org/10.1016/j.jmrt.2023.09.291

Kurwadkar S (2019) Occurrence and distribution of organic and inorganic pollutants in groundwater. Water Environ Res 91(10):1001–1008. https://doi.org/10.1002/wer.1166

Lakshmanan R, Okoli C, Boutonnet M, Järås S, Rajarao GK (2014) Microemulsion prepared magnetic NPs for phosphate removal: time efficient studies. J Environ Chem Eng 2(1):185–189. https://doi.org/10.1016/j.jece.2013.12.008

Lee Y-L et al (2022) Toxic effects and mechanisms of silver and zinc oxide NPS on zebrafish embryos in aquatic ecosystems. Nms 12(4):717. https://doi.org/10.3390/nano12040717

Li H, Wang L, Wei Y, Yan W, Feng J (2022) Preparation of templated materials and their application to typical pollutants in wastewater: a review. Front Chem 10:882876. https://doi.org/10.3389/fchem.2022.882876

Li J et al (2022) Application of α-Fe_2O_3-based heterogeneous photo-Fenton catalyst in wastewater treatment: a review of recent advances. J Environ Chem Eng 10(5):108329. https://doi.org/10.1016/j.jece.2022.108329

Li S, Li L, Zhang W (2023) Nanoscale zero-valent iron (nZVI) for heavy metal wastewater treatment: a perspective. Engineering S2095809923004009. https://doi.org/10.1016/j.eng.2023.08.012

Limousy L, Thiebault T, Brendle J (2022) New materials and technologies for wastewater treatment. Materials 15(5):1927. https://doi.org/10.3390/ma15051927

Lin Y et al (2019) Research progress on wastewater treatment methods and technologies in paper industry. IOP Conf Ser Earth Environ Sci 358(2):022042. https://doi.org/10.1088/1755-1315/358/2/022042

Lin L, Yang H, Xu X (2022) Effects of water pollution on human health and disease heterogeneity: a review. Front Environ Sci 10:880246. https://doi.org/10.3389/fenvs.2022.880246

Liu X, Demir NK, Wu Z, Li K (2015) Highly water-stable zirconium metal-organic framework UiO-66 membranes supported on alumina hollow fibers for desalination. J Am Chem Soc 137(22):6999–7002. https://doi.org/10.1021/jacs.5b02276

Liu W et al (2018) Efficient removal of hexavalent chromium from water by an adsorption–reduction mechanism with sandwiched nanocomposites. RSC Adv 8(27):15087–15093. https://doi.org/10.1039/C8RA01805G

Lone MA et al (2023) Rapid hydrothermal synthesis of mesoporous NiS_2 and $NiSe_2$ nanostructures for wastewater treatment. Heliyon 9(12):e22308. https://doi.org/10.1016/j.heliyon.2023.e22308

Makarigakis AK, Jimenez-Cisneros BE (2019) UNESCO's contribution to face global water challenges. Water 11(2), Art. no. 2. https://doi.org/10.3390/w11020388

Malhotra N et al (2020) Potential toxicity of iron oxide magnetic NPs: a review. Molecules 25(14):3159. https://doi.org/10.3390/molecules25143159

Masłoń A, Czarnota J, Szczyrba P, Szaja A, Szulżyk-Cieplak J, Łagód G (2024) Assessment of energy self-sufficiency of wastewater treatment plants—a case study from Poland. Energies 17(5):1164. https://doi.org/10.3390/en17051164

Mishra RK, Mentha SS, Misra Y, Dwivedi N (2023) Emerging pollutants of severe environmental concern in water and wastewater: a comprehensive review on current developments and future research. Water-Energy Nexus 6:74–95. https://doi.org/10.1016/j.wen.2023.08.002

Mmelesi OK, Masunga N, Kuvarega A, Nkambule TT, Mamba BB, Kefeni KK (2021) Cobalt ferrite NPs and nanocomposites: photocatalytic, antimicrobial activity and toxicity in water treatment. Mater Sci Semicond Process 123:105523. https://doi.org/10.1016/j.mssp.2020.105523

Murugan R, Ganesh Ram C (2018) Energy efficient drinking water purification system using TiO_2 solar reactor with traditional methods. Mater Today Proc 5(1):415–421. https://doi.org/10.1016/j.matpr.2017.11.100

Nam NH, Luong NH (2019) NPs: synthesis and applications. In: Materials for biomedical engineering. Elsevier, Amsterdam, pp 211–240. https://doi.org/10.1016/B978-0-08-102814-8.00008-1

Nasrollahzadeh M, Sajjadi M, Iravani S, Varma RS (2021) Carbon-based sustainable NMs for water treatment: state-of-art and future perspectives. Chemosphere 263:128005. https://doi.org/10.1016/j.chemosphere.2020.128005

Nawaz T, Sengupta S (2021) Wastewater: novel treatment technologies and source for epidemiological studies. In: Handbook of water purity and quality. Elsevier, Amsterdam, pp 293–337. https://doi.org/10.1016/B978-0-12-821057-4.00015-X

Ojha A, Tiwary D (2021) Organic pollutants in water and its health risk assessment through consumption. In: Contamination of water. Elsevier, Amsterdam, pp 237–250. https://doi.org/10.1016/B978-0-12-824058-8.00039-6

Pandey PK, Kass PH, Soupir ML, Biswas S, Singh VP (2014) Contamination of water resources by pathogenic bacteria. AMB Express 4(1):51. https://doi.org/10.1186/s13568-014-0051-x

Pourzare K, Farhadi S, Mansourpanah Y (2017) Graphene oxide/Co_3O_4 nanocomposite: synthesis, characterization, and its adsorption capacity for the removal of organic dye pollutants from water. Acta Chim Slov 945–958. https://doi.org/10.17344/acsi.2017.3642

Qasem NAA, Mohammed RH, Lawal DU (2021) Removal of heavy metal ions from wastewater: a comprehensive and critical review. Npj Clean Water 4(1):36. https://doi.org/10.1038/s41545-021-00127-0

Rathod S, Preetam S, Pandey C, Bera SP (2024) Exploring synthesis and applications of green NPs and the role of nanotechnology in wastewater treatment. Biotechnol Rep 41:e00830. https://doi.org/10.1016/j.btre.2024.e00830

Rojviroon T, Rojviroon O, Sirivithayapakorn S, Angthong S (2021) Application of TiO_2 nanotubes as photocatalysts for decolorization of synthetic dye wastewater. Water Resour Ind 26:100163. https://doi.org/10.1016/j.wri.2021.100163

Roy N, Chakraborty S (2021) ZnO as photocatalyst: an approach to waste water treatment. Mater Today Proc 46:6399–6403. https://doi.org/10.1016/j.matpr.2020.06.264

Ryu JH, Messersmith PB, Lee H (2018) Polydopamine surface chemistry: a decade of discovery. ACS Appl Mater Interfaces 10(9):7523–7540. https://doi.org/10.1021/acsami.7b19865

Saleemi MA, Hosseini Fouladi M, Yong PVC, Chinna K, Palanisamy NK, Wong EH (2021) Toxicity of carbon nanotubes: molecular mechanisms, signaling cascades, and remedies in biomedical applications. Chem. Res. Toxicol. 34(1):24–46. https://doi.org/10.1021/acs.chemrestox.0c00172

Salehi M (2022) Global water shortage and potable water safety; today's concern and tomorrow's crisis. Environ Int 158:106936. https://doi.org/10.1016/j.envint.2021.106936

Satyam S, Patra S (2024) Innovations and challenges in adsorption-based wastewater remediation: a comprehensive review. Heliyon 10(9):e29573. https://doi.org/10.1016/j.heliyon.2024.e29573

Shah SNA, Shah Z, Hussain M, Khan M (2017) Hazardous effects of titanium dioxide NPs in ecosystem. Bioinorg Chem Appl 2017:1–12. https://doi.org/10.1155/2017/4101735

Shah A, Arjunan A, Baroutaji A, Zakharova J (2023) A review of physicochemical and biological contaminants in drinking water and their impacts on human health. Water Sci. Eng. 16(4):333–344. https://doi.org/10.1016/j.wse.2023.04.003

Shivalkar S, Gautam PK, Chaudhary S, Samanta SK, Sahoo AK (2021) Recent development of autonomously driven micro/nanobots for efficient treatment of polluted water. J Environ Manage 281:111750. https://doi.org/10.1016/j.jenvman.2020.111750

Shukla S, Khan R, Daverey A (2021) Synthesis and characterization of magnetic NPs, and their applications in wastewater treatment: a review. Environ Technol Innov 24:101924. https://doi.org/10.1016/j.eti.2021.101924

Singh BJ, Chakraborty A, Sehgal R (2023) A systematic review of industrial wastewater management: evaluating challenges and enablers. J Environ Manage 348:119230. https://doi.org/10.1016/j.jenvman.2023.119230

Suryanarayana C, Prabhu B (2007) Synthesis of nanostructured materials by inert-gas condensation methods. In: Nanostructured materials. Elsevier, Amsterdam, pp 47–90. https://doi.org/10.1016/B978-081551534-0.50004-X

Tarrass F, Benjelloun M (2012) The effects of water shortages on health and human development. Perspect Public Health 132(5):240–244. https://doi.org/10.1177/1757913910391040

Toivettula A, Varis O, Vahala R, Juvakoski A (2023) Making waves: mental health impacts of inadequate drinking water services—from sidenote to research focus. Water Res 243:120335. https://doi.org/10.1016/j.watres.2023.120335

Tong Z-H, Bischoff M, Nies LF, Carroll NJ, Applegate B, Turco RF (2016) Influence of fullerene (C60) on soil bacterial communities: aqueous aggregate size and solvent co-introduction effects. Sci Rep 6(1):28069. https://doi.org/10.1038/srep28069

Tso C, Zhung C, Shih Y, Tseng Y-M, Wu S, Doong R (2010) Stability of metal oxide NPs in aqueous solutions. Water Sci Technol 61(1):127–133. https://doi.org/10.2166/wst.2010.787

Vahdat A, Ghasemi B, Yousefpour M (2019) Synthesis of hydroxyapatite and hydroxyapatite/Fe_3O_4 nanocomposite for removal of heavy metals. Environ Nanotechnol Monit Manag 12:100233. https://doi.org/10.1016/j.enmm.2019.100233

Verma B, Gumfekar SP, Sabapathy M (2022) A critical review on micro- and nanomotors: application towards wastewater treatment. Can J Chem Eng 100(3):540–558. https://doi.org/10.1002/cjce.24184

Verma P, Ratan JK (2020) Assessment of the negative effects of various inorganic water pollutants on the biosphere—an overview. In: Inorganic pollutants in water. Elsevier, Amsterdam, pp 73–96. https://doi.org/10.1016/B978-0-12-818965-8.00005-6

Vijayanand M et al (2023) Polyaromatic hydrocarbons (PAHs) in the water environment: a review on toxicity, microbial biodegradation, systematic biological advancements, and environmental fate. Environ Res 227:115716. https://doi.org/10.1016/j.envres.2023.115716

Wang L, Villa K (2021) Self-propelled micro/nanomotors for removal of insoluble water contaminants: microplastics and oil spills. Environ Sci Nano 8(12):3440–3451. https://doi.org/10.1039/D1EN00663K

Wasewar KL, Singh S, Kansal SK (2020) Process intensification of treatment of inorganic water pollutants. In: Inorganic pollutants in water. Elsevier, Amsterdam, pp 245–271. https://doi.org/10.1016/B978-0-12-818965-8.00013-5

Wilkinson JL et al (2022) Pharmaceutical pollution of the world's rivers. Proc Natl Acad Sci 119(8):e2113947119. https://doi.org/10.1073/pnas.2113947119

Yang J et al (2019) NMs for the removal of heavy metals from wastewater. Nms 9(3):424. https://doi.org/10.3390/nano9030424

Zhan S et al (2014) Efficient removal of pathogenic bacteria and viruses by multifunctional amine-modified magnetic NPs. J Hazard Mater 274:115–123. https://doi.org/10.1016/j.jhazmat.2014.03.067

Chapter 17
Current Perspectives of Nanomaterials in Wastewater Treatment for Sustainable Environment

Samim Dullah, Rahul Gogoi, Pranaba Nanda Bhattacharyya, Dhrubajit Borah, Rajani Kumari, Jugabrata Das, and Sudipta Sankar Bora

Abstract Nanotechnology is becoming an indispensable component of the Sustainable Development Goals (SDGs), particularly in the areas of environmental preservation and circular economy. Several anthropogenic activity-induced issues, like pollution, biodiversity loss, climate change, etc., have posed challenges in achieving these goals by 2030. Nanotechnology holds significant implications for global sustainability endeavors as evident from its presence in numerous scientific domains, including engineering, medicinal science, energy, and the environment. Green nanotechnology offers an interesting avenue for the synthesis of safe and cost-effective nano-materials/particles with requisite specification. Green synthesis of nanomaterials (NMs) employs bio-resources such as bacteria, fungi, herbs, and agricultural residues, which lowers the synthesis process' carbon footprint. This chapter discusses the green nanotechnology-enabled solutions for wastewater treatment under the ambit of SDG-goal 6 "clean water and sanitation." We also talk about

S. Dullah
Department of Zoology, Nanda Nath Saikia College, Titabar, Jorhat, Assam, India

R. Gogoi
Department of Agricultural Biotechnology, Assam Agricultural University, Jorhat, Assam, India

P. N. Bhattacharyya
Department of Botany and DBT-IBT-HUB, Nanda Nath Saikia College, Titabar, Assam, India

D. Borah
Department of Chemistry, Nanda Nath Saikia College, Titabar, Jorhat, Assam, India

R. Kumari
Department of Microbiology, Kaziranga University, Jorhat, Assam, India

J. Das
College of Horticulture and Farming System Research, Assam Agricultural University, Nalbari, Assam, India

S. S. Bora (✉)
DBT-North East Centre for Agricultural Biotechnology (DBT-NECAB), Assam Agricultural University, Jorhat, Assam, India
e-mail: sudip.asm@gmail.com

N. Agarwal et al. (eds.), *Nanomaterials in Wastewater Research*, Advances in Wastewater Research, https://doi.org/10.1007/978-981-96-3908-3_17

potential green NMs, their characteristics and mechanisms for wastewater management applications, benefits and drawbacks in comparison with current methods, and the challenges and future scope of research activities.

Keywords Nanotechnology · Green NMs · Wastewater treatment · SDG · Environment

17.1 Introduction

Water is a vital component of the Earth covering 70% of its surface. Of the 70%, 97.5% contribute to saline water, and only 2.5% is available as fresh water. Moreover, out of the available fresh water, i.e., 2.5%, about 68.9% are present in glaciers and as ice and snow. About 30.8% of this fresh water is present in ground and from it only a scarce amount, i.e., 0.3% is accessible for the living organisms to perform their vital activities (Üneş et al. 2017). It cannot be denied that each and every organism present on Earth is dependent on water for carrying out the important metabolic and physiological activities which is important for survival. The presence of water that is safe and clean for consumption and utilization is of utmost importance for the civilized society and also to all the other organisms ranging from plants, animals, and microbes (Acharya et al. 2020).

In the recent years, there has been a rapid surge in the global population, particularly in the cities, which has been estimated to rise by about two-third of the current population by 2050 (Plesse 2050). In one sense, this is good because, more population living in cities means that economic development will increase. But with the greater population, the demand for freshwater among the population also increases, which is a major cause of concern. Another, important concern is that the use of commodities and thus generation of wastes will also increase with the increase in population, such as plastics, chemicals present in products of personal care, and also effluents released from industries. All these pollutants are directly or indirectly discharged into the water bodies, thus creating wastewater (Rathi and Senthil Kumar 2021; Senta et al. 2020).

Wastewater possesses a serious threat to the environment and all the living organisms. Scarcity of safe as well as clean water for the living organisms has become an important goal. The focus is now shifting on treatment of wastewaters sustainably. The United Nations Sustainable Development Goal-6 (SDG-6) emphasizes on the theme-clean water for all (Baskaran 2022). It is important to note that development is required but in doing so care has to be taken so that the environment which is the basic sustenance of life is not harmed and damaged. Due to this reason, the SDGs have been formulated which can help to find ways through which development can be achieved without harming the environment. Prime focus has been given to management of wastewaters as overpopulation along with industrialization and urbanization ultimately has led to discharge of huge amount of pollutants into the water system (Farraji et al. 2016). According to the reports, management of wastewaters can help in achieving a total of 11 SDGs, by reducing pollution as well as use of materials that are hazardous; recycling of materials (Obaideen et al. 2022). Also, emphasis is given to circular economy now-a-days that helps to produce products (which include raw materials) that are available for a prolonged period. This circular economy also contributes to wastewater management by recovering the nutrients and reusing the water (Gherghel et al. 2019).

It has been observed that the traditional methods for managing the wastewater are insufficient because of increasing population and industrialization. Moreover, the traditional methods are very costly and the efficiency of wastewater treated is also considered low in some cases. Furthermore, while disinfecting the wastewater by traditional methods, disinfection by-products (DBPs) are formed which is a great threat to the environment and living organisms. Thus, it has become essential to develop new technologies that can ensure the availability of clean and safe water by treating the wastewater. One such novel technology developed for wastewater treatment is nanotechnology. The sustainable treatment of wastewaters by using NMs is currently becoming a prime topic of research and development. Use of NMs for wastewater treatment comes under green technology. The NMs find their use in removing various toxic, unwanted materials from wastewaters (e.g., heavy metals, inorganic as well as organic solvents) and also biological pathogens that are responsible in causing diseases like typhoid and cholera (Gupta et al. 2022; Saravanan et al. 2022). Use of NMs for wastewater management is considered as a great choice as the adsorption capacity is high in NMs. Moreover, NMs have high reactivity as well as sensitivity (Nasrollahzadeh et al. 2021). Wastewater treatment by NMs can be done through photocatalysis, biosorption, nanofiltration, control of pathogens, and many more This book chapter discusses the role of NMs for wastewater treatment, such as photocatalysis, heavy metal recovery, antimicrobial NMs, nanofilters, disinfection, and pathological control. It also discusses the use of waste products as NM production and application. It highlights the green synthesis of NMs and NMs from chitosan that can be used in removing dyes for wastewater treatment. Moreover, the safety, toxicity, and environmental impact of NMs have been outlined in this chapter.

17.2 Role of NMs for Wastewater Treatment

17.2.1 NMs as Photocatalysts

NMs have been extensively evaluated for their potential to act as photocatalysts for the treatment of wastewater bodies. NMs offer several advantages, including their high surface area, tunable properties, and the ability to generate reactive oxygen species (ROS) under light irradiation that can substantially degrade organic pollutants including the heavy metals present in wastewater bodies (Bhat et al. 2022; Yaqoob et al. 2020). Instances can be cited for the usages of Pd-doped ZnO NMs that have demonstrated significant photocatalytic reactivity and potential degradation kinetics against toxic contaminants in wastewater treatment (Yaqoob et al. 2020). It may be mentioned that green synthesis methods, such as the use of plant extracts, have also been recently explored for the production of efficient NMs like CuO for rapid wastewater treatment. The nanoparticles (NPs) that have been produced using green and cleaner technologies have shown excellent photocatalytic activity and promises for sustainable water treatment (Das et al. 2018). Recent findings have illustrated for the pivotal role played by NMs in wastewater treatment, with multidimensional applications not only in addressing environmental contamination but also in promoting a healthier and more sustainable environment (Hashmi et al. 2021).

17.2.2 Antimicrobial NPs

Antimicrobial NPs are tiny particles that have tremendous opportunity to use as antagonistic molecules in wastewater bodies to combat against multiple infections caused by bacteria, fungi, and viruses. Multiple and periodic treatment with efficacious NPs mitigate the chances of rapid infections in those materials which are under three "R"s (Reduce, Reuse and Recycle), for the well-being of mankind. Likewise, these molecules have also poses significant interest in agriculture as they have been suggested for their use in plant protection. In a more similar vein, a recent study suggested the application potential of inorganic NMs in plant pathogen management, thus strengthening the concept of green and sustainable alternatives in climate smart agriculture (Francesconi et al. 2022). Many important antimicrobial NPs have been extensively used for wastewater management in soil and environment. Silver NPs (AgNPs) are particularly effective at removing microbes from wastewater bodies due to their immense antimicrobial properties. They have their abilities to degrade organic matter and can be reused, making them a cost-effective and sustainable alternative in wastewater treatment (Palani et al. 2023). Metal Oxide NPs, due to their large surface area and affordable cost, are highly efficient for treating contaminated water. They can remove pollutants like pesticides and heavy metals that rapidly get exposed to surrounding water bodies (Naseem and Durrani 2021). Nanophotocatalysts are often used for wastewater purification, as they have their abilities to enhance the reactivity of catalysts due to their increased surface area (Yaqoob et al. 2020). According to a report, the nanofilters, made from nanotubes, nanoribbons, and nanofibers, are highly effective at removing dyes, heavy metals, and other contaminants from wastewater bodies (Keyvanfard et al. 2013). Antimicrobial nanocomposites, which combine NPs with other materials, are highly effective for wastewater treatment and have powerful antimicrobial properties (Naseem and Waseem 2022).

17.2.3 NMs for Heavy Metal Recovery

Every year, a variety of geological and human activities introduce different compounds known to be harmful to the wastewater bodies. According to a recent study, the ecotoxicological impacts of these chemicals and the methods employed to detoxify them have raised global concerns (Mukherjee et al. 2022). Among these, heavy metals including chromium (Cr), cadmium (Cd), mercury (Hg), lead (Pb), arsenic (As), cupric (Cu), nickel (Ni), selenium (Se), and others are prevalent and tend to accumulate in the soil and wastewater bodies and exhibit resistance to degradation (Sukthavorn et al. 2024). Anthropogenic sources of heavy metal contamination encompass activities like mining, wastewater treatment, industrial processes, and agricultural practices, all of which contribute significantly to elevated levels of heavy metals in the soil. The burning of fossil fuels releases mercury into the air, while lead is emitted through vehicle exhaust. Additionally, the use of fertilizers and insecticides in agricultural areas may lead to an increase in the concentrations of certain heavy metals in the soil, such as Cd and Zn (Masindi et al. 2018; Selvi et al. 2019). NMs like carbon nanotubes and zeolites possess a massive surface area-to-volume ratio, which increases their adsorption capacity for heavy metals. This superior adsorption ability allows for more efficient removal of toxic ions and pollutants from wastewater (Baby et al. 2022; Yang et al. 2019; Ethaib et al. 2022). Additionally, certain NMs have unique functional groups and high negative charge density, which attract and bind heavy metals effectively. Graphene-based NMs, for example, have shown great promises to remove heavy metals due to their large specific surface area, superior mechanical strength, and reactivity toward polar and nonpolar contaminants and other extraordinary properties (Yang et al. 2019). It may be mentioned that the microbots' structure consists of nanosized multilayers of graphene oxide, nickel, and platinum, each of which serves multiple functions leading to environmental detoxification. NMs can be functionalized or modified to enhance their selectivity and capacity to detoxify specific heavy metals. This adaptability allows them to be tailored for the treatment of wastewater contaminated with different types of heavy metals (Parvin et al. 2019). It may be mentioned that the process of utilizing NMs in heavy metal removal is an energy-efficient process as compared to traditional methods like chemical precipitation or ion exchange. A recent study has suggested NM-based treatments as more sustainable and cost-effective solution for heavy metal detoxification from wastewater bodies (Kolluru et al. 2021). Likewise, another study addressed the application potential of NMs in potential decontamination of heavy metal ions and dyes from rapid contaminations to wastewater bodies, highlighting their substantial potential in remediating heavy metal-contaminated water and soil (Sadegh et al. 2017).

17.2.4 Reuse and Retention of NMs

The reuse and retention of NMs in wastewater treatment plays crucial role in advancing the concept of sustainable and effective water purification technologies. NMs, due to their unique physicochemical properties, have demonstrated remarkable potential for the efficient removal of various contaminants from wastewater (Manikandan et al. 2022). However, the sustainable application of NMs in wastewater treatment requires careful strategies for their reuse and retention within the treatment process. For instance, Manikandan et al. has emphasized the importance of optimizing the design and operation of nanotechnology-based water treatment systems to enhance the retention of NMs, thereby improving the overall process of efficiency and sustainability (Manikandan et al. 2022). According to Jain et al., studies pertaining to the development of advanced membrane-based systems that can enable the selective recovery and reuse of NMs in subsequent wastewater treatment stages or even in other applications for enhancing sustainability are seemed to be inevitable (Jain et al. 2021). However, the retention of NMs in wastewater treatment processes is influenced by various factors, including the specific properties of the NMs, the characteristics of the wastewater, and the conditions of the treatment processes and further optimization. NMs with larger sizes or higher charges are generally more likely to be retained within the toxicity treatment system. Moreover, the optimization of process parameters such as pH, temperature, and flow rates can significantly impact the retention efficiency of NMs (Jain et al. 2021). Additionally, the reuse of NMs eventually offers several advantages to toxicity decontamination. Firstly, it promotes the sustainable use of these valuable resources by preventing the unnecessary consumption of new NMs. Secondly, the reuse of effective NMs can enhance the performance of wastewater treatment processes by maintaining or even improving the removal efficiency for contaminants. Lastly, the recovery and reuse of NMs contribute to the cost-effectiveness and sustainability of wastewater treatment systems (Manikandan et al. 2022). Despite the clear benefits of reusing and retaining NMs in wastewater treatment system, several challenges need to be addressed. These include developing robust and cost-effective methods for the recovery and reuse of NMs, minimizing the potential environmental and health impacts associated with the release of NMs back into the environment, thereby ensuring the long-term stability and performance of NMs within wastewater treatment systems.

17.2.5 Nanofilters

Nanofilters play a crucial role in wastewater treatment system for upgrading a sustainable environment. They are used to selectively remove ions and organic substances, making them essential in various, water and wastewater treatment processes (Abdel-Fatah 2018). In textile wastewater treatment, for instances, nanofilters are particularly effective at removing harmful contaminants (Abdel-Fatah 2018). Several studies have recently been explored to denote the performance of nano-filters in wastewater treatment. One of these studies investigated the potential of a commercial Desal 5 DK nanofiltration membrane for use as a tertiary treatment in wastewater treatment plants (Oliveira et al. 2022). The results demonstrated the membrane's effectiveness in removing ecologically toxic elements, highlighting the potential of nanofilters in addressing emerging water pollution challenges (Oliveira et al. 2022). A recent study presented a comprehensive review of the theoretical performance of nanofiltration membranes for wastewater treatment (Abdel-Fatah 2018). The research emphasized the importance of optimizing membrane design and operation conditions to enhance the retention of contaminants and improve the overall efficiency of nanofiltration-based systems. The researchers also highlighted the potential application of nanofilters in wastewater treatment, including the treatment of industrial effluents to remove dissolved impurities and color (Abdel-Fatah 2018). Overall, the reuse and retention of nanofilters in wastewater treatment are crucial for achieving sustainable environmental outcomes. By selectively removing toxic contaminants and enabling the recovery of valuable resources, nanofilters can contribute to the creation of a cleaner, eco-friendly and healthier environment.

17.2.6 Disinfection and Pathological Control

Disinfection and pathological control in wastewater treatment is crucial for sustainable development. Utilizing nanocomposites in effective wastewater treatment systems that incorporate disinfection technologies can help to prevent the spread of diseases and protect public health. Keeping possible track in minimizing the trends in human infection, a study in China during the COVID-19 pandemic highlighted the importance of proper disinfection strategies for pathological and pharmaceutical wastes (Wang et al. 2020). In addition to traditional methods, innovative techniques like vermiculture technology can be employed. This involves using earthworms to disinfect, detoxify, and enhance the nutrient content of wastewater. Such sustainable practices can handle loads that conventional sewage treatment plants (STPs) often overlook (Rajiv et al. 2010). Technological interventions, such as the delivery of disinfecting agents through oxidation processes, are also vital for sustainable wastewater treatment. A recent study emphasized that targeted delivery of these agents ensures effective disinfection while minimizing environmental impact (Singh et al. 2023). The management of medical waste, including the potential for oxidation technology to disinfect COVID-19-contaminated sewage (Olabi et al. 2024), is another critical aspect of sustainable wastewater treatment. Proper waste management aligns with the United Nations' Sustainable Development Goals (SDGs) and contributes to overall public health and environmental sustainability. Overall, effective disinfection and pathological control in wastewater treatment require a combination of proven technologies, innovative practices, and careful management. These strategies not only protect public health and the environment but also support long-term sustainable development.

17.3 Waste Products in NM Production and Application

17.3.1 Waste from Various Substances that Contains Carbon, Copper, Zinc, Helping in NM Production

Using waste materials for building value-added products has become a prevailing practice across industries. This approach provides an effective means of managing waste generated from diverse sources and an opportunity to create novel materials with significant economic value. Over the past two decades, there have been notable explorations of using important materials present in industrial waste for preparing NPs. For example, the waste of rubber/tire, which contains carbon and trace metals, serves as a starting material for producing carbon-based NPs. These NPs find applications in catalysis, sensing, and drug delivery, showcasing the potential of waste-derived materials in various industries. Similarly, waste batteries are the potential source of copper and zinc, essential elements for synthesizing metal NPs. These NPs have found applications in electronics, energy storage, and antimicrobial coatings. Moreover, plastic wastes with high carbon contents have been leveraged as potential precursors for synthesizing carbon NMs. To produce carbon NMs (CNMs), large solid polymers are broken down into small carbon materials via pyrolysis using various reactors such as quartz tubes, autoclaves, crucibles, and muffle furnaces. Notably, this process yields diverse CNMs including carbon nanotubes, carbon nanofibers, and carbon spheres sourced from various waste plastic materials containing polyolefins.

Notably, Wu and colleagues reported the synthesis of high-quality carbon nanotubes from waste polypropylene, while Kong et al. devised a solvent-free synthesis method for developing straight and helical carbon nanotubes using polyethylene-based waste plastics. Remarkably, they also successfully synthesized core–shell magnetite Fe3O4@C composites from waste polyethylene feedstock, finding applications in drug delivery, recycling of nanocatalysts, and magnetically controllable on–off reactions, highlighting the significance of waste-derived materials in various aspects of human life (Zhang et al. 2013). Additionally, the research groups lead by Zou, and Pol developed CNTs and CSs from recycled waste high-density polyethylene (HDPE) and low-density polyethylene (LDPE) and these NMs showed superior performance in lithium electrochemical cells (Pol and Thiyagarajan 2010; Zou et al. 2004).

Beyond carbon nanotubes, graphene, a versatile material with diverse applications in energy storage, coating industry, catalysis, and biomedicine, can also be derived from waste plastic materials. For example, Gong and co-workers developed a method to synthesize graphene flakes from waste polypropylene using organically modified montmorillonite (OMMT) catalysts. The typical procedure involves heating a mixture of polypropylene (89 wt%), talcum (~ 11 wt %), and OMMT in a crucible at 700 °C for 15 min, resulting in the formation of carbonized char. Subsequently, the resulting carbon char was immersed in HF and HNO_3 acid. Hydrofluoric acid helps to remove impurities while nitric acid oxidizes amorphous carbon. Finally, graphene was isolated from the solution through multiple centrifugation steps. Furthermore, waste polystyrene and plastic with 86% PE and 14% PS were also used as carbon sources for producing high-quality grapheme (Gong et al. 2014).

This ongoing exploration of waste-derived materials underscores their potential to revolutionize various industries while simultaneously addressing environmental concerns associated with waste disposal.

17.3.2 Green Synthesis of Metal NPs and Their Applications

The synthesis of metal NPs holds paramount importance due to their diverse applications in optoelectronics, biotechnology, biomedicine, and other allied disciplines. Consequently, various physical and chemical methods have been developed over the last few decades to synthesize metal NPs. However, many of these methods rely on toxic chemicals that do not align with the principles of green chemistry. Hence, recent efforts have been directed to develop green, non-toxic, and clean methods for producing metal NPs.

In this context, plant-mediated synthesis of metal NPs has emerged as an effective as well as sustainable approach. It has been observed that biomolecules found in plants such as flavones, terpenoids, ketones, alkaloids, aldehydes, vitamins, proteins, amino acids, tannins, phenolics, saponins, and polysaccharides play an imperative role in reducing metal ions, thus eliminating the need of toxic chemical reducing agents. Over the past few years, several metal NPs have been produced by using plant residues. For example, the research group led by Agnihotri developed AgNPs by using pineapple waste and investigated their efficacy in degrading methylene blue dye. This work revealed that pineapple waste–derived AgNPs can remove the dye by up to 98.04% (Agnihotri et al. 2018). Similarly, Angana and co-workers developed AgNPs by utilizing neem leaf and banana peel, showing that this NP can be used for dye degradation and antimicrobial activity (Sengupta and Sarkar 2022). Moreover, another group of workers reported the synthesis of magnetic NPs derived from polysaccharide, gum acacia and demonstrated their efficacy in removing magnetic materials from wastewater and degrading cationic dyes (Kulal et al. 2022). Additionally, a research group developed ZnO NPs using peel of *Passiflora foetida* fruit which proved to be effective in removing methylene blue and Rhodamine B from wastewater (Khan et al. 2021). Beyond dye degradation, these NPs can combat antibiotic-resistant bacteria, remove heavy metals, and treat organic and inorganic pollutants from wastewater treatment facilities (Mbarek et al. 2022).

17.3.3 NMs from Chitosan Used for Removal of Dyes

Chitosan, a bio-remediable polysaccharide first synthesized by Professor C. Rouget in 1859 through the deacetylation of chitin, plays a vital role in various industries.

With its polar amine and hydroxyl functional groups, chitosan serves as an effective adsorbent for polar and ionic impurities, making it indispensable in industrial wastewater treatment facilities. Moreover, its biodegradability and biocompatibility render it suitable for applications in food, agriculture, pharmacy, cosmetics, and more recently, in biomedical and bioremediation research (Van der Lubben et al. 2001).

NMs derived from chitosan have attracted significant interest for their potential in removing dyes from textile wastewater. For instance, a NP composite was synthesized comprising chitosan, Al_2O_3, and magnetic iron oxide, demonstrating its efficacy in removing methyl orange dye from wastewater. Additionally, the researchers reported the synthesis of magnetic chitosan and evaluated their efficiency as an adsorbent for dye removal (Tanhaei et al. 2015).

Further research delved into glutaraldehyde crosslinked magnetic chitosan NMs, prepared via a reduction precipitation process, achieving a remarkable absorption capacity of 90.06 mg/g for Acid Red 2 dye. Notably, the synthetic protocol adheres to green chemistry principles, eschewing toxic chemicals and extreme heat. The adsorption mechanism hinges on electrostatic interactions between the negatively charged groups of Acid Red 2 dye and the positively charged surface of the magnetic chitosan nanocomposite. This positive surface charge is attributed to the protonation of chitosan's amine group, highlighting the intricate interplay of molecular forces driving the adsorption process (Bhatt et al. 2023).

Overall, chitosan-based NMs offer a promising avenue for addressing environmental challenges such as dye contamination in wastewater, with their efficient adsorption capabilities and environmentally friendly synthesis methods.

The use of pesticides has become a widespread practice in modern crop protection, offering effective means to eliminate pests, weeds, and harmful substances from soil, thereby enhancing crop productivity. However, alongside their benefits, they also provide serious environmental challenges that deserve attention. Many commercial pesticides used in agriculture exhibit toxic properties and they can disrupt the aquatic ecosystem when they leach into the water body. To mitigate the environmental issue, it is imperative to develop effective adsorbents to remove harmful pesticides from the environment. In this regard, chitosan NPs have attracted significant interest as recent research showcased that they can be used for pesticide removal. For example, Dehaghi et al. demonstrated that chitosan combined with ZnO NPs can be utilized for removing pests (Dehaghi et al. 2014).

17.4 Safety, Toxicity, and Environmental Impact of NMs

In several fields today, including materials science, cosmetics, medicines, ecology, and more, NMs are being used more and more. However, using NMs has some hazardous side effects on ecology and the environment as well. In lakes, rivers, and oceans, a number of NMs are inevitably dumped. Eventually, these NMs may also have an impact on potable water. This means that aquatic life and human life may be poisoned by the pollution in aquatic habitats caused by these NMs (Keller et al.

2023). It has been observed that certain species are adversely affected by a variety of NM, whether they are distinct sorts of materials or even comparable types of materials at different sizes. Aqueous habitats are home to an enormous variety of microorganisms. Therefore, as NMs enter aqueous settings, it is common for them to interact with microbes (Jeevanandam et al. 2018).

The early 2000s saw the emergence of concerns over NM toxicity. The primary goals of the toxicological studies conducted in the first ten years were to: (1) determine which NMs were toxic to model and unique native organisms; (2) collect data on toxicity parameters, including lethal concentration (LC_{50}), median effect concentration (EC_{50}), lowest observed effect concentration (LOEC), and no observable effect concentration (NOEC); (3) investigate the impact of surface properties, texture, and size on NM toxicity; and (4) assess the toxicity of NMs at the organismal and cellular levels. The following topics are further explored in the second decade: (1) estimating and measuring the concentrations of NMs in various environmental compartments; (2) assessing the impact of environmental factors on NM toxicity; and (3) conducting mechanistic studies on the toxicity of NMs at the molecular level, particularly unique NMs.

Numerous studies on the toxicological consequences of NMs have already been published to date. Nevertheless, knowledge of the hazards that NPs pose to the environment still lags behind the introduction of new goods with nanotechnology and the creation of new applications (Gambardella and Pinsino 2022). Regarding NMs' ecotoxicity, opinions are still divided, and there are still a lot of unanswered questions regarding the safe use of NMs. To address this issue, two factors are necessary: (1) the concentrations of NMs in the actual environment and (2) the toxicity of NMs in practical situations. However, there are still a lot of differing viewpoints and opinions regarding, among other things, the validity of environmental concentration measurements and predictions in various environmental compartments, the precision of toxicological investigation techniques, and the dependability of NM toxicity in environments.

The last 20 years have seen a rise in interest in the phrase "nanotoxicology." Numerous developments have been achieved in this field since then. Initially, there is a remarkable advancement in the creation of novel NM types with unsettling physical and chemical properties as well as the large-scale manufacturing of diversified NMs. Secondly, several investigations grounded in the ever-improving nature of nanomedicines have prompted inquiries into Physics, Chemistry, with Bioengineering, resulting in fresh transdisciplinary advancements in the field of nanoscience and its attendant applications. The bioapplication of NMs, for instance, has advanced significantly (Malik et al. 2023).

At the intersection of toxicology and NMs, nanotoxicology has emerged as a topic. NMs differ from their bigger versions in that they are much smaller and have a higher surface area-to-volume ratio, which may allow for unanticipated interactions with tissues and cells (Lin et al. 2021). The field of nanotoxicology frequently draws attention to the potentially harmful interactions that NMs may have with many biological systems, including cells, tissues, and living things. Numerous studies throughout the years have demonstrated how intricately NMs interact with both the environment and living things' cells. How NMs' morphological and physicochemical characteristics may affect these interactions is still unknown (Monteiro-Riviere 2022).

NMs interact with biological cells in a very different way depending on their morphological and physicochemical characteristics, which can also affect their toxicity. Since the toxicity of NMs is largely dependent on the size of the materials, nanotoxicology is responsible for analyzing the toxic consequences of NMs. The notion of nanotoxicology is predicated on various factors, including dimensions, surface area, shape, composition, surface chemistry, and processes related to aggregation and agglomeration (Behzadi et al. 2017). Indeed, each of these variables plays a crucial role in determining the dose of the NPs and, in turn, in accurately determining their level of toxicity. In vitro and in vivo testing, however, are necessary to determine the exposure limit values of dangerous NMs (Ferdous and Nemmar 2020). Numerous methods are currently in use to investigate nanotoxicology and how NMs interact with biological systems. Early research on the toxicity of NMs was conducted in the final ten years of the twentieth century, and it was found that while materials at the nanometric scale might be somewhat poisonous, materials at the micrometric scale did not offer any toxicity (Kang et al. 2023).

Silver nanoparticles (NPs) have been found to impact environmental flora and fauna, including nematodes, algae, and fungi. Benthic species, which play a crucial role in the structure and function of marine ecosystems, are particularly at risk. When silver NPs are released into aquatic systems, benthic species face some significant threats. Studies using aquatic mesocosms have shown that silver NPs cause DNA damage and oxidative stress in Scrobicularia plana. Additionally, they induce hepatopancreas pathology and trigger early apoptosis in Mytilus galloprovincialis (Mikhailova 2020; Giri et al. 2022).

Liu et al. conducted a research utilizing a freshwater bivalve that was exposed to silver NPs coated with polyvinyl pyrrolidone (Liu et al. 2018). *Corbicula fluminea* affected the fate and modification of silver NPs in the aforementioned study, as well as their ability to move from water to sediment. Since then, a massive amount of silver collected in the sand, endangering benthic life. The test results verified that *C. fluminea* suffered from oxidative damage and physiological metabolic interference due to silver bioaccumulation. At different concentrations of silver NPs, *C. fluminea* were observed to exhibit different detoxifying pathways. Although the tissues are becoming more sensitive to low concentrations, shells are a better option for indicators of larger concentrations of silver NPs, as evidenced by the bioaccumulation of these particles in bodies and shells. Due to their longer gut retention period than other chemicals, silver NPs were not observed in the feces of *C. fluminea*. Furthermore, calcium loss and calcospherite breakdown may have been driven by larger quantities of silver NPs. Further information about the relationship among *C. fluminea* along with silver NPs is provided by the aforementioned paper, which is helpful in understanding the toxicity and eventual fate of silver NPs in natural aquatic systems (Liu et al. 2018).

According to Thakkar et al., who examined the impact of single-walled carbon nanotubes (SWCNTs) on *Dunaliella tertiolecta*, increasing the level of concentration of SWCNTs increased the harmful impacts on growth, photosynthesis, and oxidative stress (Thakkar et al. 2016). Moreover, important determinants of the toxicity associated with these NMs include the diameter, length, and structure of carbon nanotubes. This NM's toxicity will rise in tandem with the length of the carbon nanotubes. To give an example, it was demonstrated that the macrophages will take up short-length CNTs while rejecting longer-length CNTs. According to a study functionalized MWCNT in aquatic environments can reduce environmental lead toxicity (Jang and Hwang 2018). However, another study shown that carbon nanotubes may make copper more poisonous. It is impossible to accurately assess the harm that carbon nanotubes cause to the environment because they cannot survive in the real world alone and only a limited amount of these NMs can bind to other ecological pollutants (Kim et al. 2009).

Silver NP toxicity varies with water chemistry and sediment properties, most likely because of the way the particles interact with naturally occurring water colloids after they are released. The previously mentioned may have an impact on the stability and ensuing environmental behavior of silver NPs and aquatic colloids. Silver NPs may have an impact on trace metals, which are associated with colloids and particles that are important in the transportation of heavy metals.

According to Tao et al., in addition to the current state of metal pollution, silver NPs may act as a coagulant, depositing ions and colloids made of Cd, Pb, Zn, and Cu from the sediment's pore water and lowering the concentration of heavy metals (Tao et al. 2016). On the other hand, a portion of the silver NPs might stabilize in pore water and create a silver–mercury amalgam, which could indicate a substantial form of mercury transformation in the sediment. Research on mercury–silver amalgam formation has shown that it can mobilize a significant amount of colloidal mercury in surface water, which can then be carried across streams to places that are not affected. Consequently, after being released into surface water, the mercury–silver amalgam NPs may significantly improve mercury transportation over longer distances (Tao et al. 2016).

17.5 Conclusion and Future Recommendations

Due to the scarcity of clean water, it has become utmost important to convert wastewater into usable water to meet the demand of clean and safe water among the people. Researchers and stakeholders are continuously trying to develop new technologies to treat the wastewater in a sustainable way. For this development of nanotechnology which is a form of green technology and production of NMs is currently becoming a major area of research. As clean and safe water is very important both socially and economically, methods to convert wastewater into usable form can help immensely in the development of the society and also in the maintenance of the environment. Management of wastewater by developing NMs has immense potential due to their unique properties that includes increased reactivity, sensitivity, adsorption capacity, and surface-to-volume ratios. NMs have the capacity to solve the problems that arise while using the traditional methods of wastewater treatment. But an important concern to keep in mind is the adverse effect it might cause if not used in a planned and judicious manner as there are reports by some researchers who have focused on issues related to the use of NMs. The researchers are continuously studying in detail the different ways of developing NMs for wastewater treatment. But importance should be given to the use of NMs with other technologies that can be considered suitable for the treatment of wastewater.

References

Abdel-Fatah MA (2018) Nanofiltration systems and applications in wastewater treatment. Ain Shams Eng J 9(4):3077–3092
Acharya K, Blackburn A, Mohammed J, Haile AT, Hiruy AM, Werner D (2020) Metagenomic water quality monitoring with a portable laboratory. Water Res 184:116112
Agnihotri S, Sillu D, Sharma G, Arya RK (2018) Photocatalytic and antibacterial potential of silver NPs derived from pineapple waste: process optimization and modeling kinetics for dye removal. Appl Nanosci 8(8):2077–2092
Baby R, Hussein MA, Abdullah AH, Zainal Z (2022) NMs for the treatment of heavy metal contaminated water. Polymers 14(3):583
Baskaran V (2022) A systematic review on the role of geographical information systems in monitoring and achieving sustainable development goal 6: Clean water and sanitation. Sustain Dev 30(5):1417–1425
Behzadi S, Serpooshan V, Tao W, Hamaly MA, Alkawareek MY, Dreaden EC, Brown D, Alkilany AM, Farokhzad OC, Mahmoudi M (2017) Cellular uptake of NPs: journey inside the cell. Chem Soc Rev 46(14):4218–4244
Bhat SA, Sher F, Hameed M, Bashir O, Kumar R, Vo D-V, Ahmad P, Lima EC (2022) Sustainable nanotechnology based wastewater treatment strategies: achievements, challenges and future perspectives. Chemosphere 288:132606
Bhatt P, Joshi S, Bayram GMU, Khati P, Simsek H (2023) Developments and application of chitosan-based adsorbents for wastewater treatments. Environ Res 226:115530
Bondarenko O, Mortimer M, Kahru A, Feliu N, Javed I, Kakinen A, Lin S et al (2021) Nanotoxicology and nanomedicine: the Yin and Yang of nano-bio interactions for the new decade. Nano Today 39:101184
Das P, Ghosh S, Ghosh R, Dam S, Baskey M (2018) Madhuca longifolia plant mediated green synthesis of cupric oxide NPs: a promising environmentally sustainable material for waste water treatment and efficient antibacterial agent. J Photochem Photobiol, B 189:66–73
Dehaghi SM, Rahmanifar B, Moradi AM, Azar PA (2014) Removal of permethrin pesticide from water by chitosan–zinc oxide NPs composite as an adsorbent. J Saudi Chem Soc 18(4):348–355
Ethaib S, Al-Qutaifia S, Al-Ansari N, Zubaidi SL (2022) Function of NMs in removing heavy metals for water and wastewater remediation: a review. Environments 9(10):123
Farraji H, Zaman NQ, Vakili M, Faraji H (2016) Overpopulation and sustainable waste management. Int J Sustain Econ Manage (IJSEM) 5(3):13–36
Ferdous Z, Nemmar A (2020) Health impact of silver NPs: a review of the biodistribution and toxicity following various routes of exposure. Int J Mol Sci 21(7):2375
Francesconi S, Schiavi D, di Lorenzo V, Balestra GM (2022) Inorganic NMs usable in plant protection strategies. In: Nanotechnology-based sustainable alternatives for the management of plant diseases. Elsevier, Amsterdam, pp 211–231
Gambardella C, Pinsino A (2022) NM ecotoxicology in the terrestrial and aquatic environment: a systematic review. Toxics 10(7):393

Gherghel A, Teodosiu C, De Gisi S (2019) A review on wastewater sludge valorisation and its challenges in the context of circular economy. J Cleaner Prod 228:244–263

Giri AK, Jena B, Biswal B, Pradhan AK, Arakha M, Acharya S, Acharya L (2022) Green synthesis and characterization of silver NPs using Eugenia roxburghii DC. extract and activity against biofilm-producing bacteria. Sci Rep 12(1):8383

Gong J, Liu J, Jiang Z, Feng J, Chen X, Wang L, Mijowska E, Wen X, Tang T (2014) Striking influence of chain structure of polyethylene on the formation of cup-stacked carbon nanotubes/ carbon nanofibers under the combined catalysis of CuBr and NiO. Appl. Catal. B: Environ. 147:592–601

Gupta SK, Singh B, Mungray AK, Bharti R, Nema AK, Pant KK, Mulla SI (2022) Bioelectrochemical technologies for removal of xenobiotics from wastewater. Sustain Energy Technol Assess 49:101652

Hashmi SS, Shah M, Muhammad W, Ahmad A, Ullah MA, Nadeem M, Abbasi BH (2021) Potentials of phyto-fabricated NPs as ecofriendly agents for photocatalytic degradation of toxic dyes and waste water treatment, risk assessment and probable mechanism. J Ind Chem Soc 98(4):100019.

Jain K, Patel AS, Pardhi VP, Flora SJS (2021) Nanotechnology in wastewater management: a new paradigm towards wastewater treatment. Molecules 26(6):1797

Jang M-H, Hwang YS (2018) Effects of functionalized multi-walled carbon nanotubes on toxicity and bioaccumulation of lead in Daphnia magna. PLoS One 13(3):e0194935

Jeevanandam J, Barhoum A, Chan YS, Dufresne A, Danquah MK (2018) Review on NPs and nanostructured materials: history, sources, toxicity and regulations. Beilstein J Nanotechnol 9(1):1050–1074

Kang M, Weng Y, Liu Y, Wang H, Ye L, Gu Y, Bai X (2023) A review on the toxicity mechanisms and potential risks of engineered NPs to plants. Rev Environ Contamin Toxicol 261(1):5

Keller AA, Ehrens A, Zheng Y, Nowack B (2023) Developing trends in NMs and their environmental implications. Nat Nanotechnol 18(8):834–837

Keyvanfard M, Khosravi V, Karimi-Maleh H, Alizad K, Rezaei B (2013) Voltammetric determination of 6-mercaptopurine using a multiwall carbon nanotubes paste electrode in the presence of isoprenaline as a mediator. J Mol Liq 177:182–189

Khan M, Ware P, Shimpi N (2021) Synthesis of ZnO NPs using peels of *Passiflora foetida* and study of its activity as an efficient catalyst for the degradation of hazardous organic dye. SN Appl Sci 3:1–17

Kim K-T, Edgington AJ, Klaine SJ, Cho J-W, Kim SD (2009) Influence of multiwalled carbon nanotubes dispersed in natural organic matter on speciation and bioavailability of copper. Environ Sci Technol 43(23):8979–8984

Kolluru SS, Agarwal S, Sireesha S, Sreedhar I, Kale SR (2021) Heavy metal removal from wastewater using NMs-process and engineering aspects. Process Safety Environ Protect 150:323–355

Kulal P, Krishnappa PB, Badalamoole V (2022) Development of gum acacia based magnetic nanocomposite adsorbent for wastewater treatment. Polym Bull 79(11):9457–9484

Liu W, Zeng Z, Chen A, Zeng G, Xiao R, Guo Z, Yi F, Huang Z, He K, Hu L (2018) Toxicity effects of silver NPs on the freshwater bivalve *Corbicula fluminea*. J Environ Chem Eng 6(4):4236–4244

Malik S, Muhammad K, Waheed Y (2023) Nanotechnology: a revolution in modern industry. Molecules 28(2):661

Manikandan S, Subbaiya R, Saravanan M, Ponraj M, Selvam M, Pugazhendhi A (2022) A critical review of advanced nanotechnology and hybrid membrane based water recycling, reuse, and wastewater treatment processes. Chemosphere 289:132867

Masindi V, Chatzisymeon E, Kortidis I, Foteinis S (2018) Assessing the sustainability of acid mine drainage (AMD) treatment in South Africa. Sci Total Environ 635:793–802

Mbarek WB, Escoda L, Saurina J, Pineda E, Alminderej FM, Khitouni M, Suñol J-J (2022) NMs as a sustainable choice for treating wastewater: a review. Materials 15(23):8576

Mikhailova EO (2020) Silver NPs: mechanism of action and probable bio-application. J Funct Biomater 11(4):84

Monteiro-Riviere NA (2022) Perspectives of nanotoxicology: introduction. Wiley Interdiscip Rev Nanomed Nanobiotechnol 14(6)

Mukherjee A, Pal S, Das P, Mukhopadhyay SK (2022) Heavy metal exposure to a migratory waterfowl, Northern Pintail (*Anas acuta*), in two peri-urban wetlands. Sci Total Environ 851:158238

Naseem T, Durrani T (2021) The role of some important metal oxide NPs for wastewater and antibacterial applications: a review. Environ Chem Ecotoxicol 3:59–75

Naseem T, Waseem M (2022) A comprehensive review on the role of some important nanocomposites for antimicrobial and wastewater applications. Int J Environ Sci Technol 19(3):2221–2246

Nasrollahzadeh M, Sajjadi M, Iravani S, Varma RS (2021) Carbon-based sustainable NMs for water treatment: state-of-art and future perspectives. Chemosphere 263:128005

Obaideen K, Shehata N, Sayed ET, Abdelkareem MA, Mahmoud MS, Olabi AG (2022) The role of wastewater treatment in achieving sustainable development goals (SDGs) and sustainability guideline. Energy Nexus 7:100112

Olabi AG, Shehata N, Obaideen K, Sayed ET, Mahmoud M, AlMallahi MN, Abdelkareem MA (2024) COVID-19: medical waste management, impact on sustainable development goals, and bibliometric analysis. Chem Eng Technol 47(1):4–19

Oliveira M, Leonardo IC, Silva AF, Crespo JG, Nunes M, Crespo MTB (2022) Nanofiltration as an efficient tertiary wastewater treatment: elimination of total bacteria and antibiotic resistance genes from the discharged effluent of a full-scale wastewater treatment plant. Antibiotics 11(5):630

Palani G, Trilaksana H, Merlyn Sujatha R, Kannan K, Rajendran S, Korniejenko K, Nykiel M, Uthayakumar M (2023) Silver NPs for waste water management. Molecules 28(8):3520

Parvin F, Rikta SY, Tareq SM (2019) Application of NMs for the removal of heavy metal from wastewater. In: Nanotechnology in water and wastewater treatment. Elsevier, Amsterdam, pp 137–157

Plesse M (2020) Global food and water security in 2050: Demographic change and increased demand. Future Directions International

Pol VG, Thiyagarajan P (2010) Remediating plastic waste into carbon nanotubes. J Environ Monitor 12(2):455–459

Rathi BS, Senthil Kumar P (2021)Application of adsorption process for effective removal of emerging contaminants from water and wastewater. Environ Pollution 280:116995

Sadegh H, Ali GAM, Gupta VK, Makhlouf ASH, Shahryari-Ghoshekandi R, Nadagouda MN, Sillanpää M, Megiel E (2017) The role of NMs as effective adsorbents and their applications in wastewater treatment. J Nanostruct Chem 7:1–14

Saravanan A, Senthil Kumar P, Hemavathy RV, Jeevanantham S, Jawahar MJ, Neshaanthini JP, Saravanan R (2022) A review on synthesis methods and recent applications of NM in wastewater treatment: challenges and future perspectives. Chemosphere 307:135713

Selvi A, Rajasekar A, Theerthagiri J, Ananthaselvam A, Sathishkumar K, Madhavan J, Rahman PKSM (2019) Integrated remediation processes toward heavy metal removal/recovery from various environments-a review. Front Environ Sci 7:66

Sengupta A, Sarkar A (2022) Synthesis and characterization of NPs from neem leaves and banana peels: a green prospect for dye degradation in wastewater. Ecotoxicology 31(4):537–548

Senta I, Rodríguez-Mozaz S, Corominas L, Petrovic M (2020) Wastewater-based epidemiology to assess human exposure to personal care and household products—a review of biomarkers, analytical methods, and applications. Trends Environ Anal Chem 28:e00103

Singh A, Garg P, Srivastava P, Sharma VP (2023) Technological interventions for wastewater treatment: monitoring and management. In: Advanced application of nanotechnology to industrial wastewater. Springer, Singapore, pp 353–363

Sinha RK, Sunita A, Krunal C, Vinod C, Kiranbhai SB (2010) Vermiculture technology: reviving the dreams of Sir Charles Darwin for scientific use of earthworms in sustainable development programs. Technology and Investment

Sukthavorn K, Nootsuwan N, Rajendran R, Rojviroon T, Veranitisagul C, Laobuthee A (2024) Polylactic acid composite nonwoven fabric incorporating nano-silver coated titanium dioxide for photocatalytic degradation of carbaryl in water. J Polymers Environ 1–11

Tanhaei B, Ayati A, Lahtinen M, Sillanpää M (2015) Preparation and characterization of a novel chitosan/Al_2O_3/magnetite NPs composite adsorbent for kinetic, thermodynamic and isotherm studies of methyl orange adsorption. Chem Eng J 259:1–10

Tao W, Chen G, Zeng G, Yan M, Chen A, Guo Z, Huang Z, He K, Hu L, Wang L (2016)Influence of silver NPs on heavy metals of pore water in contaminated river sediments. Chemosphere 162:117–124

Thakkar M, Mitra S, Wei L (2016) Effect on growth, photosynthesis, and oxidative stress of single walled carbon nanotubes exposure to marine alga *Dunaliella tertiolecta*. J NMs

Üneş F, Demirci M, İspir E, Kaya YZ, Mamak M, Taşar B (2017) Estimation of groundwater level using artificial neural networks: a case study of Hatay-Turkey

Van der Lubben IM, Coos Verhoef J, Borchard G, Junginger HE (2001) Chitosan and its derivatives in mucosal drug and vaccine delivery. Europ J Pharmaceut Sci 14(3):201–207

Wang J, Shen J, Ye D, Yan X, Zhang Y, Yang W, Li X, Wang J, Zhang L, Pan L (2019) Disinfection technology of hospital wastes and wastewater: suggestions for disinfection strategy during coronavirus Disease 2019 (COVID-19) pandemic in China. Environ Pollut 262:114665

Yang J, Hou B, Wang J, Tian B, Bi J, Wang N, Li X, Huang X (2019) NMs for the removal of heavy metals from wastewater. NMs 9(3):424

Yaqoob AA, Parveen T, Umar K, Mohamad Ibrahim MN (2020) Role of NMs in the treatment of wastewater: a review. Water 12(2):495

Zhang J, Yan B, Wan S, Kong Q (2013) Converting polyethylene waste into large scale one-dimensional Fe3O4@ C composites by a facile one-pot process. Indus Eng Chem Res 52(16):5708–5712

Zou Y, Feng Y, Lu W, Liu X (2004) Processing and properties of MWNT/HDPE composites. Carbon 42(2):271–277

Chapter 18
Nanomaterials in Wastewater Remediation: Unveiling Cutting-Edge Solutions to Environmental Challenges

Astha Kumari and Surbhi Sinha

Abstract Nanotechnology has emerged as a cutting-edge scientific field because of its ability to address various environmental issues. Nanomaterials (NMs) possess unique properties that render them useful. For example, they have a large surface area, are highly reactive, and have strong mechanical properties. These properties are highly efficient and effective for wastewater treatment. A variety of NMs, whether single or composite, have been extensively applied in wastewater treatment for the removal of organic contaminants, microbes, and heavy metals. Therefore, the purpose of the present chapter is to examine the functions of NMs in the context of wastewater treatment, highlighting their benefits. Various NMs are discussed in this chapter, including those based on carbon, metals, metal oxides, and composite NMs. Finally, prospects for future applications of NMs in the field of water remediation are emphasized.

Keywords NMs · Wastewater treatment · Contaminants · Catalysis · Adsorption · Sustainability

18.1 Introduction

Environmental pollution from human activities is rapidly increasing, causing significant health issues and reducing access to clean water. Traditional wastewater treatment methods, such as adsorption, biological oxidation, and chemical oxidation, often face significant limitations, including high energy consumption, incomplete pollutant removal, and the creation of hazardous by-products (Ahmed et al. 2022a). These limitations highlight the need for more effective and sustainable water remediation techniques.

Nanotechnology presents a promising solution to these challenges owing to the unique properties of NMs. According to the National Nanotechnology Initiative

A. Kumari · S. Sinha (✉)
Amity Institute of Biotechnology, Amity University Uttar Pradesh, Noida, Uttar Pradesh, India
e-mail: ssinha2@amity.edu

N. Agarwal et al. (eds.), *Nanomaterials in Wastewater Research*, Advances in Wastewater Research, https://doi.org/10.1007/978-981-96-3908-3_18

(NNI), NMs are defined as materials with dimensions between 1 and 100 nm (Litter and Ahmad 2023). Their high surface area and reactivity, combined with properties, such as superior surface-to-volume ratio, photocatalytic activity, improved solubility, large surface charge, and abundant reaction sites, render them particularly effective for wastewater treatment (Shokri and Sanavi Fard 2022). In recent years, NMs have gained significant attention because of their high efficiency in removing heavy metals, microorganisms, and organic pollutants from wastewater. Various types of NMs, including carbon-based materials, metal and metal oxides, metal-organic frameworks, and hybrid NMs, have been explored and utilized for this purpose (Bhat et al. 2022). The constant demand for freshwater has driven continuous advancements in nanotechnology for wastewater treatment. Over the past two decades, novel NMs and composites have been developed using simple, green, and economical synthesis methods, leading to significant improvements in water treatment processes. NMs are classified based on their size, composition, shape, and origin, which allows the creation of materials with tailored properties (Sudha et al. 2018). They can be synthesized in different forms, such as zero-dimensional (0D) spherical particles, one-dimensional (1D) fibers and tubes, two-dimensional (2D) sheets, and three-dimensional (3D) interconnected structures (Salem et al. 2022). These morphologies endow NMs with unique properties that make them versatile tools for wastewater remediation. In wastewater treatment, NMs serve as adsorbents, photocatalysts, membranes, and chemical disinfectants, leveraging their unique properties to enhance the purification processes (Thanigaivel et al. 2022). Research in nanoscience and nanotechnology continues to progress as they offer new opportunities for developing advanced materials for environmental remediation. Sustainable wastewater treatment practices have gained significant attention because of their potential to minimize environmental impacts and promote resource recovery. Over time, advancements in science and technology have led to the development of more effective sustainable practices that consider long-term impacts on the environment. These practices emphasize the importance of minimizing pollution, conserving energy, and maximizing resource recovery by reclaiming water, nutrients, and energy from wastewater (Mo and Zhang 2013).

Sustainable wastewater treatment practices are essential to preserve the environment and safeguard public health. Innovative technologies, such as advanced oxidation processes, membrane bioreactors, and anaerobic digestion, contribute to more efficient and environmentally friendly treatment processes (Rostam and Taghizadeh 2020). Source control measures, including the reduction of pollutants at their origin and the implementation of green infrastructure, help mitigate the burden on the treatment facilities. Regulatory frameworks play a crucial role in promoting sustainability by setting stringent discharge standards and encouraging the adoption of best practices (Schellenberg et al. 2020). The successful implementation of sustainable practices has been demonstrated by numerous case studies that have shown notable achievements. For example, the use of constructed wetlands for wastewater treatment not only effectively removes contaminants, but also provides habitats for wildlife and recreational spaces for communities (Campbell and Ogden 1999). In another example, the adoption of nutrient recovery technologies in wastewater treatment plants has enabled the production of fertilizers from wastewater, reducing the need for synthetic fertilizers and closing the nutrient loop (Rosemarin et al. 2020). In future, continued advancement of sustainable wastewater treatment practices will play a critical role in ensuring a sustainable and healthy world. Integrating emerging technologies, such as NMs for pollutant removal, and biotechnologies for enhanced treatment efficiency will further drive sustainability in this field.

This chapter reviews the significant role of NMs in wastewater treatment and highlights their advantages and potential future applications. The synthesis, characterization, and application of these materials are discussed with a focus on composite NMs for environmental remediation. Ongoing research and development in this field promise innovative and sustainable solutions to address the growing challenges of water pollution and scarcity.

18.2 NMs

NMs have structural components smaller than 100 nm in at least one dimension (Asha and Narain 2020). At the nanoscale, they exhibit unique physical and chemical properties that differ significantly from those of their bulk counterparts, such as a high surface area-to-volume ratio, quantum effects, and enhanced reactivity (Onyia et al. 2020). These properties make NMs particularly useful for a variety of applications including medicine, electronics, and environmental remediation. NMs can be categorized into several types. Carbon-based NMs, such as graphene, carbon nanotubes, and fullerenes, are known for their high electrical conductivity, mechanical strength, and surface area, making them suitable for applications in electronics, energy storage, and environmental cleanup (Afreen et al. 2018). Metallic NMs, including Ag, Au, and Fe nanoparticles (NPs), exhibit high electrical and thermal conductivities and catalytic activities (Narayan et al. 2019). They are used in medical applications because of their antibacterial properties, catalysis, and environmental remediation ability to remove heavy metals and organic pollutants. Metal oxide NMs, such as titanium dioxide (TiO_2) and zinc oxide (ZnO), exhibit photocatalytic activity and chemical stability and are employed in photocatalysis for environmental cleanup, sensors, and antimicrobial agents (Danish et al. 2021). Polymeric NMs, which are composed of polymers with nanoscale dimensions, are often biocompatible and versatile, with applications in drug delivery systems and tissue engineering, and as carriers for treatment agents in environmental remediation (Ngwabebhoh and Yildiz 2019). Composite NMs, which combine different materials, such as carbon nanotubes with metal oxides or polymers, offer synergistic effects that enhance mechanical strength, reactivity, and adsorption capacity. These materials are used in advanced filtration systems and catalysis and as multifunctional materials in various industries (Huang et al. 2020). NMs exhibit several unique characteristics. Their high surface areas enhance their reactivity and ability to adsorb contaminants. Quantum effects at the nanoscale influence the optical, electrical, and magnetic properties of these materials. They also exhibit superior mechanical properties, such as increased strength, toughness, and high chemical reactivity, making them effective catalysts and adsorbents (Baig et al. 2021). NMs play a crucial role in wastewater treatment because of their ability to adsorb a wide range of pollutants, including heavy metals, dyes, and organic contaminant (Sahoo and Prelot 2020). Their catalytic properties enable them to break down complex pollutants into less harmful substances. NMs have also been incorporated into filtration systems to remove particulate matter, microorganisms, and other chemical contaminants. Additionally, certain NMs, such as silver NPs, have antimicrobial properties that help disinfect water by killing or inhibiting the growth of pathogens (Deshmukh et al. 2019). Overall, NMs are versatile and powerful tools that offer innovative solutions to some of the most pressing environmental and technological challenges.

18.3 Types of NMs Used in Wastewater Treatment

NMs have revolutionized the field of wastewater treatment owing to their unique properties and high contaminant removal efficiency. These materials can be broadly categorized into metallic NPs, carbon-based NMs, polymeric NMs, and composite NMs, each offering distinct advantages and applications (Table 18.1).

Table 18.1 Types of NMs and their properties for water remediation

NM	Properties	Applications	References
Carbon nanotubes (CNTs)	High surface area, strong mechanical properties, and excellent conductivity	Removal of heavy metals and organic pollutants	Mnif and Rebah (2020)
Graphene oxide	High surface area, tunable surface chemistry, and good dispersibility in water	Adsorption of dyes, heavy metals, and organic pollutants	Nizamuddin et al. (2019)
Metal oxides (e.g., TiO_2, ZnO)	High photocatalytic activity, stability under harsh conditions	Degradation of organic pollutants, antibacterial applications	Pedrosa et al. (2019)
Zeolites	High cation-exchange capacity, good thermal stability	Removal of heavy metals, ammonium ions, and radionuclides	Peng et al. (2005)
Magnetic NPs	Easy separation using magnetic fields, high surface area	Removal of heavy metals and organic pollutants	Pizarro et al. (2015)

18.3.1 Metallic NPs

Metallic NPs are nanoscale particles made of metals that typically range in size from 1 to 100 nm (Mody et al. 2010). These NPs exhibit unique physical and chemical properties owing to their small size and high surface area-to-volume ratio (Khan and Hossain 2022). These properties make them highly effective in various applications, including wastewater treatment. Silver NPs (AgNPs) are well known for their strong antimicrobial activity (Naseem and Waseem 2022). They can disrupt bacterial cell membranes and interfere with essential cellular processes, leading to the death of harmful microorganisms. AgNPs have been used in wastewater treatment to eliminate pathogens and prevent the spread of waterborne diseases (Ganguly et al. 2021). Their antimicrobial properties make them useful for coatings and materials that encounter water, such as filters and membranes. Gold NPs (AuNPs), another type of metallic NPs, possess excellent catalytic properties. They can catalyze chemical reactions at relatively low temperatures because of their high surface reactivity (Corma and Garcia 2008). AuNPs have been used for catalytic breakdown of organic pollutants in wastewater. They help to degrade complex organic molecules into simpler and less harmful substances, facilitating the detoxification process (Ayati et al. 2014). Zero-valent iron (ZVI) NPs are characterized by their high reactivity and large surface area (Pasinszki and Krebsz 2020). These properties make them effective in reducing and immobilizing a variety of contaminants. ZVI NPs are widely used for the removal of heavy metals and chlorinated organic compounds from wastewater. They can transform pollutants into less toxic forms and facilitate their removal via adsorption and precipitation (Liang et al. 2019). The large surface area of metallic NPs provides more reactive sites for interactions with pollutants, thereby enhancing the efficiency of treatment processes. Different types of metallic NPs can target a wide range of contaminants, including heavy metals, organic pollutants, and pathogens. Some metallic NPs, such as gold, can catalyze reactions that break down complex pollutants, making them easier to remove or neutralize (Pradeep 2009).

18.3.2 Carbon-Based NPs

Carbon-based NPs are nanoscale materials that are primarily composed of carbon atoms. These NPs exhibit unique structural, electrical, thermal, and mechanical properties that make them highly effective in various applications, including wastewater treatment (Sheoran et al. 2022). Graphene consists of a single layer of carbon atoms, arranged in a hexagonal lattice. It has an exceptionally high surface area, excellent electrical and thermal conductivities, and remarkable mechanical strength (Xin et al. 2015). Graphene is used in wastewater treatment because of its superior adsorption capability. It can adsorb a wide range of pollutants, including heavy metals, organic contaminants, and dyes owing to its large surface area and strong interactions with various molecules. CNTs, which are common carbon-based NPs, are cylindrical structures composed of carbon atoms. Depending on the number of graphene layers rolled into the tubes, they can be single walled (SWCNTs) or multiwalled (MWCNTs) (Malik et al. 2008). CNTs possess high surface area, excellent mechanical strength, and unique electrical properties. CNTs have been used for adsorption and filtration in wastewater treatment. Their high surface area allows for efficient adsorption of organic pollutants, heavy metals, and even microorganisms. CNTs can also be integrated into membranes to enhance filtration performance by providing high permeability and selectivity (Fan et al. 2015). Both graphene and CNTs have large surface areas and high surface reactivity, making them highly effective for adsorbing a wide variety of pollutants. These include heavy metals, such as lead and mercury, organic pollutants, such as pesticides and pharmaceuticals, and dyes from textile effluents. Their adsorption capacities are driven by various interactions, including van der Waals forces, π–π interactions, and electrostatic attraction (Mashkoor et al. 2020). When incorporated into membranes, carbon-based NPs enhance filtration processes by providing high flux rates and excellent contaminant removal efficiencies. For instance, graphene oxide membranes can selectively filter out ions and molecules based on their size and charge, whereas CNT-based membranes offer high permeability and strength, making them suitable for removing both organic and inorganic contaminants from water. The large surface areas of graphene and CNTs provide numerous active sites for pollutant adsorption, thereby enhancing their capacity to remove contaminants from water (Xu et al. 2018). Carbon-based NPs are chemically stable and resistant to degradation, which ensures their long-term performance in water treatment applications. The mechanical robustness of graphene and CNTs enables them to withstand high pressure and mechanical stress, making them suitable for integration into durable filtration membranes.

18.3.3 Polymeric NPs

Polymeric NPs are nanoscale particles composed of natural or synthetic polymers. These NPs typically range in size from 1 to 100 nm and are valued for their versatility, stability, and ability to be engineered for specific functions (Elsabahy and Wooley 2012). Polymeric NPs can be designed to interact with various pollutants, making them highly effective for wastewater treatment. They consist of nanoscale polymers that can be engineered to have specific chemical affinities for different pollutants, thus making them highly versatile (Khdary et al. 2023). They can be designed to form hydrogels, nanofibers, or NPs tailored for specific treatment processes, such as adsorption, and flocculation, or serving as carriers for other active agents. Their biodegradability and potential for functionalization offer sustainable solutions for water treatment (Virkutyte and Varma 2011). The nanospheres were solid spherical particles, and the entire mass was composed of a polymer matrix. They are used to adsorb pollutants onto surfaces or within structures. Nanocapsules are hollow particles with a polymer shell surrounding the core. The core contains active agents, such as biocides or catalysts that are released during the treatment process (Maia et al.

2015). Nanogels are hydrated polymer networks that can swell in water. Nanogels are effective in adsorbing pollutants and can be used for the controlled release of treatment agents. Nanofibers are fibrous structures with a large surface area. They are often used in filtration systems to capture and remove contaminants from the water (Wang and Hsiao 2016). Polymeric NPs are highly effective for wastewater treatment owing to their versatile applications. They can adsorb a wide range of pollutants, including heavy metals, such as lead and cadmium, organic pollutants, such as dyes, pharmaceuticals, and pathogens, owing to their large surface areas and customizable surface properties. Additionally, certain polymeric NPs can induce flocculation, aggregating fine particles into larger clumps that are easier to remove, thereby improving the removal of suspended solids and colloidal particles (Mohamed Noor et al. 2022). Furthermore, these NPs can serve as carriers for other treatment agents, including catalysts, antimicrobial agents, and enzymes, thereby ensuring targeted delivery and enhancing the overall efficiency of the treatment process. Many polymeric NPs are biocompatible and biodegradable, reducing their environmental impact and making them suitable for sustainable water treatment solutions. The ability to modify the surface properties and functionalities of polymeric NPs allows tailored solutions to specific water treatment challenges. Polymeric NPs are generally stable under various environmental conditions, ensuring consistent performance during treatment (Kim and Bruggen 2010).

18.3.4 Composite NMs

Composite NMs are engineered materials composed of two or more types of NMs combined to exploit the unique properties of each component (Dahl et al. 2014). This synergistic combination enhances the overall performance and efficiency of the materials, making them highly effective for wastewater treatment applications. They are designed to optimize the strengths of different NMs, resulting in superior properties and functionalities. By integrating materials, such as metals, carbon-based NMs, and polymers, composite NMs can address complex water treatment challenges more effectively than individual components (Hussain et al. 2022). Composite NMs offer enhanced properties and applications in wastewater treatment, owing to their synergistic combination. For instance, combining CNTs with metal oxides leverages the high surface area and excellent adsorption capacity of CNTs, along with the reactivity of metal oxides, effectively capturing and removing organic pollutants, heavy metals, and other contaminants (Hoang et al. 2022). Similarly, integrating gold NPs with titanium dioxide (TiO_2) harnesses the catalytic properties of gold and the photocatalytic activity of TiO_2, creating a potent material for breaking down organic pollutants under light exposure and aiding wastewater detoxification (Pattnaik et al. 2023). The exceptional mechanical strength of graphene oxide, combined with the flexibility and processability of polymeric nanofibers, results in robust filtration membranes that efficiently remove particulate matter, microorganisms, and chemical contaminants (Cui et al. 2021). Furthermore, embedding AgNPs in a polymer matrix combines antimicrobial properties with structural support, producing materials capable of both filtering and disinfecting water (Bhardwaj et al. 2021). These composites are vital in water purification systems for eliminating pathogens and ensuring the delivery of safe drinking water. The synergistic combination of different materials results in superior adsorption, catalytic, and filtration properties, making composite NMs more effective in removing a wide range of contaminants. Composite NMs can be tailored to address specific pollutants and treatment processes and offer targeted solutions for various wastewater treatment challenges. By combining complementary materials, composite NMs often require fewer resources and less energy to produce and operate, thereby contributing to more sustainable water treatment practices.

18.4 Mechanisms of NMs Action

18.4.1 Adsorption

NMs are highly effective in adsorbing contaminants from wastewater because of their unique physicochemical properties, such as high surface area, tunable surface chemistry, and the presence of abundant active sites (Saxena et al. 2020). Adsorption involves several mechanisms that collectively contribute to the removal of pollutants. Surface adsorption occurs via physisorption and chemisorption processes. Physisorption involves weak van der Waals forces between the NM surface and contaminants and is a reversible process in which pollutants adhere to the surface without forming strong bonds. This is typically influenced by factors, such as the surface area and pore size of NMs (Sahoo and Prelot 2020). In contrast, chemisorption involves the formation of strong covalent or ionic bonds between the contaminants and the NM surface, making it an irreversible process dependent on the chemical nature of both the NM and pollutants (Agboola and Benson 2021). Electrostatic interactions also play a crucial role in contaminant adsorption via electrostatic attractions. Depending on the surface charge of the NM, oppositely charged contaminants are attracted. For instance, positively charged NMs can attract and bind negatively charged contaminants, such as anions and organic pollutants, and vice versa (Tee et al. 2022). The surface charge of NMs can be adjusted by modifying their surface with functional groups. Complexation is another mechanism that is particularly effective for metal-oxide NMs, which can form complex structures with contaminants (Joo and Zhao 2017). This involves the formation of coordination bonds between the metal ions on the NM surface and contaminants, particularly heavy metals. For example, iron oxide NPs can adsorb arsenic by forming complexes between the iron ions and arsenic species. Hydrophobic interactions are significant for hydrophobic NMs, such as carbon-based NMs (e.g., graphene and carbon nanotubes), which can adsorb hydrophobic organic contaminants (Smith and Rodrigues 2015). These interactions cause hydrophobic contaminants to adhere preferentially to the hydrophobic surfaces of NMs, facilitating their removal from the aqueous phase. NMs with aromatic structures, such as graphene and carbon nanotubes, can engage in π–π interactions with aromatic contaminants (Queiroz et al. 2022). These interactions occur between the π-electrons of the aromatic rings in both NMs and contaminants, enhancing the adsorption capacity for organic pollutants, such as polycyclic aromatic hydrocarbons (PAHs) and dyes. Ion-exchange mechanisms have also been employed by some NMs, particularly those functionalized with specific ion-exchange groups (Alabi et al. 2018). This involves the exchange of ions between the NM surface and contaminants in the wastewater. For example, zeolite-based NMs can exchange cations with heavy metal ions in water, thereby removing the metals (Ethaib et al. 2022). Certain NMs, such as titanium dioxide (TiO_2), exhibit photocatalytic properties that can degrade organic contaminants (Mahlambi et al. 2015). Under light irradiation, these NMs generate reactive oxygen species (ROS), such as hydroxyl radicals and superoxide ions, which can oxidize and break down organic pollutants into less harmful substances. This mechanism adsorbs and transforms contaminants, making it highly effective for the removal of organic pollutants (Pi et al. 2018). The adsorption efficiency of NMs is affected by several factors. NMs with high surface areas and optimal pore size distributions provide more active sites for contaminant adsorption, thereby enhancing their capacity to capture pollutants. Modifying the surfaces of NMs with functional groups can significantly improve their adsorption efficiency by introducing specific interactions with the contaminants (Ahmad et al. 2020). The pH and ionic strength of wastewater can also influence the adsorption process; for example, the charge of both the NM and the contaminants can change with pH, affecting electrostatic interactions. The temperature can also affect the kinetics and thermodynamics of adsorption. Higher temperatures may enhance the adsorption capacity by increasing the mobility of contaminants and the availability of active sites (Abbas et al. 2018). Additionally, sufficient contact time between the NMs and contaminants is crucial for achieving maximum adsorption. Adsorption kinetics can vary depending on the nature of the contaminants and NMs used. Furthermore, by optimizing factors, such as surface functionalization, pH, and contact time, the efficiency of NM-based adsorption processes can be further enhanced, thereby contributing to more effective and sustainable wastewater treatment (Ahmed et al. 2022b).

18.4.2 Catalysis

Catalysis plays a crucial role in wastewater treatment by accelerating the degradation of pollutants into less harmful substances, thereby enhancing the efficiency and effectiveness of the treatment process (Saravanan et al. 2021). Traditional catalysts, while effective, often suffer from limitations, such as high energy requirements and the potential for secondary pollution. NMs offer advanced catalytic properties that address these limitations and significantly improve the degradation of pollutants. NMs exhibit unique catalytic properties because of their high surface area-to-volume ratios, quantum effects, and abundant active sites (Chadha et al. 2022). These characteristics enable NMs to facilitate various catalytic processes including photocatalysis, electrocatalysis, and chemical catalysis, which are essential for the breakdown of complex organic and inorganic pollutants in wastewater.

In photocatalysis, NMs, such as titanium dioxide (TiO_2), zinc oxide (ZnO), and graphene oxide, act as photocatalysts. These materials absorb light and generate ROS, such as hydroxyl and superoxide ions. These ROS are highly reactive and can oxidize a wide range of organic pollutants, including dyes, pharmaceuticals, and pesticides, breaking them down into less harmful compounds, such as carbon dioxide and water (Puri and Gupta 2023). When NMs like TiO_2 absorb UV light, they generate electron–hole pairs. These electron-hole pairs interact with water molecules to produce hydroxyl radicals (•OH) and superoxide anions (O_2•−). These reactive species attack the pollutant molecules, breaking their chemical bonds and converting them into less harmful substances. Electrocatalysis is another area where NMs can excel. Materials, such as platinum NPs, doped carbon NMs, and metal-organic frameworks (MOFs), enhance the electrochemical reduction of pollutants (Wang et al. 2020). NMs facilitate electron transfer reactions and convert pollutants to less harmful substances. NMs act as efficient conductors and transfer electrons to the pollutant molecules. Pollutants undergo reduction or oxidation and are transformed into less toxic forms. For example, nitrate ions (NO_3^-) can be reduced to nitrogen gas (N_2) to effectively remove pollutants (Zhang et al. 2024). Chemical catalysis using NMs involves direct interaction of pollutants with the catalyst surface, leading to their transformation into less toxic forms. NMs, such as palladium (Pd) and gold (Au) NPs, are highly effective catalysts for hydrogenation and oxidation reactions. Pollutants are adsorbed onto the surfaces of the NMs. The chemical bonds in the pollutants are broken and reformed, transforming them into less harmful substances. The transformed products are then desorbed from the catalyst surface (Ren et al. 2020).

NMs have revolutionized the catalytic degradation of pollutants during wastewater treatment. Their unique properties, such as their high surface area, quantum effects, and ability to generate reactive species, make them exceptionally effective in breaking down complex pollutants. NMs provide innovative and efficient solutions to address wastewater pollution challenges through photocatalysis, electrocatalysis, and chemical catalysis. Ongoing advancements in NM synthesis and functionalization continue to enhance catalytic performance, paving the way for more sustainable and effective wastewater treatment technologies. As research progresses, the potential of NMs to offer multifunctional, adaptive, and sustainable solutions for environmental remediation will undoubtedly expand, thereby contributing to cleaner and safer water resources.

18.4.3 Filtration

Nanofiltration (NF) is a membrane filtration process that serves as an intermediate between reverse osmosis (RO) and ultrafiltration (UF) (Tay et al. 2018). It is primarily used for water purification, especially for removing dissolved organic compounds and multivalent ions, such as calcium and magnesium. NF membranes typically have pore sizes ranging from 1 to 10 nm, allowing them to effectively filter out contaminants while retaining essential minerals, making them suitable for applications in drinking water treatment and industrial wastewater management (Bruggen et al. 2003). Table 18.2 lists some common NM-enhanced nanofiltration membranes. Nanofiltration is based on the principles of size exclusion and charge repulsion. The membranes used in NF are semi-permeable, allowing water and certain low-molecular-weight solutes to pass through while rejecting larger molecules and ions based on their size and charge. This dual mechanism ensures high efficiency in removing a broad spectrum of contaminants, including natural organic matter, synthetic organic chemicals, and certain salts. One of the key advantages of nanofiltration is its ability to remove hardness from water by selectively rejecting divalent and polyvalent ions while allowing monovalent ions, such as sodium and chloride, to pass through (Suhalim et al. 2022). This makes it particularly effective for softening hard water and desalting brackish water. Additionally, NF membranes are less susceptible to fouling than RO membranes, which reduces the operational costs and extends the lifespan of the filtration system. The efficiency of nanofiltration can be influenced by several factors, including the characteristics of the feed water, membrane properties, operating conditions, and the presence of fouling agents. To optimize performance, careful control of parameters, such as pressure, temperature, and flow rate, is essential. Pretreatment processes, such as coagulation and sedimentation, are often employed to reduce the load of suspended solids and organic matter, thereby minimizing membrane fouling and enhancing overall filtration efficiency (Farahbakhsh et al. 2004). Nanofiltration techniques enhanced with sustainable NMs represent a highly efficient and environmentally friendly approach to wastewater treatment. The integration of NMs enhances contaminant removal, water flux, antifouling properties, photocatalytic degradation, and mechanical stability (Salama et al. 2021). These improvements, combined with the sustainable aspects of resource and energy efficiency and reduced environmental impact, make NM-enhanced NF membranes a promising solution for addressing the global water purification challenges. Nanofiltration (NF) membranes enhanced with NMs exhibit remarkable efficiency in wastewater treatment, owing to their advanced physicochemical properties. Contaminant removal is significantly improved with NM-enhanced NF membranes, which demonstrate superior removal rates for a variety of contaminants, such as heavy metals, organic pollutants, and microorganisms. The high surface area and reactivity of NMs facilitate the adsorption and degradation of these pollutants, ensuring a cleaner water output. For instance, the incorporation of graphene oxide or CNTs increases adsorption capacity, targeting a broader range of contaminants with higher efficiency than traditional NF membranes (Ali et al. 2019). Water flux is another crucial aspect, in which NM integration provides significant benefits. NMs, such as graphene oxide and CNTs, enhance water permeability, allowing for a higher water flux without compromising rejection rates. This improvement is critical in large-scale applications, where maintaining high throughput without sacrificing quality is essential (Bhol et al. 2021). Enhanced water flux not only improves the efficiency of the filtration process but also reduces the operational time and costs associated with water treatment. The antifouling properties of NMs play a pivotal role in extending the lifespans and efficiencies of NF membranes. Fouling, a common issue in membrane filtration, leads to frequent cleaning cycles and increases maintenance costs. NMs, such as silver NPs, which possess antimicrobial properties, help reduce biofouling. This results in less frequent membrane cleaning, thereby reducing downtime and operational costs. The antifouling properties ensure sustained performance over prolonged usage, enhancing the overall reliability of the NF membranes (Allaoui et al. 2022). Photocatalytic degradation is another innovative aspect of NMs, such as titanium dioxide (TiO_2) and zinc oxide (ZnO). These materials exhibit photocatalytic properties that enable the degradation of organic pollutants under light exposure. When integrated into NF membranes, these NMs provide an additional purification layer that extends beyond simple filtration. Under UV light, they generate ROS that can oxidize and break down complex organic molecules into less harmful substances. The dual functionality of filtration and degradation significantly enhances the overall efficiency of wastewater treatment. The mechanical stability of the NF membranes was also significantly improved by the incorporation of the NMs. NMs, such as silica NPs, enhance the mechanical strength and durability of membranes, making them more resistant to physical and chemical stresses encountered during wastewater treatment (Goh et al. 2015). This increased stability ensures that the membranes maintain their structural integrity and performance over time, thereby reducing the need for frequent replacements. Nanofiltration is widely used in the production of potable water, treatment of industrial effluents, and recovery of valuable materials from wastewater. It is also employed in the food and beverage industry for processes, such as whey demineralization and sugar concentration (Salehi 2014). Moreover, NF plays a significant role in the pharmaceutical and biotechnology industries in the purification and concentration of bioactive compounds. Nanofiltration offers a versatile and efficient solution for water and wastewater treatment, combining the benefits of high contaminant removal efficiency, lower energy consumption than RO, and the ability to produce water with tailored mineral content suitable for various industrial and municipal applications (Qu et al. 2013).

18.4.4 Sensing and Monitoring

Nanotechnology plays a crucial role in advancing sensing and monitoring techniques for wastewater management. Owing to the unique properties of NMs, such as their high surface area, enhanced reactivity, and sensitivity, innovative sensors have been developed to detect and monitor various pollutants with high precision. These sensors address the challenges posed by the low concentrations of contaminants and the complexity and variability of wastewater matrices. The use of nanoscale materials and devices in wastewater management has improved the detection, monitoring, and removal of pollutants. The small size and high surface-to-volume ratio of NMs enable effective interactions with contaminants at the molecular level, resulting in

Table 18.2 Various types of NM-enhanced nanofiltration membranes with advantages

NM-enhanced membrane	Contaminant removal efficiency	Advantages	References
Graphene oxide (GO) membranes	High removal rates for heavy metals, dyes, and organic pollutants	Improved water flux, antifouling properties, mechanical stability	Nizamuddin et al. (2019)
CNTs membranes	Effective in removing organic pollutants and heavy metals	Enhanced mechanical properties, high water permeability	Mnif and Rebah (2020)
TiO_2 nanocomposite membranes	Photocatalytic degradation of organic pollutants under UV light	Additional purification layer, antifouling properties	Pedrosa et al. (2019)
Zeolite-infused membranes	High removal efficiency for heavy metals and ammonium ions	High cation-exchange capacity, thermal stability	Peng et al. (2005)
Magnetic NP membranes	Efficient removal of heavy metals and organic pollutants	Easy separation and recovery, antifouling properties	Pizarro et al. (2015)

more sensitive and accurate monitoring. NMs enhance the performance of electrochemical sensors by providing a high surface area for electron transfer, thus improving sensitivity and selectivity (Zhu et al. 2015). These sensors can detect very low concentrations of heavy metals, organic pollutants, and pathogens. Utilizing the unique optical properties of NMs, such as the surface plasmon resonance in metallic NPs, optical sensors can detect changes in light absorption or emission when interacting with pollutants (Philip and Kumar 2022). This method is particularly useful for the monitoring of heavy metals and organic contaminants. Colorimetric sensors employ NMs that change color in the presence of specific contaminants. This visual change was easily detectable, making it a simple and effective method for on-site water quality monitoring. Certain NMs, such as quantum dots, exhibit fluorescence that can be quenched or enhanced in the presence of pollutants. This change in fluorescence was measured to detect and quantify the contaminants (Gaviria-Arroyave et al. 2020).

NMs used in wastewater treatment must be monitored for their presence and concentration in treated water. Various analytical techniques, such as spectroscopy (UV-vis, fluorescence), chromatography (HPLC), and mass spectrometry, have been employed to detect and quantify NMs (Laborda et al. 2016). These methods ensure compliance with the regulatory limits and assess the efficacy of NM-based treatment processes. Understanding the fate and transport of NMs in wastewater and the environment is critical for assessing their environmental impacts and potential risks. Techniques, such as electron microscopy (TEM, SEM) and dynamic light scattering (DLS), are used to study the behavior of NMs in aqueous environments, including their aggregation, sedimentation, and interaction with other pollutants. Monitoring the changes in the surface properties and stability of NMs is essential for maintaining their efficacy over time in wastewater treatment. Surface analysis techniques, such as X-ray photoelectron spectroscopy (XPS) and atomic force microscopy (AFM), help characterize surface modifications and stability under different environmental conditions. Assessing the bioavailability and potential ecotoxicity of NMs released into wastewater is crucial for environmental risk assessments. Bioassays and toxicity tests using aquatic organisms have provided insights into the impact of NMs on ecosystems and human health. Advances in sensor technology have enabled real-time monitoring of NMs in wastewater. Sensors based on electrochemical, optical, and piezoelectric principles offer rapid and sensitive detection capabilities, facilitating continuous monitoring and control of NM-based treatment processes. The integration of sensor data with computational models enhances the predictive capabilities for assessing the behavior and performance of NMs in wastewater treatment systems. Computational modeling techniques simulate the transport phenomena, degradation kinetics, and environmental interactions of NMs. Monitoring NMs in wastewater align with regulatory requirements and support risk-management strategies. Compliance with the regulatory limits for NM discharge and exposure limits ensures environmental and human health protection.

The use of nanotechnology for wastewater sensing and monitoring offers significant environmental and economic benefits. Enhanced detection capabilities lead to more efficient water treatment processes, thereby reducing the release of contaminants into the environment. Moreover, the high sensitivity and selectivity of NM-based sensors lower the cost of monitoring by reducing the need for frequent sampling and laboratory analysis. Nanotechnology provides transformative solutions for sensing and monitoring wastewater contaminants. The unique properties of NMs enable the development of highly sensitive, selective, and cost-effective sensors that can detect a wide range of pollutants in real-time. These advancements not only enhance the efficiency of wastewater management but also contribute to environmental protection and public health safety.

18.4.5 Pathological Control

Waterborne pathogens, such as bacteria, viruses, fungi, protozoa, and rickettsia, are responsible for approximately 80% of diseases in developing countries, posing significant health risks and claiming millions of lives annually worldwide. Traditional disinfection methods, such as chlorine-based treatments and ozone, suffer from drawbacks, such as high energy consumption, expensive equipment requirements, and the generation of potentially harmful disinfection by-products (DBPs). This has spurred the search for efficient, sustainable, and cost-effective disinfection methods. NMs offer unique advantages for pathogen inactivation in water because of their large surface area and specific reactivity, which are not achievable with conventional methods. These materials, including AgNPs, copper oxide (CuO), zinc oxide (ZnO), titanium dioxide (TiO_2), polymeric NPs, and CNTs, exhibit potent antimicrobial properties. They act through various mechanisms, such as the generation of

ROS, which induce oxidative stress and damage cellular components, such as lipids, proteins, and DNA, ultimately leading to microbial cell death. NMs can disrupt microbial cell membranes, increase permeability, and cause structural damage, leading to cell lysis. They also inhibit essential cellular processes, such as cell wall synthesis, by interfering with peptidoglycan production and enzyme activities involved in bacterial metabolism. Furthermore, NMs possess antibiofilm properties, making them effective against biofilm-forming pathogens that are highly resistant to traditional antibiotics. NMs penetrate and disrupt biofilms, making embedded bacteria more susceptible to antimicrobial agents. Recent research has highlighted advancements in NM design and applications for water disinfection. For example, graphene-based NMs have shown enhanced antimicrobial efficacy because of their high conductivity and ability to facilitate electron transfer during ROS generation (Ji et al. 2016). Furthermore, nanocomposites combining metal NPs with biopolymers have been developed to improve stability and reduce the environmental impact of water treatment applications. Another significant advancement is the development of smart NMs that can respond to environmental stimuli. For instance, pH-responsive NPs can release antimicrobial agents in response to pH changes in wastewater, thereby ensuring targeted and efficient pathogen inactivation (Colino et al. 2021). Additionally, magnetic NPs have been employed because of their ease of separation from treated water using external magnetic fields, which reduces the risk of secondary contamination (Rasheed 2022). Hybrid NMs that combine multiple antimicrobial mechanisms are gaining traction. For example, combining photocatalytic NPs like TiO_2 with metal NPs can enhance ROS generation under light irradiation, leading to more effective pathogen inactivation (Dikshit et al. 2021). This approach leverages the strengths of different NMs to create synergistic effects that improve overall disinfection efficiency. The environmental impact and safety of NMs used in wastewater treatment are critical considerations. Researchers are actively exploring the use of biodegradable and environmentally benign NMs to minimize potential ecological risks. For instance, natural polymer-based NPs and green-synthesized metal NPs have been investigated for their effectiveness in microbial control, while ensuring sustainability (Dikshit et al. 2021). Several pilot-scale studies have demonstrated the feasibility of using NMs for treating industrial and municipal wastewater. These studies have shown that NMs can be integrated into existing wastewater treatment infrastructure, offering a sustainable and cost-effective solution for energy recovery and pollutant removal. For example, AgNPs and CuO NPs have been successfully used in membrane bioreactors to enhance microbial inactivation and reduce membrane fouling, thereby improving the overall efficiency of wastewater treatment (Thakur et al. 2024). NMs have the potential to address emerging contaminants in wastewater, such as antibiotic-resistant bacteria and viruses. By employing targeted antimicrobial mechanisms, NMs can effectively neutralize these pathogens, reduce the spread of antibiotic resistance, and improve public health.

Overall, NMs represent a promising approach for disinfection in wastewater treatment. Their diverse and multifaceted mechanisms of action make them highly effective against a wide range of pathogens, including those are resistant to conventional treatments. This versatility positions NMs as valuable tools not only in water treatment, but also in medical devices, food preservation, and other applications where microbial control is critical. As research continues to advance, the development of safe, efficient, and sustainable NMs is essential for addressing global water quality challenges and ensuring public health protection.

18.5 Sustainability Aspects of NMs

18.5.1 Environmental Impact

NMs hold significant promise for environmental applications, particularly in wastewater treatment, owing to their ability to enhance the process efficiency and effectiveness. These materials facilitate the removal of contaminants at lower concentrations and under milder conditions than those of traditional methods, thereby reducing chemical usage and energy consumption. For instance, NMs, such as TiO_2 and AgNPs, exhibit high efficiency in degrading organic pollutants and eliminating pathogens, thereby contributing to cleaner water with fewer chemical residues (Khan and Hossain 2022). However, the environmental risks associated with the use of these NMs are substantial. The potential toxicity of NPs to ecosystems, especially when they enter water bodies through effluents, poses significant concerns. Their small size and high reactivity can adversely affect aquatic organisms (Vatanpour et al. 2022). Understanding the long-term ecological effects of NMs is crucial, necessitating comprehensive risk assessments and guidelines for their safe disposal and management to mitigate their environmental impact.

18.5.2 Economic Viability

Economic viability is a key factor for the adoption of sustainable NMs. Although the initial production costs of NMs can be high owing to advanced synthesis techniques and the need for high-purity materials, their efficiency in various applications can lead to long-term cost savings. Enhanced adsorption and catalytic properties imply that less material is required, and operational costs are reduced (Turan et al. 2019). Additionally, NM-based systems often have longer lifespans and lower maintenance requirements, owing to their superior antifouling properties and mechanical strength. For example, NM-enhanced filtration membranes last longer and require fewer replacements, which results in reduced maintenance costs (Roy and Baiker 2009). As production scales up and technology matures, the cost of NMs is expected to decrease, making them economically viable for widespread use.

18.5.3 Life Cycle Analysis

A thorough life cycle analysis (LCA) of NMs evaluates their environmental and economic impacts from production to disposal (Omar et al. 2024). This includes an assessment of the energy and resources required for synthesis, performance during use, and end-of-life management. NM synthesis often requires significant energy and potentially hazardous chemicals, underscoring the need for greener synthesis methods. In the usage phase, the superior performance of NMs can offset the initial environmental costs by providing more effective pollutant removal and lower operational demands. The disposal and recycling of NMs presents challenges. Establishing safe disposal methods and exploring recycling opportunities are essential for minimizing environmental contamination and enhancing life cycle sustainability (Miseljic and Olsen 2014). NMs offer substantial environmental and economic benefits, particularly in wastewater treatment. However, their sustainability must be carefully managed through comprehensive life cycle analysis, green production methods, and safe disposal practices. Balancing these factors ensures that the advantages of NMs are maximized while minimizing potential environmental and health risks.

18.6 Challenges and Limitations

The use of NMs in wastewater treatment has several significant challenges and limitations. Technical challenges are a primary concern, particularly regarding scalability, stability, and efficiency. Scaling up from lab-scale synthesis to industrial-scale production of NMs, such as nanoscale zero-valent iron (nZVI), is complex and costly, often resulting in inconsistent quality and properties. Effective application methods are also critical; the uniform distribution of NPs in large volumes of wastewater or across extensive agricultural fields is difficult, and clumping reduces their efficiency. Additionally, integrating NMs into existing wastewater treatment infrastructure may require significant modifications or entirely new systems, which can be costly and time-consuming. Stability issues, such as agglomeration, reduce the surface area and reactivity of NPs, thereby affecting their long-term performance. Ensuring that these materials maintain their effectiveness under various environmental conditions (pH, temperature, and ionic strength) remains a challenge. Furthermore, achieving high selectivity and efficiency across a range of pollutants requires sophisticated design and precise control of NM properties, whereas cost-effective and efficient regeneration processes are essential for sustainable applications. Regulatory and safety concerns pose substantial barriers. Regulatory frameworks for the use of NMs in wastewater treatment are still evolving, with a lack of standardized guidelines and protocols for testing, approval, and monitoring. Compliance with existing and emerging regulations can increase costs and require extensive testing and documentation. The potential health risks to humans and their environmental impacts are significant considerations. NMs can pose risks through inhalation, ingestion, or dermal exposure, and studies have indicated their potential for oxidative stress, inflammation, and other adverse health effects. Ecologically, NPs released into the environment may persist and bioaccumulate, potentially harming aquatic and terrestrial ecosystems, with long-term impacts still not being fully understood. Public perception plays a crucial role in the adoption of nanotechnology. Negative perceptions or distrust in the safety of NMs can hinder their acceptance, necessitating effective communication and education regarding their benefits and risks. Ethical concerns regarding the impact of advanced technologies on future generations can influence public acceptance. Economic and social barriers further complicate the widespread adoption of NMs in wastewater treatment. The high initial costs of research and development, production, and system installation are significant economic challenges. Operational costs, including specialized equipment, regular monitoring, and the potential regeneration or replacement of NMs, add to the financial burden. Compared to established, often cheaper, conventional treatment methods require cost-effectiveness and superior performance. Social acceptance is equally critical with the need for public education and awareness to build trust and acceptance. Cultural attitudes toward new technologies and environmental practices can influence acceptance, and engaging local communities and stakeholders in decision-making processes is essential. Inclusive approaches that address community concerns and incorporate their input can enhance social acceptance.

Addressing these challenges holistically involves ongoing research and innovation to improve scalability, stability, and efficiency. Developing robust regulatory frameworks and conducting thorough risk assessments ensure their safe and effective use. Overcoming economic barriers through cost-effective solutions, along with public education and stakeholder engagement, paves the way for the successful integration of NMs into sustainable wastewater management practices. By addressing these challenges, the full potential of NMs in wastewater treatment can be realized, contributing to more effective and sustainable environmental management.

18.7 Future Perspectives and Research Directions

The future of NM-based wastewater treatment is promising and driven by innovative research, interdisciplinary approaches, and supportive policies. Emerging trends in NMs, such as graphene-based materials, metal-organic frameworks, and nanocomposites, offer enhanced contaminant removal capabilities. These materials have been explored for their exceptional adsorption capabilities, tunable structures, and improved stability and reactivity, providing better performance in the removal of heavy metals, organic pollutants, and pathogens. In addition, biogenic NMs produced by biological processes involving microorganisms or plants are eco-friendly solutions engineered to target specific pollutants. Combining nanotechnology with fields, such as biotechnology, environmental engineering, chemical engineering, and materials science, can lead to more efficient and sustainable wastewater treatment. For example, integrating NMs with biotechnological processes can enhance biodegradation and pollutant removal, whereas advances in chemical and material science can optimize the synthesis, application, and durability of NMs. Membrane bioreactors (MBRs) enhanced with NMs offer superior filtration and contaminant removal efficiency. Furthermore, the development of smart NMs that can respond to environmental stimuli, such as changes in pH or temperature, can lead to more adaptive and efficient treatment systems. Comprehensive regulatory frameworks and supportive policies are essential to realize the full potential of these advancements. Developing standardized guidelines for the production, application, and disposal of NMs, along with comprehensive risk assessment frameworks, will ensure their safe use. This includes establishing protocols for testing the environmental and health impacts of NMs and creating clear pathways for their regulatory approval. Emphasizing sustainability criteria in future policies will promote the adoption of eco-friendly and biodegradable options. Streamlining regulatory processes and providing incentives for compliance will facilitate transition to advanced NM-based systems. In addition, providing funding and incentives for research and development, raising public awareness, and fostering global collaboration are crucial for advancing and safely deploying NMs for wastewater treatment. Governments and international bodies should offer grants, tax credits, and subsidies to stimulate the innovation and commercialization of environmentally friendly technologies. Public education campaigns and transparent communication regarding the benefits and risks of NMs will help to build trust and acceptance. Engaging local communities and stakeholders in the decision-making process will further enhance social acceptance and support of new technologies. International collaboration and information sharing can accelerate the development of harmonized regulations and standards, ensuring the safe and effective use of NMs globally.

By addressing these future perspectives and research directions, the integration of NMs into wastewater treatment can contribute significantly to sustainable environmental management. These advancements not only promise to enhance the efficiency and effectiveness of wastewater treatment but also align with the broader goals of sustainability and environmental protection. The concerted efforts of researchers, policymakers, industry stakeholders, and the public are pivotal for realizing the potential of NMs to transform wastewater treatment and safeguard water resources for future generations.

18.8 Conclusion

In summary, the application of sustainable NMs in wastewater treatment offers significant benefits, including enhanced contaminant removal efficiency, the ability to target specific pollutants, and the potential for eco-friendly and biodegradable solutions. Key innovations, such as graphene-based materials, metal-organic frameworks, and biogenic NMs, have demonstrated promising capabilities for addressing complex wastewater challenges. However, the widespread adoption of these technologies is hindered by several challenges, including technical issues related to scalability, stability, and efficiency, regulatory and safety concerns, and economic and social barriers. Future directions in this field will involve continued innovative research, interdisciplinary approaches, and supportive policy developments. Emerging trends in NM synthesis, smart NMs, and advanced integration techniques have the potential to overcome the existing limitations and enhance the sustainability of wastewater treatment systems. Collaborative efforts across biotechnology, environmental engineering, chemical engineering, and material science are essential to optimize the synthesis, application, and durability of NMs.

These thoughts emphasize the critical importance of continued research and development in the field of sustainable NMs for wastewater treatment. Addressing the challenges and capitalizing on the opportunities presented by these advanced materials will require concerted efforts from researchers, policymakers, industry stakeholders, and the public. By fostering innovation, establishing comprehensive regulatory frameworks, and promoting global collaboration, the potential of NMs to revolutionize wastewater treatment can be fully realized, contributing to sustainable environmental management and the protection of vital water resources for future generations.

References

Abbas Z, Ali S, Rizwan M, Zaheer IE, Malik A, Riaz MA, Shahid MR, Rehman MZU, Al-Wabel MI (2018) A critical review of mechanisms involved in the adsorption of organic and inorganic contaminants through biochar. Arab J Geosci 11:1–23

Afreen S, Omar RA, Talreja N, Chauhan D, Ashfaq M (2018) Carbon-based nanostructured materials for energy and environmental remediation applications. In: Approaches in bioremediation: the new era of environmental microbiology and nanobiotechnology, pp 369–392

Agboola OD, Benson NU (2021) Physisorption and chemisorption mechanisms influencing micro (nano) plastics-organic chemical contaminants interactions: a review. Front Environ Sci 9:678574

Ahmad SZN, Salleh WNW, Ismail AF, Yusof N, Yusop MZM, Aziz F (2020) Adsorptive removal of heavy metal ions using graphene-based NMs: toxicity, roles of functional groups and mechanisms. Chemosphere 248:126008

Ahmed M, Mavukkandy MO, Giwa A, Elektorowicz M, Katsou E, Khelifi O, Naddeo V, Hasan SW (2022a) Recent developments in hazardous pollutants removal from wastewater and water reuse within a circular economy. NPJ Clean Water 5(1):1–25

Ahmed SF, Mofijur M, Ahmed B, Mehnaz T, Mehejabin F, Maliat D, Hoang AT, Shafiullah GM (2022b) NMs as a sustainable choice for treating wastewater. Environ Res 214:113807

Alabi A, AlHajaj A, Cseri L, Szekely G, Budd P, Zou L (2018) Review of NMs-assisted ion exchange membranes for electromembrane desalination. NPJ Clean Water 1(1):10

Ali S, Rehman SAU, Luan HY, Farid MU, Huang H (2019) Challenges and opportunities in functional carbon nanotubes for membrane-based water treatment and desalination. Sci Total Environ 646:1126–1139

Allaoui BE, Chakhtouna H, Zari N, Bouhfid R, Qaiss AEK (2022) Recent developments in functionalized polymer NF membranes for biofouling control. Emergent Mater 5(5):1345–1371

Asha AB, Narain R (2020) NMs properties. In: Polymer science and nanotechnology, pp 343–359. Elsevier

Ayati A, Ahmadpour A, Bamoharram FF, Tanhaei B, Mänttäri M, Sillanpää M (2014) A review on catalytic applications of Au/TiO_2 NPs in the removal of water pollutant. Chemosphere 107:163–174

Baig N, Kammakakam I, Falath W (2021) NMs: a review of synthesis methods, properties, recent progress, and challenges. Mater Adv 2(6):1821–1871

Bhardwaj AK, Sundaram S, Yadav KK, Srivastav AL (2021) An overview of silver NPs as promising materials for water disinfection. Environ Technol Innov 23:101721

Bhat SA, Sher F, Hameed M, Bashir O, Kumar R, Vo DVN, Ahmad P, Lima EC (2022) Sustainable nanotechnology-based wastewater treatment strategies: achievements, challenges and future perspectives. Chemosphere 288:132606

Bhol P, Yadav S, Altaee A, Saxena M, Misra PK, Samal AK (2021) Graphene-based membranes for water and wastewater treatment: a review. ACS Applied NMs 4(4):3274–3293

Campbell CS, Ogden MH (1999) Constructed wetlands in the sustainable landscape, vol 3. Wiley

Chadha U, Selvaraj SK, Ashokan H, Hariharan SP, Mathew Paul V, Venkatarangan V, Paramasivam V (2022) Complex NMs in catalysis for chemically significant applications: from synthesis and hydrocarbon processing to renewable energy applications. Adv Mater Sci Eng 2022(1):1552334

Colino CI, Lanao JM, Gutierrez-Millan C (2021) Recent advances in functionalized NMs for the diagnosis and treatment of bacterial infections. Mater Sci Eng C 121:111843

Corma A, Garcia H (2008) Supported gold NPs as catalysts for organic reactions. Chem Soc Rev 37(9):2096–2126

Cui J, Wang Y, Lu T, Liu K, Huang C (2021) High performance, environmentally friendly and sustainable nanofiber membrane filter for removal of particulate matter 1.0. J Colloid Interface Sci 597:48–55

Dahl M, Liu Y, Yin Y (2014) Composite titanium dioxide NMs. Chem Rev 114(19):9853–9889

Danish MSS, Estrella LL, Alemaida IMA, Lisin A, Moiseev N, Ahmadi M, Nazari M, Wali M, Zaheb H, Senjyu T (2021) Photocatalytic applications of metal oxides for sustainable environmental remediation. Metals 11(1):80

Deshmukh SP, Patil SM, Mullani SB, Delekar SD (2019) Silver NPs as an effective disinfectant: a review. Mater Sci Eng C 97:954–965

Dikshit PK, Kumar J, Das AK, Sadhu S, Sharma S, Singh S, Gupta PK, Kim BS (2021) Green synthesis of metallic NPs: applications and limitations. Catalysts 11(8):902

Elsabahy M, Wooley KL (2012) Design of polymeric NPs for biomedical delivery applications. Chem Soc Rev 41(7):2545–2561

Ethaib S, Al-Qutaifia S, Al-Ansari N, Zubaidi SL (2022) Function of NMs in removing heavy metals for water and wastewater remediation: a review. Environments 9(10):123

Fan X, Zhao H, Liu Y, Quan X, Yu H, Chen S (2015) Enhanced permeability, selectivity, and antifouling ability of CNTs/Al_2O_3 membrane under electrochemical assistance. Environ Sci Technol 49(4):2293–2300

Farahbakhsh K, Svrcek C, Guest RK, Smith DW (2004) A review of the impact of chemical pretreatment on low-pressure water treatment membranes. J Environ Eng Sci 3(4):237–253

Ganguly K, Dutta SD, Patel DK, Lim KT (2021) Silver NPs for wastewater treatment. In: Aquananotechnology, pp 385–401. Elsevier

Gaviria-Arroyave MI, Cano JB, Peñuela GA (2020) NM-based fluorescent biosensors for monitoring environmental pollutants: a critical review. Talanta Open 2:100006

Goh PS, Ng BC, Lau WJ, Ismail AF (2015) Inorganic NMs in polymeric ultrafiltration membranes for water treatment. Sep Purif Rev 44(3):216–249

Hoang AT, Nižetić S, Cheng CK, Luque R, Thomas S, Banh TL, Nguyen XP (2022) Heavy metal removal by biomass-derived carbon nanotubes as a greener environmental remediation: a comprehensive review. Chemosphere 287:131959

Huang Q, Chen J, Liu M, Huang H, Zhang X, Wei Y (2020) Polydopamine-based functional materials and their applications in energy, environmental, and catalytic fields: state-of-the-art review. Chem Eng J 387:124019

Hussain N, Bilal M, Iqbal HM (2022) Carbon-based NMs with multipurpose attributes for water treatment: greening the 21st-century nanostructure materials deployment. Biomater Polym Horiz 1(1):48–58

Ji H, Sun H, Qu X (2016) Antibacterial applications of graphene-based NMs: recent achievements and challenges. Adv Drug Deliv Rev 105:176–189

Joo SH, Zhao D (2017) Environmental dynamics of metal oxide NPs in heterogeneous systems: a review. J Hazard Mater 322:29–47

Khan S, Hossain MK (2022) Classification and properties of NPs. In: NP-based polymer composites. Woodhead Publishing, pp 15–54

Khdary NH, Almuarqab BT, El Enany G (2023) NP-embedded polymers and their applications: a review. Membranes 13(5):537

Kim J, Van der Bruggen B (2010) The use of NPs in polymeric and ceramic membrane structures: review of manufacturing procedures and performance improvement for water treatment. Environ Pollut 158(7):2335–2349

Laborda F, Bolea E, Cepriá G, Gómez MT, Jiménez MS, Pérez-Arantegui J, Castillo JR (2016) Detection, characterization and quantification of inorganic engineered NMs: a review of techniques and methodological approaches for the analysis of complex samples. Anal Chim Acta 904:10–32

Liang L, Xue Y, Tian G, Mao Q, Lou Z, Wu Q, Wang Q, Du J, Meng X (2019) Performance of selenate removal by biochar embedded nano zero-valent iron and the biological toxicity to: Escherichia coli. RSC Adv 9(45):26136–26141

Litter MI, Ahmad A (2023) The world of nanotechnology. In: Industrial applications of NPs. CRC Press, pp 1–15

Mahlambi MM, Ngila CJ, Mamba BB (2015) Recent developments in environmental photocatalytic degradation of organic pollutants: the case of titanium dioxide NPs—a review. J NMs 2015(1):790173

Maia F, Silva AP, Fernandes S, Cunha A, Almeida A, Tedim J, Zheludkevich ML, Ferreira MGS (2015) Incorporation of biocides in nanocapsules for protective coatings used in maritime applications. Chem Eng J 270:150–157

Malik SR, Maqbool MA, Hussain S, Irfan H (2008) Carbon nanotubes: description, properties and applications. J Pak Mater Soc 2(1):21–26

Mashkoor F, Nasar A, Inamuddin (2020) Carbon nanotube-based adsorbents for the removal of dyes from waters: a review. Environ Chem Lett 18(3):605–629

Miseljic M, Olsen SI (2014) Life-cycle assessment of engineered NMs: a literature review of assessment status. J NP Res 16:1–33

Mnif W, Rebah FB (2020) Effective heavy metals removal from water using NMs: a review. Processes 8(6):645 MDPI

Mo W, Zhang Q (2013) Energy–nutrients–water nexus: integrated resource recovery in municipal wastewater treatment plants. J Environ Manage 127:255–267

Mody VV, Siwale R, Singh A, Mody HR (2010) Introduction to metallic NPs. J Pharm Bioallied Sci 2(4):282–289

Mohamed Noor MH, Wong S, Ngadi N, Mohammed Inuwa I, Opotu LA (2022) Assessing the effectiveness of magnetic NPs coagulation/flocculation in water treatment: a systematic literature review. Int J Environ Sci Technol 19(7):6935–6956

Narayan N, Meiyazhagan A, Vajtai R (2019) Metal NPs as green catalysts. Materials 12(21):3602

Naseem T, Waseem M (2022) A comprehensive review on the role of some important nanocomposites for antimicrobial and wastewater applications. Int J Environ Sci Technol 19(3):2221–2246

Ngwabebhoh FA, Yildiz U (2019) Nature-derived fibrous NM towards biomedicine and environmental remediation: today's state and future prospects. J Appl Polym Sci 136(35):47878

Nizamuddin S et al (2019) Iron oxide NMs for the removal of heavy metals and dyes from wastewater. In: Thomas S et al (eds) Nanoscale materials in water purification. Elsevier, pp 447–472 Retrieved from Springer Link

Omar NMA, Othman MHD, Tai ZS, Kurniawan TA, Puteh MH, Jaafar J, Rahman MA, Ismail AF, Rajamohan N, Abdullah H, Wong KY (2024) Recent strategies for enhancing the performance and lifespan of low-cost ceramic membranes in water filtration and treatment processes: a review. J Water Process Eng 62:105399
Onyia AI, Ikeri HI, Chima AI (2020) Surface and quantum effects in nanosized semiconductor. Am J Nano Res Appl 8(3):35–41
Pasinszki T, Krebsz M (2020) Synthesis and application of zero-valent iron NPs in water treatment, environmental remediation, catalysis, and their biological effects. Nms 10(5):917
Pattnaik A, Sahu JN, Poonia AK, Ghosh P (2023) Current perspective of nano-engineered metal oxide based photocatalysts in advanced oxidation processes for degradation of organic pollutants in wastewater. Chem Eng Res des 190:667–686
Pedrosa M et al (2019) Metal-free graphene-based catalytic membrane for degradation of organic contaminants by persulfate activation. Chem Eng J 369:223–232
Peng X et al (2005) Ceria NPs supported on carbon nanotubes for the removal of arsenate from water. Mater Lett 59:399–403
Philip A, Kumar AR (2022) The performance enhancement of surface plasmon resonance optical sensors using NMs: a review. Coord Chem Rev 458:214424
Pi Y, Li X, Xia Q, Wu J, Li Y, Xiao J, Li Z (2018) Adsorptive and photocatalytic removal of persistent organic pollutants (POPs) in water by metal-organic frameworks (MOFs). Chem Eng J 337:351–371
Pizarro C et al (2015) Nanomagnetite-zeolite composites in the removal of arsenate from aqueous systems. J Braz Chem Soc 26:1887–1896
Pradeep T (2009) Noble metal NPs for water purification: a critical review. Thin Solid Films 517(24):6441–6478
Puri N, Gupta A (2023) Water remediation using titanium and zinc oxide NMs through disinfection and photo catalysis process: a review. Environ Res 227:115786
Qu X, Brame J, Li Q, Alvarez PJ (2013) Nanotechnology for a safe and sustainable water supply: enabling integrated water treatment and reuse. Acc Chem Res 46(3):834–843
Queiroz RN, Prediger P, Vieira MGA (2022) Adsorption of polycyclic aromatic hydrocarbons from wastewater using graphene-based NMs synthesized by conventional chemistry and green synthesis: a critical review. J Hazard Mater 422:126904
Rasheed T (2022) Magnetic NMs: greener and sustainable alternatives for the adsorption of hazardous environmental contaminants. J Clean Prod 362:132338
Ren XY, Feng XB, Cao JP, Tang W, Wang ZH, Yang Z, Zhao JP, Zhang LY, Wang YJ, Zhao XY (2020) Catalytic conversion of coal and biomass volatiles: a review. Energy Fuels 34(9):10307–10363
Rosemarin A, Macura B, Carolus J, Barquet K, Ek F, Järnberg L, Lorick D, Johannesdottir S, Pedersen SM, Koskiaho J, Haddaway NR (2020) Circular nutrient solutions for agriculture and wastewater–a review of technologies and practices. Curr Opin Environ Sustain 45:78–91
Rostam AB, Taghizadeh M (2020) Advanced oxidation processes integrated by membrane reactors and bioreactors for various wastewater treatments: a critical review. J Environ Chem Eng 8(6):104566
Roy S, Baiker A (2009) NO_x storage—reduction catalysis: from mechanism and materials properties to storage—reduction performance. Chem Rev 109(9):4054–4091
Sahoo TR, Prelot B (2020) Adsorption processes for the removal of contaminants from wastewater: the perspective role of NMs and nanotechnology. In: NMs for the detection and removal of wastewater pollutants. Elsevier, pp 161–222
Salama A, Abouzeid R, Leong WS, Jeevanandam J, Samyn P, Dufresne A, Bechelany M, Barhoum A (2021) Nanocellulose-based materials for water treatment: adsorption, photocatalytic degradation, disinfection, antifouling, and nanofiltration. Nms 11(11):3008
Salehi F (2014) Current and future applications for nanofiltration technology in the food processing. Food Bioprod Process 92(2):161–177
Salem SS, Hammad EN, Mohamed AA, El-Dougdoug W (2022) A comprehensive review of NMs: types, synthesis, characterization, and applications. Biointerface Res Appl Chem 13(1):41

Saravanan A, Kumar PS, Jeevanantham S, Karishma S, Tajsabreen B, Yaashikaa PR, Reshma B (2021) Effective water/wastewater treatment methodologies for toxic pollutants removal: processes and applications towards sustainable development. Chemosphere 280:130595

Saxena R, Saxena M, Lochab A (2020) Recent progress in NMs for adsorptive removal of organic contaminants from wastewater. ChemistrySelect 5(1):335–353

Schellenberg T, Subramanian V, Ganeshan G, Tompkins D, Pradeep R (2020) Wastewater discharge standards in the evolving context of urban sustainability–the case of India. Front Environ Sci 8:30

Sheoran K, Kaur H, Siwal SS, Saini AK, Vo DVN, Thakur VK (2022) Recent advances of carbon-based NMs (CBNMs) for wastewater treatment: synthesis and application. Chemosphere 299:134364

Shokri A, Sanavi Fard M (2022) A critical review in the features and application of photocatalysts in wastewater treatment. Chem Pap 76(9):5309–5339

Smith SC, Rodrigues DF (2015) Carbon-based NMs for removal of chemical and biological contaminants from water: a review of mechanisms and applications. Carbon 91:122–143

Sudha PN, Sangeetha K, Vijayalakshmi K, Barhoum A (2018) NMs history, classification, unique properties, production and market. In: Emerging applications of NPs and architecture nanostructures. Elsevier, pp 341–384

Suhalim NS, Kasim N, Mahmoudi E, Shamsudin IJ, Mohammad AW, Mohamed Zuki F, Jamari NLA (2022) Rejection mechanism of ionic solute removal by nanofiltration membranes: an overview. Nms 12(3):437

Tay MF, Liu C, Cornelissen ER, Wu B, Chong TH (2018) The feasibility of nanofiltration membrane bioreactor (NF-MBR)+ reverse osmosis (RO) process for water reclamation: comparison with ultrafiltration membrane bioreactor (UF-MBR)+RO process. Water Res 129:180–189

Tee GT, Gok XY, Yong WF (2022) Adsorption of pollutants in wastewater via biosorbents, NPs and magnetic biosorbents: a review. Environ Res 212:113248

Thakur N, Thakur N, Kumar A, Thakur VK, Kalia S, Arya V, Kumar A, Kumar S, Kyzas GZ (2024) A critical review on the recent trends of photocatalytic, antibacterial, antioxidant and nanohybrid applications of anatase and rutile TiO_2 NPs. Sci Total Environ 914:169815

Thanigaivel S, Priya AK, Gnanasekaran L, Hoang TK, Rajendran S, Soto-Moscoso M (2022) Sustainable applicability and environmental impact of wastewater treatment by emerging nanobiotechnological approach: future strategy for efficient removal of contaminants and water purification. Sustainable Energy Technol Assess 53:102484

Turan NB, Erkan HS, Engin GO, Bilgili MS (2019) NPs in the aquatic environment: usage, properties, transformation and toxicity—a review. Process Saf Environ Prot 130:238–249

Van Der Bruggen B, Vandecasteele C, Van Gestel T, Doyen W, Leysen R (2003) A review of pressure-driven membrane processes in wastewater treatment and drinking water production. Environ Prog 22:46–56

Vatanpour V, Ağtaş M, Abdelrahman AM, Erşahin ME, Ozgun H, Koyuncu I (2022) NMs in membrane bioreactors: recent progresses, challenges, and potentials. Chemosphere 302:134930

Virkutyte J, Varma RS (2011) Green synthesis of metal NPs: biodegradable polymers and enzymes in stabilization and surface functionalization. Chem Sci 2(5):837–846

Wang X, Hsiao BS (2016) Electrospun nanofiber membranes. Curr Opin Chem Eng 12:62–81

Wang C, Wang X, Lai F, Liu Z, Dong R, Li W et al (2020) Pt NPs supported on N-doped porous carbon derived from metal–organic frameworks for oxygen reduction. ACS Appl NMs 3(6):5698–5705

Xin G, Yao T, Sun H, Scott SM, Shao D, Wang G, Lian J (2015) Highly thermally conductive and mechanically strong graphene fibers. Science 349(6252):1083–1087

Xu J, Cao Z, Zhang Y, Yuan Z, Lou Z, Xu X, Wang X (2018) A review of functionalized carbon nanotubes and graphene for heavy metal adsorption from water: preparation, application, and mechanism. Chemosphere 195:351–364

Zhang Z, Huang Z, Li H, Wang D, Yao Y, Dong K (2024) Impact of nitrate on the removal of pollutants from water in reducing gas-based membrane biofilm reactors: a review. Membranes 14(5):109

Zhu C, Yang G, Li H, Du D, Lin Y (2015) Electrochemical sensors and biosensors based on NMs and nanostructures. Anal Chem 87(1):230–249

Chapter 19
Benefits, Risks, and Responsible Use of Nanomaterials in Wastewater Remediation

Mrutyunjay Padhiary

Abstract Nanoparticles (NPs) in wastewater treatment (WWT) are an effective strategy for addressing current environmental challenges. This book chapter investigates the use of NPs in WWT, focusing on their advantages, such as improved pollution removal, faster treatment times, and cost efficiency. Various kinds of nanomaterials (NMs), such as NPs, nanocomposites, and nanocatalysts, are investigated with a deep dive into their unique characteristics for wastewater remediation. The chapter also evaluates the risks and challenges associated with the use of NMs in environmental applications, highlighting the importance of comprehensive risk assessment systems. The importance of the appropriate and sustainable use of nanotechnology, with a specific focus on its ecological and human health impacts, is highlighted. The chapter examines the regulatory structure that governs the utilization of NMs in WWT, with a particular focus on adaptive governance frameworks. The approach promotes a comprehensive strategy that integrates the advantages of NMs with a deep understanding of the related risks, conscientious usage, and legal frameworks. The objective is to provide guidance to scholars, specialists, and decision-makers on the appropriate progression of nanotechnology for the enduring and efficient treatment of wastewater.

Keywords NMs · Wastewater remediation · Environmental risk assessment · Sustainable technology · Responsible use · Regulatory frameworks

19.1 Introduction

Urbanization, industrial growth, and the widening variety of pollutants have all greatly exacerbated the problems with WWT (Liang et al. 2019). As urban populations grow, the amount of domestic sewage also grows, putting pressure on the current

M. Padhiary (✉)
Department of Agricultural Engineering, Triguna Sen School of Technology, Assam University, Silchar, India
e-mail: mrutyu@gmail.com

N. Agarwal et al. (eds.), *Nanomaterials in Wastewater Research*, Advances in Wastewater Research, https://doi.org/10.1007/978-981-96-3908-3_19

infrastructure. The industrial expansion brings numerous chemical compounds into the water system, such as heavy metals, medicines, and synthetic organic pollutants, which present significant hazards to human health and disturb aquatic ecosystems (Karri et al. 2021). Conventional methods for treating wastewater are typically inadequate for dealing with these pollutants, which is why it is necessary to create new technologies like enhanced oxidation processes and membrane bioreactors.

NMs, such as graphene oxide and titanium dioxide, possess distinctive characteristics and exhibit strong reactivity (Tayel et al. 2018). They hold great potential for improving WWT procedures by effectively breaking down organic pollutants and eliminating heavy metals. This approach addresses both traditional and emerging contaminants, thereby facilitating the development of safer and more robust urban and industrial water management systems. Traditional treatment approaches, while helpful in many cases, are encountering difficulties in efficiently addressing emerging contaminants, such as microplastics, pharmaceutical residues, and antibiotic-resistant microorganisms (Morin-Crini et al. 2022). Nanotechnology has emerged as a groundbreaking discipline in environmental research, offering a novel method for addressing the issues associated with WWT, including the presence of emerging contaminants.

The use of NMs in WWT has become more prevalent due to their ability to greatly improve the effectiveness of pollutant removal (Singh et al. 2019). NMs have greater pollutant removal capabilities compared to standard methods because of their distinctive features, which include higher adsorption capacity and selectivity. NMs provide remarkable advantages, such as a large surface area, high reactivity, and strong catalytic abilities, that have the potential to revolutionize water purification (Yaqoob et al. 2020). Additionally, NMs can be customized to specifically target and remove pollutants in wastewater streams. For example, they can adsorb heavy metals and degrade organic pollutants through catalytic reactions.

This chapter examines the diverse range of NMs utilized in the treatment of wastewater, including well-known options, such as carbon nanotubes for the purpose of adsorption and titanium dioxide for the process of photocatalytic degradation. It emphasizes the distinctive qualities and practical applications of these materials (Nyabadza et al. 2024). The text provides a comprehensive examination of the benefits of utilizing NMs in WWT, including improved effectiveness in removing pollutants and less reliance on chemicals. It also highlights the drawbacks, such as potential negative effects on the environment and uncertainties over long-term consequences. The analysis also examines responsible procedures, regulatory concerns, and ethical considerations to offer a comprehensive perspective on the consequences of using NMs. The objective is to provide a fundamental understanding of the utilization of NMs for the purpose of eliminating pollutants, the legal framework governing their application, and the ethical factors involved in implementing nanotechnology. The purpose of this information is to assist researchers, practitioners, and policymakers in effectively navigating the dynamic field of NMs in WWT.

19.2 Overview of the WWT Process

Efficient WWT is essential to ensure the proper disposal of wastewater and adherence to regulations and laws. The aforementioned procedure is of utmost importance in safeguarding the well-being of the general population and their natural surroundings. Unprocessed wastewater has the potential to harbor detrimental contaminants that can have severe repercussions on both ecosystems and human health. The origin(s) and the chosen treatment method determine the make-up and combination of waste in a WWT plant (WWTP) (Jajac et al. 2019). The initial stage involves the characterization of the wastewater to determine its distinct components, including pH, biological oxygen demand (BOD), chemical oxygen demand (COD), alkalinity, hardness, mineral content, anionic composition, and non-ionic composition (Sillanpää et al. 2018).

The current WWT process (Fig. 19.1) consists of three main steps: preliminary treatment, primary treatment, and secondary treatment. Preliminary treatment involves the removal of larger solid materials, including plastic and paper, that have the potential to form pipe clogs and hinder the progress of the subsequent treatment procedures. Clarifiers or settling tanks are used in the primary treatment process to effectively remove a substantial amount of suspended material. Coagulants can be employed to enhance the settling of suspended solids, although they come with a high cost and necessitate recovery or reuse (Kurniawan et al. 2020a).

Secondary treatment focuses on eliminating dissolved solids that are resistant to removal by filtration or sedimentation. An instance of biological treatment is the utilization of sequencing batch reactors (SBRs), which consist of a succession of two or more tanks (Sangamnere et al. 2023). Following the initial discharge of wastewater into the first tank, blowers aerate the tank to add oxygen. The settling

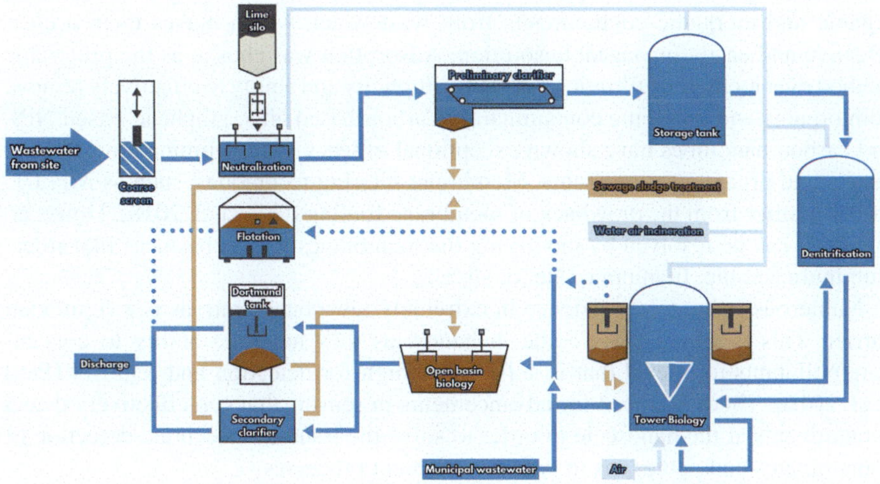

Fig. 19.1 WWT plant process flow diagram

process facilitates the separation of biomass in the tank, while a decant arm assists in the separation of clean water. While this process effectively eliminates a significant amount of organic and inorganic contaminants, it is unable to remove pollutants that are in the nanoscale range. Secondary treatment methods typically lack the ability to completely remove emerging contaminants, such as pharmaceuticals and household cleaning products, that can be detrimental to the environment when discharged as wastewater (Krzeminski et al. 2019). Nevertheless, antibiotic resistance genes (ARGs) can be more efficiently eliminated during the subsequent treatment process.

Tertiary treatment is the ultimate phase of WWTPs when the elimination of pollutants and remaining toxins from the effluent takes place. This procedure enhances the quality of the discharged effluent and complies with more stringent criteria compared to secondary treatment. The majority of tertiary treatment procedures include physicochemical approaches, including activated carbon adsorption, supplementary disinfection, and reverse osmosis. Advanced oxidation technologies, such as photocatalysis and photoelectrocatalysis, are nanotechnological methods that effectively remove contaminants from water by utilizing semiconductor catalysts in the form of NPs (Prakruthi et al. 2022).

Although current treatment procedures have been successful, large-scale applications frequently encounter difficulties. Chlorination results in an unpleasant taste and odor in the water, whereas ozonation is efficient in deactivating harmful microorganisms but can be costly due to energy expenses. Nanofiltration, reverse osmosis, microfiltration, and ultrafiltration are efficient methods for eliminating micropollutants; however, they can lead to equipment fouling (Gul et al. 2021). There is a requirement for more advanced technologies that can effectively eliminate recently discovered toxins, and nanotechnology has been regarded as a successful technology in this aspect.

NPs have the potential to enhance the quality of effluents by removing pollutants that would otherwise remain present. These particles have the ability to absorb both organic and inorganic contaminants from wastewater, which makes them a cost-effective and easily manageable solution. Adsorption was chosen as the preferable method over membrane filtration due to its simplicity and ability to effectively remove both organic and inorganic contaminants. Carbon-based NPs, graphene-based NPs, and carbon nanotubes have shown exceptional efficacy in the elimination of heavy metals and organic contaminants. Membrane filtering techniques, such as MF, UF, and NF, suffer from the drawback of membrane fouling (Goh et al. 2018). However, this issue can be resolved by modifying the membranes with substances like silica, aluminum, zeolite, titanium oxide, or silver.

Numerous pollutants' existence in extremely low concentrations is a significant barrier. This is where NMs come in handy, as they have the ability to concentrate pollutants to a level that is sufficiently high for detection and removal (Patil et al. 2016). There is a need for advancements in sensors that can effectively detect contaminants at the nanoscale in order to solve the issue of inaccurate detection of contaminants and pathogens in various treatment processes.

19.3 NMs Used in WWT

NMs are utilized in WWT to target different contaminants and enhance the efficiency of the treatment process. These materials are specifically designed with unique features to effectively handle a range of pollutants. This section examines the primary types of NMs used in wastewater cleanup, providing insight into their various properties and uses.

19.3.1 NPs

NPs are frequently employed in WWT as a prevalent kind of NMs (Fig. 19.2). These minute particles, usually smaller than 100 nm, possess a significant ratio of surface area to volume, enabling effective adsorption of pollutants. NPs, including titanium dioxide, iron oxide, and silver NPs, have demonstrated significant efficacy in eliminating heavy metals, organic compounds, and pathogens from wastewater (Naseem and Durrani 2021). Moreover, their antibacterial characteristics render them valuable for the purpose of disinfection in water treatment procedures. Scientists are actively investigating novel methods to maximize the efficiency of NPs in the process of purifying wastewater. Their primary objective is to improve the stability, reusability, and overall effectiveness of these particles in treatment systems.

NPs, specifically zero-valent iron NPs (nZVI), have garnered significant interest due to their remarkable reactivity and adsorption characteristics (Xue et al. 2018). Because it has a large surface area, nZVI has strong reducing properties that help break down pollutants faster in a number of ways, including chemical reduction and catalysis. NPs, such as nZVI, are highly effective in eliminating heavy metals and organic contaminants from wastewater, hence enhancing water quality.

19.3.2 Nanocomposites

Nanocomposites are specifically designed materials that incorporate two or more types of NMs or mix NPs with larger-scale materials in order to take advantage of the combined and enhanced effects. These materials frequently demonstrate improved characteristics in comparison with individual components, rendering them extremely adaptable in WWT. NMs made of carbon, like graphene oxide or carbon nanotubes, can improve the ability of nanocomposites to absorb and remove pollutants, making them more effective at getting rid of a wide range of pollutants (Wang et al. 2019).

Nanocomposites used in WWT include silver NPs incorporated into a polymer matrix to enhance antibacterial capabilities and titanium dioxide NPs coupled with activated carbon to enhance the photocatalytic destruction of organic contaminants. Nanocomposites provide a promising and effective option for tackling water quality

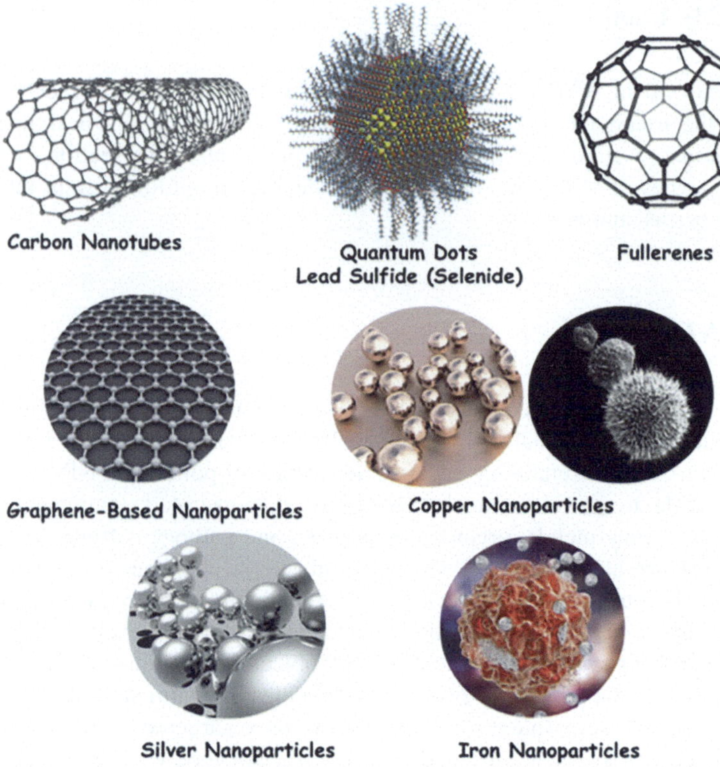

Fig. 19.2 NPs used in wastewater remediation

concerns by utilizing the distinctive characteristics of NPs with conventional materials in a sustainable manner (Elgarahy et al. 2021). One example is adding silver NPs to a polymer matrix has shown to be very effective at stopping the growth of bacteria and viruses in wastewater, making it safer to release into the environment (Ganguly et al. 2021). Also, using titanium dioxide NPs and activated carbon together can speed up the photocatalytic processes that break down organic pollutants. This enhances the purification of water sources, rendering them more suitable for human utilization.

19.3.3 Nanocatalysts

Nanocatalysts play a crucial role in expediting chemical reactions that are involved in the breakdown of pollutants, especially in the context of treating wastewater. Titanium dioxide (TiO_2) and zinc oxide (ZnO) NPs have the ability to degrade organic contaminants in wastewater under the influence of UV light (Maposa et al. 2021).

This process offers an effective method of utilizing solar energy for chemical reactions. The NPs possess a significant ratio of surface area to volume, which enables them to have more interaction with contaminants and improves their efficacy in breaking them down. These particles can be readily distributed in water, which makes them very suitable for utilization in WWT procedures.

NMs provide novel methods for the treatment of wastewater, utilizing their distinct characteristics to tackle diverse environmental issues. Notable types of NMs include metal oxides, such as TiO_2 and ZnO, which exhibit exceptional efficacy in the process of photocatalytic destruction of organic contaminants. Carbon-based NMs, such as carbon nanotubes (CNTs) and graphene oxide (GO), are known for their large surface areas and powerful adsorption abilities (Liu et al. 2022). Nano-zero-valent iron (nZVI) and other zero-valent metals are extremely reactive and have a significant surface area, which enables them to efficiently reduce pollutants as powerful reducing agents. Magnetic nanocomposites consist of magnetic cores that are coated, allowing for the convenient separation and retrieval of contaminants from water by utilizing an external magnetic field. Dendrimers are polymers with many functional groups. They are very good at getting rid of heavy metals and organic pollutants by adsorbing and complexing them. Metal-organic frameworks (MOFs), like zeolitic imidazolate frameworks (ZIF-8), are made up of metal ions connected to organic linkers (Zou et al. 2018). This arrangement results in the formation of porous structures that have large surface areas.

19.3.4 Properties and Applications

The selection of a NM for a particular application is often determined by the unique features it has to possess. For example, the strong reactivity of NPs makes them well suited for specifically removing impurities, whereas nanocomposites provide a versatile framework for customizing material characteristics (Harish et al. 2023). Nanocatalysts, due to their capacity to utilize energy for catalysis, exhibit exceptional efficacy in photodegradation processes. Comprehending the unique attributes of each category (Table 19.1) allows researchers and practitioners to choose or create NMs that correspond to the specific qualities of the wastewater and the contaminants that need to be remediated.

By taking into account the distinctive features of NMs, researchers can enhance the efficiency of WWT operations. NPs are well suited for the specific removal of contaminants because of their elevated reactivity and the ratio of their surface area to volume (Zamel et al. 2023). Nanocomposites provide a flexible framework for customizing material characteristics to suit particular uses. Nanocatalysts, due to their capacity to harness energy, are highly efficient in facilitating photodegradation reactions for the purpose of eliminating pollutants. Gaining a comprehensive understanding of the benefits and constraints associated with different types of NMs enables researchers and practitioners to make well-informed choices when it comes to selecting or creating materials for the purpose of treating wastewater.

Table 19.1 NM's properties and applications in wastewater remediation

Types/category	NM	Subcategory	Specific application in wastewater remediation	Chemical reaction	References
Metal oxides	Titanium dioxide (TiO_2)	Anatase, rutile	Photocatalytic degradation of organic pollutants, removal of heavy metals	$TiO_2 + h\nu \rightarrow TiO_2 (e^- + h^+)$	Tahir et al. (2020)
	Zinc oxide (ZnO)	Nanospheres, nanorods	Disinfection, removal of dyes and heavy metals	$ZnO + H_2O \rightarrow Zn^{2+} + 2OH^-$	Gopakumar et al. (2024)
Carbon-based	Carbon nanotubes (CNTs)	Single-walled (SWCNT), Multi-walled (MWCNT)	Adsorption of organic contaminants, heavy metals, and pathogens	CNT + pollutants (adsorption)	Marghade and Karunanidhi (2024)
	Graphene oxide (GO)	Reduced graphene oxide (rGO)	High adsorption capacity for heavy metals and organic pollutants	GO + pollutants (adsorption)	Zhao et al. (2019)
Zero-valent metals	Nano-zero-valent iron	nZVI, coated nZVI	Reduction and removal of chlorinated compounds, heavy metals	$Fe^0 + RX + H_2O \rightarrow Fe^{2+} + RH + OH^- + X^-$	Di et al. (2023)
Composites	Magnetic nanocomposites	$Fe_3O_4@SiO_2$, $Fe_3O_4@C$	Easy separation and recovery of pollutants from water using external magnetic fields	Fe_3O_4 + pollutants (adsorption)	Li et al. (2015)
Polymers	Dendrimers	PAMAM, PPI	Enhanced removal of heavy metals and organic contaminants through adsorption	Dendrimer + metal ions → complexation	Sajid et al. (2018)
Metal-organic frameworks (MOFs)	ZIF-8	Zeolitic imidazolate frameworks	Adsorption of organic pollutants, selective removal of heavy metals	ZIF-8 + pollutants (adsorption)	Li et al. (2021)

19.4 Advantages of NMs in WWT

NMs, owing to their distinctive characteristics at the nanoscale, provide numerous benefits that have catapulted them to the forefront of pioneering solutions for WWT. This section examines the various advantages of using NMs to tackle the problems of pollutant removal and treatment efficiency.

19.4.1 Improved Efficiency in Removing Pollutants

NMs possess a considerably greater ratio of surface area to volume in comparison with their larger counterparts, which allows for a more extensive interface for interactions with pollutants found in wastewater (Jain et al. 2021). The enhanced surface area of NMs improves their ability to adsorb and catalyze, resulting in more effective and comprehensive removal of contaminants. NPs, nanocomposites, and nanocatalysts, designed for specific purposes, boost the diversity and adaptability of NMs, achieving improved efficiency in removing pollutants.

19.4.2 Reduced Treatment Time and Increased Efficiency

The rapid kinetics of processes facilitated by NMs result in shorter treatment durations compared to conventional approaches. Nanocatalysts have the ability to expedite degrading reactions, resulting in faster transformation and elimination of pollutants. NMs offer a compelling solution for time-sensitive applications by improving the efficiency of WWT and reducing resource usage.

19.4.3 Cost Effectiveness and Resource Optimization

Engineered NMs can provide cost-effective solutions for treating wastewater when tailored for specific uses. The effectiveness of pollutant elimination, together with the possibility of reusing or regenerating materials, contributes to the optimization of resources and decreased operational expenses. Nanotechnology could also be used to make pollutants-targeting NPs, which means that large amounts of treatment chemicals are not needed. This makes nanotechnology more economically viable for cleaning up wastewater.

19.4.4 *Diverse Range of NMs and Their Unique Properties*

The wide array of NMs used in WWT includes NPs, nanocomposites, and nanocatalysts, each possessing unique characteristics. NPs, such as zero-valent iron (nZVI), demonstrate exceptional reactivity and adsorption capabilities (Zhao et al. 2016). Nanocomposites are made up of different materials that work better when they work together, while nanocatalysts speed up reactions that break down materials (Petronella et al. 2017). The presence of diverse NMs enables the customization of treatment procedures for different types of pollutants and wastewater compositions.

The advantages of utilizing NMs in wastewater remediation include their improved ability to remove pollutants, shortened treatment duration, cost-effectiveness, and the flexibility provided by a wide variety of NM options. Nanotechnology's benefits make it a promising and versatile alternative for tackling the changing difficulties of WWT.

19.5 Potential Hazards and Difficulties

The utilization of NMs in the process of wastewater cleanup, although showing potential, is not devoid of its inherent hazards and difficulties. This part provides a full analysis of the possible negative consequences linked to the utilization of NMs. It highlights the significance of conducting comprehensive risk evaluations and employing responsible strategies for their implementation.

19.5.1 *Environmental Fate and Persistence*

An important issue is the environmental destiny and durability of NMs once they are introduced into wastewater systems. The compact dimensions and large surface area of NPs may enhance their ability to move and disperse in aquatic environments (Peng et al. 2017). Gaining insight into the behavior of NMs in various environmental matrices is essential for evaluating their long-term effects and developing ways to mitigate undesired outcomes (Fig. 19.3).

19.5.2 *Toxicity and Potential Ecological Consequences*

The ecological impact of NMs is a worry due to their possible harm to aquatic organisms. Research on the interaction between NPs and live species, especially in aquatic habitats, is now being conducted. It is important to have a thorough grasp of nanotoxicology in order to responsibly employ NMs, as they can have harmful impacts on

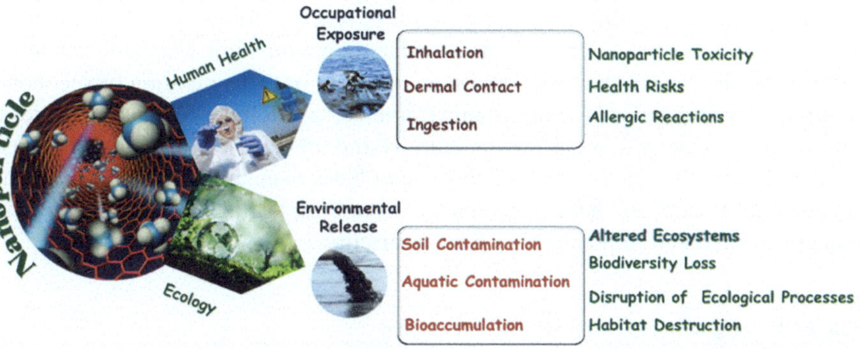

Fig. 19.3 Effects of NP use on human health and ecology

various aquatic creatures, including algae and higher trophic levels (Vasyukova et al. 2021).

19.5.3 Human Health Factors

Given the potential for NMs to enter drinking water sources or agricultural systems, it is crucial to evaluate their effects on human health. Potential health concerns may arise from inhaling, ingesting, or coming into contact with NPs (Pietroiusti et al. 2018). To make sure that everyone is safe, including people who work in WWT, it is important to do thorough studies that look at how NMs interact with living things and what health effects might happen after long-term exposure.

19.5.4 Consequences and Significance in the Long Run

Ongoing research is being conducted to investigate the enduring impacts of continual exposure to NMs on both the environment and human health. Assessing the long-term ecological and human health implications of NMs requires a thorough understanding of their persistence and bioaccumulation potential in different ecological compartments (Abbas et al. 2020).

NMs offer substantial advantages for the treatment of wastewater; however, it is crucial to effectively regulate and address their potential long-term environmental consequences and implications (Table 19.2). Titanium dioxide (TiO_2) and zinc oxide (ZnO) have the ability to build up and discharge ions, which can be harmful to aquatic habitats. Carbon nanotubes (CNTs) and graphene oxide (GO) can remain in the environment for an extended period of time, which may result in the buildup of these substances in living organisms and the disturbance of ecosystems. Nano-zero-valent iron (nZVI) has the ability to convert into less reactive forms, which necessitates careful monitoring to avoid the risk of secondary contamination. Magnetic nanocomposites have the potential to emit coatings, resulting in the generation of secondary pollution. Dendrimers and metal-organic frameworks (MOFs) provide difficulties in terms of their capacity to break down naturally and the release of metal ions, respectively. To ensure the sustainable and safe use of NMs in environmental applications, it is essential to address these concerns by using long-lasting materials, improving biodegradability, and implementing strict restrictions.

Table 19.2 Long-term effects of NPs on environment

NM	Long-term effects	Implications
Titanium dioxide (TiO_2)	Accumulation in aquatic environments	Potential toxicity to aquatic organisms, disruption of ecosystems, need for safe disposal methods
Zinc oxide (ZnO)	Release of Zn^{2+} ions into water bodies	Possible bioaccumulation, negative impact on aquatic life, stringent regulations required
Carbon nanotubes (CNTs)	Persistence in the environment	Long-term toxicity to aquatic and terrestrial organisms, concerns over human health
Graphene oxide (GO)	Stability and persistence in the environment	Potential to disrupt microbial communities, bioaccumulation, need for lifecycle analysis
Nano-zero-valent iron (nZVI)	Transformation into less reactive forms	Long-term stability issues, secondary contamination, monitoring of transformation products
Magnetic nanocomposites	Potential release of coating materials	Secondary pollution, impact on aquatic systems, requirement for durable and inert coatings
Dendrimers	Biodegradability and persistence issues	Potential for long-term environmental persistence, need for development of biodegradable variants
Metal-organic frameworks (MOFs)	Decomposition and metal ion release	Release of toxic metal ions, impact on water quality, need for stability assessments

19.5.5 The Significance of Risk Assessment

To deal with the hazards outlined before, it is necessary to create thorough frameworks for assessing those risks. These frameworks should take into account the physicochemical characteristics of NMs, the ways in which they might be exposed, and the possible effects they may have on both living and non-living parts of ecosystems. An active and flexible method of evaluating risk is essential for recognizing and reducing the possible adverse effects linked to the utilization of NPs in the treatment of wastewater.

It is crucial to incorporate ethical and sustainable methods, based on scientific data, into the development and implementation of nanotechnology for WWT in order to effectively address these risks and obstacles.

19.6 Ethical Utilization and Regulatory Factors

To effectively and safely apply NMs in WWT, it is crucial to have a comprehensive understanding of responsible usage and strong regulatory systems in place. This section explores the importance of responsible practices, ethical considerations, and the changing regulatory environment related to the use of nanotechnology in WWT.

19.6.1 Adaptive Governance Frameworks

Effective utilization of NMs requires flexible governance systems capable of adapting to the ever-changing characteristics of nanotechnology. As concerns grow, new technical advances and new insights are gained about the risks and benefits of NMs, and governance institutions should have the skills to deal with them effectively (Prasad et al. 2017). Adaptive frameworks promote a harmonious combination of stimulating innovation and protecting against potential dangers to the environment and human health.

19.6.2 Ethical Considerations in the Use of NMs

The ethical aspects of using NMs in WWT are of utmost importance. Stakeholders, such as researchers, industry participants, and politicians, need to carefully assess the possible effects on ecosystems, human well-being, and social fairness. Ensuring transparency in research, engaging the public, and conducting ethical evaluations of the wider consequences of using NMs all play a role in making responsible decisions when developing and implementing nanotechnology for wastewater cleanup.

19.6.3 Public Perception and Stakeholder Engagement

Public perception has a significant impact on the acceptance and success of NM applications. Efficient communication, involvement of stakeholders, and dissemination of information to the public are crucial for promoting comprehension and building trust within communities. To ensure the responsible and socially acceptable implementation of nanotechnology in WWT, it is crucial to address concerns, provide correct information, and involve the public in decision-making processes.

19.6.4 The Role of Regulatory Bodies in Overseeing Nanotechnology

Regulatory agencies are crucial in guaranteeing the secure and accountable utilization of NMs in the treatment of wastewater. Current environmental and occupational health standards may need to be modified in order to consider the distinct characteristics and possible hazards of NPs (León-Silva et al. 2016). Regulatory frameworks should include thorough evaluations of potential risks, standardized procedures for testing, and precise rules for the appropriate advancement and use of nanotechnology in the process of treating wastewater (Planting robots: Sowing the seeds of the agricultural future|HowToRobot, 2024).

When it comes to using NMs responsibly, a framework that includes adaptive governance structures, ethical concerns, public participation, and strong regulatory supervision is crucial. This framework promotes innovation while also protecting the environment and human health. The progress of nanotechnology will heavily depend on responsible practices and regulatory considerations in order to shape the sustainable future of using NMs in WWT.

19.6.5 Practical Applications on a Global Scale

The aforementioned area examines practical examples and uses of NPs in treating wastewater. It provides insights into successful applications, lessons learned, and the practical consequences of incorporating nanotechnology into techniques for cleaning up contaminated water.

19.7 Application of NMs in the Process of Treating Wastewater

Zero-valent iron NPs (nZVI) were used in a groundwater remediation effort to tackle the issue of chlorinated solvent contamination. The strong reactivity of nano-zero-valent iron (nZVI) enabled the efficient and quick elimination of pollutants through reduction. The project specifically showcased a decrease in trichloroethylene (TCE) levels from 500 parts per billion (ppb) to levels that cannot be detected during a span of six months (Bakke et al. 2007). The results of this study show that NMs, especially nZVI, can be very useful tools for dealing with complicated pollution problems. Compared to traditional methods, they break down contaminants quickly and effectively.

A municipal WWT facility incorporated silver NPs (AgNPs) to improve microbial disinfection. The AgNPs were very good at killing bacteria. Within 30 min of contact, the amount of E. coli dropped from 10^6 CFU/mL to levels that could not be detected (Chavan et al. 2023). This study shows that silver NPs can be used to make effluents free of pathogens, which is faster and more effective than traditional chlorine-based disinfection methods.

Arsenic (As) contamination was eliminated in an industrial WWT project with the application of nZVI. Within 48 h of using nZVI, arsenic levels dropped from 100 parts per billion (ppb) to a level below the World Health Organization's legal limit, which is 10 ppb (Chahar et al. 2023). This case exemplifies the efficacy of nZVI in effectively mitigating heavy metal pollution and adhering to rigorous regulatory requirements.

Graphene oxide and metal NPs were incorporated into nanocomposite membranes for enhanced filtering in WWT plants. These membranes demonstrated improved selectivity and permeability, allowing for the elimination of both organic and inorganic contaminants. The membranes demonstrated a 95% efficacy in removing heavy metals, such as lead (Pb) and a 90% decrease in organic dyes within a short period of operation (Khulbe and Matsuura 2018). The application's success demonstrates the adaptability of nanocomposites in offering inventive options for water purification, greatly enhancing filtration efficiency in comparison with traditional membranes.

In a desalination facility, a pilot experiment utilized carbon nanotube (CNT) membranes to achieve an 85% water recovery rate and a 99% salt rejection rate (Fajardo-Diaz et al. 2022). The CNT-based membranes exhibited greater performance compared to conventional reverse osmosis membranes, demonstrating reduced fouling and increased permeability. This emphasizes the potential of CNT-based membranes to enhance the efficiency of desalination processes.

Polymeric membranes containing silica NPs were evaluated in a WWT facility to eliminate organic contaminants, such as phenols. The membranes exhibited a 92% efficacy in removing phenolic chemicals during a 24-h timeframe (Bello and Leiviskä 2022). This example shows that polymeric nanocomposite membranes can effectively remove certain organic pollutants, offering a long-lasting answer for treating industrial wastewater.

A WWT system utilized titanium dioxide NPs (TiO_2) in a photocatalytic process to break down organic contaminants. According to a study (Kurniawan et al. 2020b), the treatment successfully achieved a degradation rate of over 99% for methylene blue dye when exposed to UV light for a period of 4 h. Nevertheless, the case study emphasized the significance of adjusting operational factors, such as NP concentration, light intensity, and reaction time, in order to achieve maximum efficiency. The study revealed that a concentration of 0.2 g/L of TiO_2, combined with a light intensity of 30 W/m^2, yielded the most favorable outcomes. This highlights the significance of precisely adjusting these parameters to achieve efficient pollutant elimination.

A textile WWTP employed zinc oxide (ZnO) NPs in a photocatalytic system to break down azo dyes (Weldegebrieal 2020). The ZnO NPs were able to degrade 85% of the Congo red dye in 180 min when exposed to visible light. This study emphasized the need to regulate factors, such as pH and NP dose in order to improve the effectiveness of photocatalysis, emphasizing the need for operational optimization in practical applications.

A project aimed at treating agricultural runoff utilized iron oxide NPs (Fe_3O_4) to effectively eliminate excessive phosphates. The initiative successfully decreased phosphate levels from 10 mg/L down below 0.1 mg/L in just one week (Mahadevan et al. 2020). This case study underscored the need to maintain the stability and reusability of NPs, emphasizing the necessity of creating durable materials that can endure several usage cycles without experiencing a substantial decline in effectiveness.

Active involvement of stakeholders was crucial in a community-oriented WWT project that included NMs. Community support was cultivated through public outreach programs, town hall meetings, and clear communication regarding the advantages and potential drawbacks of nanotechnology. According to the study (Oschlies et al. 2023), a pre- and post-engagement survey revealed a significant 30% rise in public support for the use of nanotechnology in water treatment. This case study highlights the importance of engaging stakeholders in decision-making processes to establish trust and address concerns. It demonstrates that employing good communication tactics can improve public perception and acceptance of modern technologies in environmental applications.

With the objective of educating the public about the advantages of nanotechnology in WWT through the implementation of workshops and interactive sessions, the post-campaign surveys were conducted by Ahmed et al. (2022). It was found that there was a 40% rise in awareness and positive perception of nanotechnology uses in environmental remediation. This case highlights the need for extensive instructional programs to promote public endorsement and comprehension.

The implementation of a comprehensive WWT enhancement project, which engaged the local government, scientists, and community representatives in the planning phase, led to a significant level of approval and backing from the community. The adoption of a collaborative strategy resulted in the effective implementation of nanotechnology-based solutions, resulting in an 80% reduction in pollutant levels within the initial year (Bhat et al. 2022). This study emphasizes the significance of collaborative decision-making and the involvement of stakeholders in technology adoption programs.

These above case studies highlight the concrete achievements, difficulties encountered, and insights gained from the real-world use of NPs in treating wastewater. Real-world applications offer vital insights for academics, practitioners, and policymakers who aim to exploit the potential of nanotechnology while managing associated risks and guaranteeing the appropriate utilization of these novel solutions.

19.8 Future Perspectives and Recommendations

This section explores future views and provides recommendations for expanding research, development, and responsible deployment of NMs in WWT systems.

19.8.1 Emerging Trends in NMs for WWT

NMs for resource recovery: Future advancements in nanotechnology for WWT are expected to focus more on the extraction and reuse of valuable resources. NMs, including magnetic NPs, can be designed to specifically absorb valuable metals from wastewater streams, offering possibilities for both pollution management and resource retrieval.

NMs for intelligent sensing: The progress in nanotechnology could lead to the creation of intelligent NPs that possess sensory capabilities. These materials have the potential to allow for the continuous monitoring of water quality parameters, offer a proactive method for treating wastewater, and guarantee timely actions are taken.

19.8.2 Requirements for Research and Development

An Exploration of Nanotoxicology: Additional investigation into nanotoxicology is crucial in order to uncover the possible enduring consequences of NM exposure on aquatic ecosystems and human health. This encompasses in-depth research on the accumulation, transformation, and harmful effects of diverse NMs in varying environmental circumstances.

Investigating sustainable synthesis techniques: The focus should be on creating sustainable and environmentally friendly techniques for producing NPs utilized in the treatment of wastewater. Utilizing green synthesis methods and bio-inspired materials can help decrease the environmental impact linked to the manufacturing of NMs.

19.8.3 Recommendations for Responsible and Sustainable Use

Performing thorough life cycle assessments (LCAs) for NMs utilized in WWT is essential. Life cycle assessments (LCAs) can assist in the identification of environmental and health consequences at each phase of a NM's life cycle, providing guidance to researchers and practitioners in making more sustainable and responsible decisions. Promoting international collaboration among researchers, companies, and regulatory agencies is crucial. Collaborating on knowledge, standardizing testing methods, and adopting a coordinated approach to regulatory frameworks will promote responsible worldwide advancement and implementation of nanotechnology in WWT.

19.8.4 The Functions of Academia, Industry, and Policymakers

Academia, businesses, and governments should promote interdisciplinary collaboration to address knowledge gaps and achieve a comprehensive understanding of the advantages and drawbacks of NMs in WWT. Collaborative endeavors can expedite the transformation of research discoveries into tangible implementations.

It is necessary to establish training programs and educational efforts in order to provide professionals in the water treatment sector with the necessary knowledge and skills for the proper use of NMs. This encompasses knowledge of ethical considerations, strategies for assessing risks, and adherence to ever-changing regulatory norms.

When considering the future of using NMs in WWT, it is crucial to adopt new trends, meet research requirements, and encourage ethical and sustainable practices. Stakeholders can contribute to the development of creative, effective, and ethical nanotechnology solutions for WWT by adopting a proactive approach and incorporating recommendations.

19.9 Conclusion

The incorporation of NPs into WWT has significant promise for enhancing the efficiency, selectivity, and sustainability of procedures for removing pollutants. NMs, including NPs, nanocomposites, and nanocatalysts, can be used to specifically address the complex compositions of pollutants found in various wastewater streams. Nevertheless, the prudent utilization of NMs necessitates a meticulous evaluation of possible hazards, including environmental destiny, toxicity, and enduring consequences. The development and deployment of nanotechnology in WWT necessitate careful attention to ethical considerations, public perception, and stakeholder participation. Upcoming developments involve the retrieval of resources and the use of intelligent sensing, which will lead the industry toward inventive and environmentally friendly solutions.

The primary areas of attention for research and development should be the comprehensive understanding of nanotoxicology, the exploration of sustainable synthesis methodologies, and the implementation of life cycle assessments. Responsible usage recommendations prioritize worldwide collaboration, interdisciplinary endeavors, education, and training. Collaboration between academics, industry, and policymakers is crucial for the responsible progress of NM applications in WWT. By finding a middle ground between creativity and caution, those with an interest in the matter can influence a future in which nanotechnology plays a vital role in promoting environmentally friendly water treatment methods while also safeguarding the well-being of both the ecosystem and human beings.

References

Abbas Q et al (2020) Transformation pathways and fate of engineered NPs (ENPs) in distinct interactive environmental compartments: a review. Environ Int 138:105646. https://doi.org/10.1016/j.envint.2020.105646

Ahmed SF et al (2022) Green approaches in synthesising NMs for environmental nanobioremediation: technological advancements, applications, benefits and challenges. Environ Res 204:111967. https://doi.org/10.1016/j.envres.2021.111967

Bakke B, Stewart PA, Waters MA (2007) Uses of and exposure to trichloroethylene in U.S. industry: a systematic literature review. J Occup Environ Hyg 4(5):375–390. https://doi.org/10.1080/15459620701301763

Bello A, Leiviskä T (2022) Sustainable tannin-based coagulants synthesized through Mannich reaction using melamine as an amine source for water treatment applications. Ind Crops Prod 176:114336. https://doi.org/10.1016/j.indcrop.2021.114336

Bhat SA et al (2022) Sustainable nanotechnology based WWT strategies: achievements, challenges, and future perspectives. Chemosphere 288:132606. https://doi.org/10.1016/j.chemosphere.2021.132606

Chahar M et al (2023) Recent advances in the effective removal of hazardous pollutants from wastewater by using NMs—a review. Front Environ Sci 11:1226101. https://doi.org/10.3389/fenvs.2023.1226101

Chavan P, Yadav R, Sharma P, Jaiswal AK (2023) Laser light as an emerging method for sustainable food processing, packaging, and testing. Foods 12(16):2983. https://doi.org/10.3390/foods12162983

Di L, Chen X, Lu J, Zhou Y, Zhou Y (2023) Removal of heavy metals in water using nano zero-valent iron composites: a review. J Water Process Eng 53:103913. https://doi.org/10.1016/j.jwpe.2023.103913

Elgarahy AM, Elwakeel KZ, Akhdhar A, Hamza MF (2021) Recent advances in greenly synthesized nanoengineered materials for water/wastewater remediation: an overview. Nanotechnol Environ Eng 6(1):9. https://doi.org/10.1007/s41204-021-00104-5

Fajardo-Diaz JL et al (2022) Antifouling performance of spiral wound type module made of carbon nanotubes/polyamide composite RO membrane for seawater desalination. Desalination 523:115445. https://doi.org/10.1016/j.desal.2021.115445

Ganguly K, Dutta SD, Patel DK, Lim K-T (2021) Silver NPs for WWT. In: Abd-Elsalam KA, Zahid M (eds) Aquananotechnology. Elsevier, pp 385–401. https://doi.org/10.1016/B978-0-12-821141-0.00016-1

Goh PS, Lau WJ, Othman MHD, Ismail AF (2018) Membrane fouling in desalination and its mitigation strategies. Desalination 425:130–155. https://doi.org/10.1016/j.desal.2017.10.018

Gopakumar DA, Parameswaranpillai J, George JJ, Dominic M (2024) NMs for air and water purification. Wiley

Gul A, Hruza J, Yalcinkaya F (2021) Fouling and chemical cleaning of microfiltration membranes: a mini-review. Polymers 13(6):846. https://doi.org/10.3390/polym13060846

Harish V et al (2023) Cutting-edge advances in tailoring size, shape, and functionality of NPs and nanostructures: a review. J Taiwan Inst Chem Eng 149:105010. https://doi.org/10.1016/j.jtice.2023.105010

Jain K, Patel AS, Pardhi VP, Flora SJS (2021) Nanotechnology in wastewater management: a new paradigm towards WWT. Molecules 26(6):1797. https://doi.org/10.3390/molecules26061797

Jajac N, Marović I, Rogulj K, Kilić J (2019) Decision support concept to selection of WWT plant location—the case study of Town of Kutina, Croatia. Water 11(4):717. https://doi.org/10.3390/w11040717

Karri RR, Ravindran G, Dehghani MH (2021) Wastewater—sources, toxicity, and their consequences to human health. In: Soft computing techniques in solid waste and wastewater management. Elsevier, pp 3–33. https://doi.org/10.1016/B978-0-12-824463-0.00001-X

Khulbe KC, Matsuura T (2018) Removal of heavy metals and pollutants by membrane adsorption techniques. Appl Water Sci 8(1):19. https://doi.org/10.1007/s13201-018-0661-6

Krzeminski P et al (2019) Performance of secondary WWT methods for the removal of contaminants of emerging concern implicated in crop uptake and antibiotic resistance spread: a review. Sci Total Environ 648:1052–1081. https://doi.org/10.1016/j.scitotenv.2018.08.130

Kurniawan S et al (2020a) Challenges and opportunities of biocoagulant/bioflocculant application for drinking water and WWT and its potential for sludge recovery. IJERPH 17(24):9312. https://doi.org/10.3390/ijerph17249312

Kurniawan TA et al (2020b) Functionalizing TiO2 with graphene oxide for enhancing photocatalytic degradation of methylene blue (MB) in contaminated wastewater. J Environ Manage 270:110871. https://doi.org/10.1016/j.jenvman.2020.110871

León-Silva S, Fernández-Luqueño F, López-Valdez F (2016) Silver NPs (AgNP) in the environment: a review of potential risks on human and environmental health. Water Air Soil Pollut 227(9):306. https://doi.org/10.1007/s11270-016-3022-9

Li S et al (2015) Recyclable CNTs/Fe_3O_4 magnetic nanocomposites as adsorbents to remove bisphenol A from water and their regeneration. Chem Eng J 260:231–239. https://doi.org/10.1016/j.cej.2014.09.032

Li K et al (2021) Sustainable application of ZIF-8 for heavy-metal removal in aqueous solutions. Sustainability 13(2):2. https://doi.org/10.3390/su13020984

Liang L, Wang Z, Li J (2019) The effect of urbanization on environmental pollution in rapidly developing urban agglomerations. J Clean Prod 237:117649. https://doi.org/10.1016/j.jclepro.2019.117649

Liu Z et al (2022) Synthesis of carbon-based NMs and their application in pollution management. Nanoscale Adv 4(5):1246–1262. https://doi.org/10.1039/D1NA00843A

Mahadevan H, Krishnan KA, Pillai RR, Sudhakaran S (2020) Assessment of urban river water quality and developing strategies for phosphate removal from water and wastewaters: integrated monitoring and mitigation studies. SN Appl Sci 2(4):772. https://doi.org/10.1007/s42452-020-2571-0

Mapossa AB, Mhike W, Adalima JL, Tichapondwa S (2021) Removal of organic dyes from water and wastewater using magnetic ferrite-based titanium oxide and zinc oxide nanocomposites: a review. Catalysts 11(12):1543. https://doi.org/10.3390/catal11121543

Marghade D, Karunanidhi D (2024) Properties and adsorption mechanism of biological contaminants by carbon nanotubes. In: Hadi Dehghani M, Karri RR, Mubarak NM (eds) Water treatment using engineered carbon nanotubes. Elsevier, pp 271–305. https://doi.org/10.1016/B978-0-443-18524-3.00012-X

Morin-Crini N et al (2022) Removal of emerging contaminants from wastewater using advanced treatments. A review. Environ Chem Lett 20(2):1333–1375. https://doi.org/10.1007/s10311-021-01379-5

Naseem T, Durrani T (2021) The role of some important metal oxide NPs for wastewater and antibacterial applications: a review. Environ Chem Ecotoxicol 3:59–75. https://doi.org/10.1016/j.enceco.2020.12.001

Nyabadza A et al (2024) Advanced NMs and dendrimers in water treatment and the recycling of NMs: a review. J Environ Chem Eng 12(3):112643. https://doi.org/10.1016/j.jece.2024.112643

Oschlies A et al (2023) Guide to best practices in ocean alkalinity enhancement research. Copernicus GmbH. https://doi.org/10.5194/sp-2-oae2023

Patil SS, Shedbalkar UU, Truskewycz A, Chopade BA, Ball AS (2016) NPs for environmental clean-up: a review of potential risks and emerging solutions. Environ Technol Innov 5:10–21. https://doi.org/10.1016/j.eti.2015.11.001

Peng C et al (2017) Behavior and potential impacts of metal-based engineered NPs in aquatic environments. Nms 7(1):21. https://doi.org/10.3390/nano7010021

Petronella F et al (2017) Nanocomposite materials for photocatalytic degradation of pollutants. Catal Today 281:85–100. https://doi.org/10.1016/j.cattod.2016.05.048

Pietroiusti A, Stockmann-Juvala H, Lucaroni F, Savolainen K (2018) NM exposure, toxicity, and impact on human health. Wires Nanomed Nanobiotechnol 10(5):e1513. https://doi.org/10.1002/wnan.1513

Planting robots: sowing the seeds of the agricultural future|HowToRobot (2024) https://howtorobot.com/expert-insight/planting-robots-sowing-seeds-agricultural-future. Accessed 27 May 2024

Prakruthi K, Ujwal MP, Yashas SR, Mahesh B, Kumara Swamy N, Shivaraju HP (2022) Recent advances in photocatalytic remediation of emerging organic pollutants using semiconducting metal oxides: an overview. Environ Sci Pollut Res 29(4):4930–4957. https://doi.org/10.1007/s11356-021-17361-1

Prasad R, Bhattacharyya A, Nguyen QD (2017) Nanotechnology in sustainable agriculture: recent developments, challenges, and perspectives. Front Microbiol 8:1014. https://doi.org/10.3389/fmicb.2017.01014

Sajid M, Nazal MK, Ihsanullah, Baig N, Osman AM (2018) Removal of heavy metals and organic pollutants from water using dendritic polymers based adsorbents: a critical review. Sep Purif Technol 191:400–423. https://doi.org/10.1016/j.seppur.2017.09.011

Sangamnere R, Misra T, Bherwani H, Kapley A, Kumar R (2023) A critical review of conventional and emerging WWT technologies. Sustain Water Resour Manag 9(2):58. https://doi.org/10.1007/s40899-023-00829-y

Sillanpää M, Ncibi MC, Matilainen A, Vepsäläinen M (2018) Removal of natural organic matter in drinking water treatment by coagulation: a comprehensive review. Chemosphere 190:54–71. https://doi.org/10.1016/j.chemosphere.2017.09.113

Singh S, Kumar V, Romero R, Sharma K, Singh J (2019) Applications of NPs in WWT. In: Prasad R, Kumar V, Kumar M, Choudhary D (eds) Nanobiotechnology in bioformulations. Springer International Publishing, Cham, pp 395–418. https://doi.org/10.1007/978-3-030-17061-5_17

Tahir MB, Rafique M, Rafique MS, Nawaz T, Rizwan M, Tanveer M (2020) Photocatalytic NMs for degradation of organic pollutants and heavy metals. In: Tahir MB, Rafique M, Rafique MS (eds) Nanotechnology and photocatalysis for environmental applications. Elsevier, pp 119–138. https://doi.org/10.1016/B978-0-12-821192-2.00008-5

Tayel A, Ramadan AR, El Seoud OA (2018) Titanium dioxide/graphene and titanium dioxide/graphene oxide nanocomposites: synthesis, characterization and photocatalytic applications for water decontamination. Catalysts 8(11):11. https://doi.org/10.3390/catal8110491

Vasyukova IA, Zakharova OV, Chaika VV, Pikula KS, Golokhvast KS, Gusev AA (2021) Toxic effect of metal-based NMs on representatives of marine ecosystems: a review. Nanotechnol Russia 16(2):138–154. https://doi.org/10.1134/S2635167621020178

Wang Y, Pan C, Chu W, Vipin A, Sun L (2019) Environmental remediation applications of carbon nanotubes and graphene oxide: adsorption and catalysis. Nms 9(3):439. https://doi.org/10.3390/nano9030439

Weldegebrieal GK (2020) Synthesis method, antibacterial and photocatalytic activity of ZnO NPs for azo dyes in WWT: a review. Inorg Chem Commun 120:108140. https://doi.org/10.1016/j.inoche.2020.108140

Xue W et al (2018) Performance and toxicity assessment of nanoscale zero valent iron particles in the remediation of contaminated soil: a review. Chemosphere 210:1145–1156. https://doi.org/10.1016/j.chemosphere.2018.07.118

Yaqoob AA, Parveen T, Umar K, Mohamad Ibrahim MN (2020) Role of NMs in the treatment of wastewater: a review. Water 12(2):495. https://doi.org/10.3390/w12020495

Zamel D, Khan AU, Waris A, Ebrahim A, Abd El-Sattar NEA (2023) NMs advancements for enhanced contaminant removal in WWT: NPs, nanofibers, and metal-organic frameworks (MOFs). Results Chem 6:101092. https://doi.org/10.1016/j.rechem.2023.101092

Zhao X, Liu W, Cai Z, Han B, Qian T, Zhao D (2016) An overview of preparation and applications of stabilized zero-valent iron NPs for soil and groundwater remediation. Water Res 100:245–266. https://doi.org/10.1016/j.watres.2016.05.019

Zhao C, Hong P, Li Y, Song X, Wang Y, Yang Y (2019) Mechanism of adsorption of tetracycline–Cu multi-pollutants by graphene oxide (GO) and reduced graphene oxide (rGO). J Chem Tech Biotech 94(4):1176–1186. https://doi.org/10.1002/jctb.5864

Zou D, Liu D, Zhang J (2018) From zeolitic imidazolate framework-8 to metal-organic frameworks (MOFs): representative substance for the general study of pioneering MOF applications. Energy Environ Mater 1(4):209–220. https://doi.org/10.1002/eem2.12018

The manufacturer's authorised representative in the EU is Springer Nature Customer Service Centre GmbH, Europaplatz 3, 69115 Heidelberg, Germany. If you have any concerns regarding our products, please contact ProductSafety@springernature.com

Printed and bound by CPI Group (UK) Ltd, Croydon, CR0 4YY
01/10/2026
02225922-0014